ANALYSIS AND CONTROL OF ELECTRIC DRIVES

ANALYSIS AND CONTROL OF ELECTRIC DRIVES

Simulations and Laboratory Implementation

Ned Mohan
University of Minnesota
Minneapolis, MN 55455
USA

Siddharth Raju
University of Minnesota
Minneapolis, MN 55455
USA

This edition first published 2021
© 2021 John Wiley & Sons, Inc.

All rights reserved. No part of this publication may be reproduced, stored in a retrieval system, or transmitted, in any form or by any means, electronic, mechanical, photocopying, recording or otherwise, except as permitted by law. Advice on how to obtain permission to reuse material from this title is available at http://www.wiley.com/go/permissions.

The right of Ned Mohan and Siddharth Raju to be identified as the authors of this work has been asserted in accordance with law.

Registered Office(s)
John Wiley & Sons, Inc., 111 River Street, Hoboken, NJ 07030, USA
John Wiley & Sons Ltd, The Atrium, Southern Gate, Chichester, West Sussex, PO19 8SQ, UK

Editorial Office
The Atrium, Southern Gate, Chichester, West Sussex, PO19 8SQ, UK

For details of our global editorial offices, customer services, and more information about Wiley products visit us at www.wiley.com.

Wiley also publishes its books in a variety of electronic formats and by print-on-demand. Some content that appears in standard print versions of this book may not be available in other formats.

Limit of Liability/Disclaimer of Warranty
MATLAB® is a trademark of The MathWorks, Inc. The MathWorks does not warrant the accuracy of the text or exercises in this book. This work's use or discussion of MATLAB® software or related products does not constitute endorsement or sponsorship by The MathWorks of a particular pedagogical approach or particular use of the MATLAB® software. Sciamble® Workbench is a trademark of Sciamble Corp., and is used with permission. The Sciamble Corp. does not warrant the accuracy of the text or exercises in this book. This work's use or discussion of Sciamble® Workbench software or related products does not constitute endorsement or sponsorship by Sciamble Corp. of a particular pedagogical approach or particular use of the Sciamble® Workbench software. While the publisher and authors have used their best efforts in preparing this work, they make no representations or warranties with respect to the accuracy or completeness of the contents of this work and specifically disclaim all warranties, including without limitation any implied warranties of merchantability or fitness for a particular purpose. No warranty may be created or extended by sales representatives, written sales materials or promotional statements for this work. The fact that an organization, website, or product is referred to in this work as a citation and/or potential source of further information does not mean that the publisher and authors endorse the information or services the organization, website, or product may provide or recommendations it may make. This work is sold with the understanding that the publisher is not engaged in rendering professional services. The advice and strategies contained herein may not be suitable for your situation. You should consult with a specialist where appropriate. Further, readers should be aware that websites listed in this work may have changed or disappeared between when this work was written and when it is read. Neither the publisher nor authors shall be liable for any loss of profit or any other commercial damages, including but not limited to special, incidental, consequential, or other damages.

Library of Congress Cataloging-in-Publication Data
Names: Mohan, Ned, author. | Raju, Siddharth, author.
Title: Analysis and control of electric drives : simulations and laboratory
   implementation / Ned Mohan, Siddharth Raju.
Description: Hoboken, NJ : Wiley, 2021. | Includes index.
Identifiers: LCCN 2020022305 (print) | LCCN 2020022306 (ebook) | ISBN
   9781119584537 (hardback) | ISBN 9781119584513 (adobe pdf) | ISBN
   9781119584551 (epub)
Subjects: LCSH: Electric driving–Mathematical models–Textbooks. |
   AC-to-AC converters–Textbooks. | Field orientation principle
   (Electrical engineering)–Textbooks.
Classification: LCC TK4058 .M5779 2021 (print) | LCC TK4058 (ebook) | DDC
   621.46–dc23
LC record available at https://lccn.loc.gov/2020022305
LC ebook record available at https://lccn.loc.gov/2020022306

Cover Design: Wiley
Cover Image: © Westend61/Getty Images

Set in 11/14pt TimesTen by SPi Global, Pondicherry, India

CONTENTS

Preface		xix
Acknowledgment		xxi
About the Companion Site		xxii
Part I	**Fundamentals of Electric Drives**	**1**
1	**Electric Drives: Introduction and Motivation**	**3**
1-1	The Climate Crisis and the Energy-Saving Opportunities	4
1-2	Energy Savings in Generation of Electricity	5
	1-2-1 Energy-Saving Potential in Harnessing of Wind Energy	6
1-3	Energy-Saving Potential in the End-Use of Electricity	6
	1-3-1 Energy-Saving Potential in the Process Industry	7
	1-3-2 Energy-Saving Potential in the Residential and Commercial Sectors	8
1-4	Electric Transportation	10
1-5	Precise Speed and Torque Control Applications in Robotics, Drones, and the Process Industry	10
1-6	Range of Electric Drives	11
1-7	The Multidisciplinary Nature of Drive Systems	12
1-8	Use of Simulation and Hardware Prototyping	15
1-9	Structure of the Textbook	16
1-10	Review Questions	17
	References	18
	Further Reading	18
	Problems	18

vi   CONTENTS

2 Understanding Mechanical System Requirements for Electric Drives   21

- 2-1 Introduction   21
- 2-2 Systems with Linear Motion   23
- 2-3 Rotating Systems   25
- 2-4 Friction   33
- 2-5 Torsional Resonances   35
- 2-6 Electrical Analogy   36
- 2-7 Coupling Mechanisms   39
  - 2-7-1 Conversion Between Linear and Rotary Motion   39
  - 2-7-2 Gears   41
- 2-8 Types of Loads   43
- 2-9 Four-Quadrant Operation   44
- 2-10 Steady-State and Dynamic Operations   45
- 2-11 Review Questions   45
- References   45
- Further Reading   46
- Problems   46

3 Basic Concepts in Magnetics and Electromechanical Energy Conversion   51

- 3-1 Introduction   51
- 3-2 Magnetic Circuit Concepts   52
- 3-3 Magnetic Field Produced by Current-Carrying Conductors   52
  - 3-3-1 Ampere's Law   52
- 3-4 Flux Density B and the Flux φ 54
  - 3-4-1 Ferromagnetic Materials   54
  - 3-4-2 Flux ϕ 56
  - 3-4-3 Flux Linkage   57
- 3-5 Magnetic Structures with air Gaps   58
- 3-6 Inductances   61
- 3-7 Magnetic Energy Storage in Inductors   63

CONTENTS vii

	3-8	Faraday's Law: Induced Voltage in a Coil due to Time-Rate of Change of Flux Linkage	65
		3-8-1 Relating $e(t)$, $\phi(t)$, and $i(t)$	67
	3-9	Leakage and Magnetizing Inductances	68
	3-10	Mutual Inductances	71
	3-11	Basic Principles of Torque Production and Voltage Induction	71
		3-11-1 Basic Structure of ac Machines	71
		3-11-2 Production of Magnetic Field	73
		3-11-3 Basic Principles of Torque Production and EMF Induction	76
		3-11-4 Application of the Basic Principles	80
		3-11-5 Energy Conversion	81
		3-11-6 Power Losses and Energy Efficiency	83
	3-12	Review Questions	84
		3-12-1 Magnetic Circuits	84
		3-12-2 Electromechanical Energy Conversion	86
	Further Reading		87
	Problems		87
4	**Basic Understanding of Switch-Mode Power Electronic Converters**		**95**
	4-1	Introduction	95
	4-2	Overview of Power Electronic Converters	95
		4-2-1 Switch-Mode Conversion: Switching Power-Pole as the Building Block	97
		4-2-2 PWM of the Switching Power-Pole (Constant f_s)	98
		4-2-3 Bidirectional Switching Power-Pole	99
		4-2-4 PWM of the Bidirectional Switching Power-Pole	101
	4-3	Converters for dc Motor Drives ($-V_d < \bar{v}_o < V_d$)	104
		4-3-1 Switching Waveforms in a Converter for dc Motor Drives	108
	4-4	Synthesis of Low-Frequency ac	112

viii CONTENTS

4-5	Three-Phase Inverters		113
	4-5-1	Switching Waveforms in a Three-Phase Inverter with Sine-PWM	117
4-6	Power Semiconductor Devices		118
	4-6-1	Device Ratings	119
	4-6-2	Power Diodes	119
	4-6-3	Controllable Switches	120
	4-6-4	"Smart Power" Modules Including Gate Drivers and Wide Bandgap Devices	121
4-7	Hardware Prototyping of PWM		122
4-8	Review Questions		124
References			125
Further Reading			125
Problems			126

5 Control in Electric Drives   **129**

5-1	Introduction		129
5-2	dc Motors		130
	5-2-1	Requirements Imposed by dc Machines on the PPU	134
5-3	Designing Feedback Controllers for Motor Drives		134
	5-3-1	Control Objectives	134
	5-3-2	Cascade Control Structure	139
	5-3-3	Steps in Designing the Feedback Controller	139
	5-3-4	System Representation for Small-Signal Analysis	140
5-4	Controller Design		143
	5-4-1	Proportional-Integral Controllers	143
	5-4-2	Example of a Controller Design	145
	5-4-3	The Design of the Position Control Loop	151
5-5	The Role of Feed-Forward		154
5-6	Effects of Limits		154
5-7	Anti-Windup (Non-Windup) Integration		155
5-8	Hardware Prototyping of dc Motor Speed Control		156

5-9	Review Questions	157
	References	158
	Further Reading	159
	Problems and Simulations	159

Part II Steady-State Operation of ac Machines   163

6 Using Space Vectors to Analyze ac Machines   165

6-1	Introduction	165
6-2	Sinusoidally Distributed Stator Windings	166
	6-2-1 Three-Phase, Sinusoidally Distributed Stator Windings	173
6-3	The Use of Space Vectors to Represent Sinusoidal Field Distributions in the Air Gap	175
6-4	Space-Vector Representation of Combined Terminal Currents and Voltages	180
	6-4-1 Physical Interpretation of the Stator Current Space Vector $\vec{i}_s(t)$	181
	6-4-2 Phase Components of Space Vectors $\vec{i}_s(t)$ and $\vec{v}_s(t)$	184
6-5	Balanced Sinusoidal Steady-State Excitation (Rotor Open-Circuited)	186
	6-5-1 Rotating Stator MMF Space Vector	187
	6-5-2 Rotating Stator MMF Space Vector in Multipole Machines	189
	6-5-3 The Relationship Between Space Vectors and Phasors in Balanced Three-Phase Sinusoidal Steady State ($\vec{v}_s\big|_{t=0} \Leftrightarrow \overline{V}_a$ and $\vec{i}_{ms}\big|_{t=0} \Leftrightarrow \overline{I}_{ma}$)	191
	6-5-4 Induced Voltages in Stator Windings	193
6-6	Review Questions	197
	References	199
	Further Reading	199
	Problems	199

x CONTENTS

7 Space Vector Pulse-Width-Modulated (SV-PWM) Inverters — 203

7-1 Introduction — 203
7-2 Synthesis of Stator Voltage Space Vector $\vec{v}_s^{\,a}$ — 203
7-3 Computer Simulation of SV-PWM Inverter — 208
7-4 Limit on the Amplitude $\hat{V}_s$ of the Stator Voltage Space Vector $\vec{v}_s^{\,a}$ — 211
7-5 Hardware Prototyping of Space Vector Pulse Width Modulation — 213
7-6 Summary — 214
Reference — 214
Further Reading — 214
Problems — 214

8 Sinusoidal Permanent-Magnet ac (PMAC) Drives in Steady State — 217

8-1 Introduction — 217
8-2 The Basic Structure of PMAC MACHINES — 219
8-3 Principle of Operation — 219
    8-3-1 Rotor-Produced Flux-Density Distribution — 219
    8-3-2 Torque Production — 220
    8-3-3 Mechanical System of PMAC Drives — 224
    8-3-4 Calculation of the Reference Values $i_a^*(t)$, $i_b^*(t)$, and $i_c^*(t)$ of the Stator Currents — 225
    8-3-5 Induced EMFs in the Stator Windings During Balanced Sinusoidal Steady State — 228
    8-3-6 Generator-Mode of Operation of PMAC Drives — 233
8-4 The Controller and the PPU — 233
8-5 Hardware Prototyping of PMAC Motor Hysteresis Current Control — 235
8-6 Review Questions — 238
Reference — 239
Further Reading — 239
Problems — 239

	11-4-1	Rotor-Winding Inductances (Stator Open-Circuited)	325
11-5	Mutual Inductances Between the Stator and the Rotor Phase Windings	326
11-6	Review of Space Vectors	327
	11-6-1	Relationship Between Phasors and Space Vectors in Sinusoidal Steady State	329
11-7	Flux Linkages	330
	11-7-1	Stator Flux Linkage (Rotor Open-Circuited)	330
	11-7-2	Rotor Flux Linkage (Stator Open-Circuited)	331
	11-7-3	Stator and Rotor Flux Linkages (Simultaneous Stator and Rotor Currents)	332
11-8	Stator and Rotor Voltage Equations in Terms of Space Vectors	333
11-9	Making a Case for a dq-Winding Analysis	334
11-10	Summary	338
Problems		339

12 Dynamic Analysis of Induction Machines in Terms of dq-Windings 341

12-1 Introduction 341
12-2 dq-Winding Representation 341
    12-2-1 Stator dq-Winding Representation 342
    12-2-2 Rotor dq-Windings (Along the Same dq-Axes as in the Stator) 345
    12-2-3 Mutual Inductance Between dq-Windings on the Stator and the Rotor 346
12-3 Mathematical Relationships of the dq-Windings (at an Arbitrary Speed ω_d) 347
    12-3-1 Relating dq-Winding Variables to Phase Winding Variables 349
    12-3-2 Flux Linkages of dq-Windings in Terms of Their Currents 350

	12-3-3	dq-Winding Voltage Equations	351
	12-3-4	Obtaining Fluxes and Currents with Voltages as Inputs	355
12-4		Choice of the dq-Winding Speed ω_d	355
12-5		Electromagnetic Torque	357
	12-5-1	Torque on the Rotor d-Axis Winding	357
	12-5-2	Torque on the Rotor q-Axis Winding	358
	12-5-3	Net Electromagnetic Torque T_{em} on the Rotor	359
12-6		Electrodynamics	360
12-7		d- and q-Axis Equivalent Circuits	360
12-8		Relationship Between the dq-Windings and the Per-Phase Phasor-Domain Equivalent Circuit in Balanced Sinusoidal Steady State	361
12-9		Computer Simulation	363
	12-9-1	Calculation of Initial Conditions	364
12-10		Phasor Analysis	365
12-11		Summary	373
Further Reading			373
Problems			373
Test Machine			375

13 Mathematical Description of Vector Control in Induction Machines — **377**

13-1		Introduction	377
13-2		Motor Model With the d-Axis Aligned Along the Rotor Flux Linkage $\vec{\lambda}_r$-Axis	378
	13-2-1	Calculation of ω_{dA}	379
	13-2-2	Calculation of T_{em}	380
	13-2-3	d-Axis Rotor Flux-Linkage Dynamics	380
	13-2-4	Motor Model	381
13-3		Vector Control	384
	13-3-1	Speed and Position Control Loops	385
	13-3-2	Initial Startup	388
	13-3-3	Calculating the Stator Voltages to be Applied	388
	13-3-4	Designing the PI Controllers	391

		Hardware Prototyping of Vector Control of	
	13-4	Induction Motor	393
	13-5	Summary	398
		Reference	398
		Problems	399

14 Speed-Sensorless Vector Control of Induction Motor — 401

	14-1	Introduction	401
	14-2	Open-Loop Speed Estimator	402
	14-3	Model-Reference Adaptive System (MRAS) Estimator	404
		14-3-1 Rotor Speed Estimation	407
		14-3-2 Stator d- and q-Axis Current Reference	410
		14-3-3 Estimation of ω_{dA} and θ_{da}	411
		14-3-4 Designing the PI controller	414
	14-4	Parameter Sensitivity of Open-Loop Estimator and MRAS Estimator	416
	14-5	Practical Implementation	417
	14-6	Summary	421
		References	422
		Further Reading	423
		Problems	423

14-A Appendix — 423

14-A-1 MRAS Linearized Error Function — 423

15 Analysis of Doubly Fed Generators (DFIGs) in Steady State and Their Vector Control — 427

	15-1	Introduction	427
	15-2	Steady-State Analysis	430
	15-3	Understanding DFIG Operation in dq Axis	436
		15-3-1 Stator Voltages	437
		15-3-2 Flux Linkages and Currents	437
		15-3-3 Rotor Voltages	438
		15-3-4 Stator and Rotor Power Inputs	438
		15-3-5 Electromagnetic Torque	439

		15-3-6	Relationships of Stator and Rotor Real and Reactive Powers	439
	15-4		Dynamic Analysis of DFIG	443
	15-5		Vector Control of DFIG	443
		15-5-1	Rotor Current Controller	443
		15-5-2	Rotor Speed Controller	445
		15-5-3	Stator Reactive Power Controller	446
		15-5-4	Rotor Position Estimator	446
	15-6		Summary	449
	References			450
	Further Reading			450
	Problems			450

16 Direct Torque Control (DTC) and Encoder-Less Operation of Induction Motor Drives — 453

	16-1	Introduction	453
	16-2	System Overview	453
	16-3	Principle of Encoder-Less DTC Operation	455
	16-4	Calculation of $\vec{\lambda}_s$, $\vec{\lambda}_r$, T_{em}, and ω_m	456
		16-4-1 Calculation of the Stator Flux $\vec{\lambda}_s$	456
		16-4-2 Calculation of the Rotor Flux $\vec{\lambda}_r$	456
		16-4-3 Calculation of the Electromagnetic Torque T_{em}	458
		16-4-4 Calculation of the Rotor Speed ω_m	459
	16-5	Calculation of the Stator Voltage Space Vector	460
	16-6	Direct Torque Control Using dq-Axes	464
	16-7	Summary	464
	Reference		467
	Further Reading		467
	Problems		468
	Test Machine		468

16-A Appendix — 469

16-A-1 Derivation of Torque Expressions — 469

17 Vector Control of Permanent-Magnet Synchronous Motor Drives — 473

17-1 Introduction — 473
17-2 dq-Analysis of Permanent-Magnet Synchronous Machines — 473
    17-2-1 Flux Linkages — 475
    17-2-2 Stator dq-Winding Voltages — 475
    17-2-3 Electromagnetic Torque — 476
    17-2-4 Electrodynamics — 476
17-3 Non-Salient Pole Synchronous Machines — 477
    17-3-1 Relationship Between the dq Circuits and the Per-Phase Phasor-Domain Equivalent Circuit in Balanced Sinusoidal Steady State — 477
    17-3-2 dq-Based Dynamic Controller for "Brush-less dc" Drives — 478
17-4 Salient-Pole Synchronous Machines — 481
    17-4-1 Rotor Position Estimation Using High-Frequency Injection — 483
    17-4-2 Speed-Sensorless Dynamic Controller for IPM Motor — 486
    17-4-3 Designing PID Controller — 488
    17-4-4 Electromagnetic Torque — 491
17-5 Hardware Prototyping of Vector Control of SPM Synchronous Motor — 495
17-6 Summary — 495
References — 497
Problems — 498

17-A Appendix — 498

17-A-1 Transformation of Stator Flux-Linkage From Rotating dq Frame to Stationary Frame — 498

18 Reluctance Drives: Stepper-Motors and Switched-Reluctance Drives — 501

- 18-1 Introduction — 501
- 18-2 The Operating Principle of Reluctance Motors — 502
- 18-3 Stepper-Motor Drives — 506
  - 18-3-1 Variable-Reluctance Stepper-Motors — 506
  - 18-3-2 Permanent-Magnet Stepper-Motors — 507
  - 18-3-3 Hybrid Stepper-Motors — 509
  - 18-3-4 Equivalent-Circuit Representation of a Stepper-Motor — 511
  - 18-3-5 Half-Stepping and Micro-Stepping — 512
  - 18-3-6 Power Electronic Converters for Stepper-Motors — 513
- 18-4 SRM Drives — 514
  - 18-4-1 Switched-Reluctance Motor — 514
  - 18-4-2 Electromagnetic Torque T_{em} — 515
  - 18-4-3 Induced Back-EMF e_a — 518
- 18-5 Instantaneous Waveforms — 518
- 18-6 Role of Magnetic Saturation — 521
- 18-7 Power Electronic Converters for SRM Drives — 522
- 18-8 Determining the Rotor Position for Encoder-LESS Operation — 523
- 18-9 Control in Motoring Mode — 523
- 18-10 Summary/Review Questions — 524
- References — 525
- Further Reading — 525
- Problems — 525

Index — 527

PREFACE

Electric machines and drives are used in all aspects of our life wherever electricity is used. Nearly all the electricity is generated through electric generators and drives. Almost two-thirds of this electricity is consumed by motor-driven systems in the United States. There are opportunities for energy savings by making generators and motors more efficient. In addition, substantial energy savings can be obtained by converting motors that operate at essentially constant speed to variable speed drives, where the motor speed is efficiently controlled to match the system requirements, thus resulting in substantial system-wide energy savings. There are emerging applications, such as electric vehicles, robotics, and drones, where precise speed and position control are essential. All these applications demand vector-controlled ac drives that are discussed in this textbook.

This textbook is divided into three parts. Part I of this textbook covers the fundamental principles that govern ac machines and electric drives. Using these fundamentals as the basis, the steady-state operation of ac machines and drives is analyzed in Part II. These two parts can be the basis for an undergraduate course, as we have at the University of Minnesota.

In a graduate course on this topic, students may not have the requisite background of what was covered in the undergraduate course using Parts I and II. Therefore, a quick review is warranted prior to the dynamic control of ac drives for precise speed and position, using vector control of ac drives. This is covered in Part III of this textbook.

A NEW APPROACH

This textbook is intended for a first course on the subject of electric machines and drives where no prior exposure to this subject is assumed. To do so in a single-semester course, a physics-based approach is used that not only leads to a thorough understanding of the basic principles on which electric machines operate, but also shows how they ought to be controlled for maximum efficiency. Moreover, electric machines are covered as a part of electric-drive systems, including power electronic converters and control, hence allowing relevant and interesting applications in wind turbines and electric vehicles, for example, to be discussed.

This textbook describes systems under steady-state operating conditions. However, the uniqueness of the approach used is that it seamlessly allows the discussion to be continued for analyzing and controlling of systems under dynamic conditions using vector control in a graduate-level course.

For discussion of all topics in this course, computer simulations are a necessity. These simulations utilize MATLAB/Simulink and the Sciamble Workbench (http://www.sciamble.com) – a University of Minnesota startup. The simulations in Sciamble Workbench can be seamlessly implemented to control hardware, as demonstrated by hardware experiments in an associated laboratory, to complement courses taught using this textbook.

ACKNOWLEDGMENT

The authors are greatly indebted for two grants from the University of Minnesota from the Office of Naval Research (ONR): N00014-15-1-2391 "Web-Enabled, Instructor-Taught Online Courses," and N00014-19-1-2018 "Developing WBG-Based, Extremely Low-Cost Laboratories for Power Electronics, Motor Drives, and Power System Protection and Relays for National Dissemination." These grants allowed the development of the Workbench simulation platform, which is available free-of-cost for educational purposes. These grants also allowed the development of a low-cost hardware laboratory, available from Sciamble (https://sciamble.com/) – a University of Minnesota startup.

ABOUT THE COMPANION SITE

Analysis and Control of Electric Drives: Simulations and Laboratory Implementation is accompanied by a companion website:

www.wiley.com/go/Mohan/Vectorcontrolinelectricdrives

The companion website page includes the following items that are mentioned in the textbook:

1. Links to research reports as Appendices
2. All the simulation file in MATLAB/Simulink
3. All the simulation files in Sciamble Workbench
4. Exact parameters of motors used as examples
5. Manual of the Hardware Laboratory, and
6. Solution to some select back-of-the-chapter problems.

Part I

Fundamentals of Electric Drives

1 Electric Drives: Introduction and Motivation

Electric machines and electric drives are shown by their block diagrams in Fig. 1-1a and b. Electric machines were invented more than 150 years ago and have been in use ever since in increasing numbers in a variety of applications. As shown in Fig. 1-1a, electric machines convert energy from the electrical system to the mechanical system, and vice versa. In their motoring mode, where the machine is called a motor, the electric power P_{elect} from the electrical system at the certain voltage/current magnitude and frequency get converted to the mechanical power P_{mech} to the mechanical system at corresponding torque and speed. The opposite is true for a machine in its generator mode, where power from the mechanical system gets converted and is supplied to the electrical system. In machines, as shown in Fig. 1-1a, some of the quantities (voltage/current, torque/speed) are dictated by external sources, and no attempt is made to control the others.

However, in certain applications, it is required that for given quantities on the electrical or the mechanical side, the other quantities be controlled, as in a wind turbine. This is made possible in electric drives shown by their block diagram in Fig. 1-1b. It should be noted that in

---

(Adapted from chapter 1 of *Electric Machines and Drives: A First Course* ISBN: 978-1-118-07481-7 by Ned Mohan, January 2012 and from chapter 1 of *Advanced Electric Drives: Analysis, Control, and Modeling Using MATLAB/Simulink* ISBN: 978-1-118-48548-4 by Ned Mohan, August 2014)

Analysis and Control of Electric Drives: Simulations and Laboratory Implementation, First Edition. Ned Mohan and Siddharth Raju.
© 2021 John Wiley & Sons, Inc. Published 2021 by John Wiley & Sons, Inc.
Companion website: www.wiley.com/go/Mohan/Vectorcontrolinelectricdrives

4 ELECTRIC DRIVES: INTRODUCTION AND MOTIVATION

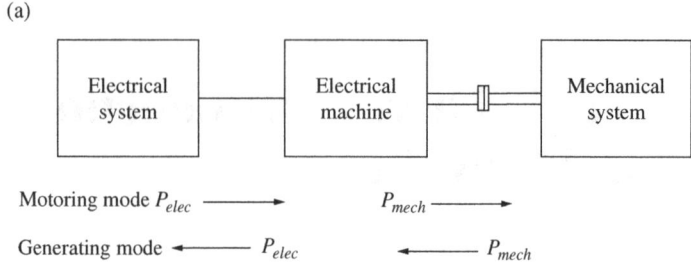

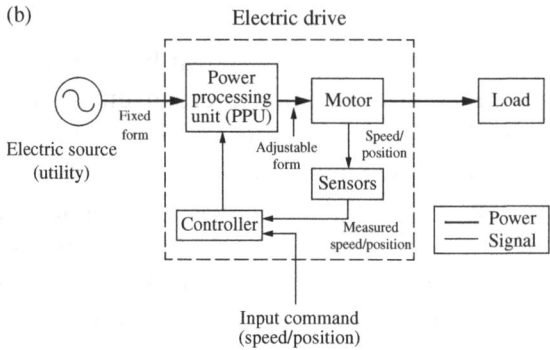

Fig. 1-1 Block diagrams of (a) electric machines and (b) electric drives (motoring mode shown).

the literature and in trade publications, electric drives sometimes refer only to the power electronic converter and its control, excluding the motor. In this textbook, however, electric drives refer to the entire block, which is shown dotted in Fig. 1-1b, that includes power electronic converter (power processing unit – PPU) and its control, as well as the electric machine, whether it is in its motoring or the generating mode. We should also note that we will be looking at only the ac machines, hence the title of the book is ac drives.

1-1 THE CLIMATE CRISIS AND THE ENERGY-SAVING OPPORTUNITIES

The climate crisis, caused by the burning of fossil fuels, is the greatest and an existential threat facing humanity. To reduce the emission of carbon dioxide, a necessary solution is first to convert our energy use to electricity, as much as possible, and then to produce that electricity using renewables such as solar and wind. As we will see in the subsequent sections in this chapter, electric drives play a significant role in generating and efficiently consuming electricity and providing ample opportunity for energy savings.

According to [1], "advances in integrated power electronics have the potential to develop a new generation of energy-efficient, high-power density, high-speed motors and generators and, in turn, save significant energy." In addition, a great deal of energy savings can be achieved by shifting from nearly constant speed motors to adjustable-speed electric drives, as explained in this chapter.

Prior to looking at the energy-saving potentials, we should understand the meaning of primary energy. According to [2], the "**Primary Energy** is energy in the form that it is first accounted for in a statistical energy balance, before any transformation to secondary or tertiary forms of energy. For example, coal can be converted to synthetic gas, which can be converted to electricity; in this example, coal is primary energy, synthetic gas is secondary energy, and electricity is tertiary energy." Often, the primary energy and the savings in the primary energy are expressed in quads, where a quad equals 10^{15} BTUs and 10 000 BTUs equal approximately 2.93 kWh.

1-2 ENERGY SAVINGS IN GENERATION OF ELECTRICITY

Nearly 99% of electricity is produced through electric machines. This percentage was nearly the same, approximately 98.6%, in the United States in 2018. According to the US Energy Information Administration [3], about 4171 billion kWh (or 4.17 trillion kWh) of electricity was generated at utility-scale electricity generation facilities in the United States in 2018. About 64% of this electricity generation was from fossil fuels (coal, natural gas, petroleum, and other gases). About 19% was from nuclear energy, and approximately 17% was from renewable energy sources. Out of the renewable energy sources, only 1.4% of the total electricity generated was by photovoltaic systems (PVs) that do not use electric machines, whereas all other sources of electricity generation use electric machines. Therefore,

6 ELECTRIC DRIVES: INTRODUCTION AND MOTIVATION

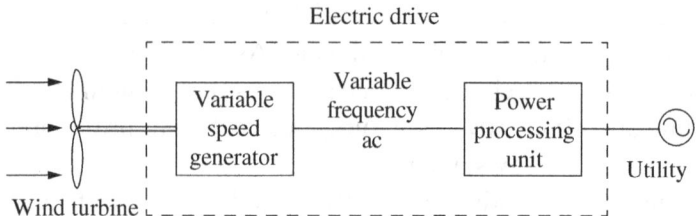

Fig. 1-2 Electric drive for wind generators.

any improvement in increasing the efficiency of machines and electric drives will be very consequential.

1-2-1 Energy-Saving Potential in Harnessing of Wind Energy

One of the significant roles of electric drives is in harnessing wind energy. The block diagram for a wind-electric system is shown in Fig. 1-2, where the variable-frequency ac produced by the wind-turbine-driven generator is interfaced with the utility system through a power electronic converter (PPU). By letting the turbine speed vary with the wind speed, it is possible to recover a higher amount of energy in the wind compared to systems where the turbine essentially rotates at a constant speed due to the generator output being directly connected to the utility grid. The harnessing of wind energy involving ac drives is crucial for generating carbon-free electricity [3], and this application is sure to grow rapidly.

1-3 ENERGY-SAVING POTENTIAL IN THE END-USE OF ELECTRICITY

According to [4], the United States consumed approximately 96 quadrillions BTU (quads) of primary energy in 2013 (it was nearly 100 quads in 2018), as shown in Fig. 1-3. Out of the total, 32% was consumed in the industrial sector and 39% in the residential and the commercial sectors combined.

Fig. 1-3 Primary energy consumption by end-use sector in the United States in 2013.

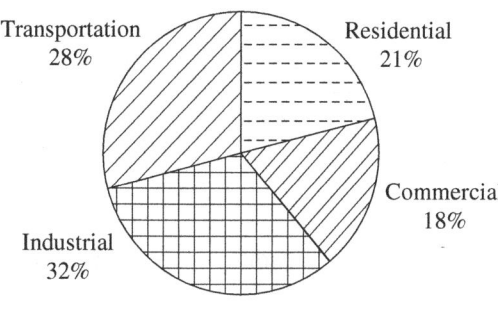

Total energy consumption = 96.1 quads

1-3-1 Energy-Saving Potential in the Process Industry

Traditionally, motors were operated uncontrolled, running at constant speeds, even in applications where efficient control over their speed could be very advantageous. For example, consider the process industry (e.g. oil refineries and chemical factories) where the flow rates of gases and fluids often need to be controlled. As Fig. 1-4a illustrates, in a pump driven at a constant speed, a throttling valve controls the flow rate. Mechanisms such as throttling valves are generally more complicated to implement in automated processes and waste large amounts of energy. In the process industry today, electronically controlled adjustable-speed drives (ASDs), shown in Fig. 1-4b, control the pump speed to match the flow requirement. Systems with ASDs are much easier to automate and offer much higher energy efficiency and lower maintenance than the traditional systems with throttling valves.

According to [1], the US industrial motor systems of all sizes and in all applications have the potential energy-saving opportunity, as a percentage of the US end-use electricity load, from 3.3 to 8.9%.

These improvements are not limited to the process industry. Electric drives for speed and position control are increasingly being used in a variety of manufacturing, heating, ventilating, and air conditioning (HVAC), and transportation systems, as we will see in the subsequent sections.

8 ELECTRIC DRIVES: INTRODUCTION AND MOTIVATION

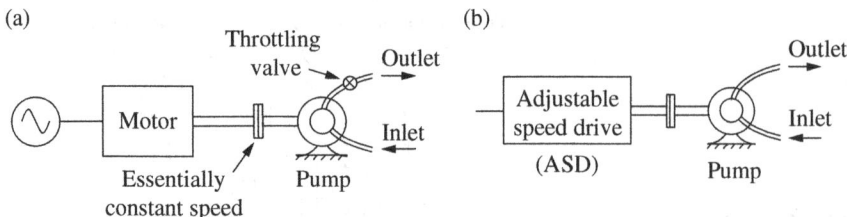

Fig. 1-4 Traditional and ASD-based flow control systems.

1-3-2 Energy-Saving Potential in the Residential and Commercial Sectors

Out of the total, the residential and commercial end-uses represent 39% of the total energy consumed, as depicted in Fig. 1-3. In the residential sector (Fig. 1-5a), the primary energy consumption of electric motor-driven systems and components is 4.73 quads. In the commercial sector (Fig. 1-5b), it is 4.87 quads. Fig. 1-5a and b provide a breakdown of motor-driven energy consumption by end-use for the residential and commercial sectors, respectively. Thus, approximately 10% of the total primary energy consumed can be attributed to electric motor-driven systems in the residential and commercial sectors.

According to [4], the technical energy-saving potential achievable through motor upgrades and variable speed technology is estimated to be 536 trillion BTU (0.54 quads) of the primary energy in the residential sector.

In the commercial sector, technical potential due to motor upgrades alone is 0.46 quads of the primary energy, whereas the potential savings resulting from the use of variable-speed drives alone is 0.53 quads of the primary energy.

Therefore, the primary energy-saving potential in the residential and the commercial sectors combined is approximately 1.53 quads. This, as a percentage of the total primary energy consumed, is approximately 1.5%. Assuming the efficiency by which the primary energy is converted to electricity to be 35%, the savings of 1.53 quads of the primary energy equals approximately 157 billion kWh of saved electricity. As a percentage of the total electricity generated in 2018 in the United States, this represents savings of 3.76%.

ENERGY-SAVING POTENTIAL IN THE END-USE OF ELECTRICITY 9

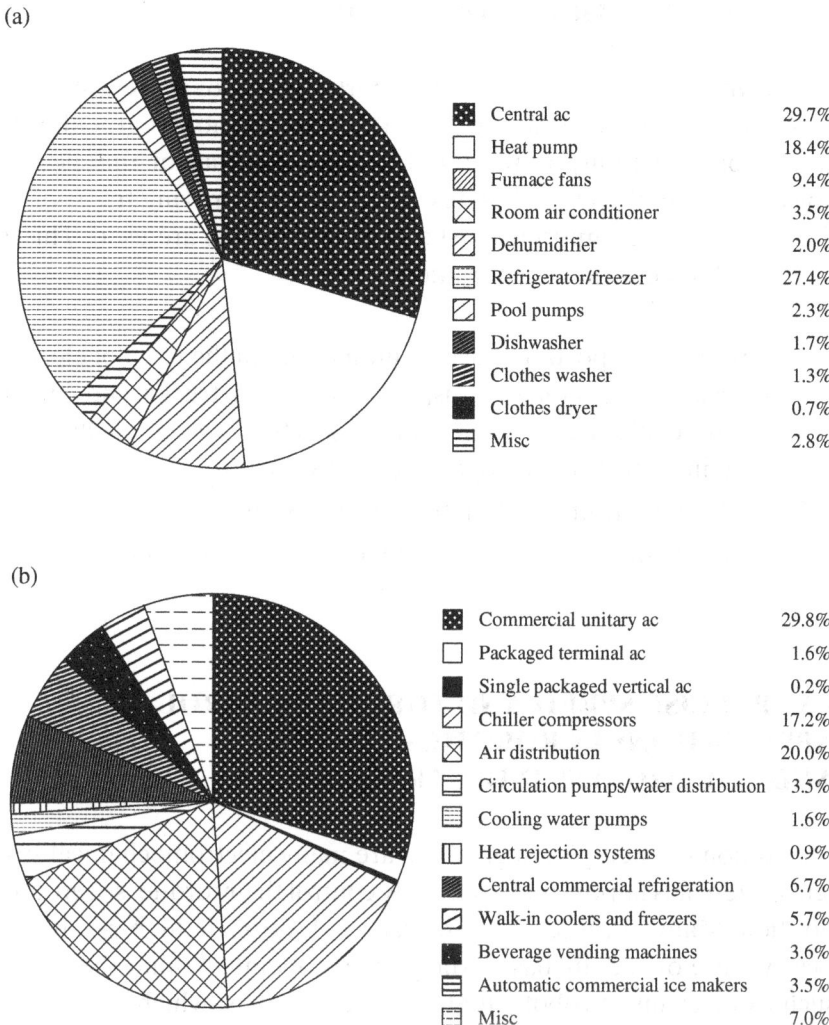

Fig. 1-5 Energy usage in (a) residential sector and (b) commercial sector.

Therefore, in the United States, industrial motor systems of all sizes and in all applications, combined with the motor systems in residential and commercial sectors, have the potential energy-saving opportunity, as a percentage of the total US end-use electricity is from approximately 7 to 12%.

1-4 ELECTRIC TRANSPORTATION

As shown earlier in Fig. 1-3, the transportation sector represents 28% of primary energy consumption. In 2016, the emission of greenhouse gases from the transportation sector surpassed that of the electric power sector in the United States. Therefore, electrifying transportation is of extreme importance where electric motors are used. This is true for all modes of transportation:

1. Ground transportation using automobiles in the form of electric vehicles for personal transport but also in trucks and buses. These could be in the form of electric, hybrid-electric, or plug-in hybrid or hydrogen fuel-cell vehicles.
2. High-speed trains and metro transit systems.
3. Aircrafts that all use electric generators and motors.

1-5 PRECISE SPEED AND TORQUE CONTROL APPLICATIONS IN ROBOTICS, DRONES, AND THE PROCESS INDUSTRY

In addition to energy savings, there are many electromechanical systems where it is important to precisely control their torque, speed, and position. Many of these systems, such as elevators in high-rise buildings, we use on a daily basis. Many others operate behind the scene, such as mechanical robots in automated factories, which are crucial for industrial competitiveness. Even in general-purpose applications of ASDs, such as pumps and compressors systems, it is possible to control ASDs in a way to increase their energy efficiency. Advanced electric drives are also needed in wind-electric systems to generate electricity at variable speed. Hybrid-electric and electric vehicles represent an important application of advanced electric drives in the immediate future. In most of these applications, increasing efficiency

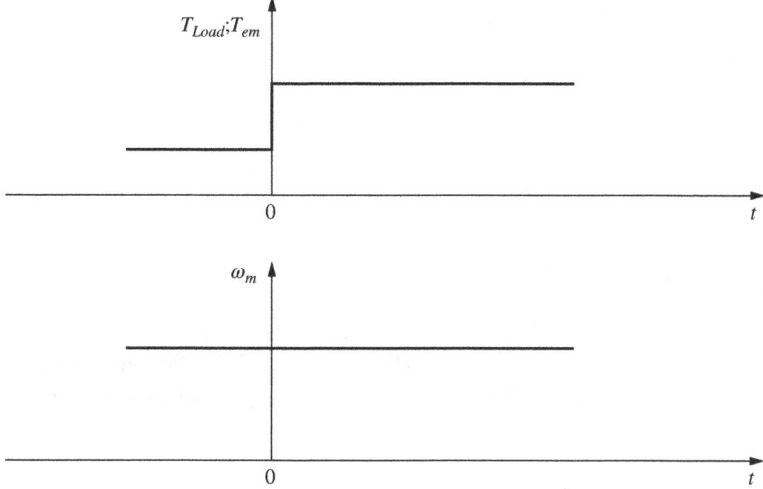

Fig. 1-6 Need for controlling the electromagnetic torque T_{em}.

requires producing maximum torque per ampere, as will be explained in this book. It also requires controlling the electromagnetic torque, as quickly and as precisely as possible. As illustrated in Fig. 1-6, the load torque T_{Load} may take a step-jump in time, in response to which the electromagnetic torque produced by the machine T_{em} must also take a step-jump if the speed ω_m of the load is to remain constant.

1-6 RANGE OF ELECTRIC DRIVES

Electric drives are increasingly being used in most sectors of the economy. Figure 1-7 shows that electric drives cover an extremely large range of power and speed – up to 100 MW in power and up to 80 000 rpm in speed.

Due to the power electronic converter, drives are not limited in speeds, unlike line-fed motors that are generally limited to 3600 rpm or so with a 60-Hz supply (3 000 rpm with a 50-Hz supply).

12 ELECTRIC DRIVES: INTRODUCTION AND MOTIVATION

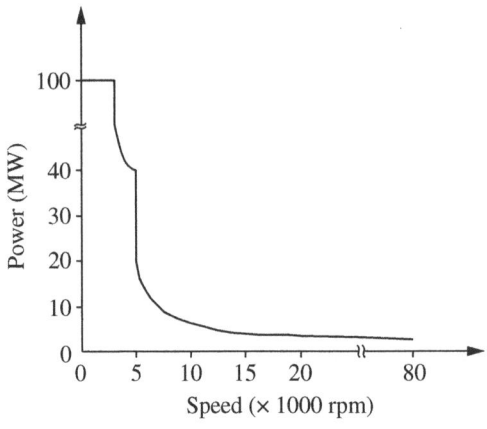

Fig. 1-7 Power and speed range of electric drives.

1-7 THE MULTIDISCIPLINARY NATURE OF DRIVE SYSTEMS

The block diagram of Fig. 1-1b points to various fields, which are essential to electric drives: electric machine theory, power electronics, analog and digital control theory, real-time application of digital controllers, mechanical system modeling, and interaction with electric power systems. A brief description of each of the fields is provided in the following subsections.

1. Theory of electric machines
   For achieving the desired motion, it is necessary to control electric motors appropriately. This requires a thorough understanding of the operating principles of various commonly used motors, such as dc, synchronous, induction, and stepper motors. The emphasis in an electric drives course needs to be different from that in traditional electric machines courses, which are oriented toward the design and application of line-fed machines.
2. Power electronics
   In Fig. 1-1b, voltages and currents from a fixed form (in frequency and magnitude) are converted to the adjustable

form best suited to the motor. It is important that the conversion takes place at a high-energy efficiency, which is realized by operating power semiconductor devices like switches.

Today, power electronics is being simplified using "Smart Power" devices, where power semiconductor switches are integrated with their protection and gate-drive circuits into a single module. Thus, the logic-level signals (such as those supplied by a digital signal processor) can directly control high-power switches in the converter. Such power-integrated modules are available with voltage handling capability approaching 4 kV and current handling capability above 1000 A. Paralleling such modules allows even higher current handling capabilities. The progress in this field has made a dramatic impact on PPUs by reducing their size and weight, while substantially increasing the number of functions that can be performed. Recently, there has been a quiet revolution where transistors based on wide bandgap materials such as SiC and GaN are commercialized. These devices have many superior characteristics compared to original Si-based devices, thus increasing the efficiency of power converters and reducing the system cost.

3. Control theory

   In the majority of applications, the speed and position of drives need not be controlled precisely. However, there is an increasing number of applications, for example, in robotics for automated factories, where accurate control of torque, speed, and position is essential. Such control is accomplished by feeding back the measured quantities, and by comparing them with their desired values, in order to achieve a fast and accurate control. In most motion control applications, it is sufficient to use a simple proportional-integral (PI) control, as discussed in this book. The task of designing and analyzing PI-type controllers is made easy due to the availability of powerful simulation tools.

14 ELECTRIC DRIVES: INTRODUCTION AND MOTIVATION

4. Real-time control using DSPs
   All modern electric drives use microprocessors and digital signal processors (DSPs) for flexibility of control, fault diagnosis, and communication with the host computer and other process computers. The use of 8-bit microprocessors is being replaced by 16-bit and even 32-bit microprocessors. DSPs are used for real-time control in applications which demand high performance or where a slight gain in the system efficiency more than pays for the additional cost of sophisticated control.
5. Mechanical system modeling
   Specifications of electric drives depend on the torque and speed requirements of the mechanical loads. Therefore, it is often necessary to model mechanical loads. Rather than considering the mechanical load and the electric drive as two separate subsystems, it is preferable to consider them together in the design process. This design philosophy is at the heart of Mechatronics.
6. Sensors
   As shown in the block diagram of electric drives in Fig. 1-1b, voltage, current, speed, and position measurements may be required. For thermal protection, the temperature needs to be sensed.
7. Interactions of drives with the utility grid
   Unlike line-fed electric motors, electric motors in drives are supplied through a power electronic interface (see Fig. 1-1b). Therefore, unless corrective action is taken, electric drives draw currents from the utility that are distorted (non-sinusoidal) in their wave shape. This distortion in line currents interferes with the utility system, degrading its power quality by distorting the utility voltages. Available technical solutions make the drive interaction with the utility harmonious, even more so than line-fed motors. The sensitivity of drives to power system disturbances such as sags swells and transient overvoltages should also be considered. Again, solutions are available to reduce or eliminate the effects of these disturbances.

1-8 USE OF SIMULATION AND HARDWARE PROTOTYPING

Through the course of this book, modeling tools are used to facilitate discussion and provide in-depth understanding of the concepts in electric drives. MATLAB/Simulink™ and Sciamble™ Workbench [5] are computer simulation tools used to demonstrate all topics in this course.

MATLAB/Simulink™ is widely accepted as the standard tool for electric drives simulation and analysis. The student version of this tool is sufficient for the purposes of this text and is reasonably priced for educational institutions. With an abundance of online training material, it is easy for a new user to become proficient in its use.

Sciamble™ Workbench is a mathematical simulation tool developed at the University of Minnesota for educational purposes. Workbench simulation software is free of cost enabling academic institutions access to advanced tools in education. All examples and key concepts explained in this book are also simulated using Workbench and the results are provided in the accompanying website.

Another noteworthy motivation for using Sciamble™ Workbench is its seamless transition between mathematical simulation and hardware prototyping modes. Hardware prototyping simplifies the development of a real-time controller enabling rapid laboratory experimentation of concepts taught in this book. Real-world experimentation enables a more in-depth and practical understanding of the contents of this book. Sciamble™'s electric drives hardware kit is used for laboratory implementation of all topics in this book.

All the simulation files using both MATLAB/Simulink™ and Sciamble™ Workbench, as well as the manual for laboratory implementation using Workbench, are available on the accompanying website.

16 ELECTRIC DRIVES: INTRODUCTION AND MOTIVATION

1-9 STRUCTURE OF THE TEXTBOOK

This book is in three parts: Part I describes the fundamental concepts required for the study of ac electric machines and drives, Part II describes the steady-state analysis of ac machines and drives, and Part III describes the dynamic analysis of ac drives and their vector control through simulations, leading to the hardware implementation of vector control. Throughout this book, the analysis leads to simulations using MATLAB/Simulink™ and Workbench of Sciamble™ (www.sciamble.com), a University of Minnesota startup. The simulations in Workbench of Sciamble™ seamlessly lead to hardware implementation, as described throughout the book. These are as follows:

1. Part I: Fundamental Concepts
   Chapter 2 deals with the modeling of mechanical systems coupled to electric drives, as well as how to determine drive specifications for various types of loads. Chapter 3 describes magnetic concepts and the laws governing electromechanical energy conversion. An introduction to PPUs is presented in Chapter 4. Chapter 5 explains the design of feedback control in vector control of ac drives. As a background to the discussion of ac motor drives, the rotating fields in ac machines are described in Chapter 6 utilizing space vectors. In controlling PPUs, the role of space vector PWM is described in Chapter 7.
2. Part II: Steady-State Analysis
   Using the space vector theory, the sinusoidal PMAC motor drives are discussed in Chapter 8. Chapter 9 analyzes induction motors and focuses on their basic principles of operation in a steady state. A concise but comprehensive discussion of controlling speed with induction-motor drives is provided in Chapter 10.
3. Part III: Dynamic Analysis
   Chapters 11 and 12 lay down the foundation on dq-based analysis of ac machines and drives. Chapter 13 describes the vector-control of induction motors. Chapter 14 is on encoder-less

vector control of induction motor drives. Chapter 15 is on doubly fed generators used in wind generators. Direct-torque control is explained in Chapter 16. The vector control is applied to the surface permanent, and the interior permanent-magnet motor drives in Chapter 17. The reluctance drives, including stepper-motors and switched-reluctance drives, are explained in Chapter 18.

1-10 REVIEW QUESTIONS

1. What is an electric drive? Draw the block diagram and explain the roles of its various components.

2. What has been the traditional approach to controlling the flow rate in the process industry? What are the major disadvantages which can be overcome by using ASDs?

3. What are the factors responsible for the growth of the adjustable-speed drive market?

4. How does an air conditioner work?

5. How does a heat pump work?

6. How do ASDs save energy in air conditioning and heat pump systems?

7. What is the role of ASDs in industrial systems?

8. There are proposals to store energy in flywheels for load leveling in utility systems. During the off-peak period for energy demand at night, these flywheels are charged to high speeds. At peak periods during the day, this energy is supplied back to the utility. How would ASDs play a role in this scheme?

9. What is the role of electric drives in electric transportation systems of various types?

10. What are the different disciplines that make up the study and design of electric-drive systems?

18 ELECTRIC DRIVES: INTRODUCTION AND MOTIVATION

REFERENCES

1. https://eere-exchange.energy.gov/fileContent.aspx?fileID=3b30e33e-9f3e-442b-b623-d724924b8581
2. https://www.eia.gov/tools/glossary/index.php?id=Primary%20energy
3. https://www.eia.gov/tools/faqs/faq.php?id=427&t=3
4. https://www.energy.gov/sites/prod/files/2014/02/f8/Motor%20Energy%20Savings%20Potential%20Report%202013-12-4.pdf
5. https://sciamble.com
6. Clark, K., Miller, N. W., and Sanchez-Gasca, J. J. (2009). Modeling of GE wind turbine-generators for grid studies. GE Energy Report, Version 4.4, 9 September 2009.
7. Johnson, K. E. (2004). Adaptive torque control of variable speed wind turbines. NREL Technical Report, August 2004.

FURTHER READING

Mohan, N. (1981). Techniques for energy conservation in AC motor driven systems. Electric Power Research Institute Final Report EM-2037, Project 1201-1213, September 1981.

Mohan, N. and Ramsey, J. (1986). Comparative study of adjustable-speed drives for heat pumps. Electric Power Research Institute Final Report EM-4704, Project 2033-4, August 1986.

PROBLEMS

1-1 A US Department of Energy report estimates that over 100 billion kWh/year can be saved in the United States by various energy conservation techniques applied to the pump-driven systems. Calculate (a) how many 1000-MW generating plants running constantly supply this wasted energy and (b) the annual savings in dollars if the cost of electricity is 0.10$/kWh.

1-2 Visit your local machine-tool shop and make a list of various electric drive types, applications, and speed/torque ranges.

1-3 Repeat Problem 1-2 for automobiles.

1-4 Repeat Problem 1-2 for household appliances.

1-5 In wind turbines, the ratio (P_{shaft}/P_{wind}) of the power available at the shaft to the power in the wind is called the coefficient of performance, C_p, which is a unit-less quantity. For informational purpose, the plot of this coefficient, as a function of λ, is shown in Fig. P1-5 [6] for various values of the blades pitch-angle θ, where λ is a constant times the ratio of the blade-tip speed and the wind speed.

The rated power is produced at the wind speed of 12 m/s where the rotational speed of the blades is 20 rpm. The cut-in wind speed is 4 m/s. Calculate the range over which the blade speed should be varied, between the cut-in and the rated wind speeds, to harness the maximum power from the wind. In this range of wind speeds, the blade's pitch-angle θ is kept at nearly zero. Note: this simple problem shows the benefit of varying the speed of wind turbines, by means of a variable-speed drive, for maximizing the harnessed energy.

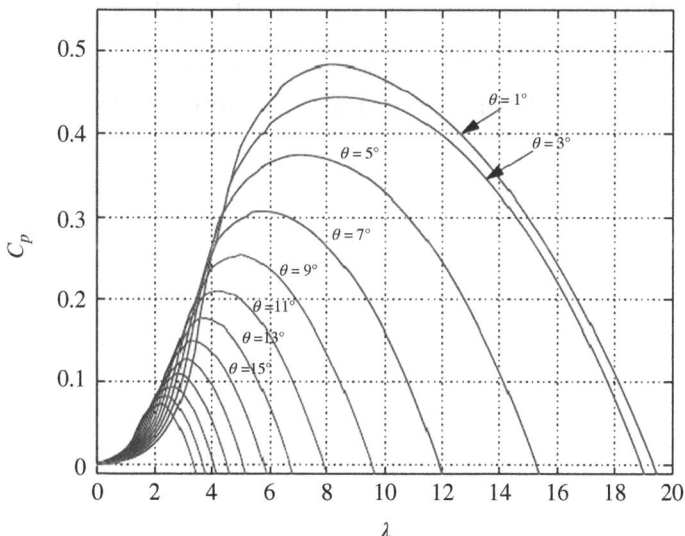

Fig. P1-5 Plot of C_p as a function of λ [6].

1-6 Read the report in Appendix 1-A in the accompanying website "Adaptive Torque Control of Variable Speed Wind Turbines" by Kathryn E. Johnson, National Renewable Energy Laboratory [7].

1-7 In wind turbines, describe the Standard Region 2 Control and describe how it works in your own words.

1-8 Read the report in Appendix 1-B in the accompanying website "Final Report on Assessment of Motor Technologies for Traction Drives of Hybrid and Electric Vehicles" and answer the following questions for HEV/EV applications:
(a) What are the types of machines considered?
(b) What type of motor is the most popular choice?
(c) What are the alternatives if NdFeB magnets are not available?
(d) What are the advantages and disadvantages of SR motors?

1-9 Read the report in Appendix 1-C in the accompanying website "Evaluation of the 2010 Toyota Prius Hybrid Synergy Drive System" and answer the following questions:
(a) What are ECVT, P, CU, and ICE?
(b) What type of motor is used in this application?

1-10 Read the report in Appendix 1-D and summarize the possibility of using SR motors in HEV/EV applications.

2 Understanding Mechanical System Requirements for Electric Drives

2-1 INTRODUCTION

Electric drives are an interface between an electrical system and a mechanical system, as Fig. 2-1 shows. They themselves consist of an electric machine and a power electronic converter. In subsequent chapters of this book, we will look at the power electronic converter and its role, as well as analyze electric machines and how they can be controlled, given the desired speed and position of the mechanical system.

In electric drives, the power flow may be in either direction. For example, in an electric vehicle in driving mode, power flows from the electric source, a battery, to the electric motor, a mechanical system, through a power electronic converter. On the other hand, while slowing down the vehicle, the roles are reversed. The kinetic energy of the moving vehicle energy is extracted (called regenerative braking), and power flows from the electric motor to the battery, again through a power electronic converter.

Electric drives must satisfy the requirements of torque and speed imposed by mechanical loads connected to them. The load in

---

(Adapted from chapter 1 of *Electric Machines and Drives: A First Course* ISBN: 978-1-118-07481-7 by Ned Mohan, January 2012)

Analysis and Control of Electric Drives: Simulations and Laboratory Implementation, First Edition. Ned Mohan and Siddharth Raju.
© 2021 John Wiley & Sons, Inc. Published 2021 by John Wiley & Sons, Inc.
Companion website: www.wiley.com/go/Mohan/Vectorcontrolinelectricdrives

22 UNDERSTANDING MECHANICAL SYSTEM REQUIREMENTS

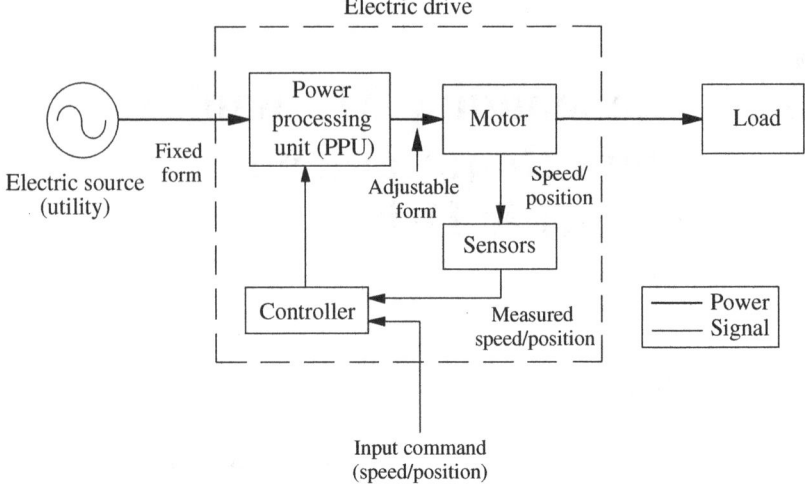

Fig. 2-1 Block diagram of adjustable speed drives.

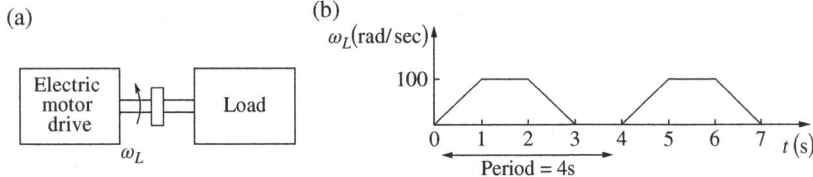

Fig. 2-2 (a) Electric drive system and (b) example of a load-speed profile requirement.

Fig. 2-2, for example, may require a trapezoidal profile for the angular speed, as a function of time. In this chapter, we will briefly review the basic principles of mechanics for understanding the requirements imposed by mechanical systems on electric drives. This understanding is necessary for selecting an appropriate electric drive for a given application.

This analysis equally applies when the load becomes the source of power, as in a wind turbine, and the electric drive generates and transfers power to the utility grid, an electric system.

2-2 SYSTEMS WITH LINEAR MOTION

We will begin by applying physical laws of motion in their simplest form, starting with linear systems. In Fig. 2-3a, a load of a constant mass M is acted upon by an external force f_e that causes it to move in the linear direction x at speed $u = dx/dt$.

This movement is opposed by the load, represented by a force f_L. The linear momentum associated with the mass is defined as $M \times u$. As shown in Fig. 2-3b, in accordance with Newton's Law of Motion, the net force $f_M (= f_e - f_L)$ equals the rate of change of momentum, which causes the mass to accelerate:

$$f_M = \frac{d}{dt}(Mu) = M\frac{du}{dt} = Ma \qquad (2\text{-}1)$$

where a is the acceleration in m/s², which from Eq. (2-1) is

$$a = \frac{du}{dt} = \frac{f_M}{M} \qquad (2\text{-}2)$$

In MKS units, a net force of 1 Newton (or 1 N), acting on a constant mass of 1 kg, results in an acceleration of 1 m/s². Integrating the acceleration with respect to time, we can calculate the speed as

$$u(t) = u(0) + \int_0^t a(\tau) \cdot d\tau \qquad (2\text{-}3)$$

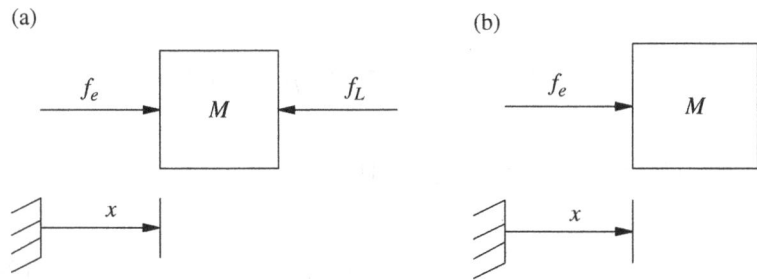

Fig. 2-3 Motion of a mass M due to the action of forces.

24 UNDERSTANDING MECHANICAL SYSTEM REQUIREMENTS

and, integrating the speed with respect to time, we can calculate the position as

$$x(t) = x(0) + \int_0^t u(\tau) \cdot d\tau \qquad (2\text{-}4)$$

where τ is a variable of integration.

The differential work dW done by the mechanism supplying the force f_e is

$$dW_e = f_e\, dx \qquad (2\text{-}5)$$

Power is the time-rate at which the work is done. Therefore, differentiating both sides of Eq. (2-5) with respect to time t, and assuming that the force f_e remains constant, the power supplied by the mechanism exerting the force f_e is

$$p_e(t) = \frac{dW_e}{dt} = f_e \frac{dx}{dt} = f_e u \qquad (2\text{-}6)$$

It takes a finite amount of energy to bring a mass to a speed from rest. Therefore, a moving mass has stored kinetic energy that can be recovered. Note that in the system of Fig. 2-3, the net force $f_M(=f_e - f_L)$ is responsible for accelerating the mass. Therefore, assuming that f_M remains constant, the net power $p_M(t)$ going into accelerating the mass can be calculated by replacing f_e in Eq. (2-6) with f_M:

$$p_M(t) = \frac{dW_M}{dt} = f_M \frac{dx}{dt} = f_M u \qquad (2\text{-}7)$$

From Eq. (2-1), substituting f_M as $M\dfrac{du}{dt}$,

$$p_M(t) = Mu\frac{du}{dt} \qquad (2\text{-}8)$$

The energy input, which is stored as kinetic energy in the moving mass, can be calculated by integrating both sides of Eq. (2-8) with respect to time. Assuming the initial speed u to be zero at time $t = 0$, the stored kinetic energy in the mass M can be calculated as

$$W_M = \int_0^t p_M(\tau)\,d\tau = M\int_0^t u\frac{du}{d\tau}d\tau = M\int_0^u u\,du = \frac{1}{2}Mu^2 \qquad (2\text{-}9)$$

where τ is a variable of integration.

2-3 ROTATING SYSTEMS

Most electric motors are of a rotating type. Consider a lever, pivoted and free to move as shown in Fig. 2-4a. When an external force f is applied in a *perpendicular* direction at a radius r from the pivot, then the torque acting on the lever is

$$\begin{array}{ccc} T & = & f \; r \\ \text{[Nm]} & & \text{[N] [m]} \end{array} \qquad (2\text{-}10)$$

which acts in a counterclockwise direction, considered here to be positive.

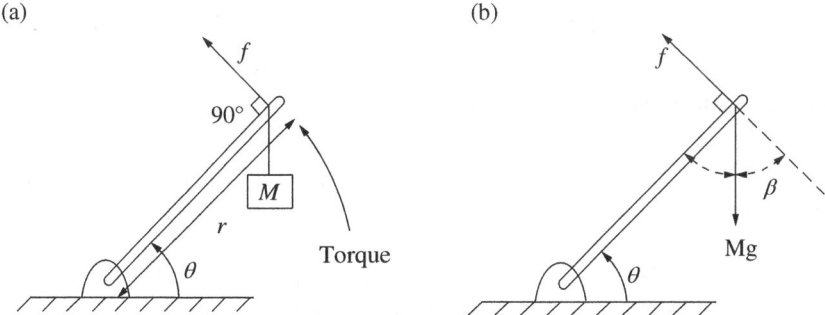

Fig. 2-4 (a) Pivoted lever and (b) holding torque for the lever.

26 UNDERSTANDING MECHANICAL SYSTEM REQUIREMENTS

EXAMPLE 2-1

In Fig. 2-4a, a mass M is hung from the tip of the lever. Calculate the holding torque required to keep the lever from turning, as a function of angle θ in the range of 0–90°. Assume that $M = 0.5$ kg and $r = 0.3$ m.

Solution

The gravitational force on the mass is shown in Fig. 2-4b. For the lever to be stationary, the net force perpendicular to the lever must be zero, i.e. $f = Mg\cos\beta$ where $g = 9.8$ m/s² is the gravitational acceleration. Note in Fig. 2-4b that $\beta = \theta$. The holding torque T_h must be $T_h = fr = Mgr\cos\theta$. Substituting the numerical values,

$$T_h = 0.5 \times 9.8 \times 0.3 \times \cos\theta = 1.47\cos\theta \text{ Nm}$$

In electric machines, the various forces shown by arrows in Fig. 2-5 are produced due to electromagnetic interactions. The definition of torque in Eq. (2-10) correctly describes the resulting electromagnetic torque T_{em} that causes the rotation of the motor and the mechanical load connected to it by a shaft.

In a rotational system, the angular acceleration, due to the net torque acting on it, is determined by its moment-of-inertia J. The example below shows how to calculate the moment-of-inertia J of a rotating solid cylindrical mass.

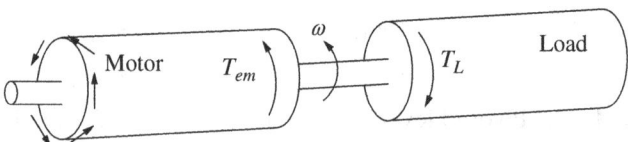

Fig. 2-5 Torque in an electric motor.

EXAMPLE 2-2

(a) Calculate the moment-of-inertia J of a solid cylinder that is free to rotate about its axis, as shown in Fig. 2-6a, in terms of its mass M and the radius r_1.

(b) Given that a solid steel cylinder has a radius $r_1 = 6$ cm, length $\ell = 18$ cm, and material density $\rho = 7.85 \times 10^3$ kg/m³, calculate its moment-of-inertia J.

Solution

(a) From Newton's Law of Motion, in Fig. 2-6a, to accelerate a differential mass dM at a radius r, the net differential force df required in a perpendicular (tangential) direction, from Eq. (2-1), is

$$(dM)\left(\frac{du}{dt}\right) = df \tag{2-11}$$

where the linear speed u in terms of the angular speed ω_m (in rad/s) is

$$u = r\,\omega_m \tag{2-12}$$

Multiplying both sides of Eq. (2-11) by the radius r, recognizing that $(r\,df)$ equals the net differential torque dT and using Eq. (2-12),

$$r^2\,dM\,\frac{d}{dt}\omega_m = dT \tag{2-13}$$

The same angular acceleration $\frac{d}{dt}\omega_m$ is experienced by all elements of the cylinder. With the help of Fig. 2-6b, the differential mass dM in Eq. (2-13) can be expressed as

$$dM = \rho\ \underbrace{r\,d\theta}_{arc}\ \underbrace{dr}_{height}\ \underbrace{d\ell}_{length} \tag{2-14}$$

where ρ is the material density in kg/m³. Substituting dM from Eq. (2-14) into Eq. (2-13),

(*Continued*)

$$\rho \left(r^3 dr\, d\theta\, d\ell\right) \frac{d}{dt}\omega_m = dT \qquad (2\text{-}15)$$

The net torque acting on the cylinder can be obtained by integrating over all differential elements in terms of r, θ, and ℓ as

$$\rho \left(\int_0^{r_1} r^3 dr \int_0^{2\pi} d\theta \int_0^{\ell} d\ell\right) \frac{d}{dt}\omega_m = T \qquad (2\text{-}16)$$

Carrying out the triple integration yields

$$\underbrace{\left(\frac{\pi}{2}\rho \ell\, r_1^4\right)}_{J_{cyl}} \frac{d}{dt}\omega_m = T \text{ or} \qquad (2\text{-}17)$$

$$J_{cyl} \frac{d\omega_m}{dt} = T \qquad (2\text{-}18)$$

where the quantity within the brackets in Eq. (2-17) is called the moment-of-inertia J, which for a solid cylinder is

$$J_{cyl} = \frac{\pi}{2}\rho \ell\, r_1^4 \qquad (2\text{-}19)$$

Since the mass of the cylinder in Fig. 2-6a is $M = \rho\left(\pi r_1^2\right)\ell$, the moment-of-inertia in Eq. (2-19) can be written as

$$J_{cyl} = \frac{1}{2} M r_1^2 \qquad (2\text{-}20)$$

(b) Substituting $r_1 = 6$ cm, length $\ell = 18$ cm, and $\rho = 7.85 \times 10^3$ kg/m³ in Eq. (2-19), the moment-of-inertia J_{cyl} of the cylinder in Fig. 2-5a is

$$J_{cyl} = \frac{\pi}{2} \times 7.85 \times 10^3 \times 0.18 \times (0.06)^4 = 0.029 \text{ kg·m}^2$$

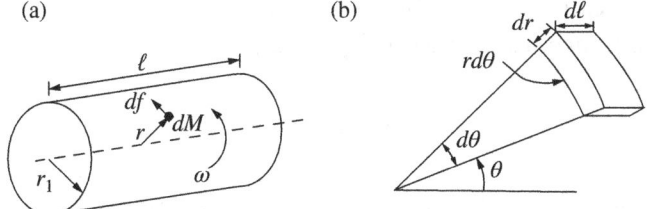

Fig. 2-6 Calculation of the inertia, J_{cyl}, of a solid cylinder.

The net torque T_J acting on the rotating body of inertia J causes it to accelerate. Similar to systems with linear motion, where $f_M = Ma$ Newton's Law in rotational systems becomes

$$T_J = J\alpha \qquad (2\text{-}21)$$

where the angular acceleration $\alpha(=d\omega/dt)$ in rad/s^2 is

$$\alpha = \frac{d\omega_m}{dt} = \frac{T_J}{J} \qquad (2\text{-}22)$$

which is similar to Eq. (2-18) in the previous example. In MKS units, a torque of 1 Nm acting on the inertia of 1 kg $\cdot$ m^2 results in an angular acceleration of 1 rad/s^2.

In systems such as the one shown in Fig. 2-7a, the motor produces an electromagnetic torque T_{em}. The bearing friction and wind resistance (drag) can be combined with the load torque T_L opposing the rotation. In most systems, we can assume that the rotating part of the motor with inertia J_M is rigidly coupled (without flexing) to the load inertia J_L. The net torque, which is the difference between the electromagnetic torque developed by the motor and the load torque

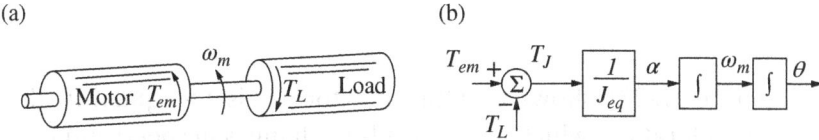

Fig. 2-7 Motor and load torque interaction with a rigid coupling.

30 UNDERSTANDING MECHANICAL SYSTEM REQUIREMENTS

opposing it, causes the combined inertias of the motor and the load to accelerate in accordance with Eq. (2-22):

$$\frac{d}{dt}\omega_m = \frac{T_J}{J_{eq}} \qquad (2\text{-}23)$$

where the net torque $T_J = T_{em} - T_L$ and the equivalent combined inertia is $J_{eq} = J_M + J_L$.

EXAMPLE 2-3

In Fig. 2-7a, each structure has the same inertia as the cylinder in Example 2-2. The load torque T_L is negligible. Calculate the required electromagnetic torque, if the speed is to increase linearly from rest to 1800 rpm in 5 s.

Solution

Using the results of Example 2-2, the combined inertia of the system is

$$J_{eq} = 2 \times 0.029 = 0.058 \text{ kg·m}^2$$

The angular acceleration is

$$\frac{d}{dt}\omega_m = \frac{\Delta \omega_m}{\Delta t} = \frac{(1800/60)2\pi}{5} = 37.7 \text{ rad/s}^2$$

Therefore, from Eq. (2-23),

$$T_{em} = 0.058 \times 37.7 = 2.19 \text{ Nm}$$

Equation (2-23) shows that the net torque is the quantity that causes acceleration, which in turn leads to changes in speed and position. Integrating the acceleration $\alpha(t)$ with respect to time,

$$\omega_m(t) = \omega_m(0) + \int_0^t \alpha(\tau)\,d\tau \qquad (2\text{-}24)$$

where $\omega_m(0)$ is the speed at $t = 0$ and τ is a variable of integration. Further integrating $\omega_m(t)$ in Eq. (2-24) with respect to time yields

$$\theta(t) = \theta(0) + \int_0^t \omega_m(\tau)\,d\tau \qquad (2\text{-}25)$$

where $\theta(0)$ is the position at $t = 0$ and τ is again a variable of integration. Equations (2-23) through (2-25) indicate that torque is the fundamental variable for controlling speed and position. Equations (2-23) through (2-25) can be represented in a block-diagram form, as shown in Fig. 2-6b.

EXAMPLE 2-4

Consider that the rotating system shown in Fig. 2-7a, with the combined inertia $J_{eq} = 2 \times 0.029 = 0.058 \text{ kg} \cdot \text{m}^2$, is required to have the angular speed profile shown in Fig. 2-1b. The load torque is zero. Calculate and plot, as functions of time, the electromagnetic torque required from the motor, and the change in position.

Solution

In the plot of Fig. 2-2b, the magnitude of the acceleration and the deceleration is 100 rad/s². During the intervals of acceleration and deceleration, since $T_L = 0$,

$$T_{em} = T_J = J_{eq}\frac{d\omega_m}{dt} = \pm 5.8 \text{ Nm}$$

as shown in Fig. 2-8.

During intervals with a constant speed, no torque is required. Since the position θ is the time-integral of speed, the resulting change of position (assuming that the initial position is zero) is also plotted in Fig. 2-8.

32 UNDERSTANDING MECHANICAL SYSTEM REQUIREMENTS

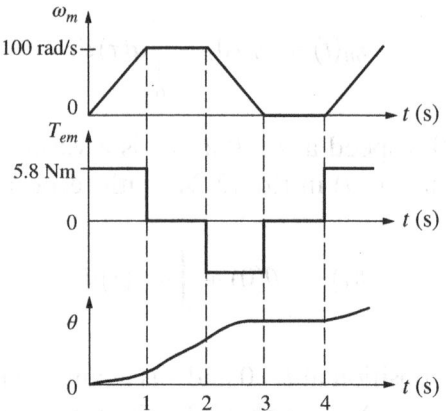

Fig. 2-8 Speed, torque, and angle variations with time.

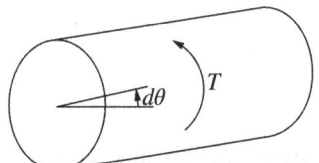

Fig. 2-9 Torque, work, and power.

In a rotational system shown in Fig. 2-9, if a net torque T causes the cylinder to rotate by a differential angle $d\theta$, the differential work done is

$$dW = T\,d\theta \qquad (2\text{-}26)$$

If this differential rotation takes place in a differential time dt, the power can be expressed as

$$p = \frac{dW}{dt} = T\frac{d\theta}{dt} = T\omega_m \qquad (2\text{-}27)$$

where $\omega_m = d\theta/dt$ is the angular speed of rotation. Substituting for T from Eq. (2-21) into Eq. (2-27),

$$p = J\frac{d\omega_m}{dt}\omega_m \qquad (2\text{-}28)$$

Integrating both sides of Eq. (2-28) with respect to time, assuming that the speed ω_m and the kinetic energy W at the time $t = 0$ are both zero, the kinetic energy stored in the rotating mass of inertia J is

$$W = \int_0^t p(\tau)\,d\tau = J\int_0^t \omega_m \frac{d\omega_m}{d\tau}\,d\tau = J\int_0^{\omega_m} \omega_m\,d\omega_m = \frac{1}{2}J\omega_m^2 \qquad (2\text{-}29)$$

This stored kinetic energy can be recovered by making the power $p(t)$ reverse direction, that is, by making $p(t)$ negative.

EXAMPLE 2-5

In Example 2-3, calculate the kinetic energy stored in the combined inertia at a speed of 1800 rpm.

Solution

From Eq. (2-29),

$$W = \frac{1}{2}(J_L + J_M)\omega_m^2 = \frac{1}{2}(0.029 + 0.029)\left(2\pi\frac{1800}{60}\right)^2 = 1030.4 \text{ J}$$

2-4 FRICTION

Friction within the motor and the load acts to oppose rotation. Friction occurs in the bearings that support rotating structures. Moreover, moving objects in air encounter windage or drag. In vehicles, this drag is a major force that must be overcome. Therefore, friction and windage can be considered as opposing forces or torque that must be overcome. The frictional torque is generally nonlinear in nature. We are

34 UNDERSTANDING MECHANICAL SYSTEM REQUIREMENTS

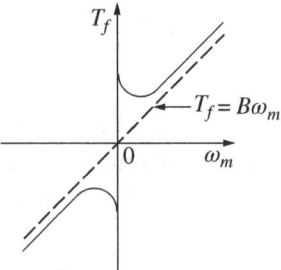

Fig. 2-10 Actual and linearized friction characteristics.

all familiar with the need for a higher force (or torque) in the beginning (from rest) to set an object in motion. This friction at zero speed is called stiction. Once in motion, the friction may consist of a component called Coulomb friction, which remains independent of speed magnitude (it always opposes rotation), as well as another component called viscous friction, which increases linearly with speed.

In general, the frictional torque T_f in a system consists of all of the aforementioned components. An example is shown in Fig. 2-10; this friction characteristic may be linearized for an approximate analysis by means of the dotted line. With this approximation, the characteristic is similar to that of viscous friction in which

$$T_f = B\omega_m \quad (2\text{-}30)$$

where B is the coefficient of viscous friction or viscous damping.

EXAMPLE 2-6

The aerodynamic drag force in automobiles can be estimated as $f_L = 0.046\, C_w\, A\, u^2$, where the coefficient 0.046 has the appropriate units, the drag force is in N, C_w is the drag coefficient (a unit-less quantity), A is the vehicle cross-sectional area in m^2, and u is the sum of the vehicle speed and headwind in km/h [1]. If $A = 1.8$ m^2 for two vehicles with $C_w = 0.3$ and $C_w = 0.5$, respectively, calculate the drag force and the power required to overcome it at the speeds of 50 and 100 km/h.

Solution

The drag force is $f_L = 0.046\, C_w A u^2$ and the power required at the constant speed, from Eq. (2-6), is $P = f_L u$ where the speed is expressed in m/s. Table 2-1 lists the drag force and the power required at various speeds for the two vehicles. Since the drag force F_L depends on the square of the speed, the power depends on the cube of the speed.

Traveling at 50 km/h, compared to 100 km/h, requires 1/8th of the power, but it takes twice as long to reach the destination. Therefore, the energy required at 50 km/h would be 1/4th that at 100 km/h.

2-5 TORSIONAL RESONANCES

In Fig. 2-7, the shaft connecting the motor with the load was assumed to be of infinite stiffness, that is, the two were rigidly connected. In reality, any shaft will twist (flex) as it transmits torque from one end to the other. In Fig. 2-11, the torque T_{shaft} available to be transmitted by the shaft is

$$T_{shaft} = T_{em} - J_M \frac{d\omega_m}{dt} \qquad (2\text{-}31)$$

This torque at the load-end overcomes the load torque and accelerates it,

TABLE 2-1 Drag Force and the Power Required

Vehicle	$u = 50$ km/h		$u = 100$ km/h	
$C_w = 0.3$	$f_L = 62.06$ N	$P = 0.86$ kW	$f_L = 248.2$ N	$P = 6.9$ kW
$C_w = 0.5$	$f_L = 103.4$ N	$P = 1.44$ kW	$f_L = 413.7$ N	$P = 11.5$ kW

36 UNDERSTANDING MECHANICAL SYSTEM REQUIREMENTS

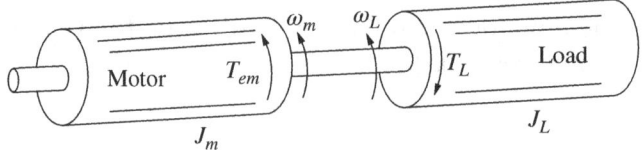

Fig. 2-11 Motor and load-torque interaction with a rigid coupling.

$$T_{shaft} = T_L + J_L \frac{d\omega_L}{dt} \quad (2\text{-}32)$$

The twisting or flexing of the shaft, in terms of the angles at the two ends, depends on the shaft torsional or the compliance coefficient K:

$$(\theta_M - \theta_L) = \frac{T_{shaft}}{K} \quad (2\text{-}33)$$

where θ_M and θ_L are the angular rotations at the two ends of the shaft. If K is infinite, $\theta_M = \theta_L$. For a shaft of finite compliance, these two angles are not equal, and the shaft acts as a spring. This compliance in the presence of energy stored in the masses, and inertias of the system, can lead to resonance conditions at certain frequencies. This phenomenon is often termed torsional resonance. Such resonances should be avoided or kept low; otherwise they can lead to fatigue and failure of the mechanical components.

2-6 ELECTRICAL ANALOGY

An analogy with electrical circuits can be very useful when analyzing mechanical systems. A commonly used analogy, though not a unique one, is to relate mechanical and electrical quantities, as shown in Table 2-2.

For the mechanical system shown in Fig. 2-11, Fig. 2-12a shows the electrical analogy, where each inertia is represented by a capacitor from its node to a reference (ground) node. In this circuit, we can write equations similar to Eqs. (2-31) through (2-33). Assuming that

ELECTRICAL ANALOGY

TABLE 2-2 Torque–Current Analogy

Mechanical system	Electrical system
Torque (T)	Current (i)
Angular speed (ω_m)	Voltage (v)
Angular displacement (θ)	Flux linkage (ψ)
Moment of inertia (J)	Capacitance (C)
Spring constant (K)	1/Inductance ($1/L$)
Damping coefficient (B)	1/Resistance ($1/R$)
Coupling ratio (n_M/n_L)	Transformer ratio (n_L/n_M)

The coupling ratio is discussed later in this chapter.

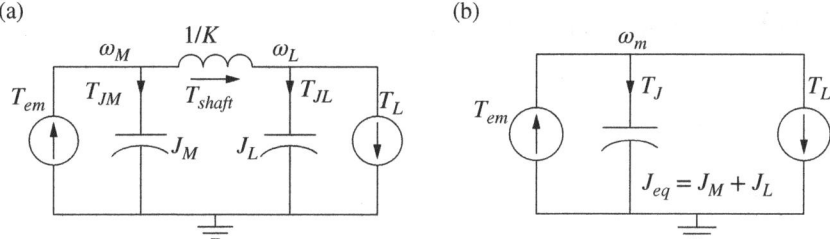

Fig. 2-12 Electrical analogy: (a) shaft of finite stiffness and (b) shaft of infinite stiffness.

the shaft is of infinite stiffness, the inductance representing it becomes zero, and the resulting circuit is shown in Fig. 2-12b, where $\omega_m = \omega_M = \omega_L$. The two capacitors representing the two inertias can now be combined to result in a single equation similar to Eq. (2-23).

EXAMPLE 2-7

In an electric-motor drive, similar to that shown in Fig. 2-7a, the combined inertia is $J_{eq} = 5 \times 10^{-3}$ kg·m². The load torque opposing rotation is mainly due to friction and can be described as $T_L = 0.5 \times 10^{-3} \omega_L$. Draw the electrical equivalent circuit and plot the electromagnetic torque required from the motor to bring the system linearly from rest to a speed of 100 rad/s in 4 s, and then to maintain that speed.

(Continued)

Solution

The electrical equivalent circuit is shown in Fig. 2-13a. The inertia is represented by a capacitor of 5 mF, and the friction by a resistance $R = 1/(0.5 \times 10^{-3}) = 2000\ \Omega$. The linear acceleration is $100/4 = 25$ rad/s², which in the equivalent electrical circuit corresponds to $dv/dt = 25$ V/s. Therefore, during the acceleration period, $v(t) = 25t$. Thus, the capacitor current during the linear acceleration interval is

$$i_c(t) = C\frac{dv}{dt} = 125.0 \text{ mA } 0 \leq t < 4 \text{ s} \qquad (2\text{-}34\text{a})$$

and the current through the resistor is

$$i_R(t) = \frac{v(t)}{R} = \frac{25t}{2000} = 12.5t \text{ mA } 0 \leq t < 4 \text{ s} \qquad (2\text{-}34\text{b})$$

Therefore,

$$T_{em}(t) = (125.0 + 12.5t) \times 10^{-3} \text{ Nm } 0 \leq t < 4 \text{ s} \qquad (2\text{-}34\text{c})$$

Beyond the acceleration stage, the electromagnetic torque is required only to overcome friction, which equals 50×10^{-3} Nm, as plotted in Fig. 2-13b.

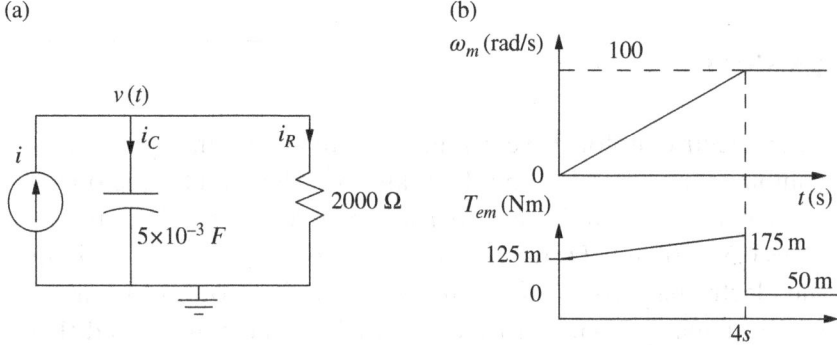

Fig. 2-13 (a) Electrical equivalent and (b) torque and speed variation.

2-7 COUPLING MECHANISMS

Wherever possible, it is preferable to couple the load directly to the motor, to avoid the additional cost of the coupling mechanism and the associated power losses. In practice, coupling mechanisms are often used for the following reasons:

- A rotary motor is driving a load which requires linear motion.
- The motors are designed to operate at higher rotational speeds (to reduce their physical size) compared to the speeds required of the mechanical loads.
- The axis of rotation needs to be changed.

There are various types of coupling mechanisms. For conversion between rotary and linear motions, it is possible to use conveyor belts (belt and pulley), rack-and-pinion, or a lead-screw type of arrangement. For rotary-to-rotary motion, various types of gear mechanisms are employed.

The coupling mechanisms have the following disadvantages:

- Additional power loss.
- Introduction of nonlinearity due to a phenomenon called backlash.
- Wear and tear.

2-7-1 Conversion Between Linear and Rotary Motion

In many systems, a linear motion is achieved by using a rotating-type motor, as shown in Fig. 2-14.

In such a system, the angular and the linear speeds are related by the radius r of the drum:

$$u = r \omega_m \quad (2\text{-}35)$$

To accelerate the mass M in Fig. 2-14, in the presence of an opposing force f_L, the force f applied to the mass, from Eq. (2-1), must be

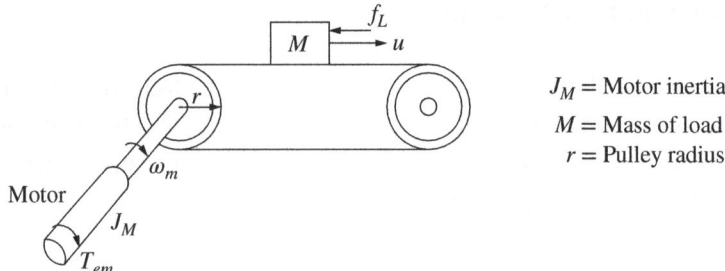

Fig. 2-14 Combination of rotary and linear motion.

$$f = M\frac{du}{dt} + f_L \qquad (2\text{-}36)$$

This force is delivered by the motor in the form of a torque T, which is related to f, using Eq. (2-35), as

$$T = r \cdot f = r^2 M \frac{d\omega_m}{dt} + r f_L \qquad (2\text{-}37)$$

Therefore, the electromagnetic torque required from the motor is

$$T_{em} = J_M \frac{d\omega_m}{dt} + \underbrace{r^2 M \frac{d\omega_m}{dt} + r f_L}_{\text{due to load}} \qquad (2\text{-}38)$$

EXAMPLE 2-8

In the vehicle of Example 2-6 with $C_w = 0.5$, assume that each wheel is powered by its own electric motor that is directly coupled to it. If the wheel diameter is 60 cm, calculate the torque and the power required from each motor to overcome the drag force when the vehicle is traveling at a speed of 100 km/h.

Solution

In Example 2-6, the vehicle with $C_w = 0.5$ presented a drag force $f_L = 413.7$ N at the speed $u = 100$ km/h. The force required from each of the four motors is $f_M = \frac{f_L}{4} = 103.4$ N. Therefore, the torque required from each motor is

$$T_M = f_M \, r = 103.4 \times \frac{0.6}{2} = 31.04 \text{ Nm}$$

From Eq. (2-35),

$$\omega_m = \frac{u}{r} = \left(\frac{100 \times 10^3}{3600}\right)\frac{1}{(0.6/2)} = 92.6 \text{ rad/s}$$

Therefore, the power required from each motor is

$$T_M \omega_m = 2.87 \text{ kW}$$

2-7-2 Gears

For matching speeds, Fig. 2-15 shows a gear mechanism where the shafts are assumed to be of infinite stiffness, and the masses of the gears are ignored. We will further assume that there is no power loss in the gears. Both gears must have the same linear speed at the point of contact. Therefore, their angular speeds are related by their respective radii r_1 and r_2 such that

$$r_1 \omega_M = r_2 \omega_L \qquad (2\text{-}39)$$

and

$$\omega_M T_1 = \omega_L T_2 \text{ (assuming no power loss)} \qquad (2\text{-}40)$$

Combining Eqs. (2-39) and (2-40),

$$\frac{r_1}{r_2} = \frac{\omega_L}{\omega_M} = \frac{T_1}{T_2} \qquad (2\text{-}41)$$

42 UNDERSTANDING MECHANICAL SYSTEM REQUIREMENTS

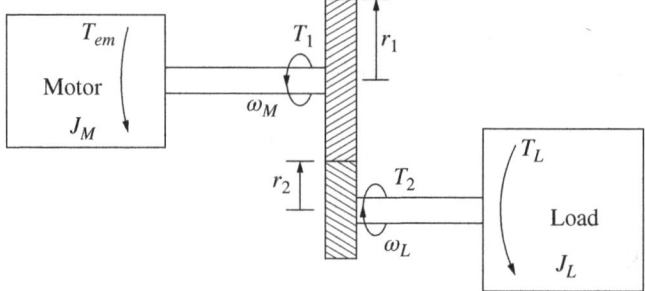

Fig. 2-15 Gear mechanism for coupling the motor to the load.

where T_1 and T_2 are the torques at the ends of the gear mechanism, as shown in Fig. 2-15. Expressing T_1 and T_2 in terms of T_{em} and T_L in Eq. (2-41),

$$\underbrace{\left(T_{em} - J_M \frac{d\omega_M}{dt}\right)}_{T_1} \frac{\omega_M}{\omega_L} = \underbrace{\left(T_L + J_L \frac{d\omega_L}{dt}\right)}_{T_2} \quad (2\text{-}42)$$

From Eq. (2-42), the electromagnetic torque required from the motor is

$$T_{em} = \underbrace{\left[J_M + \left(\frac{\omega_L}{\omega_M}\right)^2 J_L\right]}_{J_{eq}} \frac{d\omega_M}{dt} + \left(\frac{\omega_L}{\omega_M}\right) T_L \left(\text{note}: \frac{d\omega_L}{dt} = \frac{d\omega_M}{dt}\frac{\omega_L}{\omega_M}\right)$$

$$(2\text{-}43)$$

where the equivalent inertia at the motor side is

$$J_{eq} = J_M + \left(\frac{\omega_L}{\omega_M}\right)^2 J_L = J_M + \left(\frac{r_1}{r_2}\right)^2 J_L \quad (2\text{-}44)$$

Optimum Gear Ratio Equation (2-43) shows that the electromagnetic torque required from the motor to accelerate a motor-load combination depends on the gear ratio. In a basically inertial load where

T_L can be assumed to be negligible, T_{em} can be minimized, for a given load-acceleration $\dfrac{d\omega_L}{dt}$, by selecting an optimum gear ratio $(r_1/r_2)_{opt.}$. The derivation of the optimum gear ratio shows that the load inertia "seen" by the motor should equal the motor inertia, that is, in Eq. (2-44),

$$J_M = \left(\frac{r_1}{r_2}\right)^2_{opt.} J_L \quad \text{or} \quad \left(\frac{r_1}{r_2}\right)_{opt.} = \sqrt{\frac{J_M}{J_L}} \qquad (2\text{-}45a)$$

and, consequently,

$$J_{eq} = 2J_M \qquad (2\text{-}45b)$$

With the optimum gear ratio, in Eq. (2-43), using $T_L = 0$, and using Eq. (2-41),

$$(T_{em})_{opt.} = \frac{2 J_M}{\left(\dfrac{r_1}{r_2}\right)_{opt.}} \frac{d\omega_L}{dt} \qquad (2\text{-}46)$$

Similar calculations can be made for other types of coupling mechanisms (see homework problems).

2-8 TYPES OF LOADS

Load torques normally act to oppose rotation. In practice, loads can be classified into the following categories [2]:

1. Centrifugal (Squared) Torque
2. Constant Torque
3. Squared Power
4. Constant Power

Centrifugal loads, such as fans and blowers, require torque that varies with speed^2, and load power that varies with speed^3. Similarly, wind turbines produce torque that is proportional to speed^2.

44 UNDERSTANDING MECHANICAL SYSTEM REQUIREMENTS

In constant-torque loads such as conveyors, hoists, cranes, and elevators, torque remains constant with speed, and the load power varies linearly with speed. In squared-power loads such as compressors and rollers, the torque varies linearly with speed and the load power varies with speed^2. In constant-power loads, such as winders and unwinders, the torque beyond a certain speed range varies inversely with speed and the load power remains constant with speed.

2-9 FOUR-QUADRANT OPERATION

In many high-performance systems, drives are required to operate in all four quadrants of the torque-speed plane, as shown in Fig. 2-16b.

The motor drives the load in the forward direction in quadrant 1, and in the reverse direction in quadrant 3. In both of these quadrants, the average power is positive and flows from the motor to the mechanical load. In order to control the load speed rapidly, it may be necessary to operate the system in the regenerative braking mode, where the direction of power is reversed, so that it flows from the load into the motor, and usually into the utility (through the power-processing unit). In quadrant 2, the speed is positive, but the torque produced by the motor is negative. In quadrant 4, the speed is negative and the motor torque is positive.

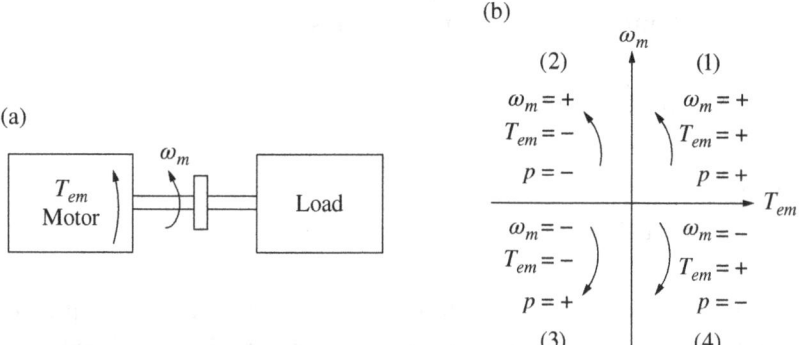

Fig. 2-16 (a) Electric drive and (d) four-quadrant operation.

2-10 STEADY-STATE AND DYNAMIC OPERATIONS

As discussed in Section 2-8, each load has its own torque-speed characteristic. For high-performance drives, in addition to the steady-state operation, the dynamic operation – how the operating point changes with time – is also important. The change of speed of the motor-load combination should be accomplished rapidly and without any oscillations (which otherwise may destroy the load). This requires a good design of the closed-loop controller, as discussed in Chapter 5, which deals with the control of drives.

2.11 REVIEW QUESTIONS

1. What are the MKS units for force, torque, linear speed, angular speed, speed, and power?

2. What is the relationship between force, torque, and power?

3. Show that torque is the fundamental variable in controlling speed and position.

4. What is the kinetic energy stored in a moving mass and a rotating mass?

5. What is the mechanism for torsional resonances?

6. What are the various types of coupling mechanisms?

7. What is the optimum gear ratio to minimize the torque required from the motor for a given load-speed profile as a function of time?

8. What are the torque-speed and the power-speed profiles for various types of loads?

REFERENCES

1. Bosch (1993). *Automotive Handbook*. Robert Bosch GmbH.
2. Nondahl, T. (1998). Proceedings of the NSF/EPRI-Sponsored Faculty Workshop on "Teaching of Power Electronics" (25–28 June 1998). University of Minnesota.

46 UNDERSTANDING MECHANICAL SYSTEM REQUIREMENTS

FURTHER READING

Gross, H. (ed.) (1983). *Electric Feed Drives for Machine Tools.* New York: Siemens and Wiley.

(1980). *DC Motors and Control ServoSystem – An Engineering Handbook*, 5e. Hopkins, MN: Electro-Craft Corporation.

Spong, M. and Vidyasagar, M. (1989). *Robot Dynamics and Control.* Wiley.

PROBLEMS

2-1 A constant torque of 5 Nm is applied to an unloaded motor at rest at time $t = 0$. The motor reaches a speed of 1800 rpm in 3 s. Assuming the damping to be negligible, calculate the motor inertia.

2-2 Calculate the inertia if the cylinder in Example 2-2 is hollow, with the inner radius $r_2 = 4$ cm.

2-3 A vehicle of mass 1500 kg is traveling at a speed of 50 km/h. What is the kinetic energy stored in its mass? Calculate the energy that can be recovered by slowing the vehicle to a speed of 10 km/h.

Belt-and-Pulley Systems

2-4 Consider the belt and pulley system in Fig. 2-13. Inertias other than that shown in the figure are negligible. The pulley radius $r = 0.09$ m and the motor inertia $J_M = 0.01$ kg·m². Calculate the torque T_{em} required to accelerate a load of 1.0 kg from rest to a speed of 1 m/s in a time of 4s. Assume the motor torque to be constant during this interval.

2-5 For the belt and pulley system shown in Fig. 2-13, $M = 0.02$ kg. For a motor with inertia $J_M = 40$ g·cm², determine the pulley radius that minimizes the torque required from the motor for a given load-speed profile. Ignore damping and the load force f_L.

Gears

2-6 In the gear system shown in Fig. 2-14, the gear ratio $n_L/n_M = 3$ where n equals the number of teeth in gear. The load and motor inertia are $J_L = 10$ kg $\cdot$ m^2 and $J_M = 1.2$ kg $\cdot$ m^2. Damping and the load-torque T_L can be neglected. For the load-speed profile shown in Fig. 2-1b, draw the profile of the electromagnetic torque T_{em} required from the motor as a function of time.

2-7 In the system of Problem 2-6, assume a triangular speed profile of the load with equal acceleration and deceleration rates (starting and ending at zero speed). Assuming a coupling efficiency of 100%, calculate the time needed to rotate the load by an angle of 30° if the magnitude of the electromagnetic torque (positive or negative) from the motor is 500 Nm.

2-8 The vehicle in Example 2-8 is powered by motors that have a maximum speed of 5000 rpm. Each motor is coupled to the wheel using a gear mechanism. (a) Calculate the required gear ratio if the vehicle's maximum speed is 150 km/h, and (b) calculate the torque required from each motor at the maximum speed.

2-9 Consider the system shown in Fig. 2-14. For $J_M = 40$ g $\cdot$ cm^2 and $J_L = 60$ g $\cdot$ cm^2, what is the optimum gear ratio to minimize the torque required from the motor for a given load-speed profile? Neglect damping and external load torque.

Lead-Screw Mechanism

2-10 Consider the lead-screw drive shown in Fig. P2-10. Derive the following equation in terms of pitch s, where $\ddot{u}_L$ = linear acceleration of the load, J_M = motor inertia, J_s = screw arrangement inertia, and the coupling ratio $n = \dfrac{s}{2\pi}$:

$$T_{em} = \frac{\ddot{u}_L}{n}\left[J_M + J_s + n^2(M_T + M_W)\right] + n F_L$$

48 UNDERSTANDING MECHANICAL SYSTEM REQUIREMENTS

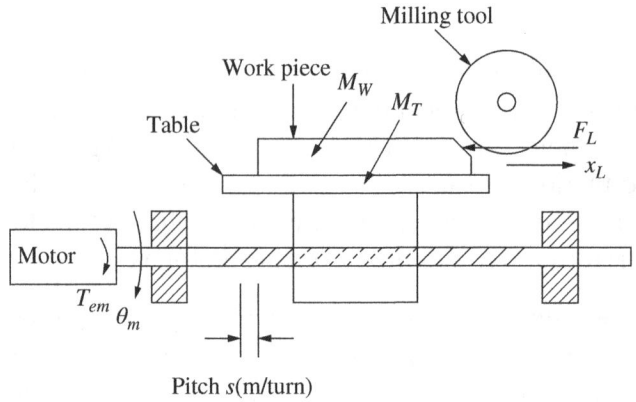

Fig. P2-10 Lead-screw system.

Wind Turbines and Electric Vehicles

2-11 In wind turbines, the shaft power available is given as follows, where the pitch-angle θ is nearly zero to "catch" all the wind energy available:

$$P_{shaft} = C_p \left(\frac{1}{2}\rho A_r V_W^3\right)$$

where C_p is the wind-turbine Coefficient of Performance (a unit-less quantity), ρ is the air density, A_r is the area swept by the rotor-blades, and V_W is the wind speed, all in MKS units. The rotational speed of the wind turbine is controlled, such that it is operating near its optimum value of the coefficient of performance with $C_p = 0.48$. Assume the combined efficiency of the gear-box, the generator, and the power electronic converter to be 90%, and the air density to be 1.2 kg/m³, $A_r = 4000$ m². Calculate the electrical power output of such a wind turbine at its rated wind speed of 13 m/s.

2-12 A wind turbine is rotating at 22 rpm in steady state at a wind speed of 13 m/s and producing 1.5 MW of power. The inertia of the mechanism is 3.4×10^6 kg·m². Suddenly, there is a short-circuit on the electric grid and the electrical output goes to zero for two seconds. Calculate the increase in speed in rpm during this interval. Assume that the shaft-torque remains constant and all other efficiencies to be 100% for the purpose of this calculation.

2-13 In an electric vehicle, each wheel is powered by its own motor. The vehicle weight is 2000 kg. This vehicle increases in its speed linearly from 0 to 60 mph in 10 seconds. The tire diameter is 70 cm. Calculate the maximum power required from each motor in kW.

2-14 In an electric vehicle, each of the four wheels is supplied by its own motor. This EV weighs 1000 kg, and the tire diameter is 50 cm. Using regenerative braking, its speed is brought from 20 m/s (72 km/h) to zero in 10 seconds, linearly with time. Neglect all losses. Calculate and plot, as a function of time for each wheel, the following: (a) the electromagnetic deceleration torque T_{em} in Nm, (b) rotation speed ω_m in rad/s, and (c) power P_m recovered in kW. Label the plots.

Simulation Problems

2-15 Making an electrical analogy, solve Problem 2-4.

2-16 Making an electrical analogy, solve Problem 2-6.

3 Basic Concepts in Magnetics and Electromechanical Energy Conversion

3-1 INTRODUCTION

Electric machines, like motors, convert electrical power input into mechanical output, as shown in Fig. 1-1a and discussed in Chapter 1. These machines may be operated solely as generators, but they also enter the generating mode when slowing down (during regenerative braking) where the power flow is reversed. In this chapter, we will briefly look at the basic structure of electric machines and the fundamental principles of the electromagnetic interactions that govern their operation. We will limit our discussion to rotating machines, although the same principles apply to linear machines.

In electric machines, torque is produced by the interaction of magnetic flux and currents. This requires a magnetic structure, therefore, we will begin with how the magnetic field and the flux are produced, resulting in torque and speed.

The purpose of this chapter is to review some of the basic concepts regarding magnetic circuits and their application for the study of ac motors and generators.

(Adapted from chapters 5 and 6 of *Electric Machines and Drives: A First Course* ISBN: 978-1-118-07481-7 by Ned Mohan, January 2012)

Analysis and Control of Electric Drives: Simulations and Laboratory Implementation, First Edition. Ned Mohan and Siddharth Raju.
© 2021 John Wiley & Sons, Inc. Published 2021 by John Wiley & Sons, Inc.
Companion website: www.wiley.com/go/Mohan/Vectorcontrolinelectricdrives

52 BASIC CONCEPTS IN MAGNETICS

3-2 MAGNETIC CIRCUIT CONCEPTS

In this section, we will look at basic magnetic concepts related to the interaction of current, flux, and speed to produce torque and to induce emf.

3-3 MAGNETIC FIELD PRODUCED BY CURRENT-CARRYING CONDUCTORS

When a current i is passed through a conductor, a magnetic field is produced. The direction of the magnetic field depends on the direction of the current. As shown in Fig. 3-1a, the current through a conductor, perpendicular and *into* the paper plane, is represented by "×"; this current produces a magnetic field in a clockwise direction. Conversely, the current *out of* the paper plane, represented by a dot, produces a magnetic field in a counterclockwise direction, as shown in Fig. 3-1b.

3-3-1 Ampere's Law

The magnetic field intensity H, produced by current-carrying conductors, can be obtained by means of Ampere's Law, which in its simplest form, states that, at any time, the line (contour) integral of the magnetic field intensity along *any* closed path equals the total current enclosed by this path. Therefore, in Fig. 3-1c,

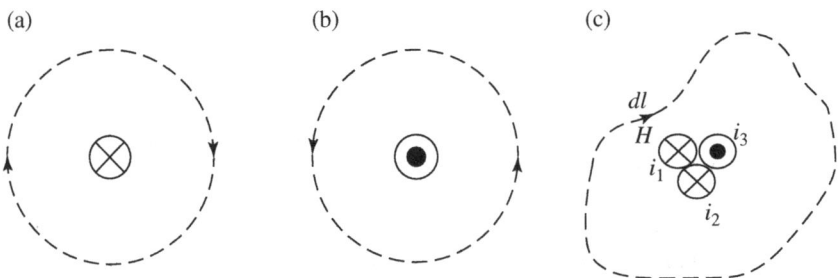

Fig. 3-1 (a and b) Magnetic field and (c) Ampere's Law.

$$\oint H d\ell = \sum i \qquad (3\text{-}1)$$

where $\oint$ represents a contour or a closed-line integration. Note that the scalar H in Eq. (3-1) is the component of the magnetic field intensity (a vector field) in the direction of the differential length $d\ell$ along the closed path. Alternatively, we can express the field intensity and the differential length to be vector quantities, which will require a dot product on the left side of Eq. (3-1).

EXAMPLE 3-1

Consider the coil in Fig. 3-2, which has $N = 25$ turns. The toroid on which the coil is wound has an inside diameter $ID = 5$ cm and an outside diameter $OD = 5.5$ cm. For a current $i = 3$ A, calculate the field intensity H along the mean-path length within the toroid.

Solution

Due to symmetry, the magnetic field intensity H_m along a circular contour within the toroid is constant. In Fig. 3-2, the mean radius $r_m = \frac{1}{2}\left(\frac{OD + ID}{2}\right)$. Therefore, the mean path of length $\ell_m (= 2\pi r_m = 0.165$ m) encloses the current i N-times, as shown in Fig. 3-2b. Therefore, from Ampere's Law in Eq. (3-1), the field intensity along this mean path is

$$H_m = \frac{Ni}{\ell_m} \qquad (3\text{-}2)$$

which for the given values can be calculated as

$$H_m = \frac{25 \times 3}{0.165} = 454.5 \text{ A/m}$$

If the width of the toroid is much smaller than the mean radius r_m, it is reasonable to assume a uniform H_m throughout the cross-section of the toroid.

54 BASIC CONCEPTS IN MAGNETICS

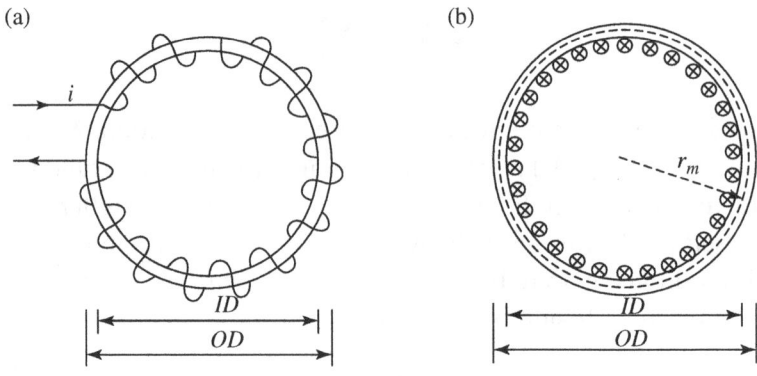

Fig. 3-2 Toroid.

The field intensity in Eq. (3-2) has the units of [A/m], noting that "turns" is a unit-less quantity. The product Ni is commonly referred to as the ampere-turns or mmf F that produces the magnetic field. The current in Eq. (3-2) may be dc, or time-varying. If the current is time-varying, the relationship in Eq. (3-2) is valid on an instantaneous basis; that is, $H_m(t)$ is related to $i(t)$ by N/ℓ_m.

3-4 FLUX DENSITY B AND THE FLUX φ

At any instant of time t for a given H-field, the density of flux lines, called the flux density B (in units of [T] for Tesla), depends on the permeability μ of the material on which this H-field is acting. In air,

$$B = \mu_o H \mu_o = 4\pi \times 10^{-7} \text{ H/m} \tag{3-3}$$

where μ_o is the permeability of air or free space.

3-4-1 Ferromagnetic Materials

Ferromagnetic materials guide magnetic fields and, due to their high permeability, require small ampere-turns (a small current for a given number of turns) to produce the desired flux density. These materials exhibit the multivalued nonlinear behavior shown by their B-H

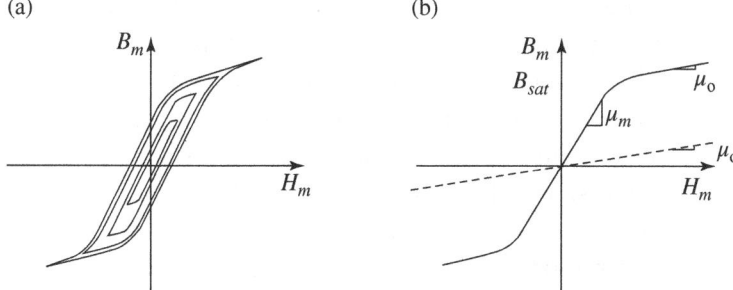

Fig. 3-3 B-H characteristics of ferromagnetic materials.

characteristics in Fig. 3-3a. Imagine that the toroid in Fig. 3-2 consists of a ferromagnetic material such as silicon steel. If the current through the coil is slowly varied in a sinusoidal manner with time, the corresponding H-field will cause one of the hysteresis loops shown in Fig. 3-3a to be traced. Completing the loop once results in a net dissipation of energy within the material, causing power loss, referred to as the hysteresis loss.

Increasing the peak value of the sinusoidally varying H-field will result in a bigger hysteresis loop. Joining the peaks of the hysteresis loop, we can approximate the B-H characteristic by the single curve shown in Fig. 3-3b. At low values of the magnetic field, the B-H characteristic is assumed to be linear with a constant slope, such that

$$B_m = \mu_m H_m \qquad (3\text{-}4a)$$

where μ_m is the permeability of the ferromagnetic material. Typically, the μ_m of a material is expressed in terms of a permeability μ_r relative to the permeability of air:

$$\left(\mu_m = \mu_r \mu_o \quad \mu_r = \frac{\mu_m}{\mu_o} \right) \qquad (3\text{-}4b)$$

In ferromagnetic materials, μ_m can be several thousand times larger than the μ_o.

In Fig. 3-3b, the linear relationship (with a constant μ_m) is approximately valid until the "knee" of the curve is reached, beyond which

56   BASIC CONCEPTS IN MAGNETICS

the material begins to saturate. Ferromagnetic materials are often operated up to a maximum flux density, slightly above the "knee" of 1.6 to 1.8 T, beyond which many more ampere-turns are required to increase flux density only slightly. In the saturated region, the incremental permeability of the magnetic material approaches μ_o, as shown by the slope of the curve in Fig. 3-3b.

In this course, we will assume that the magnetic material is operating in its linear region, and, therefore, its characteristic can be represented by $B_m = \mu_m H_m$, where μ_m remains constant.

3-4-2   Flux ϕ

Magnetic flux lines form closed paths, as shown in the toroidal magnetic core of Fig. 3-4, which is surrounded by the current-carrying coil. The flux in the toroid can be calculated by selecting a circular area A_m in a plane perpendicular to the direction of the flux lines. As discussed in Example 3-1, it is reasonable to assume a uniform H_m and hence a uniform flux density B_m throughout the core cross-section.

Substituting H_m from Eq. (3-2) into Eq. (3-4a),

$$B_m = \mu_m \frac{Ni}{\ell_m} \qquad (3-5)$$

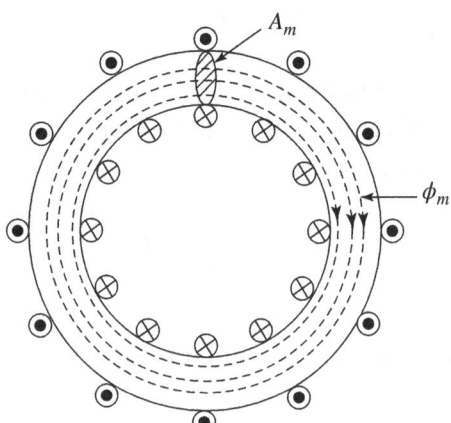

Fig. 3-4   Toroid with flux ϕ_m.

where B_m is the density of flux lines in the core. Therefore, making the assumption of a uniform B_m, the flux ϕ_m can be calculated as

$$\phi_m = B_m A_m \qquad (3\text{-}6)$$

where flux has the units of Weber [Wb]. Substituting for B_m from Eq. (3-5) into Eq. (3-6),

$$\phi_m = A_m \left(\mu_m \frac{Ni}{\ell_m} \right) = \frac{Ni}{\underbrace{\left(\dfrac{\ell_m}{\mu_m A_m} \right)}_{\mathfrak{R}_m}} \qquad (3\text{-}7)$$

where Ni equals the ampere-turns (or mmf F) applied to the core, and the term within the brackets on the right side is called the reluctance $\mathfrak{R}_m$ of the magnetic core. From Eq. (3-7),

$$\mathfrak{R}_m = \frac{\ell_m}{\mu_m A_m} \quad [\text{A/Wb}] \qquad (3\text{-}8)$$

Equation (3-7) makes it clear that the reluctance has the units [A/Wb]. Equation (3-8) shows that the reluctance of a magnetic structure, for example, the toroid in Fig. 3-4, is linearly proportional to its magnetic path length and inversely proportional to both its cross-sectional area and the permeability of its material.

Equation (3-7) shows that the amount of flux produced by the applied ampere-turns $F(=Ni)$ is inversely proportional to the reluctance $\mathfrak{R}$; this relationship is analogous to Ohm's Law ($I = V/R$) in electric circuits in dc steady state.

3-4-3 Flux Linkage

If all turns of a coil, for example, the one in Fig. 3-4, are linked by the same flux ϕ, then the coil has a flux linkage λ, where

$$\lambda = N\phi \qquad (3\text{-}9)$$

EXAMPLE 3-2

In Example 3-1, the core consists of a material with $\mu_r = 2000$. Calculate the flux density B_m and flux ϕ_m.

Solution

In Example 3-1, we calculated that $H_m = 454.5$ A/m. Using Eqs. (3-4a) and (3-4b), $B_m = 4\pi \times 10^{-7} \times 2000 \times 454.5 = 1.14$ T. The toroid width is $\frac{OD-ID}{2} = 0.25 \times 10^{-2}$ m. Therefore, the cross-sectional area of the toroid is

$$A_m = \frac{\pi}{4}\left(0.25 \times 10^{-2}\right)^2 = 4.9 \times 10^{-6} \text{ m}^2$$

Hence, from Eq. (3-6), assuming that the flux density is uniform throughout the cross-section,

$$\phi_m = 1.14 \times 4.9 \times 10^{-6} = 5.59 \times 10^{-6} \text{ Wb}$$

3-5 MAGNETIC STRUCTURES WITH AIR GAPS

In the magnetic structures of electric machines, the flux lines have to cross two air gaps. To study the effect of air gaps, let us consider the simple magnetic structure of Fig. 3-5, consisting of an N-turn coil on a magnetic core made up of iron. The objective is to establish a desired magnetic field in the air gap of length ℓ_g by controlling the coil current i. We will assume the magnetic field intensity H_m to be uniform along the mean-path length ℓ_m in the magnetic core. The magnetic field intensity in the air gap is denoted as H_g. From Ampere's Law in Eq. (3-1), the line integral along the mean path within the core and in the air gap yields the following equation:

$$H_m \ell_m + H_g \ell_g = Ni \quad (3\text{-}10)$$

MAGNETIC STRUCTURES WITH AIR GAPS 59

Applying Eq. (3-3) to the air gap and Eq. (3-4a) to the core, the flux densities corresponding to H_m and H_g are

$$B_m = \mu_m H_m \quad \text{and} \quad B_g = \mu_o H_g \tag{3-11}$$

In terms of the flux densities of Eq. (3-11), Eq. (3-10) can be written as

$$\frac{B_m}{\mu_m}\ell_m + \frac{B_g}{\mu_o}\ell_g = Ni \tag{3-12}$$

Since flux lines form closed paths, the flux crossing any perpendicular cross-sectional area in the core is the same as that crossing the air gap (neglecting the leakage flux, which is discussed later on). Therefore,

$$\phi = A_m B_m = A_g B_g \quad \text{or} \tag{3-13}$$

$$B_m = \frac{\phi}{A_m} \quad \text{and} \quad B_g = \frac{\phi}{A_g} \tag{3-14}$$

Generally, flux lines bulge slightly around the air gap, as shown in Fig. 3-5. This bulging is called the *fringing effect*, which can be accounted for by estimating the air gap area A_g, which is done by increasing each dimension in Fig. 3-5 by the length of the air gap:

$$A_g = \left(W + \ell_g\right)\left(d + \ell_g\right) \tag{3-15}$$

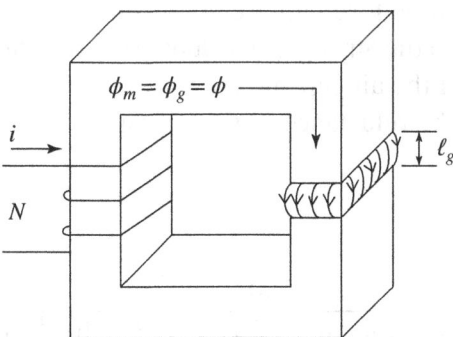

Fig. 3-5 Magnetic structure with air gap.

60 BASIC CONCEPTS IN MAGNETICS

Substituting flux densities from Eq. (3-14) into Eq. (3-12),

$$\phi \left(\frac{\ell_m}{A_m \mu_m} + \frac{\ell_g}{A_g \mu_o} \right) = Ni \tag{3-16}$$

In Eq. (3-16), we can recognize from Eq. (3-8) that the two terms within the parenthesis equal the reluctances of the core and the air gap, respectively. Therefore, the effective reluctance $\mathfrak{R}$ of the whole structure in the path of the flux lines is the sum of the two reluctances:

$$\mathfrak{R} = \mathfrak{R}_m + \mathfrak{R}_g \tag{3-17}$$

Substituting from Eq. (3-17) into Eq. (3-16), where Ni equals the applied mmf F,

$$\phi = \frac{F}{\mathfrak{R}} \tag{3-18}$$

Equation (3-18) allows ϕ to be calculated for the applied ampere-turns (mmf F). Then, B_m and B_g can be calculated from Eq. (3-14).

EXAMPLE 3-3

In the structure of Fig. 3-5, all flux lines in the core are assumed to cross the air gap. The structure dimensions are as follows: core cross-sectional area $A_m = 20$ cm^2, mean-path length $\ell_m = 40$ cm, $\ell_g = 2$ mm, and $N = 75$ turns. In the linear region, the core permeability can be assumed to be constant with $\mu_r = 4500$. The coil current i (=30 A) is below the saturation level. Ignore the flux fringing effect. Calculate the flux density in the air gap, (a) including the reluctance of the core as well as that of the air gap and (b) ignoring the core reluctance in comparison to the reluctance of the air gap.

Solution

From Eq. (3-8),

$$\mathfrak{R}_m = \frac{\ell_m}{\mu_o \mu_r A_m} = \frac{40 \times 10^{-2}}{4\pi \times 10^{-7} \times 4500 \times 20 \times 10^{-4}} \text{ and}$$

$$= 3.54 \times 10^4 \text{ A/Wb}$$

$$\mathfrak{R}_g = \frac{\ell_g}{\mu_o A_g} = \frac{2 \times 10^{-3}}{4\pi \times 10^{-7} \times 20 \times 10^{-4}} = 79.57 \times 10^4 \text{ A/Wb}$$

(a) Including both reluctances, from Eq. (3-16),

$$\phi_g = \frac{Ni}{\mathfrak{R}_m + \mathfrak{R}_g} \text{ and}$$

$$B_g = \frac{\phi_g}{A_g} = \frac{Ni}{(\mathfrak{R}_m + \mathfrak{R}_g)A_g} = \frac{75 \times 30}{(79.57 + 3.54) \times 10^4 \times 20 \times 10^{-4}}$$
$$= 1.35 \text{ T}$$

(b) Ignoring the core reluctance, from Eq. (3-16),

$$\phi_g = \frac{Ni}{\mathfrak{R}_g} \text{ and } B_g = \frac{\phi_g}{A_g} = \frac{Ni}{\mathfrak{R}_g A_g} = \frac{75 \times 30}{79.57 \times 10^4 \times 20 \times 10^{-4}}$$
$$= 1.41 \text{ T}$$

This example shows that the reluctance of the air gap dominates the flux and the flux density calculations; thus, we can often ignore the reluctance of the core in comparison to that of the air gap.

3-6 INDUCTANCES

At any instant of time in the coil of Fig. 3-6a, the flux linkage of the coil (due to flux lines entirely in the core) is related to the current i by a parameter defined as the inductance L_m:

$$\lambda_m = L_m i \qquad (3\text{-}19)$$

where the inductance $L_m(=\lambda_m/i)$ is constant if the core material is in its linear operating region.

The coil inductance in the linear magnetic region can be calculated by multiplying all the factors shown in Fig. 3-6b, which are based on earlier equations:

62   BASIC CONCEPTS IN MAGNETICS

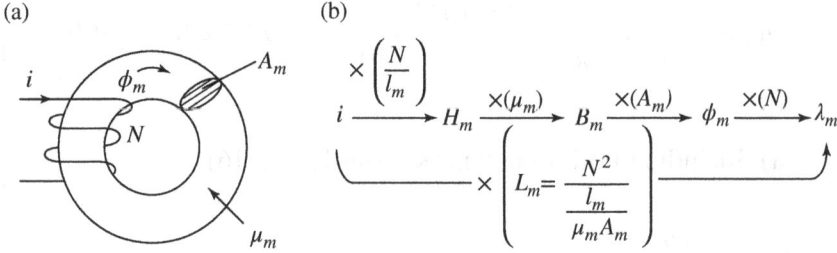

Fig. 3-6   Coil inductance.

$$L_m = \underbrace{\left(\frac{N}{\ell_m}\right)}_{\text{Eq.(3-2)}} \underbrace{\mu_m}_{\text{Eq.(3-4a)}} \underbrace{A_m}_{\text{Eq.(3-6)}} \underbrace{N}_{\text{Eq.(3-9)}} = \frac{N^2}{\left(\dfrac{\ell_m}{\mu_m A_m}\right)} = \frac{N^2}{\mathcal{R}_m} \quad (3\text{-}20)$$

Equation (3-20) indicates that the inductance L_m is strictly a property of the magnetic circuit (i.e. the core material, the geometry, and the number of turns), provided that the operation is in the linear range of the magnetic material, where the slope of its B-H characteristic can be represented by a constant μ_m.

EXAMPLE 3-4

In the rectangular toroid of Fig. 3-7, $w = 5$ mm, $h = 15$ mm, the mean-path length $\ell_m = 18$ cm, $\mu_r = 5000$, and $N = 100$ turns. Calculate the coil inductance L_m, assuming that the core is unsaturated.

Solution

From Eq. (3-8),

$$\mathcal{R}_m = \frac{\ell_m}{\mu_m A_m} = \frac{0.18}{5000 \times 4\pi \times 10^{-7} \times 5 \times 10^{-3} \times 15 \times 10^{-3}}$$

$$= 38.2 \times 10^4 \text{ A/Wb}$$

Therefore, from Eq. (3-20),

$$L_m = \frac{N^2}{\mathcal{R}_m} = 26.18 \text{ mH}$$

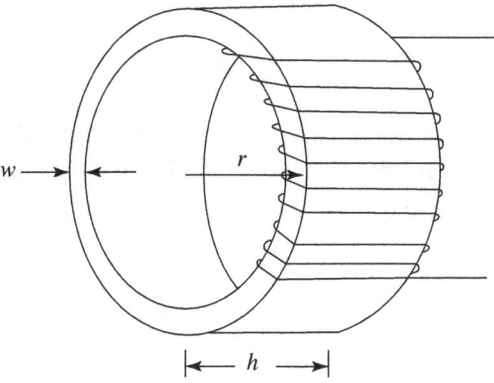

Fig. 3-7 Rectangular toroid.

3-7 MAGNETIC ENERGY STORAGE IN INDUCTORS

The energy in an inductor is stored in its magnetic field. From the study of electric circuits, we know that at any time, with a current i, the energy stored in the inductor is

$$W = \frac{1}{2} L_m i^2 \quad [\text{J}] \tag{3-21}$$

where [J], for Joules, is a unit of energy. Initially assuming a structure without an air gap, such as in Fig. 3-6a, we can express the energy storage in terms of flux density, by substituting into Eq. (3-21) the inductance from Eq. (3-20) and the current from the Ampere's Law in Eq. (3-2):

$$W_m = \frac{1}{2} \underbrace{\frac{N^2}{\ell_m/\mu_m A_m}} \underbrace{(H_m \ell_m/N)^2}_{i^2} = \frac{1}{2} \frac{(H_m \ell_m)^2}{\ell_m/\mu_m A_m} = \frac{1}{2} \frac{B_m^2}{\mu_m} \underbrace{A_m \ell_m}_{volume} \quad [\text{J}]$$

(3-22a)

where $A_m \ell_m = volume$, and in the linear region $B_m = \mu_m H_m$. Therefore, from Eq. (3-22a), the energy density in the core is

$$w_m = \frac{1}{2}\frac{B_m^2}{\mu_m} \qquad (3\text{-}22b)$$

Similarly, the energy density in the air gap depends on μ_o and the flux density in it. Therefore, from Eq. (3-22b), the energy density in any medium can be expressed as

$$w = \frac{1}{2}\frac{B^2}{\mu} \quad [\text{J/m}^3] \qquad (3\text{-}23)$$

In electric machines, where air gaps are present in the path of the flux lines, the energy is primarily stored in the air gaps. This is illustrated by the following example.

EXAMPLE 3-5

In Example 3-3 part (a), calculate the energy stored in the core and in the air gap, and compare the two.

Solution

In Example 3-3 part (a), $B_m = B_g = 1.35\,\text{T}$. Therefore, from Eq. (3-23),

$$w_m = \frac{1}{2}\frac{B_m^2}{\mu_m} = 161.1\,\text{J/m}^3 \text{ and } w_g = \frac{1}{2}\frac{B_g^2}{\mu_o} = 0.725 \times 10^6\,\text{J/m}^3$$

Therefore, $\dfrac{w_g}{w_m} = \mu_r = 4500$. Based on the given cross-sectional areas and lengths, the core volume is 200 times larger than that of the air gap. Therefore, the ratio of energy storage is

$$\frac{W_g}{W_m} = \frac{w_g}{w_m} \times \frac{(volume)_g}{(volume)_m} = \frac{4500}{200} = 22.5$$

FARADAY'S LAW   65

3-8 FARADAY'S LAW: INDUCED VOLTAGE IN A COIL DUE TO TIME-RATE OF CHANGE OF FLUX LINKAGE

In our discussion so far, we have established, in magnetic circuits, relationships between the electrical quantity i and the magnetic quantities H, B, ϕ, and λ. These relationships are valid under dc (static) conditions, as well as at any instant when these quantities are varying with time. We will now examine the voltage across the coil under time-*varying* conditions. In the coil of Fig. 3-8, Faraday's Law dictates that the time-rate of change of flux linkage equals the voltage across the coil at any instant:

$$e(t) = \frac{d}{dt}\lambda(t) = N\frac{d}{dt}\phi(t) \qquad (3\text{-}24)$$

This assumes that all flux lines link all N turns such that $\lambda = N\phi$. The polarity of the emf $e(t)$ and the direction of $\phi(t)$ in the above equation are yet to be justified.

The above relationship is valid, no matter what is causing the flux to change. One possibility is that a second coil is placed on the same core. When the second coil is supplied by a time-varying current, mutual coupling causes the flux ϕ through the coil shown in Fig. 3-8 to change with time. The other possibility is that a voltage $e(t)$ is applied across the coil in Fig. 3-8, causing the change in flux, which can be calculated by integrating both sides of Eq. (3-24) with respect to time:

$$\phi(t) = \phi(0) + \frac{1}{N}\int_0^t e(\tau)\cdot d\tau \qquad (3\text{-}25)$$

where $\phi(0)$ is the initial flux at $t = 0$ and τ is a variable of integration.

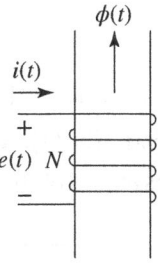

Fig. 3-8 Voltage polarity and direction of flux and current.

Recalling the Ohm's Law equation $v = Ri$, the current direction through a resistor is defined to be into the terminal chosen to be of the positive polarity. This is the passive sign convention. Similarly, in the coil of Fig. 3-8, we can establish the voltage polarity and the flux-direction in order to apply Faraday's Law, given by Eqs. (3-24) and (3-25). If the flux direction is given, we can establish the voltage polarity as follows: first, determine the direction of a hypothetical current that will produce flux in the same direction as given. Then, the positive polarity for the voltage is at the terminal which this hypothetical current is entering. Conversely, if the voltage polarity is given, imagine a hypothetical current entering the positive-polarity terminal. This current, based on how the coil is wound, for example, in Fig. 3-8, determines the flux direction for use in Eqs. (3-24) and (3-25).

Another way to determine the polarity of the induced emf is to apply Lenz's Law, which states the following: if a current is allowed to flow due to the voltage induced by an increasing flux linkage, for example, the direction of this hypothetical current will be to oppose the flux change.

EXAMPLE 3-6

In the structure of Fig. 3-8, the flux $\phi_m (= \hat{\phi}_m \sin \omega t)$ linking the coil is varying sinusoidally with time, where $N = 300$ turns, $f = 60$ Hz, and the cross-sectional area $A_m = 10$ cm^2. The peak flux density $\hat{B}_m = 1.5$ T. Calculate the expression for the induced voltage with the polarity shown in Fig. 3-8. Plot the flux and the induced voltage as functions of time.

Solution

From Eq. (3-6), $\hat{\phi}_m = \hat{B}_m A_m = 1.5 \times 10 \times 10^{-4} = 1.5 \times 10^{-3}$ Wb. From Faraday's Law in Eq. (3-24), $e(t) = \omega N \hat{\phi}_m \cos \omega t = 2\pi \times 60 \times 300 \times 1.5 \times 10^{-3} \times \cos \omega t = 169.65 \cos \omega t$ V. The waveforms are plotted in Fig. 3-9.

FARADAY'S LAW   67

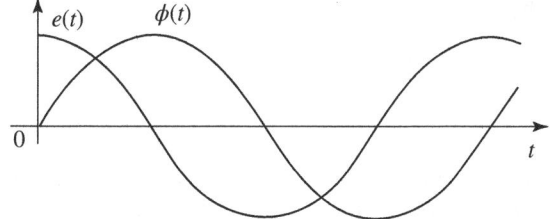

Fig. 3-9   Waveforms of flux and induced voltage.

Example 3-6 illustrates that the voltage is induced due to $d\phi/dt$, regardless of whether any current flows in that coil. In the following subsection, we will establish the relationship between $e(t)$, $\phi(t)$, and $i(t)$.

3-8-1   Relating $e(t)$, $\phi(t)$, and $i(t)$

In the coil of Fig. 3-10a, an applied voltage $e(t)$ results in $\phi(t)$, which is dictated by the Faraday's Law equation in the integral form, Eq. (3-25). However, what about the current drawn by the coil to establish this flux? Rather than going back to Ampere's Law, we

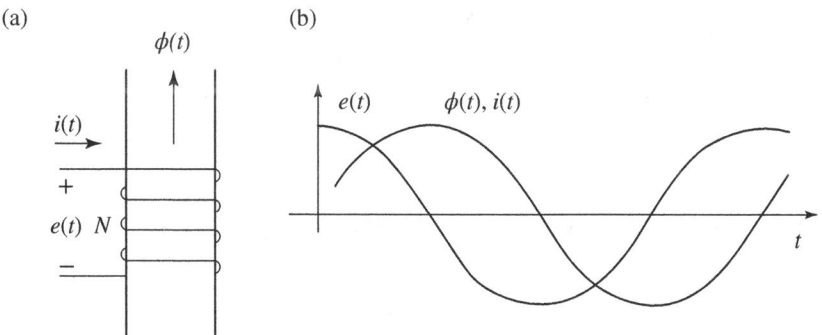

Fig. 3-10   Voltage, current, and flux.

68 BASIC CONCEPTS IN MAGNETICS

can express the coil flux linkage in terms of its inductance and current using Eq. (3-19):

$$\lambda(t) = L\,i(t) \qquad (3\text{-}26)$$

Assuming that the entire flux links all N turns, the coil flux linkage $\lambda(t) = N\phi(t)$. Substituting this into Eq. (3-26) gives

$$\phi(t) = \frac{L}{N} i(t) \qquad (3\text{-}27)$$

Substituting for $\phi(t)$ from Eq. (3-27) into Faraday's Law in Eq. (3-24) results in

$$e(t) = N\frac{d\phi}{dt} = L\frac{di}{dt} \qquad (3\text{-}28)$$

Equations (3-27) and (3-28) relate $i(t)$, $\phi(t)$, and $e(t)$; all of these are plotted in Fig. 3-10b.

EXAMPLE 3-7

In Example 3-6, the coil inductance is 50 mH. Calculate the expression for the current $i(t)$ in Fig. 3-10b.

Solution

From Eq. (3-27), $i(t) = \dfrac{N}{L}\phi(t) = \dfrac{300}{50 \times 10^{-3}} 1.5 \times 10^{-3} \sin \omega t = 9.0 \sin \omega t$ A

3-9 LEAKAGE AND MAGNETIZING INDUCTANCES

Just as conductors guide currents in electric circuits, magnetic cores guide *flux* in *magnetic circuits*. However, there is an important difference. In electric circuits, the conductivity of copper is approximately 10^{20} times higher than that of air, allowing leakage currents to be

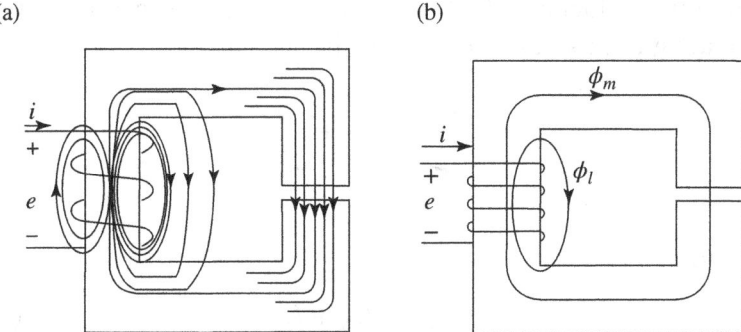

Fig. 3-11 (a) Magnetic and leakage fluxes and (b) equivalent representation of these fluxes.

neglected at dc or at low frequencies such as 60 Hz. In magnetic circuits, however, the permeabilities of magnetic materials are only around 10^4 times greater than that of air. Because of this relatively low ratio, the core window in the structure of Fig. 3-11a has "leakage" flux lines, which do not reach their intended destination – the air gap. Note that the coil shown in Fig. 3-11a is drawn schematically. In practice, the coil consists of multiple layers and the core is designed to fit as snugly to the coil as possible, thus minimizing the unused "window" area.

The leakage effect makes accurate analysis of magnetic circuits more difficult, so that it requires sophisticated numerical methods, such as finite element analysis. However, we can account for the effect of leakage fluxes by making certain approximations. We can divide the total flux ϕ into two parts: the magnetic flux ϕ_m, which is completely confined to the core and links all N turns, and the leakage flux, which is partially or entirely in air and is represented by an "equivalent" leakage flux ϕ_ℓ, which also links all N turns of the coil but does not follow the entire magnetic path, as shown in Fig. 3-11b. Thus,

$$\phi = \phi_m + \phi_\ell \qquad (3\text{-}29)$$

where ϕ is the equivalent flux, which links all N turns. Therefore, the total flux linkage of the coil is

$$\lambda = N\phi = \underbrace{N\phi_m}_{\lambda_m} + \underbrace{N\phi_\ell}_{\lambda_\ell} = \lambda_m + \lambda_\ell \qquad (3\text{-}30)$$

70 BASIC CONCEPTS IN MAGNETICS

The total inductance (called the self-inductance) can be obtained by dividing both sides of Eq. (3-30) by the current i:

$$\underbrace{\frac{\lambda}{i}}_{L_{self}} = \underbrace{\frac{\lambda_m}{i}}_{L_m} + \underbrace{\frac{\lambda_\ell}{i}}_{L_\ell} \quad (3\text{-}31)$$

Therefore,

$$L_{self} = L_m + L_\ell \quad (3\text{-}32)$$

where L_m is often called the *magnetizing inductance* due to ϕ_m in the magnetic core and L_ℓ is called the *leakage inductance* due to the leakage flux ϕ_ℓ. From Eq. (3-32), the total flux linkage of the coil can be written as

$$\lambda = (L_m + L_\ell)i \quad (3\text{-}33)$$

Hence, from Faraday's Law in Eq. (3-24),

$$e(t) = L_\ell \frac{di}{dt} + \underbrace{L_m \frac{di}{dt}}_{e_m(t)} \quad (3\text{-}34)$$

This results in the circuit of Fig. 3-12a. In Fig. 3-12b, the voltage drop due to the leakage inductance can be shown separately so that the voltage induced in the coil is solely due to the magnetizing flux.

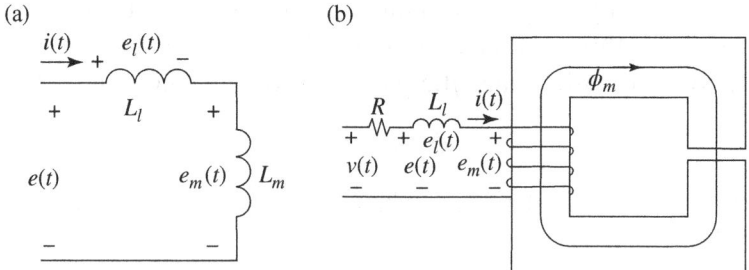

Fig. 3-12 (a) Circuit representation and (b) leakage inductance separated from the core.

The coil resistance R can then be added in series to complete the representation of the coil.

3-10 MUTUAL INDUCTANCES

Most magnetic circuits, such as those encountered in electric machines and transformers, consist of multiple coils. In such circuits, the flux established by the current in one coil partially links the other coil or coils. This phenomenon can be described mathematically by means of mutual inductances, as examined in circuit theory courses. Mutual inductances are also needed to develop mathematical models for dynamic analysis of electric machines. Since it is not the objective of this book, we will not elaborate any further on the topic of mutual inductances. Rather, we will use simpler and more intuitive means to accomplish the task at hand.

3-11 BASIC PRINCIPLES OF TORQUE PRODUCTION AND VOLTAGE INDUCTION

Having looked at how the magnetic field flux is produced in general, we will examine how it is produced in electric machines. Then, we will examine how its interaction with current produces torque, and how the voltage is induced in conductors that are moving in the presence of flux.

3-11-1 Basic Structure of ac Machines

We often describe electric machines by viewing a cross-section, as if the machine is "cut" by a plane perpendicular to the shaft axis and viewed from one side, as shown in Fig. 3-13a. Because of symmetry, this cross-section can be taken anywhere along the shaft axis. The simplified cross-section in Fig. 3-13b shows that all machines have a stationary part, called the stator, and a rotating part, called the rotor, separated by an air gap, thereby allowing the rotor to rotate freely on a shaft, supported by bearings. The stator is firmly affixed to a foundation to prevent it from turning.

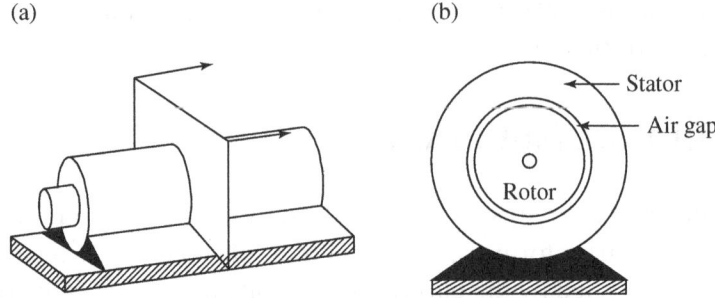

Fig. 3-13 Motor construction: (a) "cut" perpendicular to the shaft-axis and (b) cross-section as seen from one side.

In order to require small ampere-turns to create flux lines shown crossing the air gap in Fig. 3-14a, both the rotor and the stator are made up of high permeability ferromagnetic materials, and the length of the air gap is kept as small as possible. In machines with ratings under 10 kW, a typical length of the air gap is about 1 mm, which is shown highly exaggerated for ease of drawing.

The stator-produced flux distribution in Fig. 3-14a, is shown for a 2-pole machine where the field distribution corresponds to a combination of a single north pole and a single south pole. Often there are more than 2 poles, for example, 4 or 6. The flux distribution in a 4-pole machine is represented in Fig. 3-14b. Due to complete symmetry around the periphery of the air gap, it is sufficient to consider only

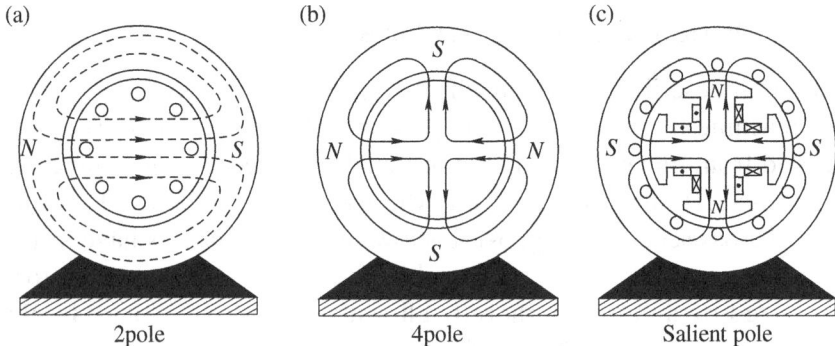

Fig. 3-14 Structure of machines.

BASIC PRINCIPLES OF TORQUE PRODUCTION 73

one pole pair consisting of adjacent north and south poles. Other pole pairs have identical conditions of magnetic fields and currents.

If the rotor and the stator are perfectly round, the air gap is uniform, and the magnetic reluctance in the path of flux lines crossing the air gap is uniform. Machines with such structures are called non-salient pole machines. Sometimes, the machines are purposely designed to have saliency so that the magnetic reluctance is unequal along various paths, as shown in Fig. 3-14c. Such saliency results in what is called the reluctance torque, which may be the primary or a significant means of producing the torque.

We should note that to reduce eddy-current losses, the stator and the rotor often consist of laminations of silicon steel, which are insulated from each other by a layer of thin varnish. These laminations are stacked together, perpendicular to the shaft axis. Conductors which run parallel to the shaft axis may be placed in slots cut into these laminations to place. Readers are urged to purchase a used dc motor and an induction motor and then take them apart to look at their construction.

3-11-2 Production of Magnetic Field

We will now examine how coils produce magnetic fields in electric machines. For illustration, a concentrated coil of N_s turns is placed in two stator slots 180° (called full-pitch) apart, as shown in Fig. 3-15a. The rotor is present without its electrical circuit. We will consider only the magnetizing flux lines that completely cross the two air gaps, and at present, ignore the leakage flux lines. The flux lines *in the air gap* are radial, that is, in a direction which goes through the center of the machine. Associated with the radial flux lines, the field intensity in the air gap is also in a radial direction; it is assumed to be positive $(+H_s)$ if it is away from the center of the machine, otherwise negative $(-H_s)$. The subscript "s" (for stator) refers to the field intensity *in the air gap* due to the stator. We will assume the permeability of iron to be infinite; hence the H-fields in the stator and the rotor are zero. Applying Ampere's Law along any of the closed paths shown in Fig. 3-15a, at any instant of time t,

(a)

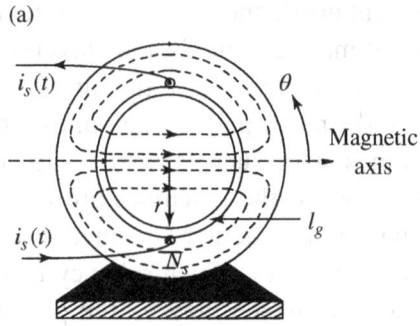

(b)

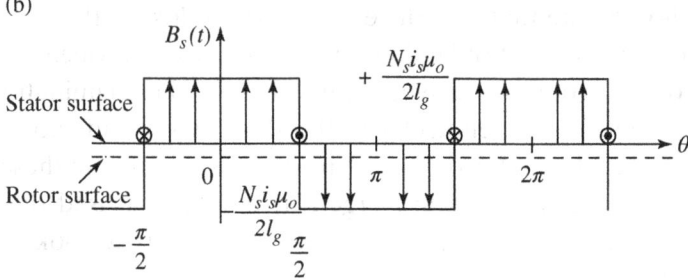

Fig. 3-15 Production of a magnetic field.

$$\underbrace{H_s \ell_g}_{\text{outward}} - \underbrace{(-H_s)\ell_g}_{\text{inward}} = N_s i_s \text{ or } H_s = \frac{N_s i_s}{2\ell_g} \quad (3\text{-}35)$$

where a negative sign is associated with the integral in the inward direction because while the path of integration is inward, the field intensity is measured outward.

The total mmf acting along any path shown in Fig. 3-15a is $N_s i_s$. Having assumed the permeability of the stator and the rotor iron to be infinite, by symmetry, half of the total ampere-turns $((N_s i_s)/2)$ are "consumed" or "acting" in making the flux lines cross each air gap length. Hence, the mmf F_s acting on each air gap is

$$F_s = \frac{N_s i_s}{2} \quad (3\text{-}36)$$

Substituting for $\dfrac{N_s i_s}{2}$ from Eq. (3-36) into Eq. (3-35),

BASIC PRINCIPLES OF TORQUE PRODUCTION 75

$$F_s = H_s \ell_g \tag{3-37}$$

Associated with H_s in the air gap is the flux density B_s, which using Eq. (3-35) can be written as

$$B_s = \mu_o H_s = \mu_o \frac{N_s i_s}{2\ell_g} \tag{3-38}$$

All field quantities (H_s, F_s, and B_s) directed away from the center of the machine are considered positive. Figure 3-15b shows the "developed" view as if the circular cross-section in Fig. 3-15a were flat. Note that the field distribution is a square wave. From Eqs. (3-35), (3-36), and (3-38), it is clear that all three stator-produced field quantities (H_s, F_s, and B_s) are proportional to the instantaneous value of the stator current $i_s(t)$ and are related to each other by constants. Therefore, in Fig. 3-15b, the square wave plot of B_s distribution at an instant of time also represents H_s and B_s distributions at that time plotted on different scales.

In the structure of Fig. 3-15a, the axis through $\theta = 0°$ is referred to as the magnetic axis of the coil or winding that is producing this field. The magnetic axis of a winding goes through the center of the machine in the direction of the flux lines produced by a positive value of the winding current and is perpendicular to the plane in which the winding is located.

EXAMPLE 3-8

In Fig 3-15a, consider a concentrated coil with $N_s = 25$ turns, and air gap length $\ell_g = 1$ mm. The mean radius (in the middle of the air gap) is $r = 15$ cm, and the length of the rotor is $\ell = 35$ cm. At an instant of time t, the current $i_s = 20$ A.

(a) Calculate the H_s, F_s, and B_s distributions in the air gap as a function of θ, and (b) calculate the total flux crossing the air gap.

(*Continued*)

76 BASIC CONCEPTS IN MAGNETICS

Solution

(a) Using Eq. (3-36), $F_s = \frac{N_s i_s}{2} = 250$ A·turns. From Eq. (3-35), $H_s = \frac{N_s i_s}{2\ell_g} = 2.5 \times 10^5$ A/m. Finally, using Eq. (3-38), $B_s = \mu_0 H_s = 0.314$ T

Plots of the field distributions are similar to those shown in Fig. 3-15b.

(b) The flux crossing the rotor is $\phi_s = \int B \cdot dA$, calculated over half of the curved cylindrical surface A. The flux density is uniform, and the area A is one-half of the circumference times the rotor length: $A = \frac{1}{2}(2\pi r)\,\ell = 0.165$ m^2. Therefore, $\phi_s = B_s \cdot A = 0.0518$ Wb.

Note that the length of the air gap in electrical machines is extremely small, typically 1–2 mm. Therefore, we will use the radius r in the middle of the air gap to also represent the radius to the conductors located in the rotor and the stator slots.

3-11-3 Basic Principles of Torque Production and EMF Induction

There are two basic principles that govern electric machines' operation to convert between electric energy and mechanical work:

1. A force is produced on a current-carrying conductor when it is subjected to an *externally established* magnetic field.
2. An emf is induced in a conductor moving in a magnetic field.

Electromagnetic Force Consider the conductor of length ℓ shown in Fig. 3-16a. The conductor is carrying a current i and is subjected to an *externally established* magnetic field of a uniform flux density B perpendicular to the conductor length. A force f_{em} is exerted on the conductor, due to the electromagnetic interaction between the external magnetic field and the conductor current. The magnitude of this force is given as

BASIC PRINCIPLES OF TORQUE PRODUCTION 77

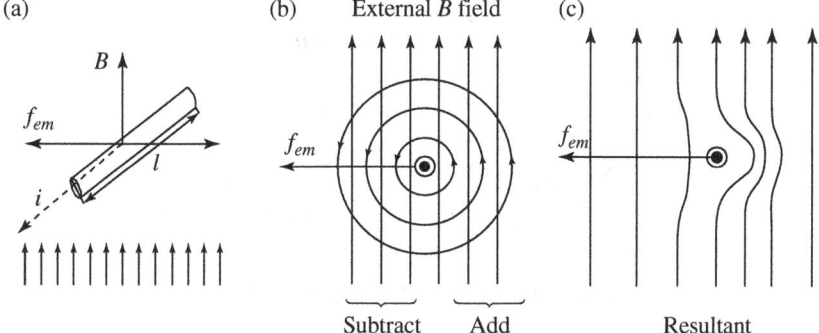

Fig. 3-16 Electric force on a current-carrying conductor in a magnetic field.

$$\underbrace{f_{em}}_{[Nm]} = \underbrace{B}_{[T]} \underbrace{i}_{[A]} \underbrace{\ell}_{[m]} \qquad (3\text{-}39)$$

As shown in Fig. 3-16a, the direction of the force is perpendicular to the directions of both *i* and *B*. To obtain the direction of this force, we will superimpose the flux lines produced by the conductor current, which are shown in Fig. 3-16b. The flux lines add up on the right side of the conductor and subtract on the left side, as shown in Fig. 3-16c. Therefore, the force f_{em} acts *from the higher concentration of flux lines to the lower concentration*, that is, from right to left in this case.

EXAMPLE 3-9

In Fig. 3-17a, the conductor is carrying a current into the paper plane in the presence of an external, uniform field. Determine the direction of the electromagnetic force.

Solution

The flux lines are clockwise and add up on the upper-right side, hence the resulting force shown in Fig. 3-17b.

78 BASIC CONCEPTS IN MAGNETICS

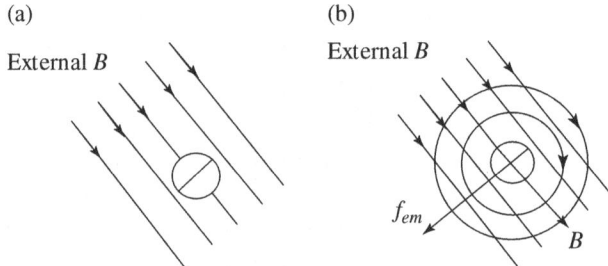

Fig. 3-17 Figure for Example 3-9.

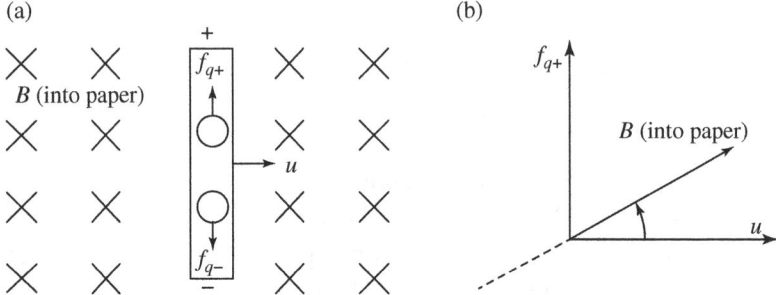

Fig. 3-18 Conductor moving in a magnetic field.

Induced EMF In Fig. 3-18a, a conductor of length ℓ is moving to the right at a speed u. The B-field is uniform and is perpendicularly directed into the paper plane. The magnitude of the induced emf at any instant of time is then given by

$$\underbrace{e}_{[V]} = \underbrace{B}_{[T]} \underbrace{l}_{[m]} \underbrace{u}_{[m/s]} \qquad (3\text{-}40)$$

The polarity of the induced emf can be established as follows: due to the conductor motion, the force on a charge q (positive, or negative in the case of an electron) within the conductor can be written as

$$f_q = q(\mathbf{u} \times \mathbf{B}) \qquad (3\text{-}41)$$

where the speed and the flux density are shown by bold letters to imply that these are vectors, and their cross product determines the force.

Since **u** and **B** are orthogonal to each other, as shown in Fig. 3-18b, the force on a positive charge is upward. Similarly, the force on an electron is downward. Thus, the upper end will have a positive potential with respect to the lower end. This induced emf across the conductor is independent of the current that would flow if a closed path were to be available (as would normally be the case). With the current flowing, the voltage across the conductor will be the induced-emf $e(t)$ in Eq. (3-40) minus the voltage drops across the conductor resistance and inductance.

EXAMPLE 3-10

In Figs. 3-19a and b, the conductors perpendicular to the paper plane are moving in the directions shown, in the presence of an external, uniform B-field. Determine the polarity of the induced emf.

Solution

The vectors representing **u** and **B** are shown. In accordance with Eq. (3-41), the top side of the conductor in Fig. 3-19a is positive. The opposite is true in Fig. 3-19b.

Magnetic Shielding of Conductors in Slots The current-carrying conductors in the stator and the rotor are often placed in slots, which shield the conductors magnetically. As a consequence, the force is

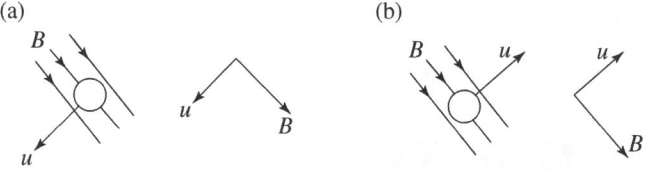

Fig. 3-19 Figure for Example 3-10.

80 BASIC CONCEPTS IN MAGNETICS

mainly exerted on the iron around the conductor. It can be shown, although we will not prove it here, that this force has the same magnitude and direction as it would in the absence of the magnetic shielding by the slot. Since our aim in this course is not to design, but rather to utilize, electric machines, we will completely ignore the effect of the magnetic shielding of conductors by slots in our subsequent discussions. The same argument applies to the calculation of induced emf and its direction using Eqs. (3-40) and (3-41).

3-11-4 Application of the Basic Principles

Consider the structure of Fig. 3-20a, where we will assume that the stator has established a uniform field B_s in the radial direction through the air gap. An N_r-turn coil is located on the rotor at a radius r. We will consider the force and the torque acting on the rotor in the counterclockwise direction to be positive.

A current i_r is passed through the rotor coil, which is subjected to a stator-established field B_s in Fig. 3-20a. The coil inductance is assumed to be negligible. The current magnitude I is constant, but its direction is controlled, such that it depends on the location δ of the coil, as plotted in Fig. 3-20b. In accordance with Eq. (3-39), the force on both sides of the coil results in an electromagnetic torque on the rotor in a counterclockwise direction, where

$$f_{em} = B_s(N_r I)\ell \qquad (3\text{-}42)$$

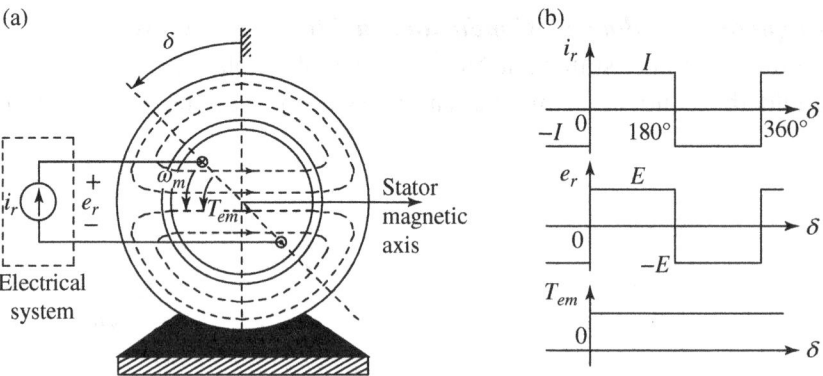

Fig. 3-20 Motoring mode.

BASIC PRINCIPLES OF TORQUE PRODUCTION

Thus,

$$T_{em} = 2f_{em}r = 2B_s(N_r I)\ell r \tag{3-43}$$

As the rotor turns, the current direction is changed every half-cycle, resulting in a torque that remains constant, as given by Eq. (3-43). This torque will accelerate the mechanical load connected to the rotor shaft, resulting in a speed ω_m. Note that an equal but opposite torque is experienced by the stator. This is precisely the reason for affixing the stator to the foundation – to prevent the stator from turning.

Due to the conductors moving in the presence of the stator field, in accordance with Eq. (3-40), the magnitude of the induced emf at any instant of time in each conductor of the coil is

$$E_{cond} = B_s \ell \underbrace{r\omega_m}_{u} \tag{3-44}$$

Thus, the magnitude of the induced emf in the rotor coil with $2N_r$ conductors is

$$E = 2N_r B_s \ell r \omega_m \tag{3-45}$$

The waveform of the emf e_r, with the polarity indicated in Fig. 3-20a, is similar to that of i_r, as plotted in Fig. 3-20b.

3-11-5 Energy Conversion

In this idealized system which has no losses, we can show that the electrical input power P_{el} is converted into the mechanical output power P_{mech}. Using the waveforms of i_r, e_r, and T_{em} in Fig. 3-20b at any instant of time,

$$P_{el} = e_r i_r = (2N_r B_s \ell r \omega_m) I \tag{3-46}$$

and

$$P_{mech} = T_{em}\omega_m = (2B_s N_r I \ell r)\omega_m \tag{3-47}$$

82 BASIC CONCEPTS IN MAGNETICS

Thus,

$$P_{mech} = P_{el} \qquad (3\text{-}48)$$

The above relationship is valid in the presence of losses. The power drawn from the electrical source is P_{el}, in Eq. (3-46), plus the losses in the electrical system. The mechanical power available at the shaft is P_{mech}, in Eq. (3-47), minus the losses in the mechanical system.

EXAMPLE 3-11

The machine shown in Fig. 3-20a has a radius of 15 cm, and the length of the rotor is 35 cm. The rotor coil has $N_r = 15$ turns, and $B_s = 1.3$ T (uniform). The current i_r, as plotted in Fig. 3-20b, has a magnitude $I = 10$ A. $\omega_m = 100$ rad/s. Calculate and plot T_{em} and the induced emf e_r. Also, calculate the power being converted.

Solution

Using Eq. (3-43), the electromagnetic torque on the rotor will be in a counterclockwise direction and of a magnitude

$$T_{em} = 2B_s(N_r I)\ell r = 2 \times 1.3 \times 15 \times 10 \times 0.35 \times 0.15 = 20.5 \text{ Nm}$$

The electromagnetic torque will have the waveform shown in Fig. 3-20b. At a speed of $\omega_m = 100$ rad/s, the electrical power absorbed for conversion into mechanical power is

$$P = \omega_m T_{em} = 100 \times 20.5 \simeq 2 \text{ kW}$$

Regenerative Braking At a speed ω_m, the rotor inertia, including that of the connected mechanical load, has stored kinetic energy. This energy can be recovered and fed back into the electrical system shown in Fig. 3-21a.

BASIC PRINCIPLES OF TORQUE PRODUCTION 83

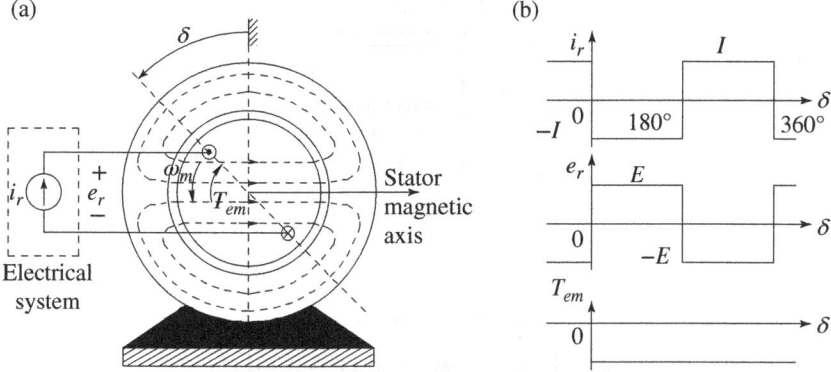

Fig. 3-21 Regenerative braking mode.

In so doing, the current is so controlled as to have the waveform plotted in Fig. 3-21b, as a function of the angle δ. Notice that the waveform of the induced voltage remains unchanged. In this regenerative case, due to the reversal of the current direction (compared to that in the motoring mode), the torque T_{em} is in the clockwise direction (opposing the rotation), and shown to be negative in Fig. 3-10b. Now the input power P_{mech} from the mechanical side equals the output power P_{el} into the electrical system. This direction of power flow represents the generator mode of operation.

3-11-6 Power Losses and Energy Efficiency

As indicated in Fig. 3-22, any electric drive has inherent power losses, which are converted into heat. These losses are complex functions of the speed and the torque of the machine. If the output power of the drive is P_o, then the input power to the motor in Fig. 3-22 is

$$P_{in,\,motor} = P_0 + P_{loss,\,motor} \tag{3-49}$$

At any operating condition, let $P_{loss,\,motor}$ equal all of the losses in the motor. Then the energy efficiency of the motor is

$$\eta_{motor} = \frac{P_o}{P_{in,\,motor}} = \frac{P_o}{P_o + P_{loss,\,motor}} \tag{3-50}$$

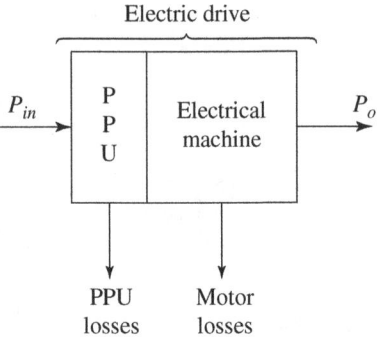

Fig. 3-22 Power losses and energy efficiency.

In the power-processing unit (PPU) of an electric drive, power losses occur due to current conduction and switching within the power semiconductor devices. Similar to Eq. (3-50), we can define the energy efficiency of the PPU as η_{PPU}. Therefore, the overall efficiency of the drive is such that

$$\eta_{drive} = \eta_{motor} \times \eta_{PPU} \qquad (3\text{-}51)$$

Energy efficiencies of drives depend on many factors. The energy efficiency of small- to medium-sized electric motors range from 85 to 93%, while that of PPUs range from 93 to 97%. Thus, from Eq. (3-51), the overall energy efficiency of drives is in the approximate range of 80 to 90%.

3-12 REVIEW QUESTIONS

3-12-1 Magnetic Circuits

1. What is the role of magnetic circuits? Why are magnetic materials with very high permeabilities desirable? What is the permeability of air? What is the typical range of the relative permeabilities of ferromagnetic materials like iron?

2. Why can "leakage" be ignored in electric circuits but not in magnetic circuits?

3. What is Ampere's Law, and what quantity is usually calculated by using it?

4. What is the definition of the mmf F?

5. What is meant by "magnetic saturation"?

6. What is the relationship between ϕ and B?

7. How can magnetic reluctance $\mathcal{R}$ be calculated? What field quantity is calculated by dividing the mmf F by the reluctance $\mathcal{R}$?

8. In magnetic circuits with an air gap, what usually dominates the total reluctance in the flux path: the air gap or the rest of the magnetic structure?

9. What is the meaning of the flux linkage λ of a coil?

10. Which law allows us to calculate the induced emf? What is the relationship between the induced voltage and the flux linkage?

11. How is the polarity of the induced emf established?

12. Assuming sinusoidal variations with time at a frequency f, how are the rms value of the induced emf, the peak of the flux linking a coil, and the frequency of variation f related?

13. How does the inductance L of a coil relate Faraday's Law to Ampere's Law?

14. In a linear magnetic structure, define the inductance of a coil in terms of its geometry.

15. What is leakage inductance? How can the voltage drop across it be represented separately from the emf induced by the main flux in the magnetic core?

16. In linear magnetic structures, how is energy storage defined? In magnetic structures with air gaps, where is energy mainly stored?

17. What is the meaning of "mutual inductance"?

86 BASIC CONCEPTS IN MAGNETICS

3-12-2 Electromechanical Energy Conversion

18. What is the role of electric machines? What do the motoring mode and the generating mode of operations mean?

19. What are the definitions of stator and rotor?

20. Why do we use high permeability ferromagnetic materials for stators and rotors in electric machines? Why are these constructed by stacking laminations together, rather than as a solid structure?

21. What is the approximate air gap length in machines with less than 10 kW ratings?

22. What are multipole machines? Why can such machines be analyzed by considering only one pair of poles?

23. Assuming the permeability of iron to be infinite, where is the mmf produced by machine coils "consumed"? What law is used to calculate the field quantities, such as flux density, for a given current through a coil? Why is it important to have a small air gap length?

24. What are the two basic principles of operation for electric machines?

25. What is the expression for force acting on a current-carrying conductor in an externally established B-field? What is its direction?

26. What is slot shielding, and why can we choose to ignore it?

27. How do we express the induced emf in a conductor "cutting" an externally established B-field? How do we determine the polarity of the induced emf?

28. How do electrical machines convert energy from one form to another?

29. What are various loss mechanisms in electric machines?

30. How is electrical efficiency defined, and what are typical values of efficiencies for the machines, the PPUs, and the overall drives?

31. What is the end-result of power losses in electric machines?

32. What is meant by the various ratings on the name-plates of machines?

FURTHER READING

Fitzgerald, A.E., Kingsley, C., and Umans, S. (1990). *Electric Machinery*, 5e. McGraw-Hill, Inc.

Slemon, G.R. (1992). *Electric Machines and Drives*. Addison-Wesley.

PROBLEMS

Magnetic Circuits

3-1 In Example 3-1, calculate the field intensity within the core: (a) very close to the inside diameter and (b) very close to the outside diameter. (c) Compare the results with the field intensity along the mean path.

3-2 In Example 3-1, calculate the reluctance in the path of flux lines if $\mu_r = 2000$.

3-3 Consider the core of dimensions given in Example 3-1. The coil requires an inductance of 25 µH. The maximum current is 3 A, and the maximum flux density is not to exceed 1.3 T. Calculate the number of turns N and the relative permeability μ_r of the magnetic material that should be used.

3-4 In Problem 3-3, assume the permeability of the magnetic material to be infinite. To satisfy the conditions of maximum flux density and the desired inductance, a small air gap is introduced. Calculate the length of this air gap (neglecting the effect of flux fringing) and the number of turns N.

88 BASIC CONCEPTS IN MAGNETICS

3-5 In Example 3-4, calculate the maximum current beyond which the flux density in the core will exceed 0.3 T.

3-6 The rectangular toroid of Fig. 3-7 in Example 3-4 consists of a material whose relative permeability can be assumed to be infinite. The other parameters are as given in Example 3-4. An air gap of 0.05 mm length is introduced. Calculate (a) the coil inductance L_m assuming that the core is unsaturated and (b) the maximum current beyond which the flux density in the core will exceed 0.3 T.

3-7 In Problem 3-6, calculate the energy stored in the core and the air gap at the flux density of 0.3 T.

3-8 In the structure of Fig. 3-11a, $L_m = 200$ mH, $L_i = 1$ mH, and $N = 100$ turns. Ignore the coil resistance. A steady-state voltage is applied, where $\overline{V} = \sqrt{2} \times 120 \angle 0$ V at a frequency of 60 Hz. Calculate the current $\overline{I}$ and $i(t)$.

Electromechanical Energy Conversion

3-9 Assume the field distribution produced by the stator in the machine shown in Fig. P3-9 to be radially uniform. The magnitude

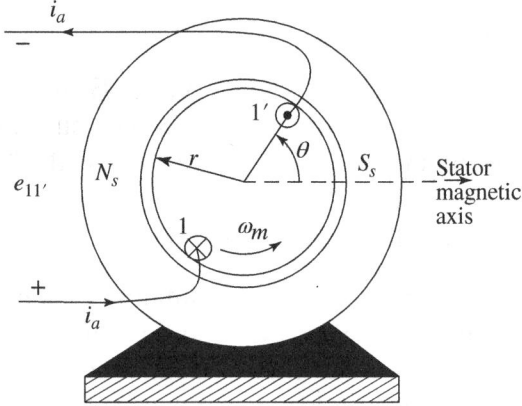

Fig. P3-9 Problem 3-9.

of the air-gap flux density is B_s, the rotor length is ℓ, and the rotational speed of the motor is ω_m.

(a) Plot the emf $e_{11'}$ induced in the coil as a function of θ for two values of i_a: 0 and 10 A. (b) In the position shown, the current i_a in the coil $11'$ equals I_a. Calculate the torque acting on the coil in this position for two values of instantaneous speed ω_m: 0 and 100 rad/s.

3-10 Figure P3-10 shows a primitive machine with a rotor producing a uniform magnetic field such that the air-gap flux density in the radial direction is of the magnitude B_r. Plot the induced emf $e_{11'}$ as a function of θ. The length of the rotor is ℓ and the radius at the air gap is r.

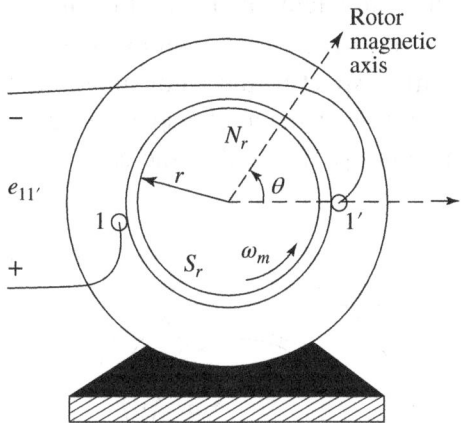

Fig. P3-10 Problem 3-10.

3-11 In the primitive machine shown in Fig. P3-11, the air-gap flux density B_s has a sinusoidal distribution given by $B_s = \hat{B}\cos\theta$. The rotor length is ℓ. (a) Given that the rotor is rotated at a speed of ω_m, plot as a function of θ the emf $e_{11'}$ induced and the torque T_{em} acting on the coil if $i_a = I$. (b) In the position shown, the current i_a in the coil equals I. Calculate P_{el}, the electrical power input to the machine, and P_{mech}, the mechanical power output of the machine, if $\omega_m = 60$ rad/s.

90 BASIC CONCEPTS IN MAGNETICS

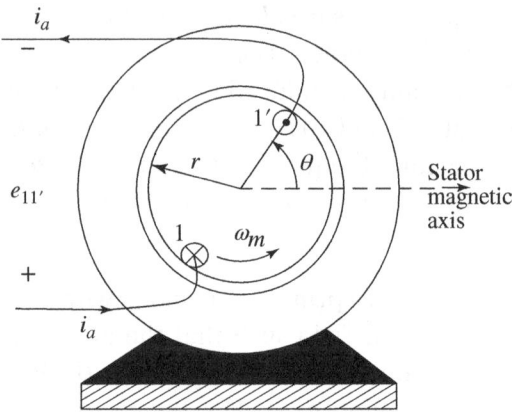

Fig. P3-11 Problem 3-11.

3-12 In the machine shown in Fig. P3-12, the air-gap flux density B_r has a sinusoidal distribution given by $B_r = \hat{B} \cos \alpha$, where α is measured with respect to the rotor magnetic axis. Given that the rotor is rotating at an angular speed ω_m and the rotor length is ℓ, plot the emf $e_{11'}$ induced in the coil as a function of θ.

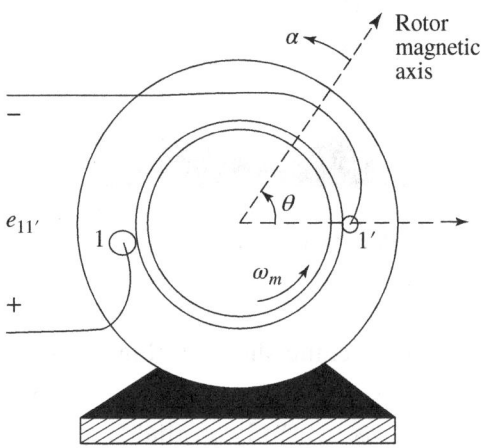

Fig. P3-12 Problem 3-12.

3-13 In the machine shown in Fig. P3-13, the air-gap flux density B_s is constant and equal to B_{max} in front of the pole faces, and is zero elsewhere. The direction of the B-field is from left

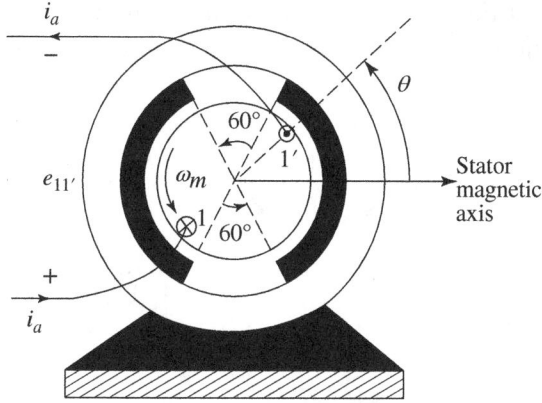

Fig. P3-13 Problem 3-13.

(north pole) to the right (south pole). The rotor is rotating at an angular speed of ω_m, and the length of the rotor is ℓ. Plot the induced emf $e_{11'}$ as a function of θ. What should be the waveform of i_a that produces an optimum electromagnetic torque T_{em}?

3-14 As shown in Fig. P3-14, a rod in a uniform magnetic field is free to slide on two rails. The resistance of the rod and the rails are negligible. Electrical continuity between the rails and the rod is assumed, so that a current can flow through the rod. A damping force, F_d, tending to slow down the rod, is proportional to the square of the rod's speed as follows: $F_d = k_f u^2$ where $k_f = 1500$. Assume that the inductance in this circuit can be ignored. Find the steady-state speed u of the rod, assuming the system extends endlessly to the right.

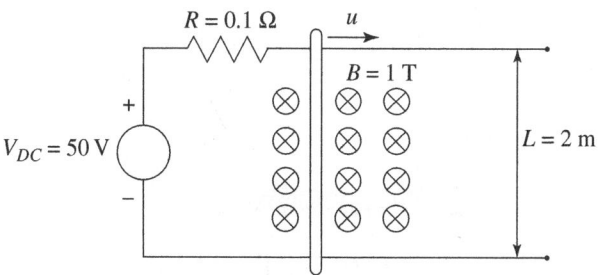

Fig. P3-14 Problem 3-14.

3-15 Consider Fig. P3-15. Plot the mmf distribution in the air gap as a function of θ for $i_a = I$. Assume each coil has a single turn.

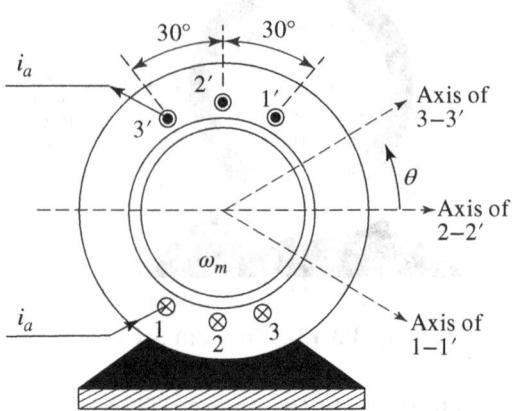

Fig. P3-15 Problem 3-15.

3-16 In Fig. P3-16, the stator coil has N_s turns, and the rotor coil has N_r turns. Each coil produces in the air gap a uniform, radial flux density B_s, and B_r, respectively. In the position shown, calculate the torque experienced by both the rotor coil and the stator coil, due to the currents i_s and i_r flowing through these coils. Show that the torque on the stator is equal in magnitude, but opposite in direction, to that experienced by the rotor.

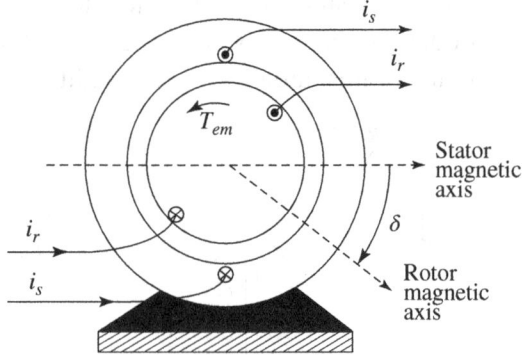

Fig. P3-16 Problem 3-16.

3-17 Figure P3-17 shows the cross-section, seen from the front, of squirrel-cage induction machines, which are discussed later on. In such machines, the magnetizing-flux density distribution produced by the stator windings is represented by a vector $\vec{B}_{ms}$ that implies that at this instant, for example, the radially oriented flux density in the air gap is in a vertically downward direction where the peak occurs, and it is co-sinusoidally distributed elsewhere. This means that, at this instant, the flux density in the air gap is zero at $\theta = 90°$ (and at $\theta = 270°$). At $\theta = 180°$, the flux density again peaks but has a negative value. This flux density distribution is rotating at a speed ω_{syn}. The rotor consists of bars, as shown, along its periphery. This rotor is rotating at a speed of ω_m. The voltages induced in the front-ends of the rotor-bars have the polarities as shown in Fig. P3-17, with respect to their back-ends. Determine if ω_m is greater or less than ω_{syn}.

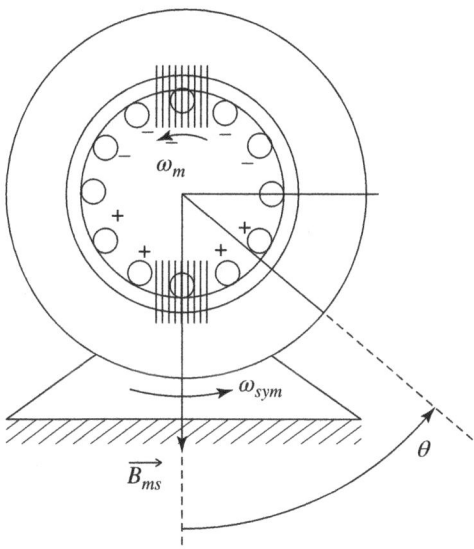

Fig. P3-17 Problem 3-17.

4 Basic Understanding of Switch-Mode Power Electronic Converters

4-1 INTRODUCTION

Electric drives require power electronic converters to efficiently match electrical-system voltages and currents to the speed and the torque of the mechanical system. Some of the sustainability-related applications are increasing the efficiency of motor-driven systems, harnessing of energy in the wind, and electric transportation of various types, as discussed in Chapter 1. Similar to linear amplifiers, power electronic converters amplify the input control signals. However, unlike linear amplifiers, power electronic converters in electric drives use switch-mode power electronics principles to achieve high energy efficiency and low cost, size, and weight. In this chapter, we will examine the basic switch-mode principles, topologies, and control for the processing of electrical power in an efficient and controlled manner.

4-2 OVERVIEW OF POWER ELECTRONIC CONVERTERS

The discussion in this chapter is excerpted from [1], to which the reader is referred for a systematic discussion. In many applications such as wind turbines, the voltage-link structure of Fig. 4-1 is used.
(Adapted from chapter 4 of *Electric Machines and Drives: A First Course* ISBN: 978-1-118-07481-7 by Ned Mohan, January)

Analysis and Control of Electric Drives: Simulations and Laboratory Implementation, First Edition. Ned Mohan and Siddharth Raju.
© 2021 John Wiley & Sons, Inc. Published 2021 by John Wiley & Sons, Inc.
Companion website: www.wiley.com/go/Mohan/Vectorcontrolinelectricdrives

96 BASIC UNDERSTANDING OF SWITCH-MODE

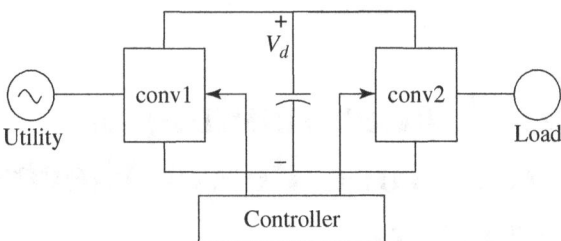

Fig. 4-1 Voltage-link system.

To provide the needed functionality to the converters in Fig. 4-1, transistors and diodes which can block voltage only of one polarity have led to this commonly used voltage-link structure.

This structure consists of two separate converters, one on the electrical side and the other on the mechanical side. The dc ports of these two converters are connected with a parallel capacitor forming a dc-link, across which the voltage polarity does not reverse, thus allowing unipolar voltage-blocking transistors to be used within these converters. The capacitor in parallel with the two converters forms a dc voltage-link; hence, this structure is called a *voltage-link* (or a *voltage-source*) structure. This structure is used in a very large power range, from a few tens of watts to several megawatts, even extending to hundreds of megawatts in utility applications. Therefore, we will mainly focus on this voltage-link structure in this book, although there are other structures as well. It should be remembered that the power flow in Fig. 4-1 can reverse.

To understand how converters in Fig. 4-1 operate, our emphasis will be to discuss how the mechanical-side converter, with the dc voltage as input, synthesizes dc or low-frequency sinusoidal voltage outputs. Functionally, this converter operates as a linear amplifier, amplifying a control signal, dc in case of dc-motor drives, and ac in case of ac-motor drives. The power flow through this converter should be reversible.

The dc voltage V_d (assumed to be constant) is used as the input voltage to the mechanical-side switch-mode converter in Fig. 4-1. The task of this converter, depending on the machine type, is to

OVERVIEW OF POWER ELECTRONIC CONVERTERS 97

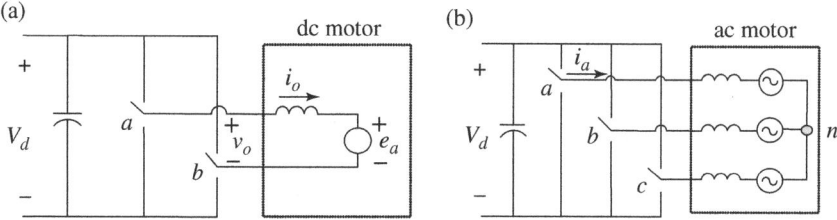

Fig. 4-2 Switch mode converters for (a) dc- and (b) ac-machine drives.

deliver an adjustable-magnitude dc or sinusoidal ac to the machine by amplifying the signal from the controller by a constant gain. The power flow through the switch-mode converter must be able to reverse. In switch-mode converters, as their name implies, transistors are operated as switches: either fully on or fully off. The switch-mode converters used for dc- and ac-machine drives can be simply illustrated, as in Fig. 4-2a and b, respectively, where each bi-positional switch constitutes a pole. The dc–dc converter for dc-machine drives in Fig. 4-2a consists of two such poles, whereas the dc-to-three-phase ac converter shown in Fig. 4-2b for ac-machine drives consists of three such poles.

Typically, the power converter efficiencies exceed 95% and can exceed 98% in very large power ratings.

4-2-1 Switch-Mode Conversion: Switching Power-Pole as the Building Block

Achieving high energy efficiency requires switch-mode conversion, where in contrast to linear power electronics, transistors (and diodes) are operated as switches, either on or off. This switch-mode conversion can be explained by its basic building block, a switching power-pole a, as shown in Fig. 4-3a.

It effectively consists of a bi-positional switch, which forms a two-port: a voltage-port across a capacitor with a voltage V_d that cannot change instantaneously, and a current-port due to the series inductor through which the current cannot change instantaneously. For now, we will assume the switch to be ideal with two positions: up or down,

98 BASIC UNDERSTANDING OF SWITCH-MODE

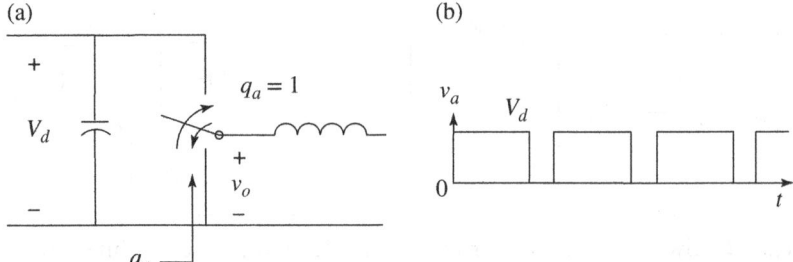

Fig. 4-3 Switching power-pole as the building block in converters.

dictated by a switching signal q_a which takes on two values: 1 and 0, respectively.

The bi-positional switch "chops" the input dc voltage V_d into a train of high-frequency voltage pulses, shown by v_a the waveform in Fig. 4-3b, by switching up or down at a high repetition rate, called the switching frequency f_s. Controlling the pulse width within a switching cycle allows control over the switching-cycle-averaged value of the pulsed output, and this pulse-width modulation (PWM) forms the basis of synthesizing adjustable dc and low-frequency sinusoidal ac outputs, as described in the next section. A switch-mode converter consists of one or more such switching power-poles.

4-2-2 PWM of the Switching Power-Pole (Constant f_s)

The objective of the switching power-pole, redrawn in Fig. 4-4a, is to synthesize the output voltage such that its *switching-cycle-average* is of the desired value: dc or ac that varies sinusoidally at a low frequency, compared to f_s. Switching at a constant switching-frequency f_s produces a train of voltage pulses in Fig. 4-4b that repeat with a constant switching time-period T_s, equal to $1/f_s$.

Within each switching cycle with the time-period T_s ($=1/f_s$) in Fig. 4-4b, the switching-cycle-averaged value $\bar{v}_a$ of the waveform is controlled by the pulse width T_{up} (during which the switch is in the up position and v_a equals V_d), as a ratio of T_s:

$$\bar{v}_a = \frac{T_{up}}{T_s} V_d = d_a V_d \quad 0 \le d_a \le 1 \qquad (4\text{-}1)$$

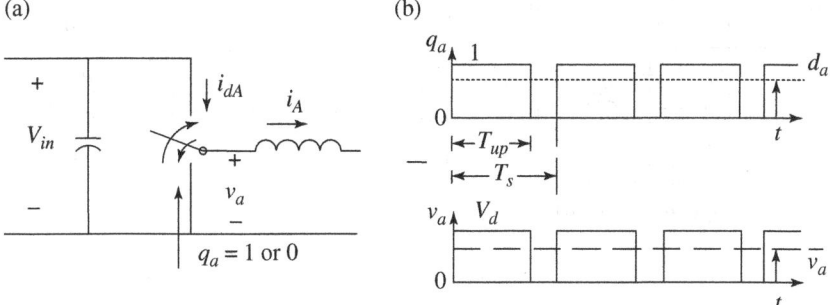

Fig. 4-4 PWM of the switching power-pole.

where $d_a(=T_{up}/T_s)$, which is the average of the q_a waveform shown in Fig. 4-4b is defined as the duty-ratio of the switching power-pole a, and the switching-cycle-averaged voltage is indicated by a "-" on top. The switching-cycle-averaged voltage and the switch duty-ratio are expressed by lowercase letters since they may vary as functions of time. The control over the switching-cycle-averaged value of the output voltage is achieved by adjusting or modulating the pulse width, which later on will be referred to as PWM. This switching power-pole and the control of its output by PWM set the stage for switch-mode conversion with high energy efficiency.

We should note that $\bar{v}_a$ and d_a in the above discussion are discrete quantities and their values, calculated over a kth switching cycle, for example, can be expressed as $\bar{v}_{a,k}$ and $d_{a,k}$. However, in practical applications, the pulse-width T_{up} changes very slowly over many switching cycles, and hence we can consider them analog quantities as $\bar{v}_a(t)$ and $d_a(t)$ that are continuous functions of time. For simplicity, we may not show their time dependence explicitly.

4-2-3 Bidirectional Switching Power-Pole

A bidirectional switching power-pole, through which the power flow can be in either direction, is implemented, as shown in Fig. 4-5a. In such a bidirectional switching power-pole, the positive inductor current i_L, as shown in Fig. 4-5b, represents a Buck-mode of operation (where the power flow is from the higher voltage to the lower voltage), where only the transistor and the diode associated with the Buck

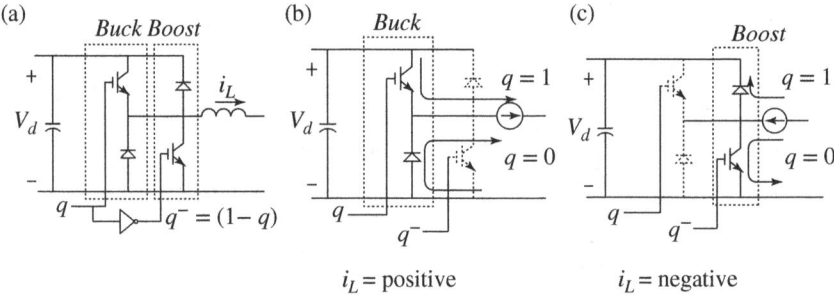

Fig. 4-5 Bidirectional power flow through a switching power-pole.

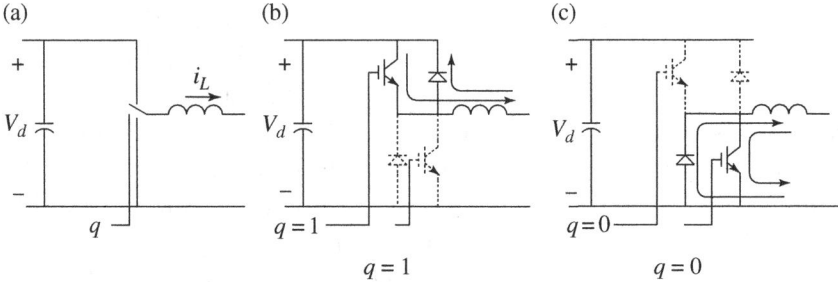

Fig. 4-6 Bidirectional switching power-pole.

converter take part; the transistor conducts during $q = 1$; otherwise, the diode conducts.

However, as shown in Fig. 4-5c, the negative inductor current represents a Boost-mode of operation (where the power flow is from the lower voltage to the higher voltage), where only the transistor and the diode associated with the Boost converter take part; the transistor conducts during $q = 0$ ($\bar{q} = 1$); otherwise, the diode conducts during $q = 1$ ($\bar{q} = 0$).

Figures 4-5b and c show that the combination of devices in Fig. 4-5a renders it to be a switching power-pole that can carry i_L in either direction. This is shown as an equivalent switch in Fig. 4-6a that is effectively in the "up" position when $q = 1$ as shown in Fig. 4-6b, and in the "down" position when $q = 0$ as shown in Fig. 4-6c, regardless of the direction of i_L.

OVERVIEW OF POWER ELECTRONIC CONVERTERS 101

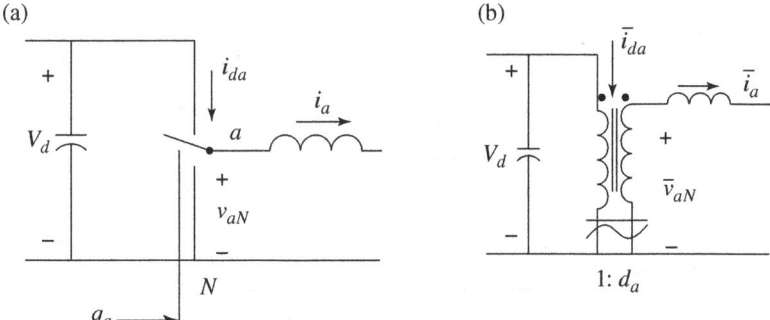

Fig. 4-7 Switching-cycle-averaged representation of the bidirectional power-pole.

The bidirectional switching power-pole of Fig. 4-6a is repeated in Fig. 4-7a for pole-a, with its switching signal identified as q_a. In response to the switching signal, it behaves as follows: "up" when $q_a = 1$; otherwise, "down." Therefore, its switching-cycle-averaged representation is an ideal transformer, shown in Fig. 4-7b, with a turns-ratio $1 : d_a(t)$.

The switching-cycle-averaged values of the variables at the voltage-port and the current-port in Fig. 4-7b are related by $d_a(t)$ as follows:

$$\overline{v}_{aN} = d_a V_d \qquad (4\text{-}2)$$

$$\overline{i}_{da} = d_a \overline{i}_a \qquad (4\text{-}3)$$

4-2-4 PWM of the Bidirectional Switching Power-Pole

The voltage of a switching power-pole at the current-port is always of positive polarity. However, the output voltages of converters in Fig. 4-2 for electric drives must be reversible in polarity. This is achieved by introducing a common-mode voltage in each power-pole as discussed below and taking the differential output between the power-poles.

To obtain the desired switching-cycle-averaged voltage $\overline{v}_{aN}$ in Fig. 4-7b that includes the common-mode voltage, requires the following power-pole duty-ratio from Eq. (4-2):

$$d_a = \frac{\overline{v}_{aN}}{V_d} \qquad (4\text{-}4)$$

102 BASIC UNDERSTANDING OF SWITCH-MODE

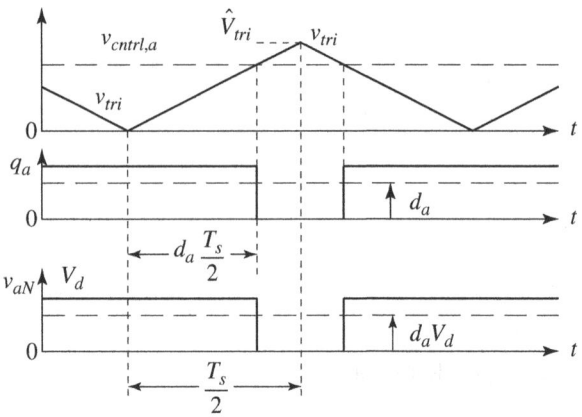

Fig. 4-8 Waveforms for PWM in a switching power-pole.

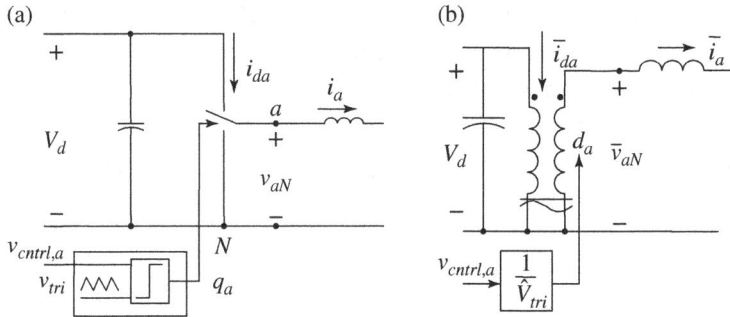

Fig. 4-9 Switching power-pole and its duty-ratio control.

where V_d is the dc-bus voltage. To obtain the switching signal q_a to deliver this duty-ratio, a control voltage $v_{cntrl,\,a}$ is compared with a triangular-shaped carrier waveform of the switching-frequency f_s and amplitude $\hat{V}_{tri}$, as shown in Fig. 4-8. Because of symmetry, only $T_s/2$, one-half of the switching time-period needs be considered. The switching-signal $q_a = 1$ if $v_{cntrl,a} > v_{tri}$; otherwise 0. Therefore, in Fig. 4-8,

$$v_{cntrl,a} = d_a \hat{V}_{tri} \qquad (4\text{-}5)$$

The switching-cycle-averaged representation of the switching power-pole in Fig. 4-9a is shown by a controllable turns-ratio ideal

OVERVIEW OF POWER ELECTRONIC CONVERTERS   103

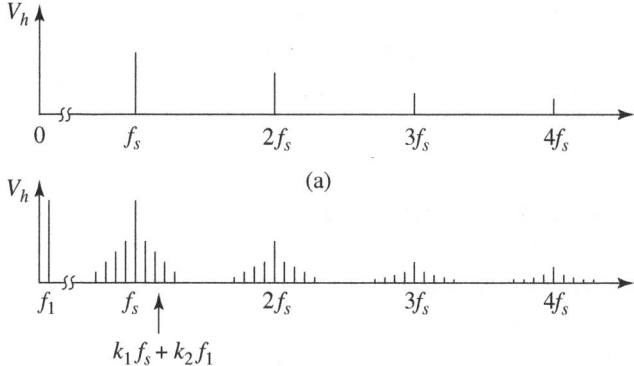

Fig. 4-10   Harmonics in the output of a switching power-pole.

transformer in Fig. 4-9b, where the switching-cycle-averaged representation of the duty-ratio control is as per Eq. (4-5).

The Fourier spectrum of the switching waveform v_{aN} is shown in Fig. 4-10, which depends on the nature of the control signal. If the control voltage is dc, the output voltage has harmonics at the multiples of the switching frequency, that is at f_s, $2f_s$, etc., as shown in Fig. 4-10a. If the control voltage varies at a low-frequency f_1, as in electric drives, then the harmonics of significant magnitudes appear in the sidebands of the switching frequency and its multiples, as shown in Fig. 4-10b, where

$$f_h = k_1 f_s + \underbrace{k_2 f_1}_{sidebands} \tag{4-6}$$

in which k_1 and k_2 are constants that can take on values 1, 2, 3, ... Some of these harmonics associated with each power-pole are canceled from the converter output voltages, where two or three such power-poles are used.

In the power-pole shown in Fig. 4-9, the output voltage v_{aN} and its switching-cycle-averaged $\overline{v}_{aN}$ are limited between 0 and V_d. To obtain an output voltage $\overline{v}_{an}$ (where "n" may be a fictitious node) that can become both positive and negative, a common-mode offset $\overline{v}_{com}$ is introduced in each power-pole so that the pole output voltage is

$$\overline{v}_{aN} = \overline{v}_{com} + \overline{v}_{an} \tag{4-7}$$

104 BASIC UNDERSTANDING OF SWITCH-MODE

where $\overline{v}_{com}$ allows $\overline{v}_{an}$ to become both positive and negative around the common-mode voltage $\overline{v}_{com}$. In the differential output, when two or three power-poles are used, the common-mode voltage gets eliminated.

4-3 CONVERTERS FOR dc MOTOR DRIVES ($-V_d < \overline{v}_o < V_d$)

Converters for dc motor drives consist of two power-poles, as shown in Fig. 4-11a, where

$$\overline{v}_o = \overline{v}_{aN} - \overline{v}_{bN} \qquad (4\text{-}8)$$

and $\overline{v}_o$ can assume both positive and negative values. Since the output voltage is desired to be in a full-range, from $-V_d$ to $+V_d$, pole-a is assigned to produce $\overline{v}_o/2$, and pole-b is assigned to produce $-\overline{v}_o/2$ toward the output:

$$\overline{v}_{an} = \frac{\overline{v}_o}{2} \text{ and } \overline{v}_{bn} = -\frac{\overline{v}_o}{2} \qquad (4\text{-}9)$$

where "n" is a fictitious node, as shown in Fig. 4-11a, chosen to define the contribution of each pole toward $\overline{v}_o$.

To achieve equal excursions in positive and negative values of the switching-cycle-averaged output voltage, the switching-

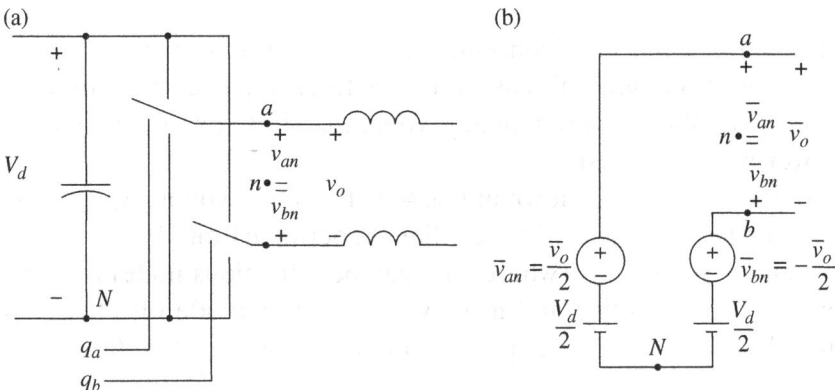

Fig. 4-11 Converter for a dc-motor drive.

cycle-averaged common-mode voltage in each pole is chosen to be one-half the dc-bus voltage

$$\bar{v}_{com} = \frac{V_d}{2} \qquad (4\text{-}10)$$

Therefore, from Eq. (4-7),

$$\bar{v}_{aN} = \frac{V_d}{2} + \frac{\bar{v}_o}{2} \text{ and } \bar{v}_{bN} = \frac{V_d}{2} - \frac{\bar{v}_o}{2} \qquad (4\text{-}11)$$

The switching-cycle-averaged output voltages of the power-poles and the converter are shown in Fig. 4-11b. From Eqs. (4-4) and (4-11),

$$d_a = \frac{1}{2} + \frac{1}{2}\frac{\bar{v}_o}{V_d} \text{ and } d_b = \frac{1}{2} - \frac{1}{2}\frac{\bar{v}_o}{V_d} \qquad (4\text{-}12)$$

and from Eq. (4-12),

$$\bar{v}_o = (d_a - d_b)V_d \qquad (4\text{-}13)$$

EXAMPLE 4-1

In a dc-motor drive, the dc-bus voltage is $V_d = 350\,\text{V}$. Determine the following: $\bar{v}_{com}$, $\bar{v}_{aN}$, and d_a for pole-a and similarly for pole-b, if the output voltage required is (a) $\bar{v}_0 = 300\,\text{V}$ and (b) $\bar{v}_0 = -300\,\text{V}$.

Solution

From Eq. (4-10), $\bar{v}_{com} = \frac{V_d}{2} = 175\,\text{V}$.

(a) For $\bar{v}_0 = 300\,\text{V}$, from Eq. (4-9), $\bar{v}_{an} = \bar{v}_o/2 = 150\,\text{V}$ and $\bar{v}_{bn} = -\bar{v}_o/2 = -150\,\text{V}$. From Eq. (4-11), $\bar{v}_{aN} = 325\,\text{V}$ and $\bar{v}_{bN} = 25\,\text{V}$. From Eq. (4-12), $d_a \simeq 0.93$ and $d_b \simeq 0.07$.

(b) For $\bar{v}_0 = -300\,\text{V}$, $\bar{v}_{an} = \bar{v}_o/2 = -150\,\text{V}$ and $\bar{v}_{bn} = -\bar{v}_o/2 = 150\,\text{V}$. Therefore, from Eq. (4-11), $\bar{v}_{aN} = 25\,\text{V}$ and $\bar{v}_{bN} = 325\,\text{V}$. From Eq. (4-12), $d_a \simeq 0.07$ and $d_b \simeq 0.93$.

106 BASIC UNDERSTANDING OF SWITCH-MODE

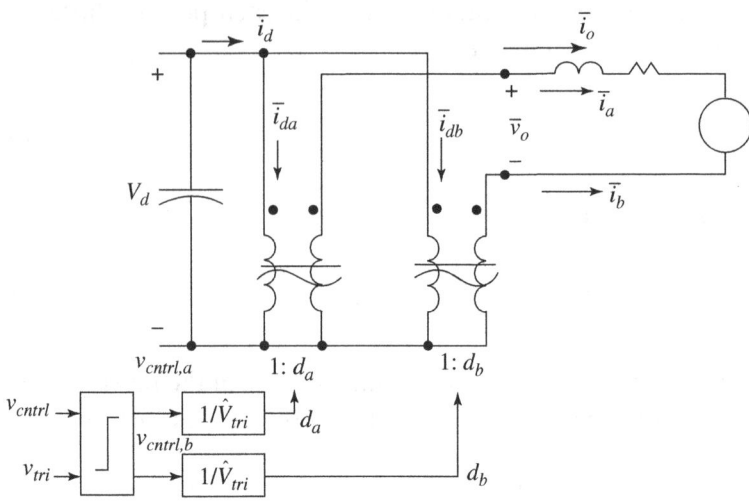

Fig. 4-12 Switching-cycle-averaged representation of the converter for dc drives.

The switching-cycle-averaged representation of the two power-poles, along with the pulse-width-modulator, in a block-diagram form is shown in Fig. 4-12.

In each power-pole of Fig. 4-12, the switching-cycle-averaged dc-side current is related to its output current by the pole duty-ratio:

$$\bar{i}_{da} = d_a \bar{i}_a \text{ and } \bar{i}_{db} = d_b \bar{i}_b \tag{4-14}$$

By Kirchhoff's current law, the total switching-cycle-averaged dc-side current is

$$\bar{i}_d = \bar{i}_{da} + \bar{i}_{db} = d_a \bar{i}_a + d_b \bar{i}_b \tag{4-15}$$

Recognizing the directions with which the currents i_a and i_b are defined,

$$\bar{i}_a(t) = -\bar{i}_b(t) = \bar{i}_o(t) \tag{4-16}$$

CONVERTERS FOR dc MOTOR DRIVES $(-V_d < \bar{v}_o < V_d)$ 107

Thus, substituting currents from Eq. (4-15) into Eq. (4-16),

$$\bar{i}_d = (d_a - d_b)\bar{i}_o \qquad (4\text{-}17)$$

EXAMPLE 4-2

In the dc-motor drive of Example 4-1, the output current into the motor is $\bar{i}_o = 15\text{A}$. Calculate the power delivered from the dc-bus and show that it is equal to the power delivered to the motor (assuming the converter to be lossless), if $\bar{v}_0 = 300\text{ V}$.

Solution

Using the values for d_a and d_b from part (a) of Example 4-1, and $\bar{i}_o = 15\text{A}$, from Eq. (4-17), $\bar{i}_d(t) = 12.9$ A and therefore the power delivered by the dc-bus is $P_d = 4.515\text{ kW}$. Power delivered by the converter to the motor is $P_o = \bar{v}_o \bar{i}_o = 4.5\text{ kW}$, which is equal to the input power (neglecting the round-off errors).

Using Eqs. (4-5) and (4-12), the control voltages for the two poles are as follows:

$$v_{cntrl,a} = \frac{\hat{V}_{tri}}{2} + \frac{\hat{V}_{tri}}{2}\left(\frac{\bar{v}_o}{V_d}\right) \quad \text{and} \quad v_{cntrl,b} = \frac{\hat{V}_{tri}}{2} - \frac{\hat{V}_{tri}}{2}\left(\frac{\bar{v}_o}{V_d}\right) \qquad (4\text{-}18)$$

In Eq. (4-18), defining the second term in the two control voltages above as one-half the control voltage, that is,

$$\frac{v_{cntrl}}{2} = \frac{\hat{V}_{tri}}{2}\left(\frac{\bar{v}_o}{V_d}\right) \qquad (4\text{-}19)$$

Equation (4-19) simplifies as follows:

$$\bar{v}_o = \underbrace{\left(\frac{V_d}{\hat{V}_{tri}}\right)}_{k_{PWM}} v_{cntrl} \qquad (4\text{-}20)$$

where $\left(V_d/\hat{V}_{tri}\right)$ is the converter gain k_{PWM}, from the feedback control signal to the switching-cycle-averaged voltage output, as shown in Fig. 4-13 in a block-diagram form.

4-3-1 Switching Waveforms in a Converter for dc Motor Drives

We will look further into the switching details of the converter in Fig. 4-11a. To produce a positive output voltage, the control voltages are as shown in Fig. 4-14. Only one-half the time-period, $T_s/2$, needs to be considered due to symmetry.

The pole output voltages v_{aN} and v_{bN} have the same waveform as the switching signals except for their amplitude. The output voltage v_o

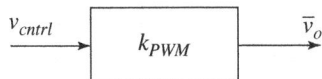

Fig. 4-13 Gain of the converter for dc drives.

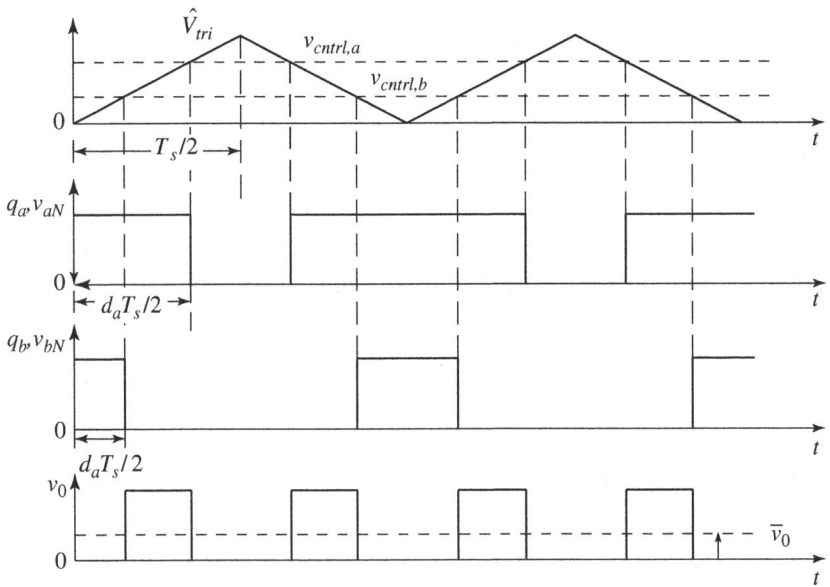

Fig. 4-14 Switching voltage waveforms in a converter for dc drive.

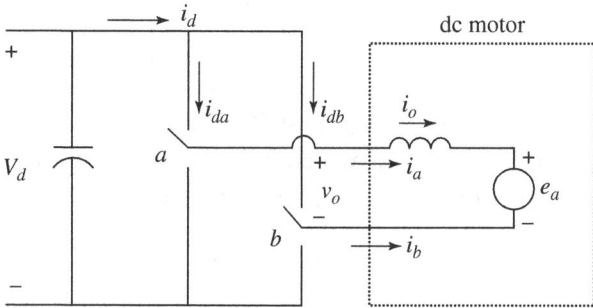

Fig. 4-15 Currents defined in the converter for dc-motor drives.

waveform shows that the effective switching frequency at the output is twice the original. That is, within the time-period of the switching-frequency f_s with which the converter devices are switching, there are two complete cycles of repetition. Therefore, the harmonics in the output are at $(2f_s)$ and at its multiples. If the switching frequency is selected sufficiently large, the motor inductance may be enough to keep the ripple in the output current within acceptable range without the need for an external inductor in series.

Next, we will look at the currents associated with this converter, repeated in Fig. 4-15. The pole currents $i_a = i_o$ and $i_b = -i_o$. The dc-side current $i_d = i_{da} + i_{db}$. The waveforms for these currents are shown by means of Example 4-3.

EXAMPLE 4-3

In the dc-motor drive of Fig. 4-15, assume the operating conditions are as follows: $V_d = 350$ V, $e_a = 236$ V(dc), and $\bar{i}_o = 4$ A. The switching-frequency f_s is 20 kHz. Assume that the series resistance R_a associated with the motor is 0.5 Ω. Calculate the series inductance L_a necessary to keep the peak–peak ripple in the output current to be 1.0 A at this operating condition. Assume that $\hat{V}_{tri} = 1$ V. Plot v_o, $\bar{v}_o$, i_o, and i_d.

(*Continued*)

Solution

As seen from Fig. 4-14, the output voltage v_o is a pulsating waveform which consists of a dc switching-cycle-averaged $\bar{v}_o$ plus a ripple component $v_{o,ripple}$, which contains subcomponents which are at very high frequencies (the multiples of $2f_s$):

$$v_o = \bar{v}_o + v_{o,ripple} \qquad (4\text{-}21)$$

Therefore, resulting current i_o consists of a switching-cycle-averaged dc component $\bar{i}_o$ and a ripple component $i_{o,\,ripple}$:

$$i_o = \bar{i}_o + i_{o,ripple} \qquad (4\text{-}22)$$

For a given v_o, we can calculate the output current employing superposition, by considering the circuit at dc and the ripple frequency (the multiples of $2f_s$), as shown in Figs. 4-16a and b, respectively. In the dc circuit, the series inductance has no effect and hence is omitted from Fig. 4-16a. In the ripple-frequency circuit of Fig. 4-16b, the back-emf e_a, that is dc, is suppressed along with the series resistance R_a, which generally is negligible compared to the inductive reactance of L_a at the very frequencies associated with the ripple. From the circuit of Fig. 4-16a,

$$\bar{v}_o = e_a + R_a \bar{i}_o = 238 \text{ V} \qquad (4\text{-}23)$$

The switching waveforms are shown in Fig. 4-17, which is based on Fig. 4-14, where the details are shown for the first half-cycle. The output voltage v_o pulsates between 0 and $V_d = 350$ V, where

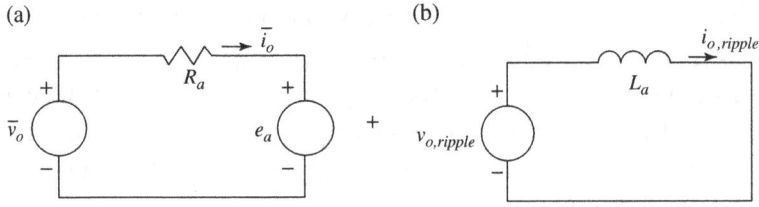

Fig. 4-16 Superposition of dc and ripple-frequency variables.

CONVERTERS FOR dc MOTOR DRIVES $(-V_d < \bar{v}_o < V_d)$ 111

from Eq. (4-12), $d_a = 0.84$ and $d_b = 0.16$. At $f_s = 20$ kHz, $T_s = 50$ μs. Using Eqs. (4-21) and (4-23), the ripple voltage waveform is also shown in Fig. 4-17, where during $\dfrac{d_a - d_b}{2} T_s$ (= 17.0 μs), the ripple voltage in the circuit of Fig. 4-16b is 112 V. Therefore, during this time interval, the peak-to-peak ripple ΔI_{p-p} in the inductor current can be related to the ripple voltage as follows:

$$L_a \dfrac{\Delta I_{p-p}}{(d_a - d_b)T_s/2} = 112 \text{ V} \quad (4\text{-}24)$$

Substituting the values in the equation above with $\Delta I_{p-p} = 1$ A, $L_a = 1.9$ mH. As shown in Fig. 4-17, the output current increases linearly during $(d_a - d_b)T_s/2$, and its waveform is symmetric around the switching-cycle-averaged value, that is, it crosses the

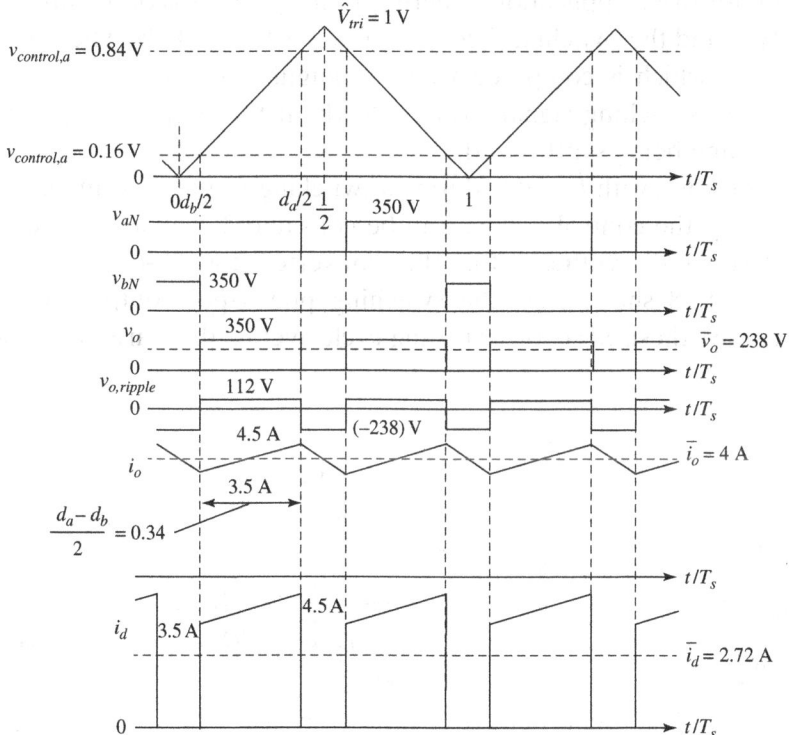

Fig. 4-17 Switching current waveforms in Example 4-3.

(*Continued*)

switching-cycle-averaged value at the midpoint of this interval. The ripple waveform in other intervals can be found by symmetry. The dc-side current i_d flows only during $(d_a - d_b)T_s/2$ interval; otherwise, it is zero as shown in Fig. 4-17. Averaging over $T_s/2$, the switching-cycle-averaged dc-side current $\bar{i}_d = 2.72$ A.

4-4 SYNTHESIS OF LOW-FREQUENCY ac

The principle of synthesizing a dc voltage for dc-motor drives can be extended for synthesizing low-frequency ac voltages, so long as the frequency f_1 of the ac being synthesized is two or three orders of magnitude smaller than the switching frequency f_s. This is the case in most ac-motor drive applications where f_1 is at 60 Hz (or is of the order of 60 Hz), and the switching frequency is a few tens of kHz. The control voltage, which is compared with a triangular waveform voltage to generate switching signals, varies slowly at the frequency f_1 of the ac voltage being synthesized.

Therefore, with $f_1 < < f_s$, during a switching-frequency time-period $T_s(=1/f_s)$, the control voltage can be considered pseudo-dc, and the analysis and synthesis for the converter for dc-drives apply. Figure 4-18 shows how the switching power-pole output voltage can be synthesized so, on switching-cycle averaged, it varies as shown

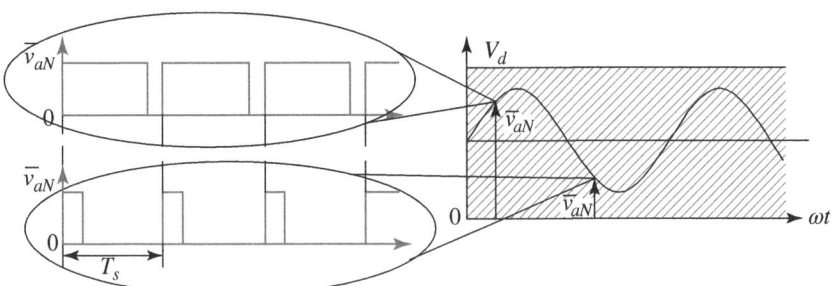

Fig. 4-18 Waveforms of a switching power-pole to synthesize low-frequency ac.

at the low-frequency f_1, where at any instant "under the microscope" shows the switching waveform with the duty-ratio that depends on the switching-cycle-averaged voltage being synthesized. The limit on switching-cycle-averaged power-pole voltage is between 0 and V_d, as in the case of converters for dc drives.

The switching-cycle-averaged representation of the switching power-pole in Fig. 4-7a is, as shown earlier in Fig. 4-7b, by an ideal transformer with controllable turns-ratio. The harmonics in the output of the power-pole in a general form were shown earlier by Fig. 4-10b. In the following section, three switching power-poles are used to synthesize three-phase ac for motor drives.

4-5 THREE-PHASE INVERTERS

Converters for three-phase outputs consist of three power-poles, as shown in Fig. 4-19a. The switching-cycle-averaged representation is shown in Fig. 4-19b.

In Fig. 4-19, $\bar{v}_{an}$, $\bar{v}_{bn}$, and $\bar{v}_{cn}$ are the desired balanced three-phase switching-cycle-averaged voltages to be synthesized: $\bar{v}_{an} = \hat{V}_{ph} \sin(\omega_1 t)$, $\bar{v}_{bn} = \hat{V}_{ph} \sin(\omega_1 t - 120°)$, and $\bar{v}_{cn} = \hat{V}_{ph} \sin(\omega_1 t - 240°)$. In series with these, common-mode voltages are added such that

$$\bar{v}_{aN} = \bar{v}_{com} + \bar{v}_{an} \qquad \bar{v}_{bN} = \bar{v}_{com} + \bar{v}_{bn} \qquad \bar{v}_{cN} = \bar{v}_{com} + \bar{v}_{cn} \qquad (4\text{-}25)$$

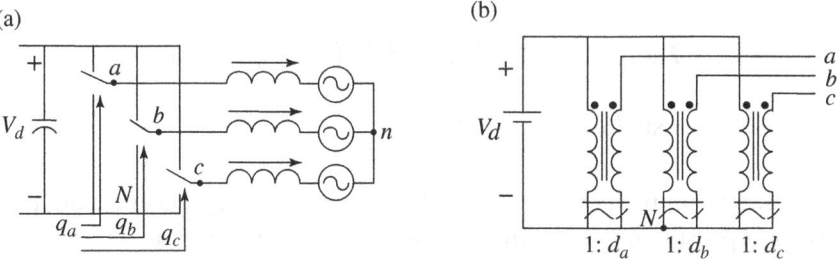

Fig. 4-19 Three-phase converter.

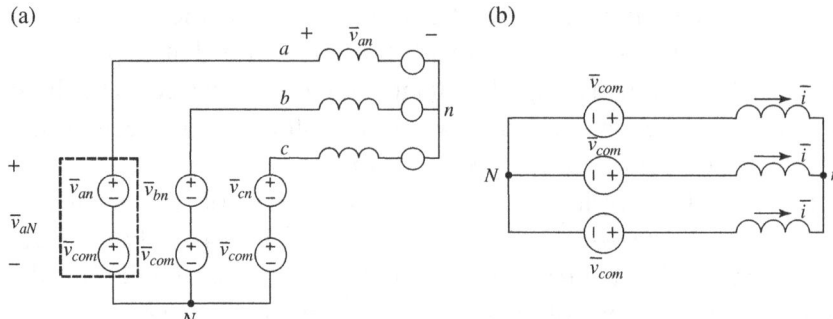

Fig. 4-20 Switching-cycle-averaged output voltages in a three-phase converter.

These voltages are shown in Fig. 4-20a. The common-mode voltages do not appear across the load; only $\bar{v}_{an}$, $\bar{v}_{bn}$, and $\bar{v}_{cn}$ appear across the load with respect to the load-neutral. This can be illustrated by applying the principle of superposition to the circuit of Fig. 4-20a.

By "suppressing" $\bar{v}_{an}$, $\bar{v}_{bn}$, and $\bar{v}_{cn}$, only equal common-mode voltages are present in each phase, as shown in Fig. 4-20b. If the current in one phase is i, then it will be the same in the other two phases. By Kirchhoff's current law at the load-neutral, $3i = 0$ and hence $i = 0$, and, therefore, the common-mode voltages do not appear across the load phases.

To obtain the switching-cycle-averaged currents drawn from the voltage-port of each switching power-pole, we will assume the currents drawn by the motor load in Fig. 4-19b to be sinusoidal but lagging with respect to the switching-cycle-averaged voltages in each phase by an angle ϕ_1, where $\bar{v}_{an}(t) = \hat{V}_{ph} \sin \omega_1 t$, and so on:

$$\bar{i}_a(t) = \hat{I} \sin(\omega_1 t - \phi_1), \; \bar{i}_b(t) = \hat{I} \sin(\omega_1 t - \phi_1 - 120°), \; \bar{i}_c(t)$$
$$= \hat{I} \sin(\omega_1 t - \phi_1 - 240°) \quad (4\text{-}26)$$

Assuming that the ripple in the output currents is negligibly small, the average power output of the converter can be written as

$$P_o = \bar{v}_{aN}\bar{i}_a + \bar{v}_{bN}\bar{i}_b + \bar{v}_{cN}\bar{i}_c \quad (4\text{-}27)$$

THREE-PHASE INVERTERS   115

Equating the average output power to the power input from the dc-bus and assuming the converter to be lossless

$$\bar{i}_d(t)V_d = \bar{v}_{aN}\bar{i}_a + \bar{v}_{bN}\bar{i}_b + \bar{v}_{cN}\bar{i}_c \quad (4\text{-}28)$$

Making use of Eq. (4-26) into Eq. (4-28),

$$\bar{i}_d(t)V_d = \bar{v}_{com}(\bar{i}_a + \bar{i}_b + \bar{i}_c) + \bar{v}_{an}\bar{i}_a + \bar{v}_{bn}\bar{i}_b + \bar{v}_{cn}\bar{i}_c \quad (4\text{-}29)$$

By Kirchhoff's current law at the load-neutral, the sum of all three phase currents within brackets in Eq. (4-29) is zero

$$\bar{i}_a + \bar{i}_b + \bar{i}_c = 0 \quad (4\text{-}30)$$

Therefore, from Eq. (4-29),

$$\bar{i}_d(t) = \frac{1}{V_d}\left(\bar{v}_{an}\bar{i}_a + \bar{v}_{bn}\bar{i}_b + \bar{v}_{cn}\bar{i}_c\right) \quad (4\text{-}31)$$

In Eq. (4-31), the sum of the products of phase voltages and currents is the three-phase power being supplied to the motor. Substituting for phase voltages and currents in Eq. (4-31),

$$\bar{i}_d(t) = \frac{\hat{V}_{ph}\hat{I}}{V_d}\left\{\begin{array}{l}\sin(\omega_1 t)\sin(\omega_1 t - \phi_1) + \sin(\omega_1 t - 120°)\sin(\omega_1 t - \phi_1 - 120°) \\ + \sin(\omega_1 t - 240°)\sin(\omega_1 t - \phi_1 - 240°)\end{array}\right\}$$

(4-32)

which simplifies to a dc current, as it should, in a three-phase circuit:

$$\bar{i}_d(t) = I_d = \frac{3}{2}\frac{\hat{V}_{ph}\hat{I}}{V_d}\cos\phi_1 \quad (4\text{-}33)$$

In three-phase converters, there are two methods of synthesizing sinusoidal output voltages, out of which we will only discuss Sine-PWM. In Sine-PWM, the switching-cycle-averaged output of power-poles, $\bar{v}_{aN}$, $\bar{v}_{bN}$, and $\bar{v}_{cN}$, have a constant dc common-mode

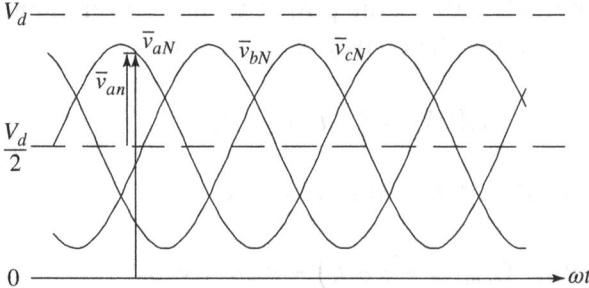

Fig. 4-21 Switching-cycle-averaged voltages due to Sine-PWM.

voltage $\overline{v}_{com} = \frac{V_d}{2}$, similar to that in dc-motor drives, around which $\overline{v}_{an}$, $\overline{v}_{bn}$, and $\overline{v}_{cn}$ can vary sinusoidally as shown in Fig. 4-21:

$$\overline{v}_{aN} = \frac{V_d}{2} + \overline{v}_{an} \qquad \overline{v}_{bN} = \frac{V_d}{2} + \overline{v}_{bn} \qquad \overline{v}_{cN} = \frac{V_d}{2} + \overline{v}_{cn} \qquad (4\text{-}34)$$

In Fig. 4-21, using Eq. (4-4), the plots of $\overline{v}_{aN}$, $\overline{v}_{bN}$, and $\overline{v}_{cN}$, each divided by V_d, are also the plots of d_a, d_b, and d_c within the limits of 0 and 1:

$$d_a = \frac{1}{2} + \frac{\overline{v}_{an}}{V_d} \qquad d_b = \frac{1}{2} + \frac{\overline{v}_{bn}}{V_d} \qquad d_c = \frac{1}{2} + \frac{\overline{v}_{cn}}{V_d} \qquad (4\text{-}35)$$

These power-pole duty-ratios define the turns-ratio in the ideal transformer representation of Fig. 4-19b. As can be seen from Fig. 4-21, at the limit, $\overline{v}_{an}$ can become a maximum of $\frac{V_d}{2}$ and hence the maximum allowable value of the phase-voltage peak is

$$\left(\hat{V}_{ph}\right)_{max} = \frac{V_d}{2} \qquad (4\text{-}36)$$

Therefore, using the properties of three-phase circuits where the line-line voltage magnitude is $\sqrt{3}$ times the phase-voltage magnitude, the maximum amplitude of the line-line voltage in Sine-PWM is limited to

$$\left(\hat{V}_{LL}\right)_{max} = \sqrt{3}\left(\hat{V}_{ph}\right)_{max} = \frac{\sqrt{3}}{2}V_d \simeq 0.867\, V_d \qquad (4\text{-}37)$$

4-5-1 Switching Waveforms in a Three-Phase Inverter with Sine-PWM

In Sine-PWM, three sinusoidal control voltages equal the duty-ratios, given in Eq. (4-35), multiplied by $\hat{V}_{tri}$. These are compared with a triangular waveform signal to generate the switching signals. These switching waveforms for Sine-PWM are shown by an example below.

EXAMPLE 4-4

In a three-phase converter of Fig. 4-19a, a Sine-PWM is used. The parameters and the operating conditions are as follows: $V_d = 350$ V, $f_1 = 60$ Hz, $\bar{v}_{an} = 160 \cos \omega_1 t$ volts, etc., and the switching frequency $f_s = 25$ kHz. $\hat{V}_{tri} = 1$ V. At $\omega_1 t = 15°$, calculate and plot the switching waveforms for one cycle of the switching frequency.

Solution

At $\omega_1 t = 15°$, $\bar{v}_{an} = 154.55$ V, $\bar{v}_{bn} = -41.41$ V, and $\bar{v}_{cn} = -113.14$ V. Therefore, from Eq. (4-34), $\bar{v}_{aN} = 329.55$ V, $\bar{v}_{bN} = 133.59$ V, and $\bar{v}_{cN} = 61.86$ V. From Eq. (4-35), the corresponding power-pole duty-ratios are $d_a = 0.942$, $d_b = 0.382$, and $d_c = 0.177$. For $\hat{V}_{tri} = 1$ V, these duty-ratios also equal the control voltages in volts. The switching time-period $T_s = 50$ μs. Based on this, the switching waveforms are shown in Fig. 4-22.

It should be noted that there is another approach called Space-Vector PWM (SV-PWM), described in Reference [1], which fully utilizes the available dc-bus voltage, and results in the ac output which can be approximately 15% higher than that possible by using the Sine-PWM approach, both in a linear range where no lower-order harmonics appear. Sine-PWM is limited to $\left(\hat{V}_{LL}\right)_{max} \simeq 0.867 V_d$, as given by Eq. (4-37), because it synthesizes output voltages on a per-pole basis, which does not take advantage of the three-phase

118 BASIC UNDERSTANDING OF SWITCH-MODE

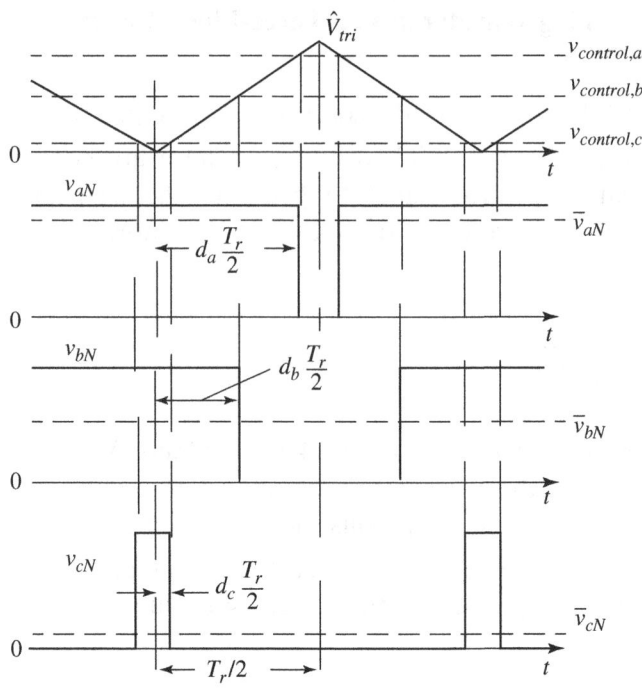

Fig. 4-22 Switching waveforms in Example 4-4.

properties. However, by considering line-line voltages, it is possible to get $\left(\hat{V}_{LL}\right)_{max} = V_d$ in SV-PWM.

4-6 POWER SEMICONDUCTOR DEVICES [2]

Electric drives owe their market success, in part, to rapid improvements in power semiconductor devices and control ICs. Switch-mode power electronic converters require diodes and transistors, which are controllable switches that can be turned on and off by applying a small voltage to their gates. These power devices are characterized by the following quantities:

1. *Voltage Rating* is the maximum voltage that can be applied across a device in its *off*-state, beyond which the device "breaks down" and irreversible damage occurs.

POWER SEMICONDUCTOR DEVICES   119

2. *Current Rating* is the maximum current (expressed as instantaneous, average, and/or rms) that a device can carry, beyond which excessive heating within the device destroys it.
3. *Switching Speeds* are the speeds with which a device can make a transition from its *on*-state to *off*-state, or vice versa. Small switching times associated with fast-switching devices result in low switching losses, or considering it differently, fast-switching devices can be operated at high switching frequencies.
4. *On-State Voltage* is the voltage drop across a device during its on-state while conducting a current. The smaller this voltage is, the smaller the on-state power loss.

4-6-1   Device Ratings

Available power devices range in voltage ratings of several kV (up to 9 kV) and current ratings of several kA (up to 5 kA). Moreover, these devices can be connected in series and parallel to satisfy any voltage and current requirements. Their switching speeds range from a fraction of a microsecond to a few microseconds, depending on their other ratings. In general, higher power devices switch more slowly than their low-power counterparts. The *on*-state voltage is usually in the range of 1–3 V.

4-6-2   Power Diodes

Power diodes are available in voltage ratings of several kV (up to 9 kV) and current ratings of several kA (up to 5 kA). The on-state voltage drop across these diodes is usually of the order of 1 V. Switch-mode converters used in motor drives require fast-switching diodes. On the other hand, the diode rectification of line-frequency ac can be accomplished by slower switching diodes, which have a slightly lower *on*-state voltage drop.

4-6-3 Controllable Switches

Transistors are controllable switches which are available in several forms: Bipolar-Junction Transistors (BJTs), metal-oxide-semiconductor field-effect transistors (MOSFETs), Gate Turn Off (GTO) thyristors, and insulated-gate bipolar transistors (IGBTs). In switch-mode converters for motor-drive applications, there are two devices that are primarily used: MOSFETs at low power levels and IGBTs in power ranges extending to MW levels. The following subsections provide a brief overview of their characteristics and capabilities.

MOSFETs In applications at voltages below 200 V and switching frequencies in excess of 50 kHz, MOSFETs are clearly the device of choice because of their low *on*-state losses in low voltage ratings, their fast switching speeds, and their ease of control. The circuit symbol of an n-channel MOSFET is shown in Fig. 4-23a. It consists of three terminals: drain (D), source (S), and gate (G). The main current flows between the drain and the source terminals. MOSFET i–v characteristics for various gate voltage values are shown in Fig. 4-23b; it is fully *off* and approximates an open switch when the gate-source voltage is zero. To turn the MOSFET *on* completely, a positive gate-to-source voltage, typically in the range of 10–15 V, must be applied. This gate-source voltage should be continuously applied in order to keep the MOSFET in its *on*-state.

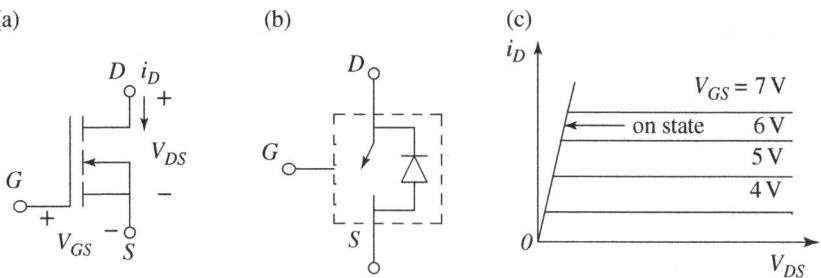

Fig. 4-23 MOSFET characteristics.

POWER SEMICONDUCTOR DEVICES   121

Insulated-Gate Bipolar Transistors   IGBTs combine the ease of control of MOSFETs with low *on*-state losses, even at fairly high voltage ratings. Their switching speeds are sufficiently fast for switching frequencies up to 30 kHz. Therefore, they are used in a vast voltage and power range – from a fractional kW to many MW.

The circuit symbol for an IGBT is shown in Fig. 4-24a and the *i–v* characteristics are shown in Fig. 4-24b. Similar to MOSFETs, IGBTs have a high impedance gate, which requires only a small amount of energy to switch the device. IGBTs have a small *on*-state voltage, even in devices with large blocking-voltage ratings (for example, V_{on} is approximately 2 V in 1200-V devices). IGBTs can be designed to block negative voltages, but most commercially available IGBTs, by design to improve other properties, cannot block any appreciable reverse-polarity voltage (similar to MOSFETs).

IGBTs have turn-on and turn-off times on the order of 1 μs and are available as modules in ratings as large as 3.3 kV and 1200 A. Voltage ratings of up to 5 kV are projected.

4-6-4   "Smart Power" Modules Including Gate Drivers and Wide Bandgap Devices

A gate-drive circuitry, shown as a block in Fig. 4-25, is required as an intermediary to interface the control signal coming from a microprocessor or an analog control integrated circuit (IC) to the power

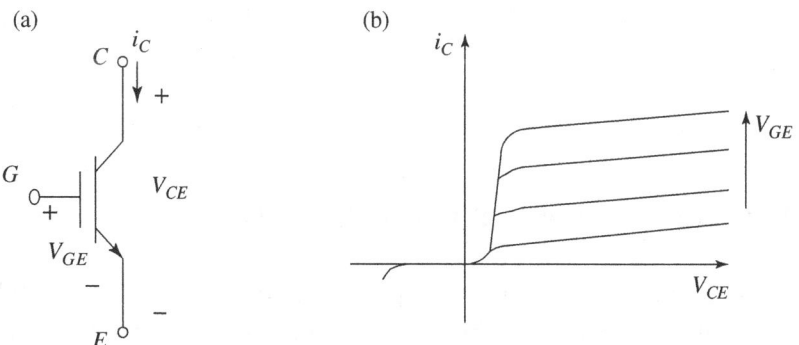

Fig. 4-24   IGBT symbol and characteristics.

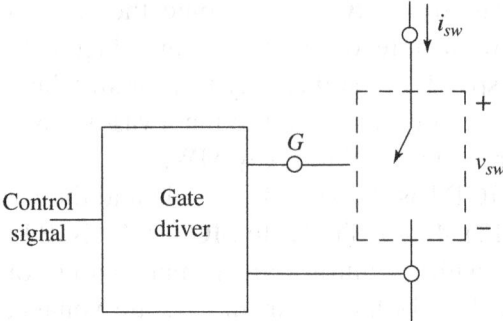

Fig. 4-25 Block diagram of a gate-drive circuit.

semiconductor switch. Such gate-drive circuits require many components, passive as well as active. Electrical isolation may also be needed between the control-signal circuit and the circuit in which the power switch is connected. The gate-driver ICs, which include all of these components in one package, have been available for some time.

Also, "Smart Power" modules, also called Power Integrated Modules (PIMs), have become available. Smart power modules combine more than one power switch and diode, along with the required gate-drive circuitry, into a single module. These modules also include fault protection and diagnostics. Such modules immensely simplify the design of power electronic converters.

In recent years, there has been a quiet revolution in the field of power electronics, where tremendous progress has been made in using new SiC- and GaN-based wide bandgap devices that have much better switching characteristics. These devices promise to revolutionize the field of power electronics and the converters based on these devices will open them to a large number of applications.

4-7 HARDWARE PROTOTYPING OF PWM

In this section, the hardware implementation of PWM for a full-bridge converter shown in Fig. 4-11 is presented. The experiment is performed on a 42 V full-bridge converter hardware and Workbench

HARDWARE PROTOTYPING OF PWM   123

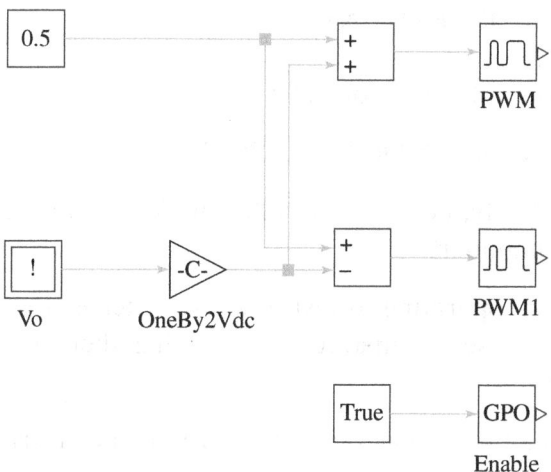

Fig. 4-26 Real-time full-bridge converter pulse-width modulation.

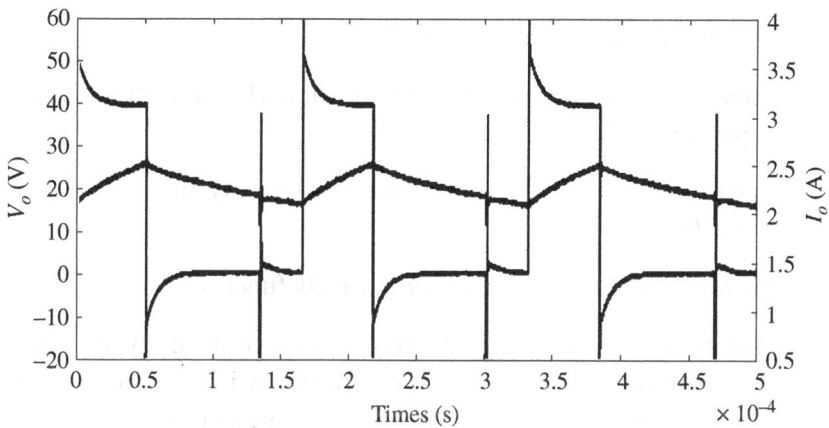

Fig. 4-27 Full-bridge converter output voltage and output current.

software as explained step-by-step in [2]. The Workbench model to perform real-time hardware implementation of the system presented in Eq. (4-12) is as shown in Fig. 4-26.

The carrier triangle frequency is set to 6 kHz and the result of running the above model at duty of 33% is shown in Fig. 4-27.

4-8 REVIEW QUESTIONS

1. What is the function of PPUs?
2. What are the sub-blocks of PPUs?
3. Qualitatively, how does a switch-mode amplifier differ from a linear amplifier?
4. Why do operating transistors as switches result in much smaller losses compared to operating them in their linear region?
5. How is a bi-positional switch realized in a converter pole?
6. What is the gain of each converter pole?
7. How does a switch-mode converter pole approach the output of a linear amplifier?
8. What is the meaning of $\bar{v}_{aN}(t)$?
9. How is the pole output voltage made linearly proportional to the input control signal?
10. What is the physical significance of the duty-ratio, for example $d_a(t)$?
11. How is PWM achieved, and what is its function?
12. Instantaneous quantities on the two sides of the converter pole, for example, pole-a, are related by the switching signal $q_a(t)$. What relates the average quantities on the two sides?
13. What is the equivalent model of a switch-mode pole in terms of its average quantities?
14. How is a switch-mode dc–dc converter which can achieve an output voltage of either polarity and an output current flowing in either direction realized?

15. What is the frequency content of the output voltage waveform in dc–dc converters?

16. In a dc-drive converter, how is it possible to keep the ripple in the output current small, despite the output voltage pulsating between 0 and V_d, or 0 and $-V_d$, during each switching cycle?

17. What is the frequency content of the input dc current? Where does the pulsating ripple component of the dc-side current flow through?

18. How is bidirectional power flow achieved through a converter pole?

19. What makes the average of the dc-side current in a converter pole related to the average of the output current by its duty-ratio?

20. How are three-phase, sinusoidal ac output voltages synthesized from a dc voltage input?

21. What are the voltage and current ratings and the switching speeds of various power semiconductor devices?

REFERENCES

1. Mohan, N. (2011). *Power Electronics: A First Course.* New York: Wiley.
2. https://sciamble.com/Resources/pe-drives-lab/basic-drives/switched-mode-dc

FURTHER READING

N. Mohan, *Power Electronics:* Computer Simulation, Analysis, and Education Using PSpice, January 1998. www.mnpere.com.

N. Mohan, T. Undeland, and W. P. Robbins, *Power Electronics: Converters, Applications, and Design,* 3rd edition, 2003, Wiley, New York.

PROBLEMS

4-1 In a switch-mode converter pole-a in Fig. 4-4a, $V_d = 150\,\text{V}$, $\hat{V}_{tri} = 5\,\text{V}$, and $f_s = 20\,\text{kHz}$. Calculate the values of the control signal $v_{cntrl,\,a}$ and the pole duty-ratio d_a during which the switch is in its top position, for the following values of the average output voltage: $\bar{v}_{aN} = 125\,\text{V}$ and $\bar{v}_{aN} = 50\,\text{V}$.

4-2 In Problem 4-1, assume the $i_a(t)$ waveform to be dc with a magnitude of 10 A. Draw the waveform of $i_{da}(t)$ for the two values of $\bar{v}_{aN}$.

dc–dc Converters (Four-Quadrant Capability)

4-3 A switch-mode dc–dc converter uses a PWM-controller IC, which has a triangular waveform signal at 25 kHz with $\hat{V}_{tri} = 3\,\text{V}$. If the input dc source voltage $V_d = 150\,\text{V}$, calculate the gain k_{PWM} of this switch-mode amplifier.

4-4 In a switch-mode dc–dc converter of Fig. 4-11a, $\dfrac{v_{cntrl}}{\hat{V}_{tri}} = 0.8$ with a switching frequency $f_s = 20\,\text{kHz}$ and $V_d = 150\,\text{V}$. Calculate and plot the ripple in the output voltage $v_o(t)$.

4-4 A switch-mode dc–dc converter is operating at a switching frequency of 20 kHz, and $V_d = 150\,\text{V}$. The average current being drawn by the dc motor is 8.0 A. In the equivalent circuit of the dc motor, $E_a = 100\,\text{V}$, $R_a = 0.25\,\Omega$, and $L_a = 4\,\text{mH}$, all in series. (a) Plot the output current and calculate the peak-to-peak ripple and (b) plot the dc-side current.

4-5 In Problem 4-5, the motor goes into regenerative braking mode. The average current being supplied by the motor to the converter during braking is 8.0 A. Plot the voltage and current waveforms on both sides of this converter. Calculate the average power flow into the converter.

4-6 In Problem 4-5, calculate $\bar{i}_{da}$, $\bar{i}_{db}$, and $\bar{i}_d (= I_d)$.

4-7 Repeat Problem 4-5 if the motor is rotating in the reverse direction, with the same current drawn and the same induced emf E_a value of the opposite polarity.

4-8 Repeat Problem 4-8 if the motor is braking while it has been rotating in the reverse direction. It supplies the same current and produces the same induced emf E_a value of the opposite polarity.

4-9 Repeat Problem 4-5 if a bipolar voltage switching is used in the dc–dc converter. In such a switching scheme, the two bi-positional switches are operated in such a manner that when switch-a is in the top position, switch-b is in its bottom position, and vice versa. The switching signal for pole-a is derived by comparing the control voltage (as in Problem 4-5) with the triangular waveform.

dc-to-Three-Phase ac Inverters

4-11 Plot $d_a(t)$ if the output voltage of the converter pole-a is $\bar{v}_{aN}(t) = \frac{V_d}{2} + 0.85 \frac{V_d}{2} \sin(\omega_1 t)$, where $\omega_1 = 2\pi \times 60$ rad/s.

4-12 In the three-phase dc–ac inverter of Fig. 4-19, $V_d = 300$ V, $\hat{V}_{tri} = 1$ V, $\bar{v}_{an}(t) = 90 \sin(\omega_1 t)$, and $f_1 = 45$ Hz. Calculate and plot $d_a(t)$, $d_b(t)$, $d_c(t)$, $\bar{v}_{aN}(t), \bar{v}_{bN}(t), \bar{v}_{cN}(t)$, and $\bar{v}_{an}(t)$, $\bar{v}_{bn}(t)$, and $\bar{v}_{cn}(t)$.

4-13 In the balanced three-phase dc–ac inverter shown in Fig. 4-19, the phase-a average output voltage is $\bar{v}_{an}(t) = \frac{V_d}{2} 0.75 \sin(\omega_1 t)$, where $V_d = 300$ V and $\omega_1 = 2\pi \times 45$ rad/s. The inductance L in each phase is 5 mH. The ac-motor internal voltage in phase-a can be represented as $e_a(t) = 106.14 \sin(\omega_1 t - 6.6°)$ V, assuming this internal voltage to be purely sinusoidal. (a) Calculate and plot $d_a(t)$, $d_b(t)$, and $d_c(t)$, (b) sketch $\bar{i}_a(t)$, and (c) sketch $\bar{i}_{da}(t)$.

4-14 In Problem 4-13, calculate and plot $\bar{i}_d(t)$, which is the average dc current drawn from the dc capacitor bus in Fig. 4-19b.

Simulation Problems

4-15 Simulate a two-quadrant pole of Fig. 4-7a in dc steady state. The nominal values are as follows: $V_d = 200$ V, and the output has in series $R_a = 0.37 \, \Omega$, $L_a = 1.5$ mH, and $E_a = 136$ V.

128 BASIC UNDERSTANDING OF SWITCH-MODE

$\hat{V}_{tri} = 1$ V. The switching frequency $f_s = 20$ kHz. In dc steady state, the average output current is $I_a = 10$ A. (a) Obtain the plot of $v_{aN}(t)$, $i_a(t)$, and $i_{da}(t)$, (b) obtain the peak–peak ripple in $i_a(t)$ and compare it with its value obtained analytically, and (c) obtain the average values of $i_a(t)$ and $i_{da}(t)$, and show that these two averages are related by the duty-ratio d_a.

4-16 Repeat Problem 4-15 by calculating the value of the control voltage such that the converter pole is operating in the boost mode with $I_a = -10$ A.

dc–dc Converters

4-17 Simulate the dc–dc converter of Fig. 4-11a in dc steady state. The nominal values are as follows: $V_d = 200$ V and the output consists of in series $R_a = 0.37\,\Omega$, $L_a = 1.5$ mH, and $E_a = 136$ V. $\hat{V}_{tri} = 1$ V. The switching frequency $f_s = 20$ kHz. In the dc steady state, the average output current is $I_a = 10$ A. (a) Obtain the plot of $v_o(t)$, $i_o(t)$, and $i_d(t)$, (b) obtain the peak–peak ripple in $i_o(t)$ and compare it with its value obtained analytically, and (c) obtain the average values of $i_o(t)$ and $i_d(t)$, and show that these two averages are related by the duty-ratio d in Eq. (4-17).

4-18 In Problem 4-17, apply a step-increase at 0.5 ms in the control voltage to reach the output current of 15 A (in steady state) and observe the output current response.

4-19 Repeat Problem 4-18 with each converter pole represented on its average basis.

dc-Three-Phase ac Inverters

4-20 Simulate the three-phase ac inverter on an average basis for the system described in Problem 4-13. Obtain the various waveforms.

4-21 Repeat Problem 4-20 for a corresponding switching circuit and compare the switching waveforms with the average waveforms in Problem 4-20.

5 Control in Electric Drives

5-1 INTRODUCTION

Many applications, such as robotics and factory automation, require precise control of speed and position. In such applications, a feedback control, as illustrated by Fig. 5-1, is used. This feedback control system consists of a power-processing unit (PPU), a motor, and a mechanical load. The output variables such as torque and speed are sensed and are fed back to be compared with the desired (reference) values. The error between the reference and the actual values are amplified to control the PPU to minimize or eliminate this error. A properly designed feedback controller makes the system insensitive to disturbances and changes in the system parameters.

The objective of this chapter is to discuss the design of motor-drive controllers. A dc-motor drive is used as an example, although the same design concepts can be applied in controlling brushless-dc-motor drives and vector-controlled induction-motor drives. In the following discussion, it is assumed that the PPU is of a switch-mode type and has a very fast response time. A permanent-magnet dc machine with a constant field flux ϕ_f is assumed.

(Adapted from chapter 7 of *Electric Machines and Drives: A First Course* ISBN: 978-1-118-07481-7 by Ned Mohan, January 2012 and from chapter 11 of *Power Electronics: A First Course* ISBN: 978-1-118-07480-0 by Ned Mohan, October 2011)

Analysis and Control of Electric Drives: Simulations and Laboratory Implementation, First Edition. Ned Mohan and Siddharth Raju.
© 2021 John Wiley & Sons, Inc. Published 2021 by John Wiley & Sons, Inc.
Companion website: www.wiley.com/go/Mohan/Vectorcontrolinelectricdrives

130 CONTROL IN ELECTRIC DRIVES

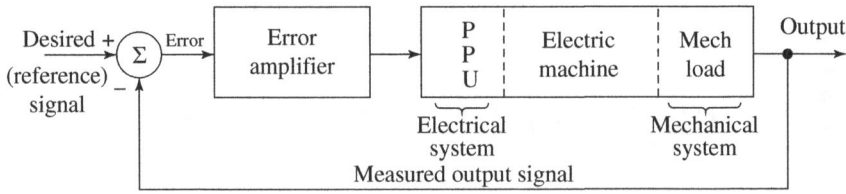

Fig. 5-1 Feedback-controlled drive.

The feedback control in this chapter is described using dc motors, such that the same principles can be used in controlling ac motors. Therefore, Section 5-2 is devoted to briefly describing dc machines. For a detailed description of dc machines, the reader is referred to chapter 7 of [1].

5-2 dc MOTORS

dc motors were widely used in the past for all types of drive applications, and they continue to be used for controlling speed and position. There are two designs of dc machines: stators consisting of either permanent magnets or a field winding. In this section, we will describe permanent-magnet dc motors, which are usually supplied by switch-mode power electronic converters.

Figure 5-2 shows a cutaway view of a dc motor. It shows a permanent-magnet stator, a rotor that carries a winding, a commutator, and the brushes. In dc machines, the stator establishes a uniform flux ϕ_f in the air gap in the radial direction (the subscript "f" is for field). If permanent magnets like those shown in Fig. 5-2 are used, the air gap flux density established by the stator remains constant. A field winding whose current can be varied can be used to achieve an additional degree of control over the air gap flux density.

As shown in Fig. 5-2, the rotor slots contain a winding, called the armature winding, which handles electrical power for conversion to (or from) mechanical power at the rotor shaft. Also, there is a commutator affixed to the rotor. On its outer surface, the commutator contains copper segments, which are electrically insulated from each other by means of mica or plastic. The coils of the armature winding

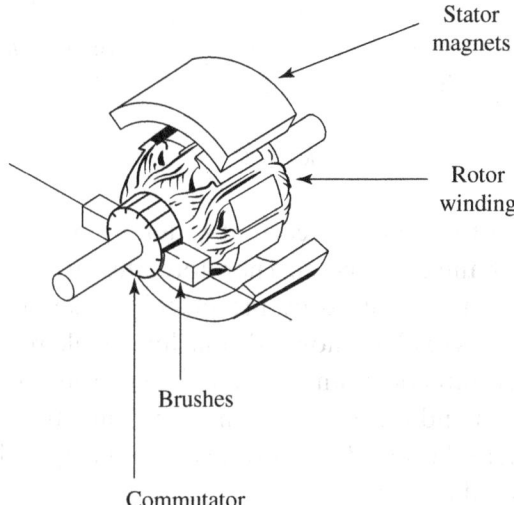

Fig. 5-2 Exploded view of a dc motor (*Source:* Electro-Craft Corporation).

are connected to these commutator segments so that a stationary dc source can supply voltage and current to the rotating commutator by means of stationary carbon brushes which rest on top of the commutator. The wear due to the mechanical contact between the commutator and the brushes requires periodic maintenance, which is the main drawback of dc machines.

In a dc machine, the magnitude and the direction of the electromagnetic torque depend on the armature current i_a. Therefore, in a permanent-magnet dc machine, the electromagnetic torque produced by the machine is linearly related to a machine torque-constant k_T, which is unique to a given machine and is specified in its data sheets:

$$T_{em} = k_T i_a \tag{5-1}$$

Similarly, the induced emf e_a depends only on the rotational speed ω_m, and can be related to it by a voltage-constant k_E, which is unique to a given machine and specified in its data sheets:

$$e_a = k_E \omega_m \tag{5-2}$$

132 CONTROL IN ELECTRIC DRIVES

Equating the mechanical power ($\omega_m T_{em}$) to the electrical power ($e_a i_a$), the torque-constant k_T and the voltage-constant k_E are exactly the same numerically in MKS units:

$$k_T = k_E \qquad (5\text{-}3)$$

The direction of the armature current i_a determines the direction of currents through the conductors. Therefore, the direction of the electromagnetic torque produced by the machine also depends on the direction of i_a. This explains how a dc machine while rotating in a forward or reverse direction can be made to go from motoring mode (where the speed and torque are in the same direction) to its generator mode (where the speed and torque are in the opposite direction) by reversing the direction of i_a.

Applying a reverse-polarity dc voltage to the armature terminals makes the armature current flow in the opposite direction. Therefore, the electromagnetic torque is reversed, and after some time, reversing the direction of rotation and the polarity of induced emfs in conductors, which depends on the direction of rotation.

The above discussion shows that a dc machine can easily be made to operate as a motor or as a generator in forward or reverse direction of rotation.

It is convenient to discuss a dc machine in terms of its equivalent circuit of Fig. 5-3, which shows the conversion between electrical and mechanical power. The armature current i_a produces the electromagnetic torque $T_{em}(=k_T i_a)$ necessary to rotate the mechanical load at a speed. Across the armature terminals, the rotation at the speed of ω_m induces a voltage, called the back-emf $e_a(=k_E \omega_m)$, represented by a dependent voltage-source.

The applied voltage v_a from the PPU overcomes the back-emf e_a and causes the current i_a to flow. Recognizing that there is a voltage drop across both the armature winding resistance R_a (which includes the voltage drop across the carbon brushes) and the armature winding inductance L_a, we can write the equation of the electrical side as

$$v_a = e_a + R_a i_a + L_a \frac{di_a}{dt} \qquad (5\text{-}4)$$

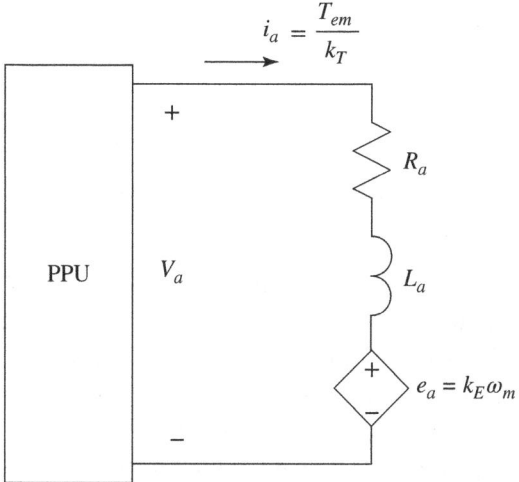

Fig. 5-3 dc motor equivalent circuit.

On the mechanical side, the electromagnetic torque produced by the motor overcomes the mechanical-load torque T_L to produce acceleration:

$$\frac{d\omega_m}{dt} = \frac{1}{J_{eq}}(T_{em} - T_L) \quad (5\text{-}5)$$

where $J_{eq}(=J_m + J_L)$ is the total effective value of the combined inertia of the dc machine and the mechanical load.

Note that the electric and the mechanical systems are coupled. The torque T_{em} in the mechanical system (Eq. (5-1)) depends on the electrical current i_a. The back-emf e_a in the electrical system (Eq. (5-2)) depends on the mechanical speed ω_m. The electrical power absorbed from the electrical source by the motor is converted into mechanical power and vice versa. In a dc steady state, with a voltage V_a applied to the armature terminals and a load-torque $T_L(=T_{em})$ being supplied to the load, the inductance L_a in the equivalent circuit of Fig. 5-3 does not play a role. Hence, in Fig. 5-3,

$$I_a = \frac{T_{em}(=T_L)}{k_T} \quad (5\text{-}6)$$

134 CONTROL IN ELECTRIC DRIVES

$$\omega_m = \frac{E_a}{k_E} = \frac{V_a - R_a I_a}{k_E} = \frac{V_a - R_a(T_{em}/k_T)}{k_E} \quad (5\text{-}7)$$

5-2-1 Requirements Imposed by dc Machines on the PPU

Based on Eqs. (5-6) and (5-7), the steady-state torque-speed characteristics for various values of V_a are plotted in Fig. 5-4a. Neglecting the voltage drop across the armature resistance in Eq. (5-7), the terminal voltage V_a varies linearly with speed ω_m, as plotted in Fig. 5-4b. At low values of speed, the voltage drop across the armature resistance can be a significant component of the terminal voltage, and hence the relationship in Fig. 5-4a in this region is shown dotted.

In summary, in dc-motor drives, the voltage rating of the PPU is dictated by the maximum speed, and the current rating is dictated by the maximum load torque being supplied.

5-3 DESIGNING FEEDBACK CONTROLLERS FOR MOTOR DRIVES

5-3-1 Control Objectives

The control system in Fig. 5-1 shown earlier is shown simplified in Fig. 5-5, where $G_p(s)$ is the Laplace-domain transfer function of the plant consisting of the PPU, the motor, and the mechanical load.

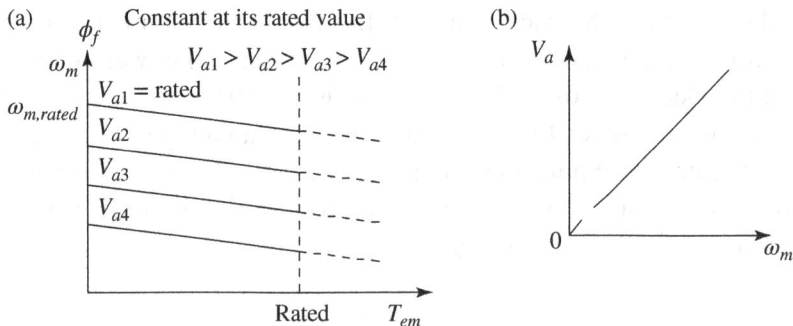

Fig. 5-4 (a) Torque-speed characteristics and (b) V_a versus ω_m.

DESIGNING FEEDBACK CONTROLLERS FOR MOTOR DRIVES 135

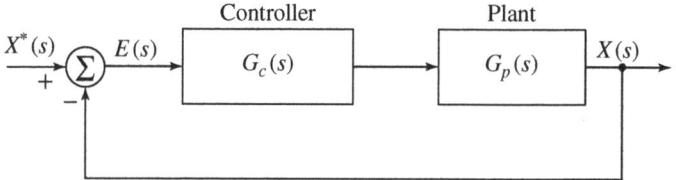

Fig. 5-5 Simplified control system representation.

$G_c(s)$ is the controller transfer function. In response to a desired (reference) input $X^*(s)$, the output of the system is $X(s)$, which (ideally) equals the reference input. The controller $G_c(s)$ is designed with the following objectives in mind:

- A zero steady-state error.
- A good dynamic response (which implies both a fast transient response, for example, to a step-change in the input, and small settling time with very little overshoot).

To keep the discussion simple, a unity feedback will be assumed. The open-loop transfer function (including the forward path and the unity feedback path) $G_{OL}(s)$ is

$$G_{OL}(s) = G_c(s) G_p(s) \qquad (5\text{-}8)$$

The closed-loop transfer function $\dfrac{X(s)}{X^*(s)}$ in a unity feedback system is

$$G_{CL}(s) = \frac{G_{OL}(s)}{1 + G_{OL}(s)} \qquad (5\text{-}9)$$

In order to define a few necessary control terms, we will consider a generic Bode plot of the open-loop transfer function $G_{OL}(s)$ in terms of its magnitude and phase angle, shown in Fig. 5-6a as a function of frequency.

The frequency at which the gain equals unity (that is $|G_{OL}(s)| = 0$ db) is defined as the crossover frequency f_c (angular frequency ω_c). At the crossover frequency, the phase delay introduced by the open-loop

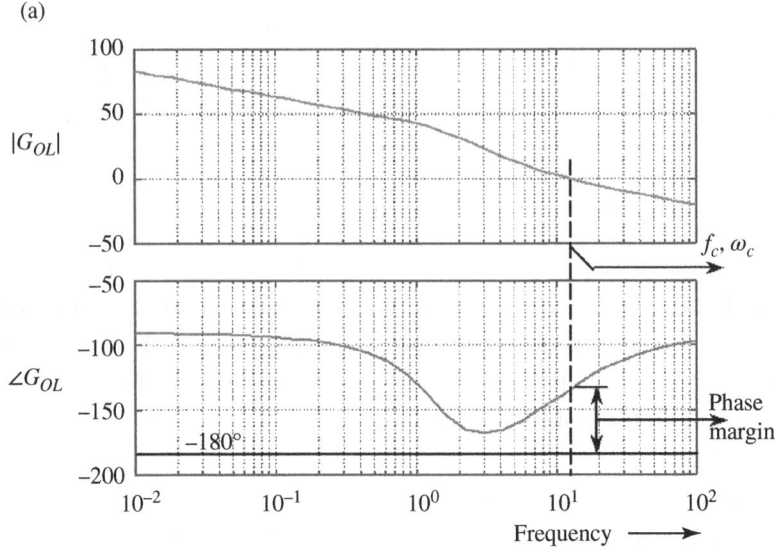

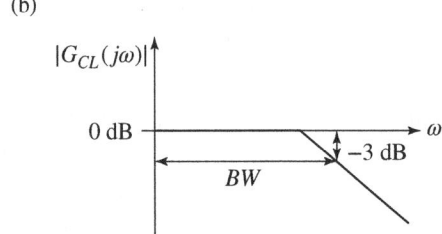

Fig. 5-6 (a) Phase margin and (b) bandwidth.

transfer function must be less than 180° in order for the closed-loop feedback system to be stable. Therefore, at f_c, the phase angle $\phi_{OL}|_{f_c}$ of the open-loop transfer function measured with respect to −180° is defined as the phase margin (PM):

$$\text{Phase margin (PM)} = \phi_{OL}|_{f_c} - (-180°) = \phi_{OL}|_{f_c} + 180° \quad (5\text{-}10)$$

Note that $\phi_{OL}|_{f_c}$ has a negative value. For a satisfactory dynamic response without oscillations, the PM should be greater than 45°, preferably close to 60°. The magnitude of the closed-loop transfer function is plotted in Fig. 5-6b (idealized by the asymptotes), in which the

DESIGNING FEEDBACK CONTROLLERS FOR MOTOR DRIVES 137

bandwidth is defined as the frequency at which the gain drops to (–3 dB). As a first-order approximation in many practical systems,

$$\text{Closed} - \text{loop bandwidth} \approx f_c \quad (5\text{-}11)$$

For a fast transient response by the control system, for example, a response to a step-change in the input, the bandwidth of the closed loop should be high. From Eq. (5-11), this requirement implies that the crossover frequency f_c (of the open-loop transfer function shown in Fig. 5-6a) should be designed to be high.

EXAMPLE 5-1

In a unity feedback system, the open-loop transfer function is given as $G_{OL}(s) = \dfrac{k_{OL}}{s}$, where $k_{OL} = 2 \times 10^3$ rad/s. (a) Plot the open-loop transfer function. What is the crossover frequency? (b) Plot the closed-loop transfer function and calculate the bandwidth. (c) Calculate and plot the time-domain closed-loop response to a step-change in the input.

Solution

(a) The open-loop transfer function is plotted in Fig. 5-9a, which shows that the crossover frequency $\omega_c = k_{OL} = 2 \times 10^3$ rad/s.

(b) The closed-loop transfer function, from Eq. (5-2), is $G_{CL}(s) = \dfrac{1}{1 + s/k_{OL}}$. This closed-loop transfer function is plotted in Fig. 5-7b, which shows that the bandwidth is exactly equal to the ω_c calculated in part *a*.

(c) For a step-change, $X^*(s) = \dfrac{1}{s}$. Therefore,

$$X(s) = \frac{1}{s} \frac{1}{1 + s/k_{OL}} = \frac{1}{s} - \frac{1}{1 + s/k_{OL}}$$

(*Continued*)

138 CONTROL IN ELECTRIC DRIVES

The Laplace-inverse transform yields

$$x(t) = \left(1 - e^{-t/\tau}\right) u(t) \quad \text{where} \quad \tau = \frac{1}{k_{OL}} = 0.5 \text{ ms}$$

The time response is plotted in Fig. 5-7c. We can see that a higher value of k_{OL} results in a higher bandwidth and a smaller time-constant τ, leading to a faster response.

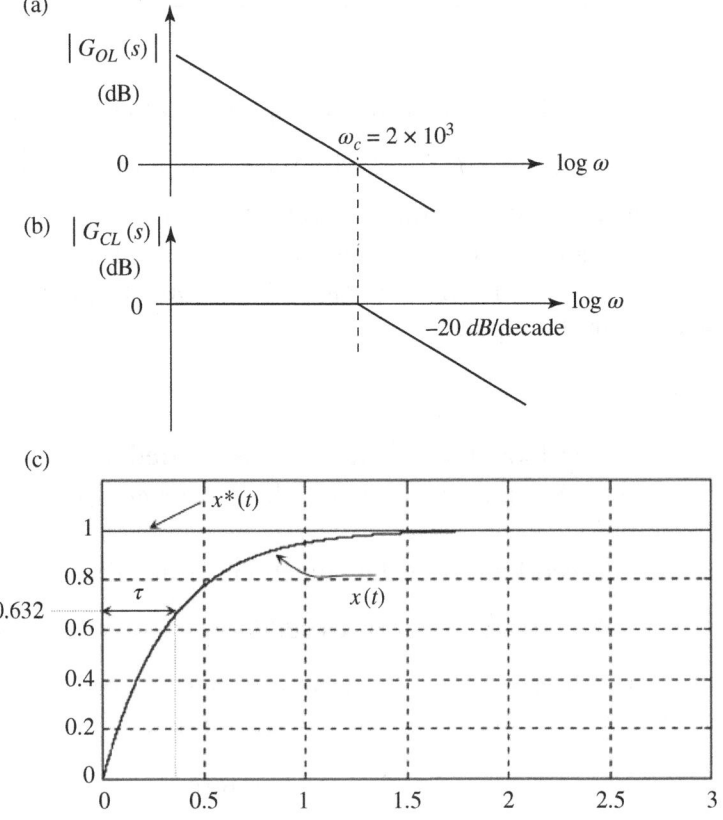

Fig. 5-7 (a) Gain magnitude of a first-order system open loop, (b) the gain magnitude of a closed loop, and (c) step response.

5-3-2 Cascade Control Structure

In the following discussion, a cascade control structure such as that shown in Fig. 5-8 is used. The cascade control structure is commonly used for motor drives because of its flexibility. It consists of distinct control loops; the innermost current (torque) loop is followed by the speed loop. If the position needs to be controlled accurately, the outermost position loop is superimposed on the speed loop. Cascade control requires that the bandwidth (speed of response) increases toward the inner loop, with the torque loop being the fastest and the position loop being the slowest. The cascade control structure is widely used in the industry.

5-3-3 Steps in Designing the Feedback Controller

Motion control systems must often respond to large changes in the desired (reference) values of the torque, speed, and position. They must reject large, unexpected load disturbances. For large changes, the overall system is often nonlinear. This nonlinearity comes about because the mechanical load is often highly nonlinear. Additional nonlinearity is introduced by voltage and current limits imposed by the PPU and the motor. In view of the above, the following steps for designing the controller are suggested:

1. The first step is to assume that, around the steady-state operating point, the input reference changes and the load disturbances are all small. In such a small-signal analysis, the overall system

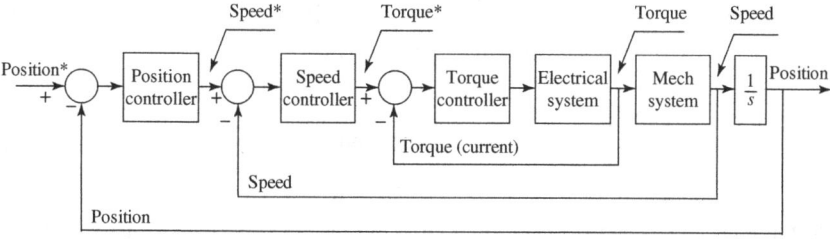

Fig. 5-8 Cascade control of a motor drive.

can be assumed to be linear around the steady-state operating point, thus allowing the basic concepts of linear control theory to be applied.

2. Based on the linear control theory, once the controller has been designed, the entire system can be simulated on a computer under large-signal conditions to evaluate the adequacy of the controller. The controller must be "adjusted" as appropriate.

5-3-4 System Representation for Small-Signal Analysis

For ease of the analysis described below, the system in Fig. 5-8 is assumed to be linear, and the steady-state operating point is assumed to be zero for all of the system variables. This linear analysis can then be extended to nonlinear systems and to steady-state operating conditions other than zero. The control system in Fig. 5-8 is designed with the highest bandwidth (associated with the torque loop), which is one or two orders of magnitude smaller than the switching frequency f_s. As a result, in designing the controller, the switching-frequency components in various quantities are of no consequence. Therefore, we will use the average variables discussed in Chapter 4, where the switching-frequency components were eliminated.

The Average Representation of the PPU For the purpose of designing the feedback controller, we will assume that the dc-bus voltage V_d within the PPU shown in Fig. 5-9a is constant. Following the averaging analysis in Chapter 4, the average representation of the switch-mode converter is shown in Fig. 5-9b.

In terms of the dc-bus voltage V_d and the triangular-frequency waveform peak $\hat{V}_{tri}$, the average output voltage $\bar{v}_a(t)$ of the converter is linearly proportional to the control voltage:

$$\bar{v}_a(t) = k_{PWM} v_c(t) \quad \left(k_{PWM} = \frac{V_d}{\hat{V}_{tri}} \right) \quad (5\text{-}12)$$

where k_{PWM} is the gain constant of the PWM converter. Therefore, in the Laplace domain, the PWM controller and the dc–dc switch-mode

DESIGNING FEEDBACK CONTROLLERS FOR MOTOR DRIVES 141

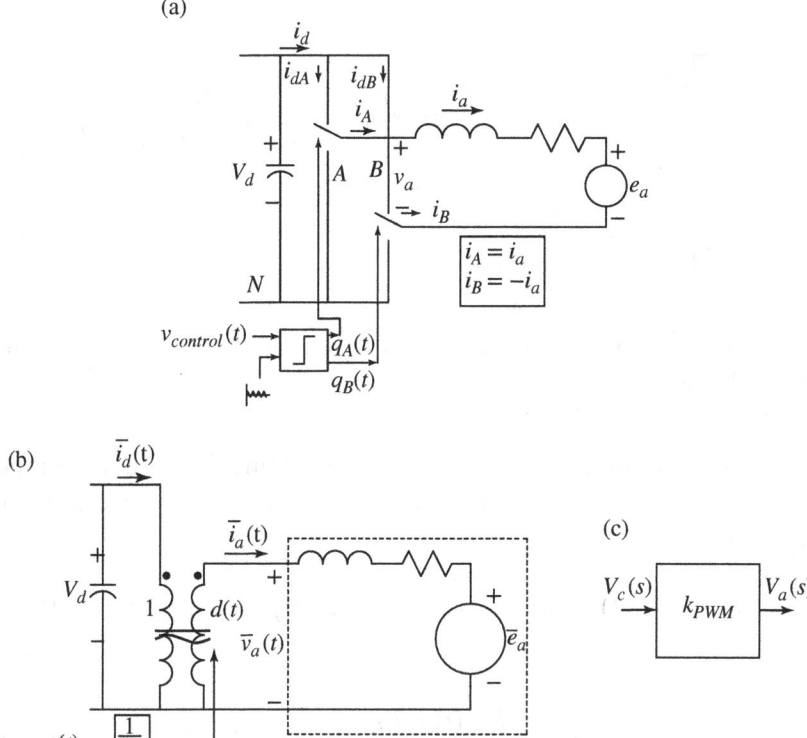

Fig. 5-9 (a) Switch-mode converter for motor drives, (b) average model of the switching-mode converter, and (c) linearized representation.

converter can be represented simply by a gain-constant k_{PWM}, as shown in Fig. 5-9c:

$$V_a(s) = k_{PWM} V_c(s) \qquad (5\text{-}13)$$

where $V_a(s)$ is the Laplace transform of $\bar{v}_a(t)$, and $V_c(s)$ is the Laplace transform of $v_c(t)$. The above representation is valid in the linear range, where $-\hat{V}_{tri} \leq v_c \leq \hat{V}_{tri}$.

The Modeling of the dc Machine and the Mechanical Load The dc motor and the mechanical load are modeled, as shown by the equivalent circuit in Fig. 5-10a, in which the speed $\omega_m(t)$ and the back-emf

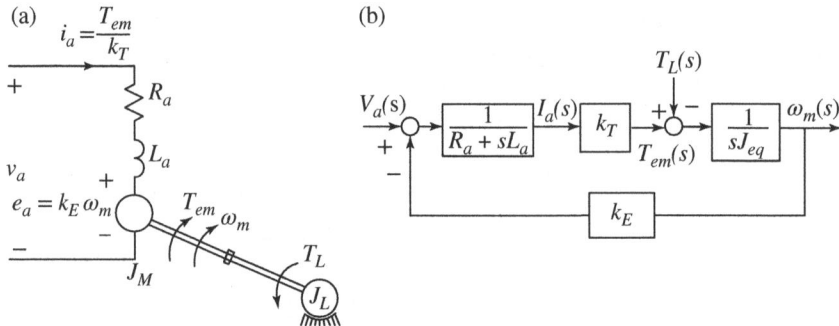

Fig. 5-10 dc motor and mechanical load: (a) equivalent circuit and (b) block diagram.

$e_a(t)$ are assumed not to contain switching-frequency components. The electrical and mechanical equations corresponding to Fig. 5-10a are

$$\overline{v}_a(t) = e_a(t) + R_a \overline{i}_a(t) + L_a \frac{d}{dt} \overline{i}_a(t), e_a(t) = k_E \omega_m(t) \quad (5\text{-}14)$$

and

$$\frac{d}{dt} \omega_m(t) = \frac{\overline{T}_{em}(t) - T_L}{J_{eq}}, \overline{T}_{em}(t) = k_T \overline{i}_a(t) \quad (5\text{-}15)$$

where the equivalent load inertia J_{eq} ($=J_M + J_L$) is the sum of the motor inertia and the load inertia, and the damping is neglected (it could be combined with the load torque T_L).

In the simplified procedure presented here, the controller is designed to follow the changes in the torque, speed, and position reference values (and hence the load torque in Eq. (5-15) is assumed to be absent). Equations (5-14) and (5-15) can be expressed in the Laplace domain as

$$V_a(s) = E_a(s) + (R_a + sL_a)I_a(s) \quad (5\text{-}16)$$

or

$$I_a(s) = \frac{V_a(s) - E_a(s)}{R_a + sL_a} \quad (5\text{-}17)$$

$$E_a(s) = k_E \omega_m(s)$$

CONTROLLER DESIGN 143

We can define the Electrical Time Constant τ_e as

$$\tau_e = \frac{L_a}{R_a} \qquad (5\text{-}18)$$

Therefore, Eq. (5-17) can be written in terms of τ_e as

$$I_a(s) = \frac{1/R_a}{1 + \frac{s}{1/\tau_e}} \{ V_a(s) - E_a(s) \}$$

$$E_a(s) = k_E \omega_m(s) \qquad (5\text{-}19)$$

From Eq. (5-15), assuming the load torque to be absent in the design procedure,

$$\omega_m(s) = \frac{T_{em}(s)}{s J_{eq}}$$

$$T_{em}(s) = k_T I_a(s) \qquad (5\text{-}20)$$

Equations (5-17) and (5-20) can be combined and represented in block-diagram form, as shown in Fig. 5-10b.

5-4 CONTROLLER DESIGN

The controller in the cascade control structure shown in Fig. 5-8 is designed with the objectives discussed in Section 5-2-1 in mind. In the following section, a simplified design procedure is described.

5-4-1 Proportional-Integral Controllers

Motion control systems often utilize a proportional-integral (PI) controller, as shown in Fig. 5-11. The input to the controller is the error $E(s) = X^*(s) - X(s)$, which is the difference between the reference input and the measured output.

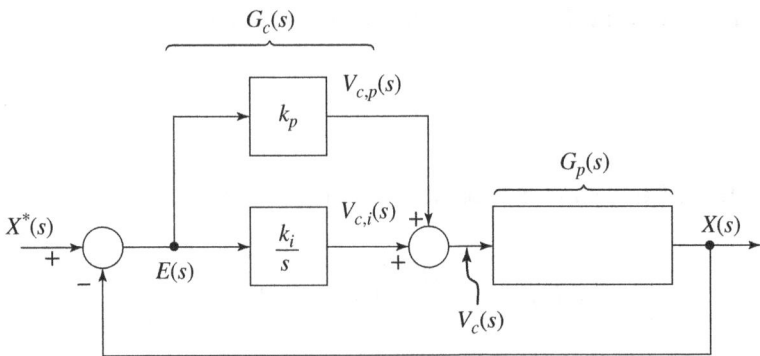

Fig. 5-11 PI controller.

In Fig. 5-11, the proportional controller produces an output proportional to the error input:

$$V_{c,p}(s) = k_p E(s) \qquad (5\text{-}21)$$

where k_p is the proportional-controller gain. In torque and speed loops, proportional controllers, if used alone, result in a steady-state error in response to step-change in the input reference. Therefore, they are used in combination with the integral controller described below.

In the integral controller shown in Fig. 5-11, the output is proportional to the integral of the error $E(s)$, expressed in the Laplace domain as

$$V_{c,i}(s) = \frac{k_i}{s} E(s) \qquad (5\text{-}22)$$

where k_i is the integral-controller gain. Such a controller responds slowly because its action is proportional to the time integral of the error. The steady-state error goes to zero for a step-change in input because the integrator action continues for as long as the error is not zero.

In motion control systems, the P controllers in the position loop and the PI controllers in the speed and torque loop are often adequate.

CONTROLLER DESIGN   145

Therefore, we will not consider differential (D) controllers. As shown in Fig. 5-11, $V_c(s) = V_{c,p}(s) + V_{c,i}(s)$. Therefore, using Eqs. (5-21) and (5-22), the transfer function of a PI controller is

$$\frac{V_c(s)}{E(s)} = \left(k_p + \frac{k_i}{s}\right) = \frac{k_i}{s}\left[1 + \frac{s}{k_i/k_p}\right] \qquad (5\text{-}23)$$

5-4-2 Example of a Controller Design

In the following discussion, we will consider the example of a permanent-magnet dc-motor supplied by a switch-mode PWM dc–dc converter. The system parameters are given in Table 5-1.

We will design the torque, speed, and position feedback controllers (assuming a unity feedback) based on the small-signal analysis, in which the load nonlinearity and the effects of the limiters can be ignored.

Design of the Torque (Current) Control Loop   As mentioned earlier, we will begin with the innermost loop in Fig. 5-12a (utilizing the transfer function block diagram of Fig. 5-10b to represent the motor-load combination, Fig. 5-9c to represent the PPU, and Fig. 5-11 to represent the PI controller).

TABLE 5-1   dc-Motor Drive System

System parameter	Value
R_a	2.0 Ω
L_a	5.2 mH
J_{eq}	152×10^{-6} kg·m^2
B	0
K_E	0.1 V/(rad/s)
k_T	0.1 Nm/A
V_d	60 V
$\hat{V}_{tri}$	5 V
f_s	33 kHz

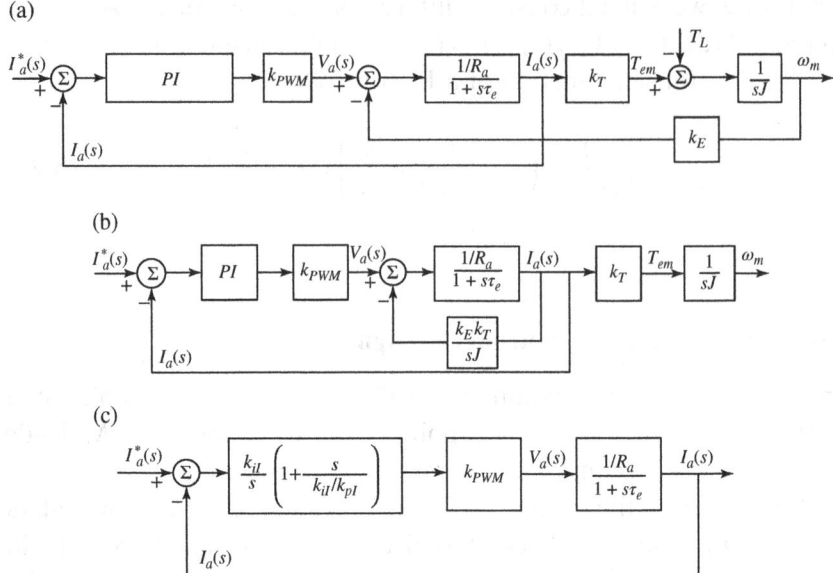

Fig. 5-12 Design of the torque control loop.

In permanent-magnet dc motors in which ϕ_f is constant, the current and the torque are proportional to each other, related by the torque-constant k_T. Therefore, we will consider the current to be the control variable because it is more convenient to use. Notice that there is a feedback in the current loop from the output speed. This feedback dictates the induced back-emf. Neglecting T_L, and considering the current to be the output, $E_a(s)$ can be calculated in terms of $I_a(s)$ in Fig. 5-12a as $E_a(s) = \dfrac{k_T k_E}{s J_{eq}} I_a(s)$. Therefore, Fig. 5-12a can be redrawn, as shown in Fig. 5-12b. Notice that the feedback term depends inversely on the inertia J_{eq}. Assuming that the inertia is sufficiently large to justify neglecting the feedback effect, we can simplify the block diagram, as shown in Fig. 5-12c.

The current-controller in Fig. 5-12c is a PI error amplifier with the proportional gain k_{pI} and the integral gain k_{iI}. Its transfer function is given by Eq. (5-23). The subscript "I" refers to the current loop. The open-loop transfer function $G_{I,OL}(s)$ of the simplified current loop in Fig. 5-12c is

$$G_{I,OL}(s) = \underbrace{\frac{k_{iI}}{s}\left[1 + \frac{s}{k_{iI}/k_{pI}}\right]}_{PI-controller} \underbrace{k_{PWM}}_{PPU} \underbrace{\frac{1/R_a}{1 + \frac{s}{1/\tau_e}}}_{motor} \quad (5\text{-}24)$$

To select the gain constants of the PI controller in the current loop, a simple design procedure, which results in a PM of 90°, is suggested as follows:

- Select the zero (k_{iI}/k_{pI}) of the PI controller to cancel the motor pole at ($1/\tau_e$) due to the electrical time-constant τ_e of the motor. Under these conditions,

$$\frac{k_{iI}}{k_{pI}} = \frac{1}{\tau_e} \quad \text{or} \quad k_{pI} = \tau_e k_{iI} \quad (5\text{-}25)$$

Cancellation of the pole in the motor transfer function renders the open-loop transfer function to be

$$G_{I,OL}(s) = \frac{k_{I,OL}}{s} \quad (5\text{-}26a)$$

$$k_{I,OL} = \frac{k_{iI} k_{PWM}}{R_a} \quad (5\text{-}26b)$$

- In the open-loop transfer function of Eq. (5-26a), the crossover frequency $\omega_{cI} = k_{I,OL}$. We will select the crossover frequency $f_{cI}(=\omega_{cI}/2\pi)$ of the current open loop to be approximately one to two orders of magnitude smaller than the switching frequency of the PPU in order to avoid interference in the control loop from the switching-frequency noise. Therefore, at the selected crossover frequency, from Eq. (5-26b),

$$k_{iI} = \omega_{cI} R_a / k_{PWM} \quad (5\text{-}27)$$

This completes the design of the torque (current) loop, as illustrated by the example below, where the gain constants k_{pI} and k_{iI} can be calculated from Eqs. (5-25) and (5-27).

EXAMPLE 5-2

Design the current loop for the example system of Table 5-1, assuming that the crossover frequency is selected to be 1 kHz.

Solution

From Eq. (5-27), for $\omega_{cI} = 2\pi \times 10^3$ rad/s, $k_{iI} = \omega_{cI} R_a / k_{PWM} = 1050.0$ and from Eq. (5-25), $k_{pI} = k_{iI}\tau_e = k_{iI}(L_a/R_a) = 2.73$.

The open-loop transfer function is plotted in Fig. 5-13a, which shows that the crossover frequency is 1 kHz, as assumed previously. The closed-loop transfer function is plotted in Fig. 5-13b.

The Design of the Speed Loop We will select the bandwidth of the speed loop to be one order of magnitude smaller than that of the current (torque) loop. Therefore, the closed-current loop can be assumed to be ideal for design purposes and represented by unity, as shown in Fig. 5-14. The speed controller is of the PI type. The resulting open-loop transfer function $G_{\Omega,OL}(s)$ of the speed loop in the block diagram of Fig. 5-14 is as follows, where the subscript "Ω" refers to the speed loop:

$$G_{\Omega,OL}(s) = \underbrace{\frac{k_{i\Omega}}{s}\left[1 + s/\left(k_{i\Omega}/k_{p\Omega}\right)\right]}_{\text{PI controller}} \underbrace{1}_{\text{current loop}} \underbrace{\frac{k_T}{sJ_{eq}}}_{\text{torque + inertia}} \qquad (5\text{-}28)$$

Equation (5-28) can be rearranged as

$$G_{\Omega,OL}(s) = \left(\frac{k_{i\Omega}k_T}{J_{eq}}\right)\frac{1 + s/\left(k_{i\Omega}/k_{p\Omega}\right)}{s^2} \qquad (5\text{-}29)$$

This shows that the open-loop transfer function consists of a double pole at the origin. At low frequencies in the Bode plot, this double pole at the origin causes the magnitude to decline at the rate

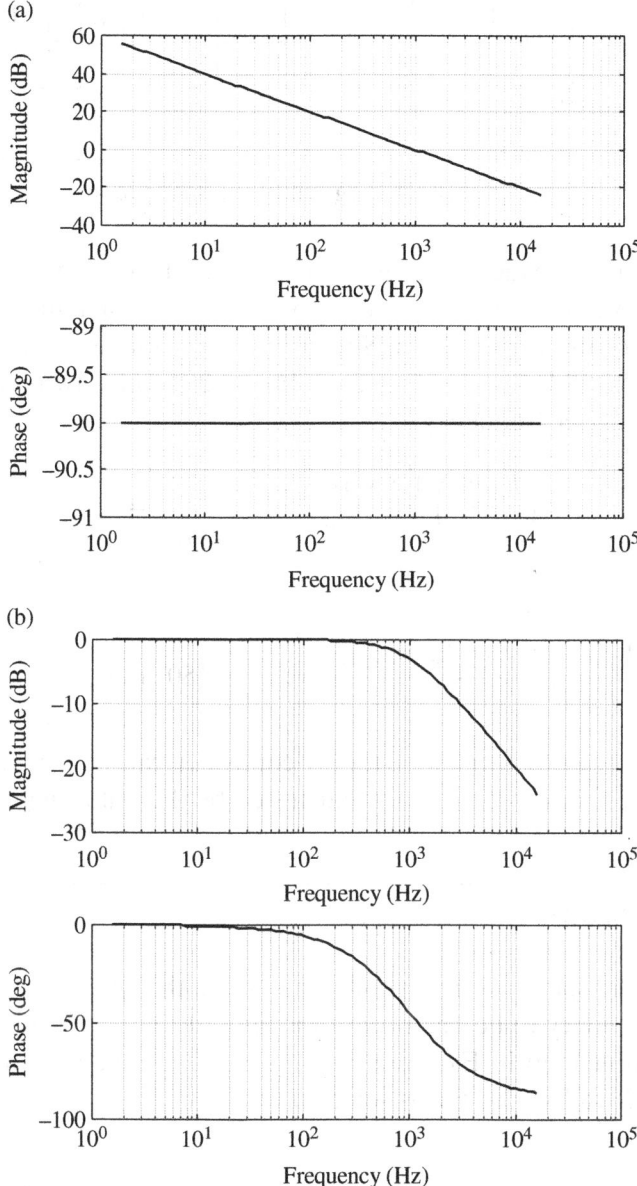

Fig. 5-13 Frequency response of the current loop: (a) open loop and (b) closed loop.

150 CONTROL IN ELECTRIC DRIVES

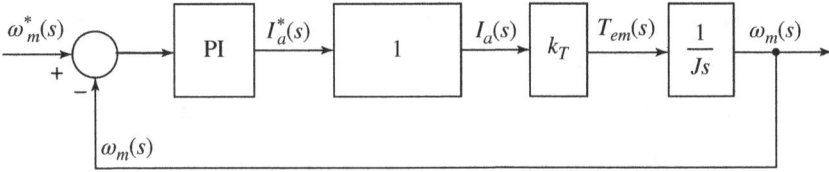

Fig. 5-14 Block diagram of the speed loop.

of -40 db per decade while the phase angle is at $-180°$. We can select the crossover frequency $\omega_{c\Omega}$ to be one order of magnitude smaller than that of the current loop. Similarly, we can choose a reasonable value of the PM $\phi_{pm,\,\Omega}$. Therefore, Eq. (5-29) yields two equations at the crossover frequency:

$$\left| \left(\frac{k_{i\Omega} k_T}{J_{eq}}\right) \frac{1 + s/(k_{i\Omega}/k_{p\Omega})}{s^2} \right|_{s = j\omega_{c\Omega}} = 1 \qquad (5\text{-}30)$$

and

$$\angle \left(\frac{k_{i\Omega} k_T}{J_{eq}}\right) \frac{1 + s/(k_{i\Omega}/k_{p\Omega})}{s^2} \bigg|_{s = j\omega_{c\Omega}} = -180° + \phi_{pm,\Omega} \qquad (5\text{-}31)$$

The two gain constants of the PI controller can be calculated by solving these two equations, as illustrated by the following example.

EXAMPLE 5-3

Design the speed-loop controller, assuming the speed-loop crossover frequency to be one order of magnitude smaller than that of the current loop in Example 5-2; that is, $f_{c\Omega} = 100\,\text{Hz}$, and thus $\omega_{c\Omega} = 628\,\text{rad/s}$. The PM is selected to be $60°$.

Solution

In Eqs. (5-30) and (5-31), substituting $k_T = 0.1\,\text{Nm/A}$, $J_{eq} = 152 \times 10^{-6}\,\text{kg} \cdot \text{m}^2$, and $\phi_{PM,\,\Omega} = 60°$ at the crossover frequency, where $s = j\omega_{c\Omega} = j628$, we can calculate that $k_{p\Omega} = 0.827$ and $k_{i\Omega} = 299.7$. The open- and the closed-loop transfer functions are plotted in Fig. 5-15a and b.

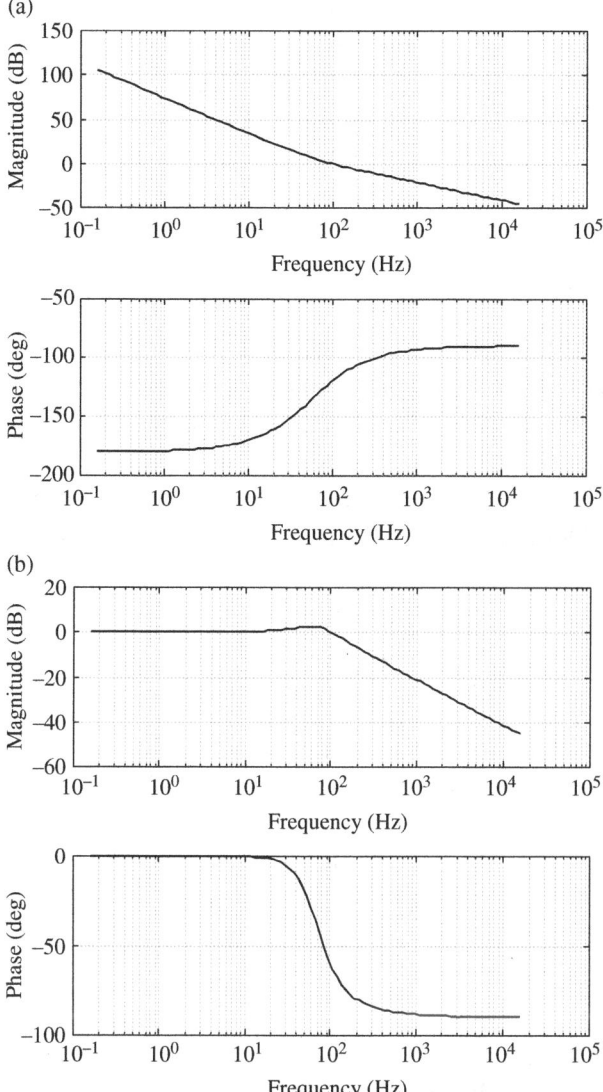

Fig. 5-15 Speed loop response: (a) open loop and (b) closed loop.

5-4-3 The Design of the Position Control Loop

We will select the bandwidth of the position loop to be one order of magnitude smaller than that of the speed loop. Therefore, the speed loop can be idealized and represented by unity, as shown in Fig. 5-16.

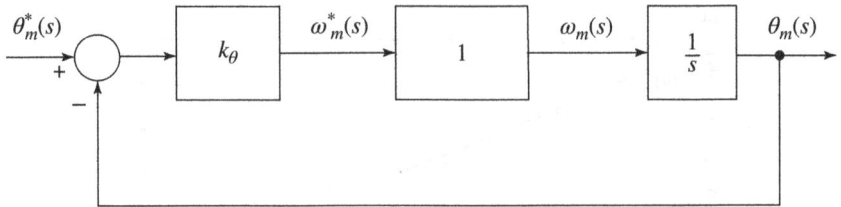

Fig. 5-16 Block diagram of position loop.

For the position controller, it is adequate to have only a proportional gain $k_{p\theta}$ because of the presence of a true integrator $\left(\frac{1}{s}\right)$ in Fig. 5-16 in the open-loop transfer function. This integrator will reduce the steady-state error to zero for a step-change in the reference position. With this choice of the controller, and with the closed-loop response of the speed loop assumed to be ideal, the open-loop transfer function $G_{\theta,\,OL}(s)$ is

$$G_{\theta,OL}(s) = \frac{k_\theta}{s} \qquad (5\text{-}32)$$

Therefore, selecting the crossover frequency $\omega_{c\theta}$ of the open loop allows k_θ to be calculated as

$$k_\theta = \omega_{c\theta} \qquad (5\text{-}33)$$

EXAMPLE 5-4

For the example system of Table 5-1, design the position-loop controller, assuming the position-loop crossover frequency to be one order of magnitude smaller than that of the speed loop in Example 5-3 (that is, $f_{c\theta} = 10$ Hz and $\omega_{c\theta} = 62.8$ rad/s).

Solution

From Eq. (5-33), $k_\theta = \omega_{c\theta} = 62.8$ rad/s. The open- and closed-loop transfer functions are plotted in Figs. 5-17a and b.

CONTROLLER DESIGN   153

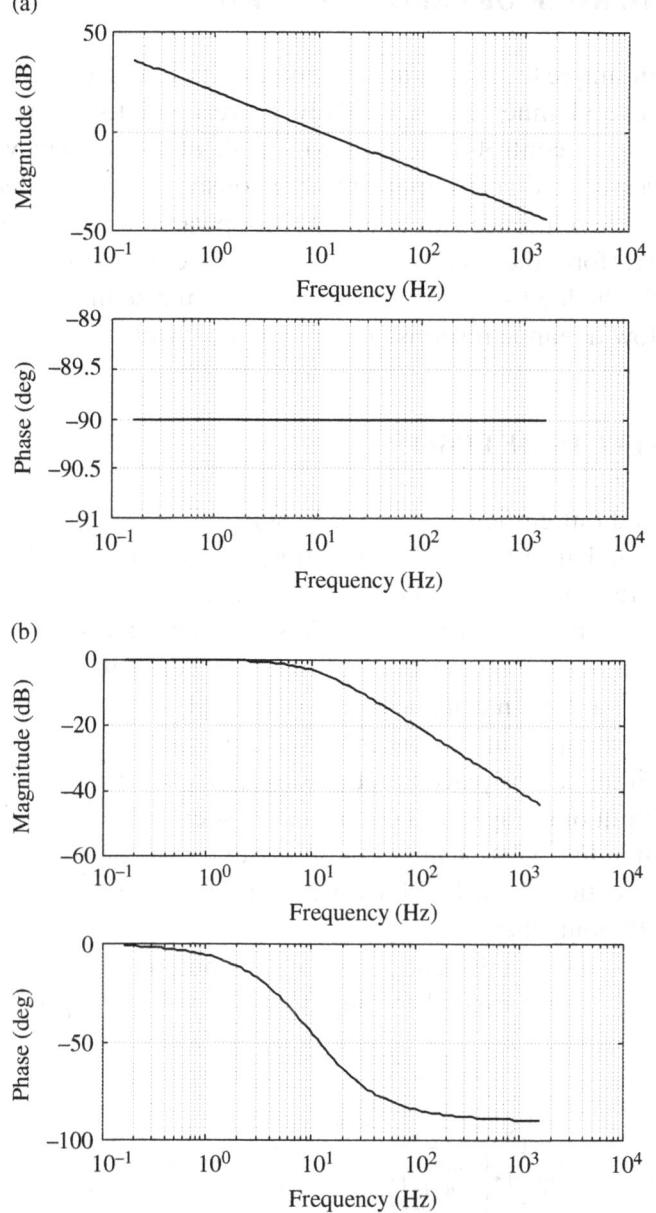

Fig. 5-17   Position loop response: (a) open-loop and (b) closed loop.

5-5 THE ROLE OF FEED-FORWARD

Although simple to design and implement, a cascaded control consisting of several inner loops is likely to respond to changes more slowly than a control system in which all of the system variables are processed and acted upon simultaneously. In industrial systems, approximate reference values of inner-loop variables are often available. Therefore, these reference values are fed forward, as shown in Fig. 5-18. The feed-forward operation can minimize the disadvantage of the slow dynamic response of cascaded control.

5-6 EFFECTS OF LIMITS

As pointed out earlier, one of the benefits of cascade control is that the intermediate variables such as torque (current) and the control signal to the PWM-IC can be limited to acceptable ranges by putting limits on their reference values. This provides safety of operation for the motor, for the power electronics converter within the power processor, and the mechanical system as well.

As an example, in the original cascade control system discussed earlier, limits can be placed on the torque (current) reference, which is the output of the speed PI controller, as seen in Fig. 5-18. Similarly, as shown in Fig. 5-19a, a limit inherently exists on the control voltage (applied to the PWM-IC chip), which is the output of the torque/current PI controller.

Similarly, a limit inherently exists on the output of the PPU, whose magnitude cannot exceed the input dc-bus voltage V_d. For a large

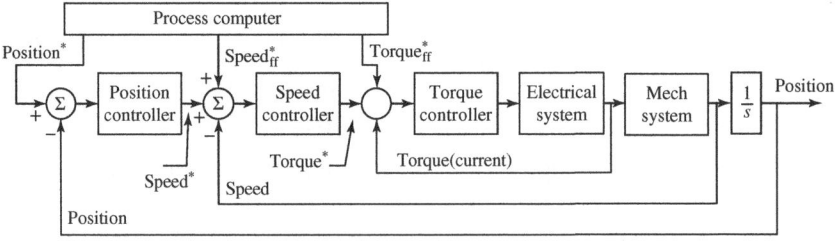

Fig. 5-18 Control system with feed-forward.

ANTI-WINDUP (NON-WINDUP) INTEGRATION 155

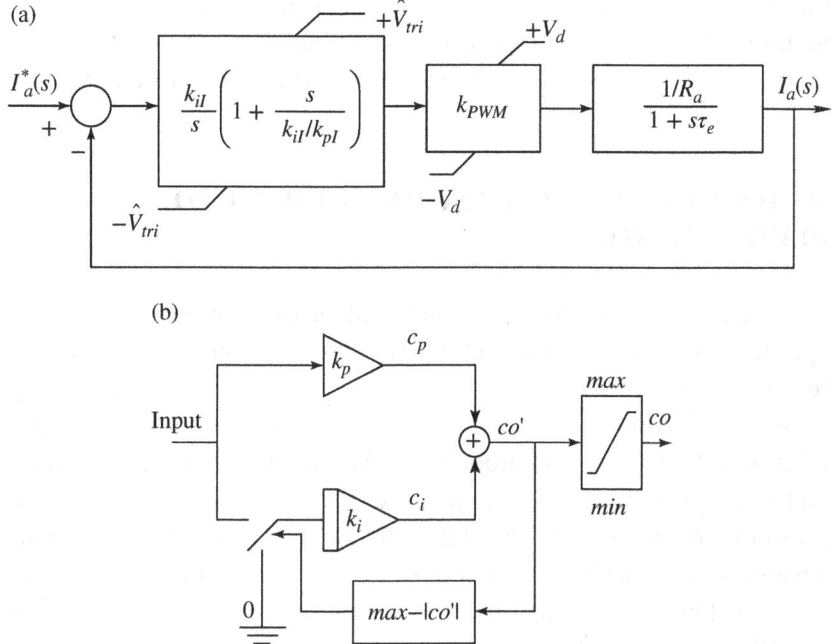

Fig. 5-19 (a) Limits on the PI controller and (b) PI with anti-windup.

change in reference or a large disturbance, the system may reach such limits. This makes the system nonlinear and introduces further delay in the loop when the limits are reached. For example, a linear controller may demand a large motor current in order to meet a sudden load torque increase, but the current limit will cause the current loop to meet this increased load torque demand slower than is otherwise possible. This is the reason that after the controller is designed based on the assumptions of linearity, its performance in the presence of such limits should be thoroughly simulated.

5-7 ANTI-WINDUP (NON-WINDUP) INTEGRATION

In order for the system to maintain stability in the presence of limits, special attention should be paid to the controllers with integrators, such as the PI controller shown in Fig. 5-19b. In the anti-windup integrator of Fig. 5-19b, if the controller output reaches its limit, then the integrator action is turned off by shorting the input of the integrator to ground if the saturation increases in the same direction.

5-8 HARDWARE PROTOTYPING OF dc MOTOR SPEED CONTROL

In this chapter, closed-loop speed control of dc-motor drives was explained in detail. In this section, the hardware implementation of the same is analyzed.

For the purposes of this comparison, a commercially available tabletop 42V dc motor is modeled in Workbench, and its hardware results are presented. Although the exact motor parameters of the dc motor are not necessary, this is available on the accompanying website. The Workbench file of the modeled system is shown in Fig. 5-20. The speed controller has a step reference input from 0 to 100 rad/s at time $t = 1$ s.

The speed result from running this model in hardware is shown in Fig. 5-21. The measured dc motor speed closely tracks that of the reference speed.

The step-by-step procedure for recreating the above comparison is presented in [2].

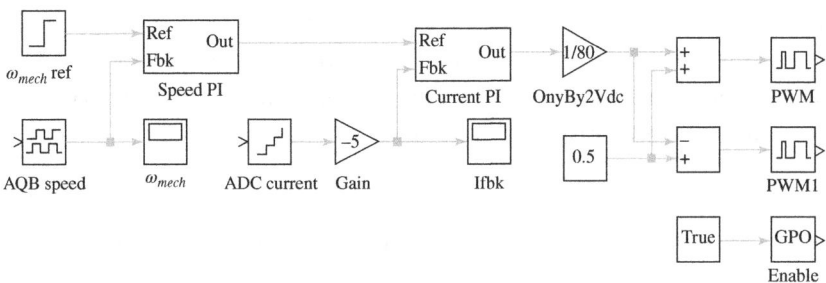

Fig. 5-20 Real-time dc motor speed control.

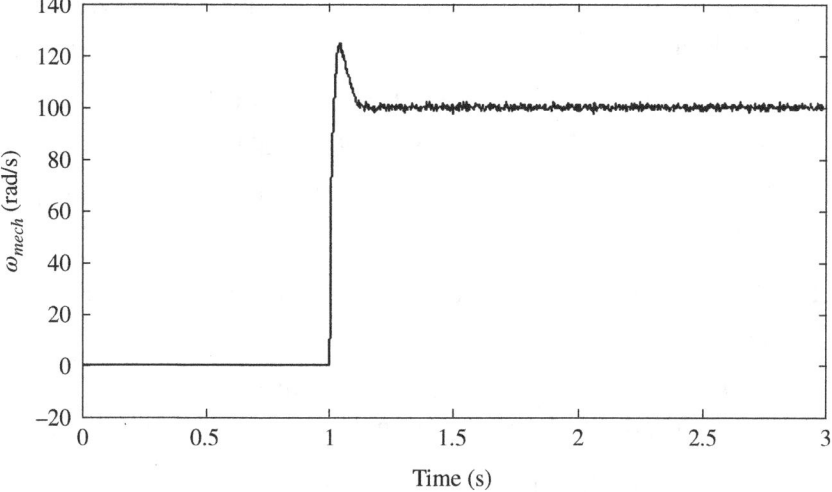

Fig. 5-21 Hardware speed control.

5-9 REVIEW QUESTIONS

1. What are the various blocks of a motor drive?

2. What is a cascaded control, and what are its advantages?

3. Draw the average models of a PWM controller and a dc–dc converter.

4. Draw the dc-motor equivalent circuit and its representation in the Laplace domain. Is this representation linear?

5. What is the transfer function of a proportional-integral (PI) controller?

6. Draw the block diagram of the torque loop.

7. What is the rationale for neglecting the feedback from speed in the torque loop?

8. Draw the simplified block diagram of the torque loop.

9. Describe the procedure for designing the PI controller in the torque loop.

10. How would we have designed the PI controller of the torque loop if the effect of the speed were not ignored?

11. What allows us to approximate the closed torque loop by unity in the speed loop?

12. What is the procedure for designing the PI controller in the speed loop?

13. How would we have designed the PI controller in the speed loop if the closed torque loop were not approximated by unity?

14. Draw the position-loop block diagram.

15. Why do we only need a P controller in the position loop?

16. What allows us to approximate the closed speed loop by unity in the position loop?

17. Describe the design procedure for determining the controller in the position loop.

18. How would we have designed the position controller if the closed speed loop was not approximated by unity?

19. Draw the block diagram with feed-forward. What are its advantages?

20. Why are limiters used, and what are their effects?

21. What is the integrator windup, and how can it be avoided?

REFERENCES

1. *Electric Machines and Drives: A First Course* ISBN: 978-1-118-07481-7 by Ned Mohan January 2012
2. https://sciamble.com/Resources/pe-drives-lab/basic-drives/dc-speed-control

FURTHER READING

Kazmierkowski, M. and Tunia, H. (1994). *Automatic Control of Converter-Fed Drives.* Elsevier 559 pages.

Kazmierkowski, M., Krishnan, R., and Blaabjerg, F. (2002). *Control of Power Electronics.* Academic Press 518 pages.

Leonard, W. (1985). *Control of Electric Drives.* New York: Springer-Verlag.

PROBLEMS AND SIMULATIONS

dc Motors

5-1 A permanent-magnet dc motor has the following parameters: $R_a = 0.35\,\Omega$ and $k_E = k_T = 0.5$ in MKS units. For a torque of up to 8 Nm, plot its steady-state torque-speed characteristics for the following values of V_a: 100, 75, and 50 V.

5-2 Consider the dc motor of Problem 5-1, whose moment-of-inertia $J_m = 0.02\,\text{kg}\cdot\text{m}^2$. Its armature inductance L_a can be neglected for slow changes. The motor is driving a load of inertia $J_L = 0.04\,\text{kg}\cdot\text{m}^2$. The steady-state operating speed is 300 rad/s. Calculate and plot the terminal voltage $v_a(t)$ that is required to bring this motor to a halt as quickly as possible, without exceeding the armature current of 12 A.

Controller Design

5-3 In a unity feedback system, the open-loop transfer function is of the form $G_{OL}(s) = \dfrac{k}{1 + s/\omega_p}$. Calculate the bandwidth of the closed-loop transfer function. How does the bandwidth depend on k and ω_p?

5-4 In a feedback system, the forward path has a transfer function of the form $G(s) = k/(1 + s/\omega_p)$, and the feedback path has a gain of k_{fb}, which is less than unity. Calculate the bandwidth

160 CONTROL IN ELECTRIC DRIVES

of the closed-loop transfer function. How does the bandwidth depend on k_{fb}?

5-5 In designing the torque loop of Example 5-2, include the effect of the back-emf, shown in Fig. 5-9a. Design a PI controller for the same open-loop crossover frequency and a phase margin of 60°. Compare your results with those in Example 5-2.

5-6 In designing the speed loop of Example 5-3, include the torque loop by a first-order transfer function based on the design in Example 5-2. Design a PI controller for the same open-loop crossover frequency and the same phase margin as in Example 5-3 and compare the results.

5-7 In designing the position loop of Example 5-4, include the speed loop by a first-order transfer function based on the design in Example 5-3. Design a P-type controller for the same open-loop crossover frequency as in Example 5-4 and for a phase margin of 60°. Compare your results with those in Example 5-4.

5-8 In an actual system in which there are limits on the voltage and current that can be supplied, why and how does the initial steady-state operating point make a difference for large-signal disturbances?

5-9 Obtain the time response of the system designed in Example 5-3, in terms of the change in speed, for a step-change of the load-torque disturbance.

5-10 Obtain the time response of the system designed in Example 5-4, in terms of the change in position, for a step-change of the load-torque disturbance.

5-11 In the example system of Table 5-1, the maximum output voltage of the dc–dc converter is limited to 60 V. Assume that the current is limited to 8 A in magnitude. How do these two limits impact the response of the system to a large step-change in the reference value?

5-12 In Example 5-3, design the speed-loop controller, without the inner current loop, as shown in Fig. P5-12, for the same crossover frequency and phase margin as in Example 5-3. Compare results with the system of Example 5-3.

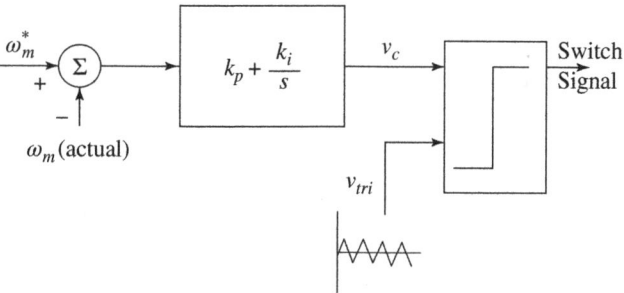

Fig. P5-12 Speed controller.

Part II
Steady-State Operation of ac Machines

6 Using Space Vectors to Analyze ac Machines

6-1 INTRODUCTION

The purpose of this chapter is to introduce space vectors, which are very useful in analyzing the operation of ac machines. Generally, three-phase ac voltages and currents supply all of these machines. The stators of the induction and the synchronous machines are similar and consist of six-phase windings. However, the rotor construction makes the operation of these two machines different. In the stator of these machines, each phase winding (a winding consists of a number of coils connected in series) produces a sinusoidal field distribution in the air gap. The field distributions, due to three phases, are displaced by 120° ($2\pi/3$ rad) in space with respect to each other, as indicated by their magnetic axes (defined in Chapter 6 for a concentrated coil) in the cross-section of Fig. 6-1 for a 2-pole machine, the simplest case. In this chapter, we will learn to represent sinusoidal field distributions in the air gap with space vectors, which will greatly simplify our analysis.

(Adapted from chapter 9 of *Electric Machines and Drives: A First Course* ISBN: 978-1-118-07481-7 by Ned Mohan, January 2012)

Analysis and Control of Electric Drives: Simulations and Laboratory Implementation, First Edition. Ned Mohan and Siddharth Raju.
© 2021 John Wiley & Sons, Inc. Published 2021 by John Wiley & Sons, Inc.
Companion website: www.wiley.com/go/Mohan/Vectorcontrolinelectricdrives

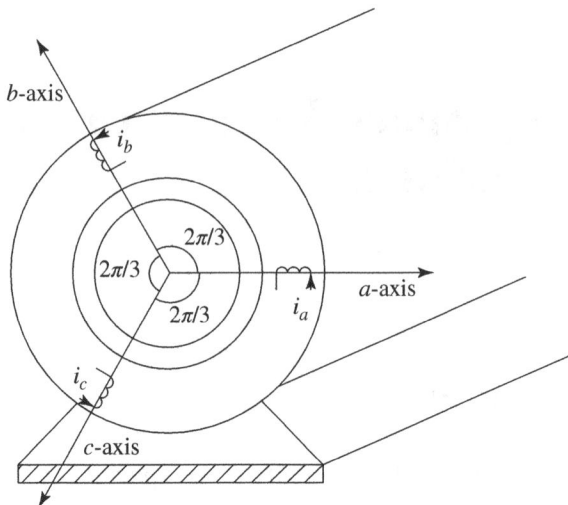

Fig. 6-1 Magnetic axes of the three phases in a 2-pole machine.

6-2 SINUSOIDALLY DISTRIBUTED STATOR WINDINGS

In the following description, we will assume a 2-pole machine (with $p = 2$). This analysis is later generalized to multipole machines by means of Example 6-2.

In ac machines, windings for each phase ideally should produce a sinusoidally distributed, radial field (F, H, and B) in the air gap. Theoretically, this requires a sinusoidally distributed winding in each phase. In practice, this is approximated in a variety of ways discussed in Refs. [1, 2]. To visualize this sinusoidal distribution, consider the winding for phase a, shown in Fig. 6-2a, where, in the slots, the number of turns-per-coil for phase-a progressively increases away from the magnetic axis, reaching a maximum at $\theta = 90°$. Each coil, such as the coil with sides 1 and 1', spans 180° where the current into coil-side 1 returns in 1' through the end-turn at the back of the machine. This coil (1, 1') is connected in series to coil-side 2 of the next coil (2, 2'), and so on. Graphically, such a winding for phase-a can be drawn, as shown in Fig. 6-2b, where bigger circles represent higher conductor densities, noting that all of the conductors in the winding are in series and hence carry the same current.

SINUSOIDALLY DISTRIBUTED STATOR WINDINGS 167

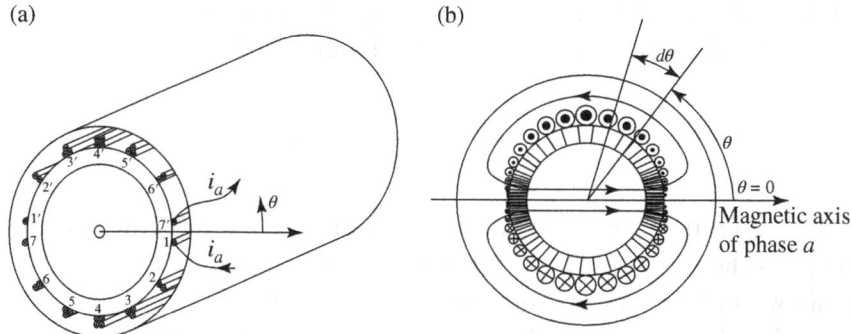

Fig. 6-2 Sinusoidally distributed winding for phase-a.

In Fig. 6-2b, in phase a, the conductor density $n_s(\theta)$, in terms of the number of conductors per radian angle, is a sinusoidal function of the angle θ, and can be expressed as

$$n_s(\theta) = \hat{n}_s \sin\theta \;\;[\text{no. of conductors/rad}] \;\; 0 < \theta < \pi \quad (6\text{-}1)$$

where $\hat{n}_s$ is the maximum conductor density, which occurs at $\theta = \frac{\pi}{2}$. If the phase winding has a total of N_s turns (that is, $2N_s$ conductors), then each winding-half, from $\theta = 0$ to $\theta = \pi$, contains N_s conductors. To determine $\hat{n}_s$ in Eq. (6-1) in terms of N_s, note that a differential angle $d\theta$ at θ in Fig. 6-2b contains $n_s(\theta) \cdot d\theta$ conductors. Therefore, the integral of the conductor density in Fig. 6-2b, from $\theta = 0$ to $\theta = \pi$, equals N_s conductors:

$$\int_0^\pi n_s(\theta)\, d\theta = N_s \quad (6\text{-}2)$$

Substituting the expression for $n_s(\theta)$ from Eq. (6-1), the integral in Eq. (6-2) yields

$$\int_0^\pi n_s(\theta)\, d\theta = \int_0^\pi \hat{n}_s \sin\theta\, d\theta = 2\hat{n}_s \quad (6\text{-}3)$$

Equating the right sides of Eqs. (6-2) and (6-3),

$$\hat{n}_s = \frac{N_s}{2} \quad (6\text{-}4)$$

Substituting $\hat{n}_s$ from Eq. (6-4) into Eq. (6-1) yields the sinusoidal conductor-density distribution in the phase-a winding as

$$n_s(\theta) = \frac{N_s}{2} \sin\theta \quad 0 \le \theta \le \pi \tag{6-5}$$

In a multipole machine (with $p > 2$), the peak conductor density remains the same, $N_s/2$, as in Eq. (6-5) for a 2-pole machine. (This is shown in Example 6-2 and the Problem 6-4.)

Rather than restricting the conductor density expression to a region $0 < \theta < \pi$, we can interpret the negative of the conductor density in the region $\pi < \theta < 2\pi$ in Eq. (6-5) as being associated with carrying the current in the opposite direction, as indicated in Fig. 6-2b.

To obtain the air gap field (mmf, flux density, and the magnetic field intensity) distribution caused by the winding current, we will make use of the symmetry in Fig. 6-3.

The radially oriented fields in the air gap at angles θ and $(\theta + \pi)$ are equal in magnitude but opposite in direction. We will assume the field direction away from the center of the machine to be positive. Therefore, the magnetic field intensity *in the air gap*, established by the current i_a (hence the subscript "a") at positions θ and $(\theta + \pi)$ will be equal in magnitude but of opposite sign: $H_a(\theta + \pi) = -H_a(\theta)$. To exploit this symmetry, we will apply Ampere's Law to a closed path shown in

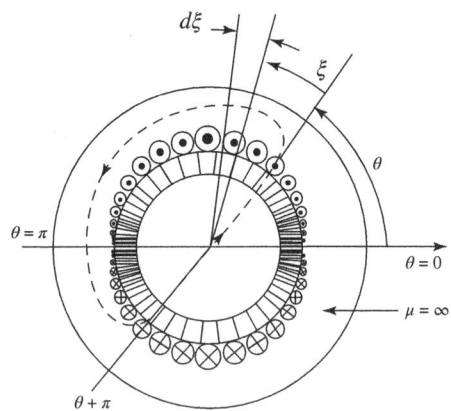

Fig. 6-3 Calculation of air gap field distribution.

Fig. 6-3 through angles θ and $(\theta+\pi)$. We will assume the magnetic permeability of the rotor and the stator iron to be infinite, and the H-field in iron to be zero. In terms of $H_a(\theta)$, application of Ampere's Law along the closed path in Fig. 6-3, at any instant of time t, results in

$$\underbrace{H_a \ell_g}_{outward} - \underbrace{(-H_a)\ell_g}_{inward} = \int_0^\pi i_a \cdot n_s(\theta + \xi) \cdot d\xi \qquad (6\text{-}6)$$

where ℓ_g is the length of each air gap, and a negative sign is associated with the integral in the inward direction because, while the path of integration is inward, the field intensity is measured outwardly. On the right side of Eq. (6-6), $n_s(\xi) \cdot d\xi$ is the number of turns enclosed in the differential angle $d\xi$ at an angle ξ, as measured in Fig. 6-3. In Eq. (6-6), integration from 0 to π yields the total number of conductors enclosed by the chosen path, including the "negative" conductors that carry current in the opposite direction. Substituting the conductor density expression from Eq. (6-5) into Eq. (6-6),

$$2H_a(\theta)\ell_g = \frac{N_s}{2} i_a \int_0^\pi \sin(\theta + \xi) \cdot d\xi = N_s i_a \cos\theta \text{ or}$$
$$H_a(\theta) = \frac{N_s}{2\ell_g} i_a \cos\theta \qquad (6\text{-}7)$$

Using Eq. (6-7), the radial flux density $B_a(\theta)$ and the mmf $F_a(\theta)$ acting on the air gap at an angle θ can be written as

$$B_a(\theta) = \mu_o H_a(\theta) = \left(\frac{\mu_o N_s}{2\ell_g}\right) i_a \cos\theta \text{ and} \qquad (6\text{-}8)$$

$$F_a(\theta) = \ell_g H_a(\theta) = \frac{N_s}{2} i_a \cos\theta \qquad (6\text{-}9)$$

The co-sinusoidal field distributions in the air gap, due to a positive value of i_a (with the direction as defined in Fig. 6-2a and b), given by Eqs. (6-7) through (6-9), are plotted in the developed view of Fig. 6-4a. The angle θ is measured in the counterclockwise direction with respect

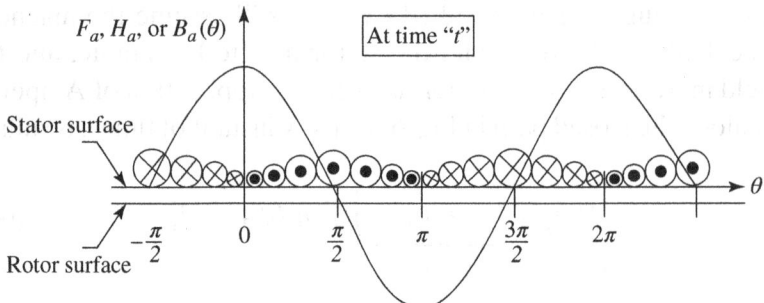

Fig. 6-4 Developed view of the field distribution in the air gap.

to the phase-*a* magnetic axis. The radial field distributions in the air gap peak along the phase-*a* magnetic axis, and at any instant of time, their amplitudes are linearly proportional to the value of i_a at that time. Figure 6-4b shows field distributions in the air gap due to positive and negative values of i_a at various times. Notice that regardless of the positive or the negative current in phase-*a*, the flux-density distribution produced by it in the air gap always has its peak (positive or negative) along the phase-*a* magnetic axis.

EXAMPLE 6-1

In the sinusoidally distributed winding of phase-*a*, shown in Fig. 6-3, $N_s = 100$ and the current $i_a = 10$ A. The air gap length $\ell_g = 1$ mm. Calculate the ampere-turns enclosed and the corresponding F, H, and B fields for the following Ampere's Law integration paths: (a) through θ equal to 0° and 180° as shown in Fig. 6-5a and (b) through θ equal to 90° and 270° as shown in Fig. 6-5b.

Solution

(a) At $\theta = 0°$, from Eqs. (6-7) through (6-9),

$$H_a\big|_{\theta=0} = \frac{N_s}{2\ell_g} i_a \cos(\theta) = 5 \times 10^5 \text{ A/m},$$

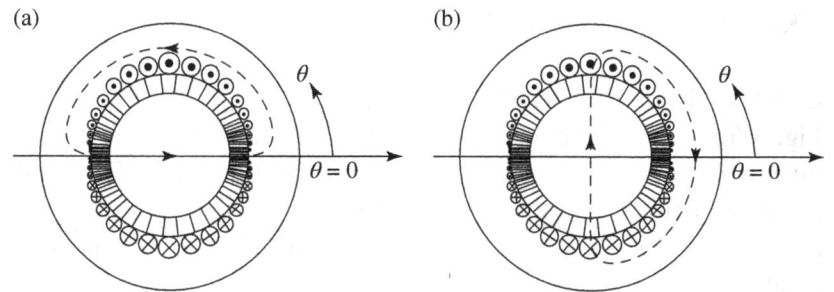

Fig. 6-5 Paths corresponding to Example 6-1.

$$B_a|_{\theta = 0} = \mu_o H_a|_{\theta = 0} = 0.628 \text{ T, and}$$

$$F_a|_{\theta = 0} = \ell_g H_a|_{\theta = 0} = 500 \text{ A turns}$$

All the field quantities reach their maximum magnitude at $\theta = 0°$ and $\theta = 180°$, because the path through them encloses all of the conductors that are carrying current in the same direction.
(b) From Eqs. (6-7) through (6-9), at $\theta = 90°$,

$$H_a|_{\theta = 90°} = \frac{N_s}{2\ell_g} i_a \cos(\theta) = 0 \text{ A/m}, \quad B_a|_{\theta = 90°} = 0, \quad \text{and} \quad F_a|_{\theta = 90°} = 0$$

Half of the conductors enclosed by this path, as shown in Fig. 6-5b, carry current in a direction opposite to that of the other half. The net effect is the cancellation of all of the field quantities in the air gap at 90° and 270°.

We should note that there is a limited number of total slots along the stator periphery, and each phase is allotted only a fraction of the total slots. In spite of these limitations, the field distribution can be made to approach a sinusoidal distribution in space, as in the ideal case discussed above. Since machine design is not our objective, we will leave the details for the interested reader to investigate in Refs. [1, 2].

EXAMPLE 6-2

Consider the phase-*a* winding for a 4-pole stator ($p = 4$) as shown in Fig. 6-6a. All of the conductors are in series. Just like in a 2-pole machine, the conductor density is a sinusoidal function. The total number of turns per-phase is N_s. Obtain the expressions for the conductor density and the field distribution, both as functions of position.

Solution

We will define an electrical angle θ_e in terms of the actual (mechanical) angle θ:

$$\theta_e = \frac{p}{2}\theta \qquad (6\text{-}10)$$

where $\theta_e = 2\theta$ ($p = 4$ poles).

Skipping a few steps (left as Problem 6-4), we can show that, in terms of θ_e, the conductor density in phase-*a* of a *p*-pole stator ideally should be

$$n_s(\theta_e) = \frac{N_s}{p}\sin\theta_e \qquad (p \geq 2) \qquad (6\text{-}11)$$

To calculate the field distribution, we will apply Ampere's Law along the path through θ_e and $(\theta_e + \pi)$, shown in Fig. 6-6a, and we will make use of symmetry. The procedure is similar to that used for a 2-pole machine (the intermediate steps are skipped here and left as Problem 6-5). The results for a multipole machine ($p \geq 2$) are as follows:

$$H_a(\theta_e) = \frac{N_s}{p\ell_g}i_a\cos\theta_e \qquad (6\text{-}12a)$$

SINUSOIDALLY DISTRIBUTED STATOR WINDINGS 173

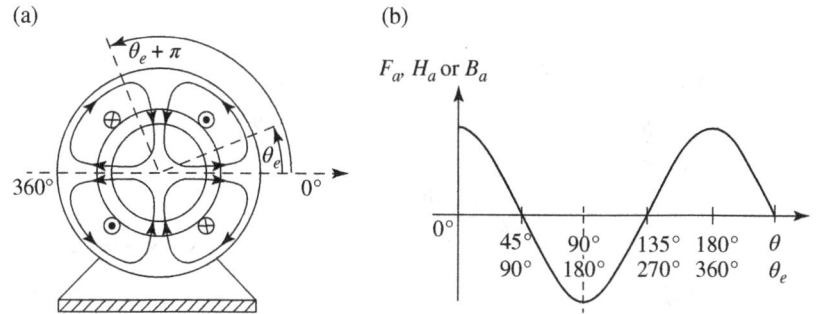

Fig. 6-6 Phase a of a 4-pole machine.

$$B_a(\theta_e) = \mu_o H_a(\theta_e) = \left(\frac{\mu_o N_s}{p\,\ell_g}\right) i_a \cos\theta_e \qquad (6\text{-}12b)$$

and

$$F_a(\theta_e) = \ell_g H_a(\theta_e) = \frac{N_s}{p} i_a \cos\theta_e \qquad (6\text{-}12c)$$

These distributions are plotted in Fig. 6-6b for a 4-pole machine. Notice that one complete cycle of distribution spans 180 mechanical degrees; therefore, this distribution is repeated twice around the periphery in the air gap.

6-2-1 Three-Phase, Sinusoidally Distributed Stator Windings

In the previous section, we focused only on phase-a, which has its magnetic axis along $\theta = 0°$. There are two more identical sinusoidally distributed windings for phases b and c, with magnetic axes along $\theta = 120°$ and $\theta = 240°$, respectively, as represented in Fig. 6-7a. These three windings are generally connected in a wye arrangement by connecting terminals a', b', and c' together, as shown in Fig. 6-7b.

Field distributions in the air gap due to currents i_b and i_c, are identical in shape to those in Fig. 6-4a and b, due to i_a, but they peak along their respective phase-b and phase-c magnetic axes. By Kirchhoff's Current Law, in Fig. 6-7b,

$$i_a(t) + i_b(t) + i_c(t) = 0 \qquad (6\text{-}13)$$

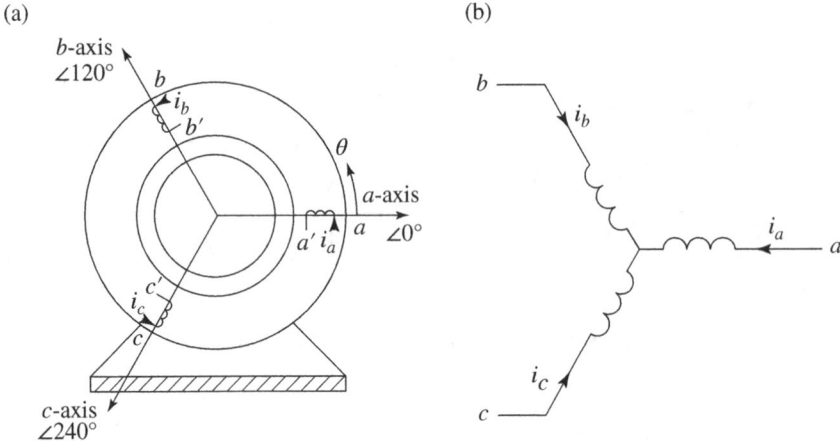

Fig. 6-7 Three-phase windings.

EXAMPLE 6-3

At any instant of time t, the stator windings of the 2-pole machine shown in Fig. 6-7b have $i_a = 10$ A, $i_b = -7$ A, and $i_c = -3$ A. The air gap length $\ell_g = 1$ mm and each winding has $N_s = 100$ turns. Plot the flux density as a function of θ, produced by each current, and the resultant flux density $B_s(\theta)$ in the air gap due to the combined effect of the three stator currents at this time. Note that the subscript "s" (which refers to the stator) includes the effect of all three stator phases on the air gap field distribution.

Solution

From Eq. (6-8), the peak flux density produced by any phase current i is

$$\hat{B} = \frac{\mu_o N_s}{2\ell_g} i = \frac{4\pi \times 10^{-7} \times 100}{2 \times 1 \times 10^{-3}} i = 0.0628\, i \quad [\text{T}]$$

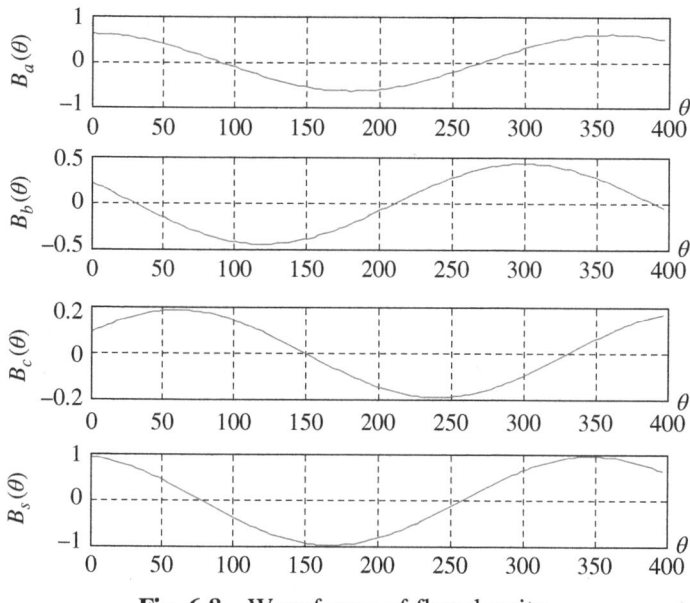

Fig. 6-8 Waveforms of flux density.

The flux-density distributions are plotted as functions of θ in Fig. 6-8 for the given values of the three-phase currents. Note that B_a has its positive peak at $\theta = 0°$, B_b has its negative peak at $\theta = 120°$, and B_c has its negative peak at $\theta = 240°$. Applying the principle of superposition under the assumption of a linear magnetic circuit, adding together the flux-density distributions produced by each phase at every angle θ yields the combined stator-produced flux density distribution $B_s(\theta)$, plotted in Fig. 6-8.

6-3 THE USE OF SPACE VECTORS TO REPRESENT SINUSOIDAL FIELD DISTRIBUTIONS IN THE AIR GAP

In linear ac circuits in a sinusoidal steady state, all voltages and currents vary sinusoidally with time. These sinusoidally time-varying voltages and currents are represented by phasors $\overline{V}$ and $\overline{I}$ for ease of calculations. These phasors are expressed by complex numbers, as discussed in Chapter 3.

176 USING SPACE VECTORS TO ANALYZE ac MACHINES

Similarly, in ac machines, at any instant of time t, sinusoidal space distributions of fields (B, H, F) in the air gap can be represented by space vectors. At any instant of time t, in representing a field distribution in the air gap with a space vector, we should note the following:

- The peak of the field distribution is represented by the amplitude of the space vector.
- Where the field distribution has its *positive* peak, the angle θ, measured with respect to the phase-a magnetic axis (by convention chosen as the reference axis), is represented by the orientation of the space vector.

Similar to phasors, space vectors are expressed by complex numbers. The space vectors are denoted by a "→" on top, and their time dependence is explicitly shown.

Let us first consider phase-a. In Fig. 6-9a, at any instant of time t, the mmf produced by the sinusoidally distributed phase-a winding has a co-sinusoidal shape (distribution) in space; that is, this distribution always peaks along the phase-a magnetic axis, and elsewhere it varies with the cosine of the angle θ away from the magnetic axis.

The amplitude of this co-sinusoidal spatial distribution depends on the phase current i_a, which varies with time. Therefore, as shown in Fig. 6-9a, at any time t, the mmf distribution due to i_a can be represented by a space vector $\overrightarrow{F_a}(t)$:

$$\overrightarrow{F_a}(t) = \frac{N_s}{2} i_a(t) \angle 0° \quad (6\text{-}14)$$

The amplitude of $\overrightarrow{F_a}(t)$ is $(N_s/2)$ times $i_a(t)$, and $\overrightarrow{F_a}(t)$ is always oriented along the phase-a magnetic axis at the angle of 0°. The phase-a magnetic axis is always used as the reference axis. A representation similar to the mmf distribution can be used for the flux-density distribution.

In a similar manner, at any time t, the mmf distributions produced by the other two phase windings can also be represented by space vectors oriented along their respective magnetic axes at and 240°, as

THE USE OF SPACE VECTORS TO REPRESENT 177

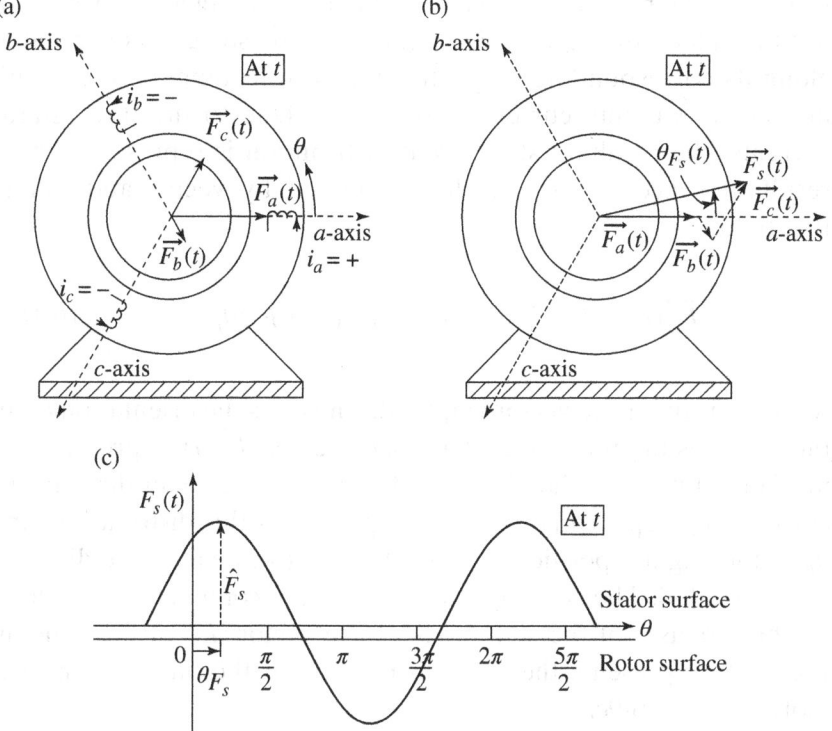

Fig. 6-9 Representation of MMF space vector in a machine.

shown in Fig. 6-9a, for negative values of i_b and i_c. In general, at any instant of time, we have the following three space vectors representing the respective mmf distributions:

$$\vec{F_a}(t) = \frac{N_s}{2} i_a(t) \angle 0°$$

$$\vec{F_b}(t) = \frac{N_s}{2} i_b(t) \angle 120° \qquad (6\text{-}15)$$

$$\vec{F_c}(t) = \frac{N_s}{2} i_c(t) \angle 240°$$

Note that the sinusoidal distribution of mmf in the air gap at any time t is a consequence of the sinusoidally distributed windings. As shown in Fig. 6-9a, for a positive value of i_a and negative values of i_b and i_c (such that $i_a + i_b + i_c = 0$), each of these vectors is pointed along its corresponding magnetic axis, with its amplitude depending on the winding current at that time. Due to the three stator currents, the resultant stator mmf distribution is represented by a resultant space vector, which is obtained by vector addition in Fig. 6-9b:

$$\vec{F_s}(t) = \vec{F_a}(t) + \vec{F_b}(t) + \vec{F_c}(t) = \hat{F}_s \angle \theta_{F_s} \qquad (6\text{-}16a)$$

where $\hat{F}_s$ is the space vector amplitude and θ_{F_s} is the orientation (with the a-axis as the reference). The space vector $\vec{F_s}(t)$, represents the mmf distribution in the air gap at this time t due to all three-phase currents; $\hat{F}_s$ represents the peak amplitude of this distribution, and θ_{F_s} is the angular position at which the positive peak of the distribution is located. The subscript "s" refers to the combined mmf due to all three phases of the stator. The space vector $\vec{F_s}$ at this time, in Fig. 6-9b, represents the mmf distribution in the air gap, which is plotted in Fig. 6-9c.

Expressions similar to $\vec{F_s}(t)$ in Eq. (6-16a) can be derived for the space vectors representing the combined-stator flux-density and the field-intensity distributions:

$$\vec{B_s}(t) = \vec{B_a}(t) + \vec{B_b}(t) + \vec{B_c}(t) = \hat{B}_s \angle \theta_{B_s} \qquad (6\text{-}16b)$$

and

$$\vec{H_s}(t) = \vec{H_a}(t) + \vec{H_b}(t) + \vec{H_c}(t) = \hat{H}_s \angle \theta_{H_s} \qquad (6\text{-}16c)$$

How are these three field distributions, represented by space vectors defined in Eqs. (6-16a) through (6-16c), related to each other? This question is answered by Eqs. (6-21a) and (6-21b) in Section 6-4-1.

THE USE OF SPACE VECTORS TO REPRESENT 179

EXAMPLE 6-4

In a 2-pole, three-phase machine, each of the sinusoidally distributed windings has $N_s = 100$ turns. The air gap length $\ell_g = 1.5$ mm. At a time t, $i_a = 10$ A, $i_b = -10$ A, and $i_c = 0$ A. Using space vectors, calculate and plot the resultant flux density distribution in the air gap at this time.

Solution

From Eqs. (6-15) and (6-16), noting that mathematically $1 \angle 0° = \cos \theta + j \sin \theta$,

$$\vec{F_s}(t) = \frac{N_s}{2}(i_a \angle 0° + i_b \angle 120° + i_c \angle 240°)$$

$$= 50 \times \{10 + (-10)(\cos 120° + j \sin 120°) + (0)(\cos 240° + j \sin 240°)\}$$

$$= 50 \times 17.32 \angle -30° = 866 \angle -30° \text{ A turns}$$

From Eqs. (6-8) and (6-9), $B_a(\theta) = (\mu_0/\ell_g)F_a(\theta)$. The same relationship applies to the field quantities due to all three stator phase currents being applied simultaneously; that is, $B_s(\theta) = (\mu_0/\ell_g)F_s(\theta)$. Therefore, at any instant of time t,

$$\vec{B_s}(t) = \frac{\mu_0}{\ell_g}\vec{F_s}(t) = \frac{4\pi \times 10^{-7}}{1.5 \times 10^{-3}}866 \angle -30° = 0.73 \angle -30° \text{ T}$$

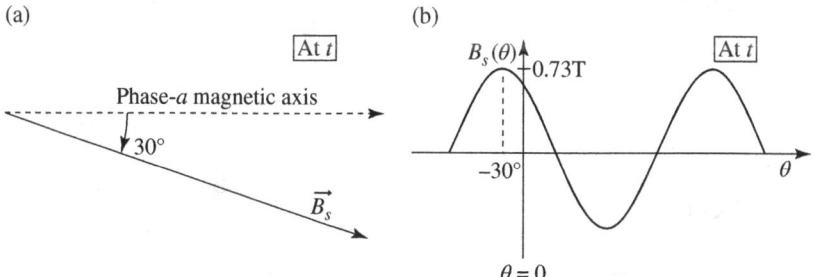

Fig. 6-10 (a) Resultant flux-density space vector and (b) flux-density distribution.

(*Continued*)

180 USING SPACE VECTORS TO ANALYZE ac MACHINES

This space vector is drawn in Fig. 6-10a. The flux density distribution has a peak value of 0.73 T, and the positive peak is located at $\theta = -30°$, as shown in Fig. 6-10b. Elsewhere, the radial flux density in the air gap, due to the combined action of all three-phase currents, is co-sinusoidally distributed.

6-4 SPACE-VECTOR REPRESENTATION OF COMBINED TERMINAL CURRENTS AND VOLTAGES

At any time t, we can measure the phase quantities, such as the voltage $v_a(t)$ and the current $i_a(t)$, at the terminals. Since there is no easy way to show that phase currents and voltages are distributed in space at any given time, we will NOT assign space vectors to physically represent these phase quantities. Rather, at any instant of time t, we will define space vectors to mathematically represent the combination of phase voltages and phase currents. These space vectors are defined to be the sum of their phase components (at that time) multiplied by their respective phase-axis orientations. Therefore, at any instant of time t, the stator current and the stator voltage space vectors are defined, in terms of their phase components (shown in Fig. 6-11a), as

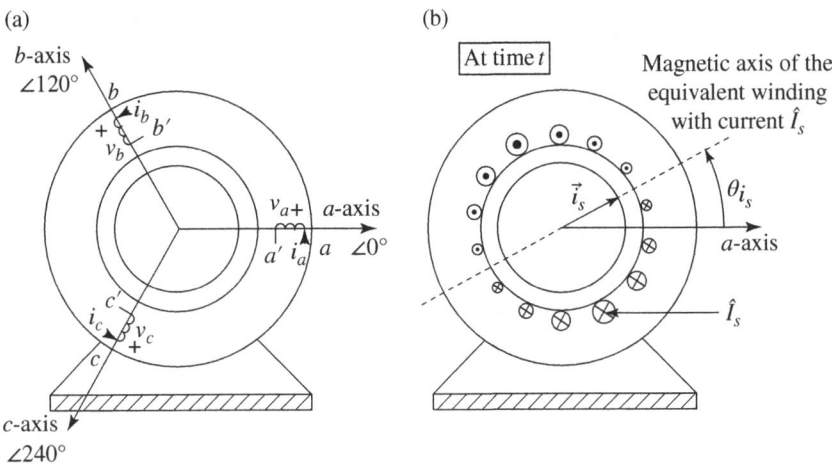

Fig. 6-11 (a) Phase voltages and currents and (b) physical interpretation of the current space vector.

SPACE-VECTOR REPRESENTATION 181

$$\vec{i_s}(t) = i_a(t)\angle 0° + i_b(t)\angle 120° + i_c(t)\angle 240° = \hat{I}_s(t)\angle \theta_{i_s}(t) \quad (6\text{-}17)$$

and

$$\vec{v_s}(t) = v_a(t)\angle 0° + v_b(t)\angle 120° + v_c(t)\angle 240° = \hat{V}_s(t)\angle \theta_{v_s}(t) \quad (6\text{-}18)$$

where the subscript "s" refers to the combined quantities of the stator. We will see later on that this mathematical description is of immense help in understanding the operation and control of ac machines.

6-4-1 Physical Interpretation of the Stator Current Space Vector $\vec{i_s}(t)$

The stator current space vector $\vec{i_s}(t)$ can be easily related to the stator mmf space vector $\vec{F_s}$. Multiplying both sides of Eq. (6-17) by $(N_s/2)$ gives

$$\frac{N_s}{2}\vec{i_s}(t) = \underbrace{\frac{N_s}{2} i_a(t)\angle 0°}_{\vec{F_a}(t)} + \underbrace{\frac{N_s}{2} i_b(t)\angle 120°}_{\vec{F_b}(t)} + \underbrace{\frac{N_s}{2} i_c(t)\angle 240°}_{\vec{F_c}(t)} \quad (6\text{-}19a)$$

Using Eq. (6-16), the sum of the mmf space vectors for the three phases is the resultant stator space vector. Therefore,

$$\frac{N_s}{2}\vec{i_s}(t) = \vec{F_s}(t) \quad (6\text{-}19b)$$

Thus,

$$\vec{i_s}(t) = \frac{\vec{F_s}(t)}{(N_s/2)} \quad (6\text{-}20)$$

where $\hat{I}_s(t) = \dfrac{\hat{F}_s(t)}{(N_s/2)}$ and $\theta_{i_s}(t) = \theta_{F_s}(t)$.

Equation (6-20) shows that the vectors $\vec{i_s}(t)$ and $\vec{F_s}(t)$ are related only by a scalar constant $(N_s/2)$. Therefore, they have the same orientation, and their amplitudes are related by $(N_s/2)$. At any instant of time t, Eq. (6-20) has the following interpretation:

The combined mmf distribution in the air gap produced by i_a, i_b, and i_c flowing through their respective sinusoidally distributed phase windings (each with N_s turns), is the same as that produced in Fig. 6-11b by a current $\hat{I}_s$ flowing through an equivalent sinusoidally distributed stator winding with its axis oriented at $\theta_{i_s}(t)$. This equivalent winding also has N_s turns.

As we will see later on, the above interpretation is very useful – it allows us to obtain, at any instant of time t, the combined torque acting on all three-phase windings by calculating the torque acting on this single equivalent winding with a current $\hat{I}_s$.

Next, we will use $\vec{i_s}(t)$ to relate the field quantities produced due to the combined effects of the three stator phase winding currents. Equations (6-7) through (6-9) show that the field distributions H_a, B_a, and F_a, produced by i_a flowing through the phase-a winding, are related by scalar constants. This will also be true for the combined fields in the air gap caused by the simultaneous flow of i_a, i_b, and i_c, since the magnetic circuit is assumed to be unsaturated and the principle of superposition applies. Therefore, we can write expressions for $\vec{B_s}(t)$ and $\vec{H_s}(t)$ in terms of $\vec{i_s}(t)$, which are similar to Eq. (6-19b) for $\vec{F_s}(t)$ (which is repeated below),

$$\vec{F_s}(t) = \frac{N_s}{2} \vec{i_s}(t)$$

$$\vec{H_s}(t) = \frac{N_s}{2\ell_g} \vec{i_s}(t) \quad \text{(rotor – circuit electrically open – circuited)}$$

$$\vec{B_s}(t) = \frac{\mu_o N_s}{2\ell_g} \vec{i_s}(t) \qquad (6\text{-}21a)$$

The relationships in Eq. (6-21a) show that these stator space vectors (with the rotor circuit electrically open-circuited) are collinear (that is, they point in the same direction) at any instant of time. Equation (6-21) also yields the relationship between the peak values as

SPACE-VECTOR REPRESENTATION

$$\hat{F}_s = \frac{N_s}{2}\hat{I}_s$$

$$\hat{H}_s = \frac{N_s}{2\ell_g}\hat{I}_s \quad \text{(rotor – circuit electrically open – circuited)} \quad (6\text{-}21b)$$

$$\hat{B}_s = \frac{\mu_o N_s}{2\ell_g}\hat{I}_s$$

EXAMPLE 6-5

For the conditions in an ac machine in Example 6-4 at a given time t, calculate $\vec{i}_s(t)$. Show the equivalent winding and the current necessary to produce the same mmf distribution as the three-phase windings combined.

Solution

In Example 6-4, $i_a = 10$ A, $i_b = -10$ A, and $i_c = 0$ A. Therefore, from Eq. (6-17),

$$\vec{i}_s = i_a\angle 0° + i_b\angle 120° + i_c\angle 240° = 10 + (-10)\angle 120° + (0)\angle 240°$$
$$= 17.32\angle -30° \text{ A.}$$

The space vector $\vec{i}_s$ is shown in Fig. 6-12a. Since the $\vec{i}_s$ vector is oriented at $\theta = -30°$, with respect to the phase-a magnetic axis, the equivalent sinusoidally distributed stator winding has its magnetic axis at an angle of $-30°$, with respect to the phase-a winding, as shown in Fig. 6-12b. The current required in the equivalent stator winding to produce the equivalent mmf distribution is the peak current $\hat{I}_s = 17.32$ A.

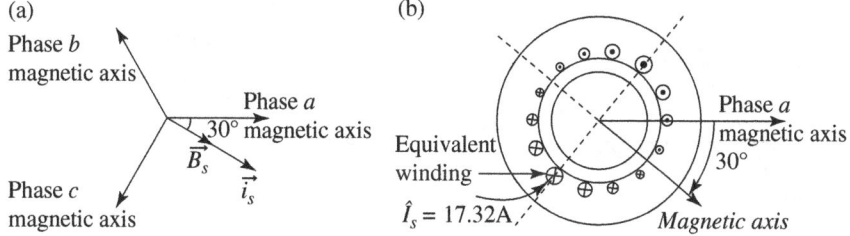

Fig. 6-12 (a) Stator current space vector and (b) the equivalent winding.

6-4-2 Phase Components of Space Vectors $\vec{i_s}(t)$ and $\vec{v_s}(t)$

If the three stator windings in Fig. 6-13a are connected in a wye arrangement, the sum of their currents is zero at any instant of time t by Kirchhoff's Current Law: $i_a(t) + i_b(t) + i_c(t) = 0$.

Therefore, as shown in Fig. 6-13b, at any time t, a space vector is constructed from a unique set of phase components, which can be obtained by multiplying the projection of the space vector along the three axes by 2/3. (We should note that if the phase currents were not required to add up to zero, there would be an infinite number of phase component combinations.)

This graphical procedure is based on the mathematical derivations described below. First, let us consider the relationship

$$1 \angle \theta = e^{j\theta} = \cos\theta + j\sin\theta \qquad (6\text{-}22)$$

The real part in the above equation is

$$\text{Re}(1 \angle \theta) = \cos\theta \qquad (6\text{-}23)$$

Therefore, mathematically, we can obtain the phase components of a space vector, such as $\vec{i_s}(t)$ as follows: multiply both sides of the $\vec{i_s}(t)$ expression in Eq. (6-17) by $1 \angle 0°$, $1 \angle -120°$, and $1 \angle -240°$,

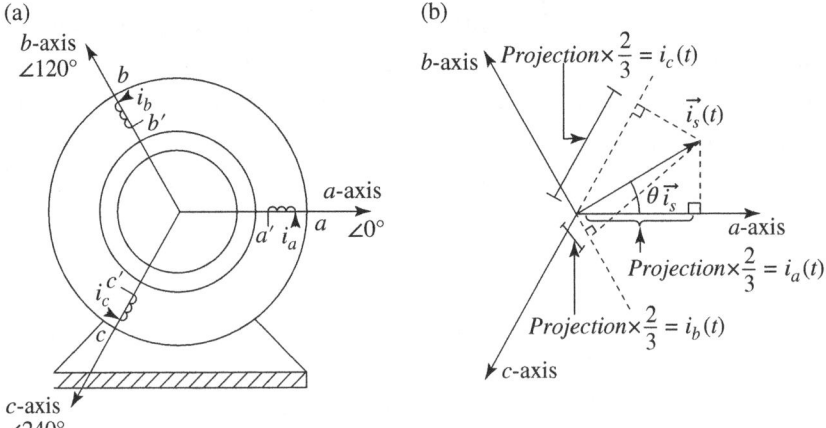

Fig. 6-13 Phase components of a space vector.

respectively. Equate the real parts on both sides and use the condition that $i_a(t) + i_b(t) + i_c(t) = 0$.

To obtain i_a : $\mathrm{Re}\left[\vec{i_s}\angle 0°\right] = i_a + \underbrace{\mathrm{Re}\,[i_b \angle 120°]}_{-\frac{1}{2}i_b} + \underbrace{\mathrm{Re}\,[i_c \angle 240°]}_{-\frac{1}{2}i_c} = \frac{3}{2}i_a s$

Therefore,

$$i_a = \frac{2}{3}\mathrm{Re}\left[\vec{i_s}\angle 0°\right] = \frac{2}{3}\mathrm{Re}\left[\hat{I}_s \angle \theta_{i_s}\right] = \frac{2}{3}\hat{I}_s \cos\theta_{i_s} \quad (6\text{-}24a)$$

To obtain i_b : $\mathrm{Re}\left[\vec{i_s}\angle -120°\right] = \underbrace{\mathrm{Re}\,[i_a \angle -120°]}_{-\frac{1}{2}i_a} + i_b + \underbrace{\mathrm{Re}\,[i_c \angle 120°]}_{-\frac{1}{2}i_c} = \frac{3}{2}i_b$

Therefore,

$$i_b = \frac{2}{3}\mathrm{Re}\left[\vec{i_s}\angle -120°\right] = \frac{2}{3}\mathrm{Re}\left[\hat{I}_s \angle(\theta_{i_s} -120°)\right] = \frac{2}{3}\hat{I}_s \cos(\theta_{i_s} -120°)$$
$$(6\text{-}24b)$$

To obtain i_c : $\mathrm{Re}\left[\vec{i_s}\angle -240°\right] = \underbrace{\mathrm{Re}\,[i_a \angle -240°]}_{-\frac{1}{2}i_a} + \underbrace{\mathrm{Re}\,[i_b \angle -120°]}_{-\frac{1}{2}i_b} + i_c = \frac{3}{2}i_c$

Therefore,

$$i_c = \frac{2}{3}\mathrm{Re}\left[\vec{i_s}\angle -240°\right] = \frac{2}{3}\mathrm{Re}\left[\hat{I}_s \angle(\theta_{i_s} -240°)\right] = \frac{2}{3}\hat{I}_s \cos(\theta_{i_s} -240°)$$
$$(6\text{-}24c)$$

Since $i_a(t) + i_b(t) + i_c(t) = 0$, it can be shown that the same uniqueness applies to components of all space vectors such as $\vec{v_s}(t)$, $\vec{B_s}(t)$, and so on for both the stator and the rotor.

EXAMPLE 6-6

In an ac machine at a given time, the stator voltage space vector is given as $\vec{v_s} = 254.56 \angle 30°$ V. Calculate the phase voltage components at this time.

Solution

From Eq. (6-24),

$$v_a = \frac{2}{3} \text{Re}\{\vec{v_s} \angle 0°\} = \frac{2}{3} \text{Re}\{254.56 \angle 30°\} = \frac{2}{3} \times 254.56 \cos 30° = 146.97 \text{ V},$$

$$v_b = \frac{2}{3} \text{Re}\{\vec{v_s} \angle -120°\} = \frac{2}{3} \text{Re}\{254.56 \angle -90°\}$$
$$= \frac{2}{3} \times 254.56 \cos(-90°) = 0 \text{ V, and}$$

$$v_c = \frac{2}{3} \text{Re}\{\vec{v_s} \angle -240°\} = \frac{2}{3} \text{Re}\{254.56 \angle -210°\} = \frac{2}{3} \times 254.56 \cos(-210°)$$
$$= -146.97 \text{ V}$$

6-5 BALANCED SINUSOIDAL STEADY-STATE EXCITATION (ROTOR OPEN-CIRCUITED)

So far, our discussion has been in very general terms where voltages and currents are not restricted to any specific form. However, we are mainly interested in the normal mode of operation, that is, balanced three-phase, sinusoidal steady-state conditions. Therefore, we will assume that a balanced set of sinusoidal voltages at a frequency $f\left(=\frac{\omega}{2\pi}\right)$ in steady state is applied to the stator, with the rotor assumed to be open-circuited. We will initially neglect the stator winding resistances R_s and the leakage inductances $L_{\ell s}$.

In steady state, applying voltages to the windings in Fig. 6-14a (under rotor open-circuit condition) results in magnetizing currents.

BALANCED SINUSOIDAL STEADY-STATE   187

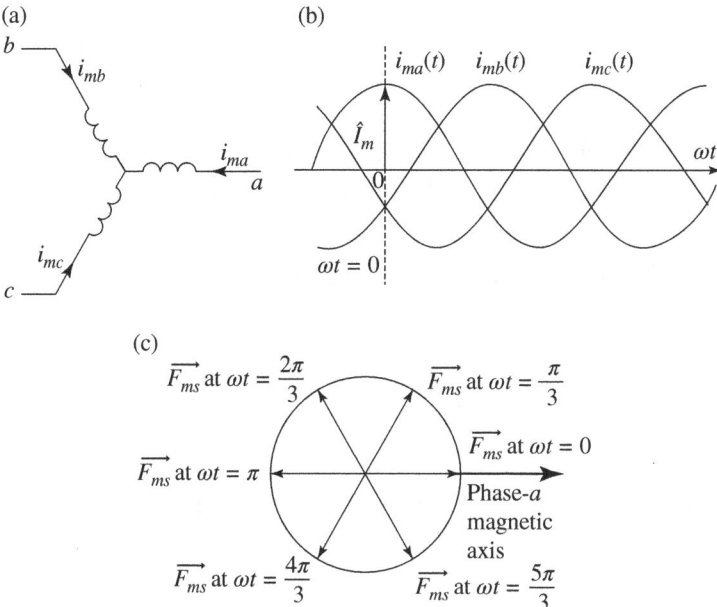

Fig. 6-14 (a) Windings, (b) magnetizing currents, and (c) rotating mmf space vector.

These magnetizing currents are indicated by adding "m" to the subscripts in the following equation and are plotted in Fig. 6-14b:

$$i_{ma} = \hat{I}_m \cos \omega t$$
$$i_{mb} = \hat{I}_m \cos (\omega t - 2\pi/3) \quad (6\text{-}25)$$
$$i_{mc} = \hat{I}_m \cos (\omega t - 4\pi/3)$$

where $\hat{I}_m$ is the peak value of the magnetizing currents and the time origin is chosen to be at the positive peak of $i_{ma}(t)$.

6-5-1 Rotating Stator MMF Space Vector

Substituting into Eq. (6-17) the expressions in Eq. (6-25), for the magnetizing currents varying sinusoidally with time, the stator magnetizing current space vector is

$$\overrightarrow{i_{ms}}(t) = \hat{I}_m [\cos \omega t \angle 0° + \cos(\omega t - 2\pi/3) \angle 120° + \cos(\omega t - 4\pi/3) \angle 240°] \quad (6\text{-}26)$$

The expression within the square bracket in Eq. (6-26) simplifies to $\frac{3}{2} \angle \omega t$ (see Problem 6-8) and Eq. (6-26) becomes

$$\overrightarrow{i_{ms}}(t) = \underbrace{\frac{3}{2} \hat{I}_m}_{\hat{I}_{ms}} \angle \omega t = \hat{I}_{ms} \angle \omega t \text{ where } \hat{I}_{ms} = \frac{3}{2} \hat{I}_m \quad (6\text{-}27)$$

From Eq. (6-21a),

$$\overrightarrow{F_{ms}}(t) = \frac{N_s}{2} \overrightarrow{i_{ms}}(t) = \hat{F}_{ms} \angle \omega t \text{ where } \hat{F}_{ms} = \frac{N_s}{2} \hat{I}_{ms} = \frac{3}{2} \frac{N_s}{2} \hat{I}_m \quad (6\text{-}28)$$

Similarly, using Eq. (6-21a) again,

$$\overrightarrow{B_{ms}}(t) = \left(\frac{\mu_o N_s}{2\ell_g}\right) \overrightarrow{i_{ms}}(t) \text{ where } \hat{B}_{ms} = \left(\frac{\mu_o N_s}{2\ell_g}\right) \hat{I}_{ms} = \frac{3}{2}\left(\frac{\mu_o N_s}{2\ell_g}\right) \hat{I}_m$$
$$(6\text{-}29)$$

Note that if the peak flux density $\hat{B}_{ms}$ in the air gap is to be at its rated value in Eq. (6-29), then the $\hat{I}_{ms}$, and hence the peak value of the magnetizing current $\hat{I}_m$, in each phase must also be at their rated values.

Under sinusoidal steady-state conditions, the stator-current, the stator-mmf, and the air gap flux-density space vectors have constant amplitudes ($\hat{I}_{ms}$, $\hat{F}_{ms}$, and $\hat{B}_{ms}$). As shown by $\overrightarrow{F_{ms}}(t)$ in Fig. 6-14c, all of these space vectors rotate with time at a constant speed, called the synchronous speed ω_{syn}, in the counterclockwise direction, which in a 2-pole machine is equal to the frequency ω ($=2\pi f$) of the voltages and currents applied to the stator:

$$\omega_{syn} = \omega \quad (p = 2) \quad (6\text{-}30)$$

BALANCED SINUSOIDAL STEADY-STATE   189

EXAMPLE 6-7

With the rotor electrically open-circuited in a 2-pole ac machine, voltages are applied to the stator and result in the magnetizing currents plotted in Fig. 6-15a. Sketch the direction of the flux lines at the instants $\omega t = 0°, 60°, 120°, 180°, 240°$, and $300°$. Show that one electrical cycle results in the rotation of the flux orientation by one revolution, in accordance with Eq. (6-30) for a 2-pole machine.

Solution

At $\omega t = 0$, $i_{ma} = \hat{I}_m$, and $i_{mb} = i_{mc} = -(1/2)\hat{I}_m$. The current directions for the three windings are indicated in Fig. 6-15b, where the circles for phase-a are shown larger, due to twice as much current in them compared to the other two phases. The resulting flux orientation is shown as well. A similar procedure is followed at other instants, as shown in Fig. 6-15c through g. These drawings clearly show that in a 2-pole machine, the electrical excitation through one cycle of the electrical frequency $f(=\omega/2\pi)$ results in the rotation of the flux orientation, and hence of the space vector $\vec{B}_{ms}$, by one revolution in space. Therefore, $\omega_{syn} = \omega$, as expressed in Eq. (6-30).

6-5-2 Rotating Stator MMF Space Vector in Multipole Machines

In the previous section, we considered a 2-pole machine. In general, in a p-pole machine, a balanced sinusoidal steady state, with currents and voltages at a frequency $f(=\omega/2\pi)$, results in an mmf space vector that rotates at a speed

$$\omega_{syn} = \frac{\omega}{p/2} \left(\frac{p}{2} = \text{pole-pairs}\right) \qquad (6\text{-}31)$$

This can be illustrated by considering a p-pole machine and repeating the procedure outlined in Example 6-7 for a 2-pole machine (this is left as Problem 6-11).

190 USING SPACE VECTORS TO ANALYZE ac MACHINES

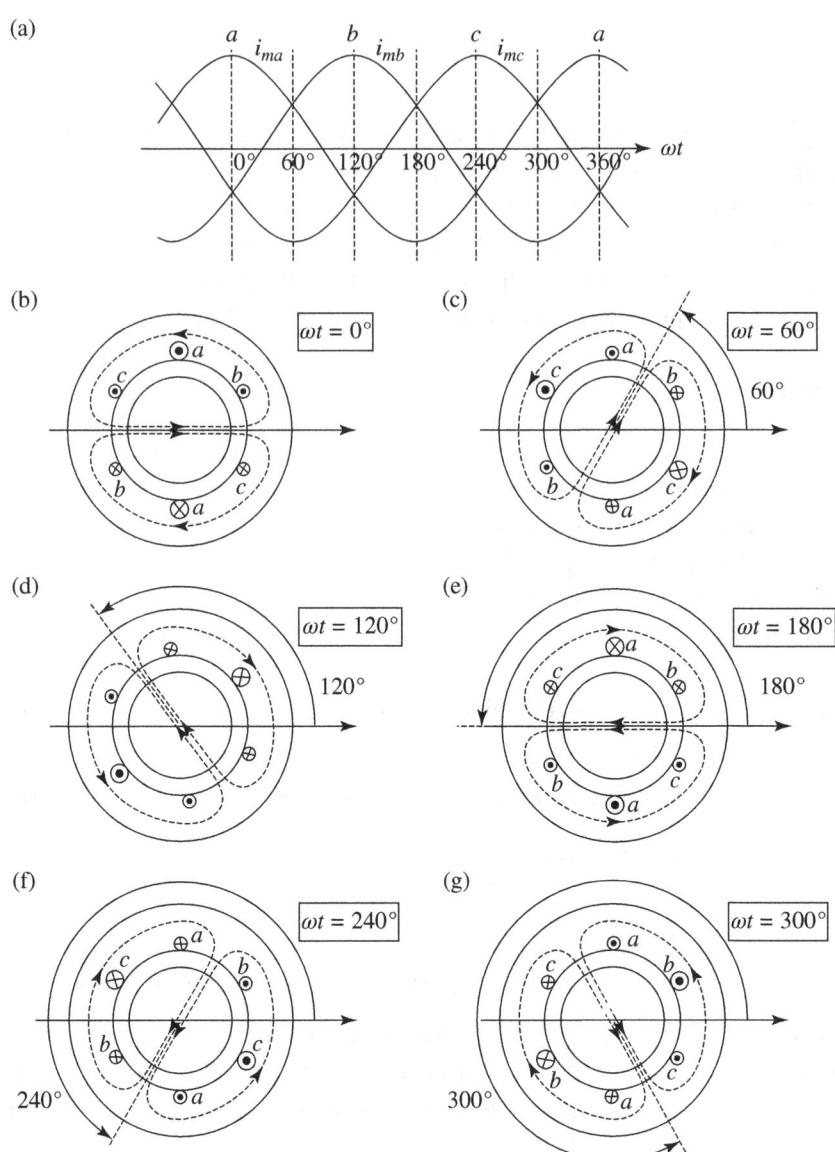

Fig. 6-15 Example 6-7.

In the space vectors for multipole machines, the three magnetic axes can be drawn as in a 2-pole machine (similar to the space vector diagrams of Fig. 6-9b or 6-13b, for example), except now the axes are separated by 120° (electrical), where the electrical angles are defined

BALANCED SINUSOIDAL STEADY-STATE   191

by Eq. (6-10). Therefore, one complete cycle of electrical excitation causes the space vector, at the synchronous speed given in Eq. (6-31), to rotate by 360° (electrical); that is, in the space vector diagram, the space vector returns to the position that it started from. This corresponds to a rotation by an angle of 360/(p/2) mechanical degrees, which is exactly what happens within the machine. However, in general (special situations will be pointed out), since no additional insight is gained by this multipole representation, it is best to analyze a multipole machine as if it were a 2-pole machine.

6-5-3 The Relationship Between Space Vectors and Phasors in Balanced Three-Phase Sinusoidal Steady State ($\vec{v_s}|_{t=0} \Leftrightarrow \overline{V}_a$ and $\vec{i_{ms}}|_{t=0} \Leftrightarrow \overline{I}_{ma}$)

In Fig. 6-14b, note that at $\omega t = 0$, the magnetizing current i_{ma} in phase-a, is at its positive peak. Corresponding to this time $\omega t = 0$, the space vectors $\vec{i_{ms}}$, $\vec{F_{ms}}$, and $\vec{B_{ms}}$ are along the a-axis in Fig. 6-14c. Similarly, at $\omega t = 2\pi/3$ rad or $120°$, i_{mb} in phase-b reaches its positive peak. Correspondingly, the space vectors $\vec{i_{ms}}$, $\vec{F_{ms}}$, and $\vec{B_{ms}}$ are along the b-axis, 120° ahead of the a-axis. Therefore, we can conclude that under a balanced three-phase sinusoidal steady state, when a phase voltage (or a phase current) is at its positive peak, the combined stator voltage (or current) space vector will be oriented along that phase axis. This can also be stated as follows: when a combined stator voltage (or current) space vector is oriented along the magnetic axis of any phase, at that time, that phase voltage (or current) is at its positive peak value.

We will make use of the information in the above paragraph. Under a balanced three-phase sinusoidal steady state, let us arbitrarily choose some time as the origin $t = 0$ in Fig. 6-16a such that the current i_{ma} reaches its positive peak at a later time $\omega t = \alpha$. The phase-a current can be expressed as

$$i_{ma}(t) = \hat{I}_m \cos(\omega t - \alpha) \qquad (6\text{-}32)$$

which is represented by a phasor below and shown in the phasor diagram of Fig. 6-16b:

192 USING SPACE VECTORS TO ANALYZE ac MACHINES

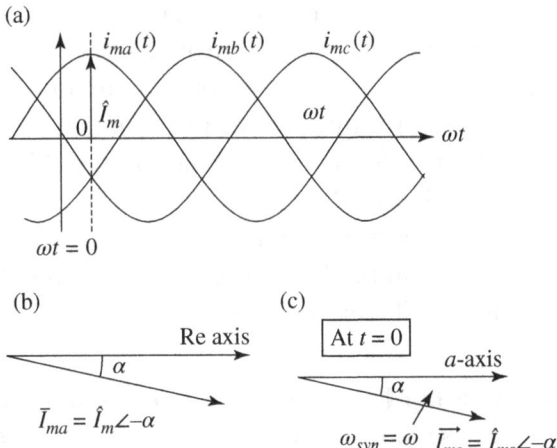

Fig. 6-16 Relationship between space vectors and phasors in balanced sinusoidal steady state.

$$\bar{I}_{ma} = \hat{I}_m \angle -\alpha \qquad (6\text{-}33a)$$

The phase-a current $i_{ma}(t)$ reaches its positive peak at $\omega t = \alpha$. Therefore, at the time $t = 0$, the $\vec{i}_{ms}$ space vector will be as shown in Fig. 6-16c, behind the magnetic axis of phase-a by an angle α, so that it will be along the a-axis at a later time $\omega t = \alpha$, when i_{ma} reaches its positive peak. Therefore, at the time $t = 0$,

$$\left. \vec{i}_{ms} \right|_{t=0} = \hat{I}_{ms} \angle -\alpha \text{ where } \hat{I}_{ms} = \frac{3}{2}\hat{I}_m \qquad (6\text{-}33b)$$

Combining Eqs. (6-33a) and (6-33b),

$$\left. \vec{i}_{ms} \right|_{t=0} = \frac{3}{2}\bar{I}_{ma} \qquad (6\text{-}34)$$

where the left side mathematically represents the combined current space vector at the time $t = 0$, and on the right side $\bar{I}_{ma}$ is the phase-a current phasor representation. In sinusoidal steady state, Eq. (6-34) illustrates an important relationship between space vectors and phasors which we will use very often:

BALANCED SINUSOIDAL STEADY-STATE   193

1. The orientation of the phase-a voltage (or current) phasor is the same as the orientation of the combined stator voltage (or current) space vector at a time $t = 0$.
2. The amplitude of the combined stator voltage (or current) space vector is larger than that of the phasor amplitude by a factor of 3/2.

Note that knowing the phasors for phase-a is sufficient, as the other phase quantities are displaced by 120° with respect to each other and have equal magnitudes. This concept will be used in the following section.

6-5-4   Induced Voltages in Stator Windings

In the following discussion, we will ignore the resistance and the leakage inductance of the stator windings, shown wye-connected in Fig. 6-17a. Neglecting all losses, under the condition that there is no electrical circuit or excitation in the rotor, the stator windings appear purely inductive. Therefore, in each phase, the phase voltage and the magnetizing current are related as

$$e_{ma} = L_m \frac{di_{ma}}{dt}, e_{mb} = L_m \frac{di_{mb}}{dt}, \text{ and } e_{mc} = L_m \frac{di_{mc}}{dt} \quad (6\text{-}35)$$

where L_m is the magnetizing inductance of the three-phase stator, which in terms of the machine parameters can be calculated as (see Problems 6-13 and 6-14)

$$L_m = \frac{3}{2} \left[\frac{\pi \mu_o r \ell}{\ell_g} \left(\frac{N_s}{2} \right)^2 \right] \quad (6\text{-}36)$$

where r is the radius, ℓ is the rotor length, and ℓ_g is the air gap length. The combination of quantities within the square bracket is the single-phase self-inductance $L_{m,1\text{-}phase}$ of each of the stator phase windings in a 2-pole machine:

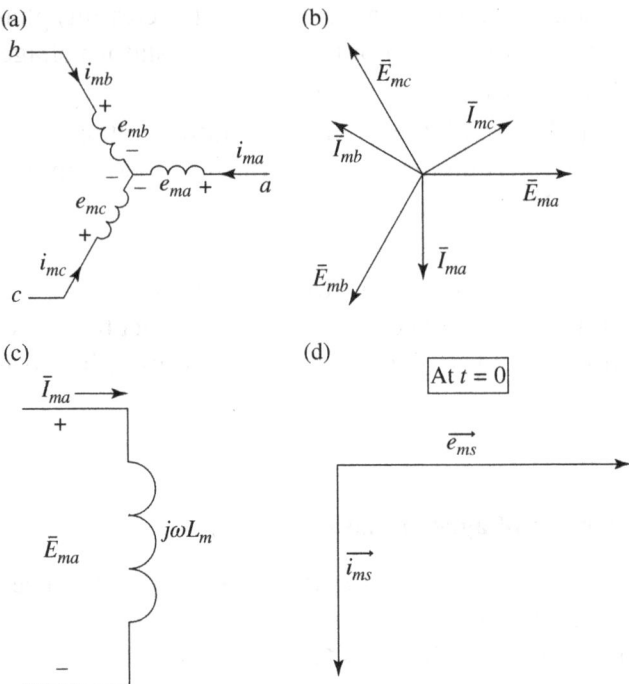

Fig. 6-17 Winding current and induced emf: (a) individual windings, (b) phasors, (c) per-phase equivalent circuit, and (d) space vectors.

$$L_{m,\text{1-phase}} = \frac{\pi \mu_o r \ell}{\ell_g} \left(\frac{N_s}{2}\right)^2 \quad (6\text{-}37)$$

Due to mutual coupling between the three phases, L_m given in Eq. (6-36) is larger than $L_{m,\text{1-phase}}$ by a factor of 3/2:

$$L_m = \frac{3}{2} L_{m,\text{1-phase}} \quad (6\text{-}38)$$

Under a balanced sinusoidal steady state, assuming that i_{ma} peaks at $\omega t = 90°$, we can draw the three-phase phasor diagram shown in Fig. 6-17b, where

$$\overline{E}_{ma} = (j\omega L_m)\overline{I}_{ma} \quad (6\text{-}39)$$

The phasor-domain circuit diagram for phase-a is shown in Fig. 6-17c, and the corresponding combined space vector diagram for $\overrightarrow{e_{ms}}$ and $\overrightarrow{i_{ms}}$ at $t = 0$ is shown in Fig. 6-17d. In general, at any time t,

BALANCED SINUSOIDAL STEADY-STATE 195

$$\overrightarrow{e_{ms}}(t) = (j\omega L_m)\overrightarrow{i_{ms}}(t) \quad (6\text{-}40)$$

where $\hat{E}_{ms} = (\omega L_m)\hat{I}_{ms} = \frac{3}{2}(\omega L_m)\hat{I}_m$.

In Eq. (6-40), substituting for $\overrightarrow{i_{ms}}(t)$ in terms of $\overrightarrow{B_{ms}}(t)$ from Eq. (6-21a) and substituting for L_m from Eq. (6-36),

$$\overrightarrow{e_{ms}}(t) = j\omega\left(\frac{3}{2}\pi r\ell \frac{N_s}{2}\right)\overrightarrow{B_{ms}}(t) \quad (6\text{-}41)$$

Equation (6-41) shows an important relationship: the induced voltages in the stator windings can be interpreted as back-emfs induced by the rotating flux-density distribution. This flux-density distribution, represented by $\overrightarrow{B_{ms}}(t)$, is rotating at a speed ω_{syn} (which equals ω in a 2-pole machine) and is "cutting" the stationary conductors of the stator phase windings. A similar expression can be derived for a multipole machine with $p > 2$ (see Problem 6-17).

EXAMPLE 6-8

In a 2-pole machine in a balanced sinusoidal steady state, the applied voltages are 208 V (L-L, rms) at a frequency of 60 Hz. Assume the phase-a voltage to be the reference phasor. The magnetizing inductance $L_m = 55$ mH. Neglect the stator winding resistances and leakage inductances and assume the rotor to be electrically open-circuited. (a) Calculate and draw the $\overline{E}_{ma}$ and $\overline{I}_{ma}$ phasors. (b) Calculate and draw the space vectors $\overrightarrow{e_{ms}}$ and $\overrightarrow{i_{ms}}$ at $\omega t = 0°$ and $\omega t = 60°$. (c) If the peak flux density in the air gap is 1.1 T, draw the $\overrightarrow{B_{ms}}$ space vector in part (b) at the two instants of time.

Solution

(a) With the phase-a voltage as the reference phasor,

$$\overline{E}_{ma} = \frac{208\sqrt{2}}{\sqrt{3}} \angle 0° = 169.83 \angle 0° \text{ V}$$

and

$$\overline{I}_{ma} = \frac{\overline{E}_{ma}}{j\omega L_m} \angle 0° = \frac{169.83}{2\pi \times 60 \times 55 \times 10^{-3}} \angle -90° = 8.19 \angle -90° \text{ A}$$

(*Continued*)

These two phasors are drawn in Fig. 6-18a.

(b) At $\omega t = 0°$, from Eq. (6-34), as shown in Fig. 6-18b,

$$\vec{e_{ms}}\big|_{\omega t = o} = \frac{3}{2}\overline{E}_{ma} = \frac{3}{2}169.83\angle 0° = 254.74\angle 0° \text{ V}$$

and

$$\vec{i_{ms}}\big|_{\omega t = 0} = \frac{3}{2}\overline{I}_{ma} = \frac{3}{2}8.19\angle -90° = 12.28\angle -90° \text{ A}$$

(c) At $\omega t = 60°$, both space vectors have rotated by an angle of 60° in a counterclockwise direction, as shown in Fig. 6-18b. Therefore,

$$\vec{e_{ms}}\big|_{\omega t = 60°} = \vec{e_{ms}}\big|_{\omega t = 0}(1\angle 60°) = 254.74\angle 60° \text{ V}$$

and

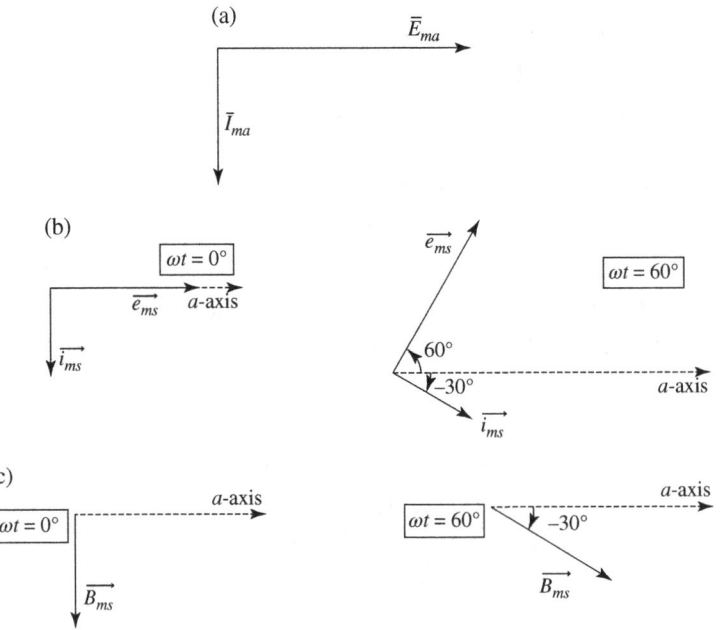

Fig. 6-18 Example 6.8.

$$\left.\vec{i_{ms}}\right|_{\omega t = 60°} = \left.\vec{i_{ms}}\right|_{\omega t = 0}(1\angle 60°) = 12.28\angle -30° \text{ A}$$

In this example, at any time, the stator flux density space vector $\vec{B_{ms}}$ is oriented in the same direction as the $\vec{i_{ms}}$ space vector. Therefore, as plotted in Fig. 6-18c,

$$\left.\vec{B_{ms}}\right|_{\omega t = 0°} = 1.1\angle -90° \text{ T and } \left.\vec{B_{ms}}\right|_{\omega t = 60°} = 1.1\angle -30° \text{ T}$$

6-6 REVIEW QUESTIONS

1. Draw the three-phase axis in the motor cross-section. Also, draw the three phasors $\overline{V}_a$, $\overline{V}_b$, and $\overline{V}_c$ in a balanced sinusoidal steady state. Why is the phase-b axis ahead of the phase-a axis by 120°, but $\overline{V}_b$ lags $\overline{V}_a$ by 120°?

2. Ideally, what should be the field (F, H, and B) distributions produced by each of the three stator windings? What is the direction of this field in the air gap? What direction is considered positive, and what is considered negative?

3. What should the conductor-density distribution in a winding be in order to achieve the desired field distribution in the air gap? Express the conductor-density distribution $n_s(\theta)$ for phase-a.

4. How is a sinusoidal distribution of conductor density in a phase winding approximated in practical machines with only a few slots available to each phase?

5. How are the three field distributions (F, H, and B) related to each other, assuming that there is no magnetic saturation in the stator and the rotor iron?

6. What is the significance of the magnetic axis of any phase winding?

7. Mathematically express the field distributions in the air gap due to i_a as a function of θ. Repeat this for i_b and i_c.

8. What do the phasors $\overline{V}$ and $\overline{I}$ denote? What are the meanings of the space vectors $\vec{B_a}(t)$ and $\vec{B_s}(t)$ at a time t, assuming that the rotor circuit is electrically open-circuited?

9. What is the constraint on the sum of the stator currents?

10. What are physical interpretations of various stator winding inductances?

11. Why is the per-phase inductance L_m greater than the single-phase inductance $L_{m,\,1\text{-}phase}$ by a factor of 3/2?

12. What are the characteristics of space vectors which represent the field distributions $F_s(\theta)$, $H_s(\theta)$, and $B_s(\theta)$ at a given time? What notations are used for these space vectors? Which axis is used as a reference to express them mathematically in this chapter?

13. Why does a dc current through a phase winding produce a sinusoidal flux-density distribution in the air gap?

14. How are the terminal phase voltages and currents combined for representation by space vectors?

15. What is the physical interpretation of the stator current space vector $\vec{i_s}(t)$?

16. With no excitation or currents in the rotor, are all of the space vectors associated with the stator $\vec{i_{ms}}(t), \vec{F_{ms}}(t)$, and $\vec{B_{ms}}(t)$ collinear (oriented in the same direction)?

17. In ac machines, a stator space vector $\vec{v_s}(t)$ or $\vec{i_s}(t)$ consists of a unique set of phase components. What is the condition on which these components are based?

18. Express the phase voltage components in terms of the stator voltage space vector.

19. Under three-phase balanced sinusoidal condition with no rotor currents, and neglecting the stator winding resistances R_s and the leakage inductance $L_{\ell s}$ for simplification, answer the following questions: (a) What is the speed at which all of the space vectors rotate? (b) How is the peak flux density related to the magnetizing currents? Does this relationship depend on the frequency f of the excitation? If the peak flux density is at its rated value, then what about the peak value of the magnetizing currents? (c) How do the magnitudes of the applied voltages depend on the frequency of excitation, in order to keep the flux density constant (at its rated value, for example)?

20. What is the relationship between space vectors and phasors under balanced sinusoidal operating conditions?

REFERENCES

1. A. E. Fitzgerald, C. Kingsley, and S. D. Umans, *Electric Machinery*, 5th edition, McGraw Hill, 1990.
2. Slemon, G.R. (1992). *Electric Machines, and Drives*. Addison-Wesley Publishing Company, Inc.

FURTHER READING

Kovacs, P.K. (1984). *Transient Phenomena in Electrical Machines*. Elsevier.

PROBLEMS

6-1 In a three-phase, 2-pole ac machine, assume that the neutral of the wye-connected stator windings is accessible. The rotor is electrically open-circuited. The phase-a is applied a current $i_a(t) = 10 \sin \omega t$. Calculate $\vec{B_a}$ at the following instants of ωt: $0°, 90°, 135°,$ and $210°$. Also, plot the $B_a(\theta)$ distribution at these instants.

6-2 In the sinusoidal conductor-density distribution shown in Fig. 6-3, make use of the symmetry at θ and at $(\pi - \theta)$ to calculate the field distribution $H_a(\theta)$ in the air gap.

6-3 In ac machines, why is the stator winding for phase-b placed 120° ahead of phase-a (as shown in Fig. 6-1), whereas the phasors for phase-b (such as $\overline{V}_b$) lag behind the corresponding phasors for phase-a?

6-4 In Example 6-2, derive the expression for $n_s(\theta_e)$ for a 4-pole machine. Generalize it for a multipole machine.

6-5 In Example 6-2, obtain the expressions for $H_a(\theta_e)$, $B_a(\theta_e)$, and $F_a(\theta_e)$.

6-6 In a 2-pole, three-phase machine with $N_s = 100$, calculate $\vec{i_s}$ and $\vec{F_s}$ at a time t if at that time the stator currents are as follows: (a) $i_a = 10$ A, $i_b = -5$ A, and $i_c = -5$ A; (b) $i_a = -5$ A, $i_b = 10$ A, and $i_c = -5$ A; (c) $i_a = -5$ A, $i_b = -5$ A, and $i_c = 10$ A.

6-7 In a wye-connected stator, at a time t, $\vec{v_s} = 150\angle -30°$ V. Calculate v_a, v_b, and v_c at that time.

6-8 Show that the expression in the square brackets of Eq. (6-26) simplifies to $\frac{3}{2}\angle\omega t$.

6-9 In a 2-pole, three-phase ac machine, $\ell_g = 1.5$ mm and $N_s = 100$. During a balanced, sinusoidal, 60-Hz steady state with the rotor electrically open-circuited, the peak of the magnetizing current in each phase is 10 A. Assume that at $t = 0$, the phase-a current is at its positive peak. Calculate the flux-density distribution space vector as a function of time. What is the speed of its rotation?

6-10 In Problem 6-9, what would be the speed of rotation if the machine had 6 poles?

6-11 By means of drawings similar to Example 6-7, show the rotation of the flux lines, and hence the speed, in a 4-pole machine.

6-12 In a three-phase ac machine, $\overline{V}_a = 120\sqrt{2}\angle 0°$ V and $\overline{I}_{ma} = 5\sqrt{2}\angle -90°$ A. Calculate and draw $\overrightarrow{e_{ms}}$ and $\overrightarrow{i_{ms}}$ space vectors at $t = 0$. Assume a balanced, sinusoidal, three-phase steady-state operation at 60 Hz. Neglect the resistance and the leakage inductance of the stator phase windings.

6-13 Show that in a 2-pole machine, $L_{m,1\text{-}phase} = \frac{\pi\mu_o r \ell}{\ell_g}\left(\frac{N_s}{2}\right)^2$.

6-14 Show that $L_m = \frac{3}{2}L_{m,1\text{-}phase}$.

6-15 In a three-phase ac machine, $\overline{V}_a = 120\sqrt{2}\angle 0°$ V. The magnetizing inductance $L_m = 75$ mH. Calculate and draw the three magnetizing current phasors. Assume a balanced, sinusoidal, three-phase steady-state operation at 60 Hz.

6-16 In a 2-pole, three-phase ac machine, $\ell_g = 1.5$ mm, $\ell = 24$ cm, $r = 6$ cm, and $N_s = 100$. Under a balanced, sinusoidal, 60-Hz steady state, the peak of the magnetizing current in each phase is 10 A. Assume that at $t = 0$, the current in phase-a is at its positive peak. Calculate the expressions for the induced back-emfs in the three-stator phases.

6-17 Recalculate Eq. (6-41) for a multipole machine with $p > 2$.

6-18 Calculate L_m in a p-pole machine ($p \geq 2$).

6-19 Combine the results of Problems 6-17 and 6-18 to show that for $p \geq 2$, $\overrightarrow{e}_{ms}(t) = j\omega L_m \overrightarrow{i}_{ms}(t)$.

6-20 In Fig. 6-15 of Example 6-7, plot the flux-density distributions produced by each of the phases in parts (b) through (g).

6-21 At some instant of time, $\overrightarrow{B_s}(t) = 1.1\angle 30°$ T. Calculate and plot the flux-density distribution produced by each of the phases as a function of θ.

6-22 In Eq. (6-41), the expression for $\vec{e}_{ms}(t)$ is obtained by using the expression of inductance in Eq. (6-36). Instead of following this procedure, calculate the voltages induced in each of the stator phases due to the rotating $\vec{B}_{ms}$ to confirm the expression of $\vec{e}_{ms}(t)$ in Eq. (6-41).

7 Space Vector Pulse-Width-Modulated (SV-PWM) Inverters

7-1 INTRODUCTION

In Chapter 4, we discussed sinusoidal-PWM to generate three-phase sinusoidal voltages. These desired voltages can also be supplied to the motor, by calculating and then applying appropriate voltages, which can be generated based on the sinusoidal pulse-width-modulation principles discussed in Chapter 4. The availability of digital signal processors in control of electric drives, provides an excellent opportunity to improve upon this sinusoidal pulse-width modulation by a procedure described in this chapter, which is termed space vector pulse-width modulation (SV-PWM). We will simulate such an inverter for use in ac drives.

7-2 SYNTHESIS OF STATOR VOLTAGE SPACE VECTOR $\vec{v}_s^a$

In terms of the instantaneous stator phase voltages, the stator space voltage vector is

$$\vec{v}_s^a(t) = v_a(t)e^{j0} + v_b(t)e^{j2\pi/3} + v_c(t)e^{j4\pi/3} \qquad (7\text{-}1)$$

(Chapter 7 of *Advanced Electric Drives: Analysis, Control, and Modeling Using MATLAB/Simulink* ISBN: 978-1-118-48548-4 by Ned Mohan, August 2014)

Analysis and Control of Electric Drives: Simulations and Laboratory Implementation, First Edition. Ned Mohan and Siddharth Raju.
© 2021 John Wiley & Sons, Inc. Published 2021 by John Wiley & Sons, Inc.
Companion website: www.wiley.com/go/Mohan/Vectorcontrolinelectricdrives

204 SPACE VECTOR PULSE-WIDTH-MODULATED

In the circuit of Fig. 7-1, in terms of the inverter output voltages with respect to the negative dc bus and hypothetically assuming the stator neutral as a reference ground,

$$v_a = v_{aN} + v_N; \quad v_b = v_{bN} + v_N; \quad v_c = v_{cN} + v_N \qquad (7\text{-}2)$$

Substituting Eq. (7-2) into Eq. (7-1) and recognizing that

$$e^{j0} + e^{j2\pi/3} + e^{j4\pi/3} = 0 \qquad (7\text{-}3)$$

the instantaneous stator voltage space vector can be written in terms of the inverter output voltages as

$$\vec{v}_s^a(t) = v_{aN}e^{j0} + v_{bN}e^{j2\pi/3} + v_{cN}e^{j4\pi/3} \qquad (7\text{-}4)$$

A switch in an inverter pole of Fig. 7-1 is in the "up" position if the pole switching function $q = 1$, otherwise in the "down" position if $q = 0$. In terms of the switching functions, the instantaneous voltage space vector can be written as

$$\vec{v}_s^a(t) = V_d\left(q_a e^{j0} + q_b e^{j2\pi/3} + q_c e^{j4\pi/3}\right) \qquad (7\text{-}5)$$

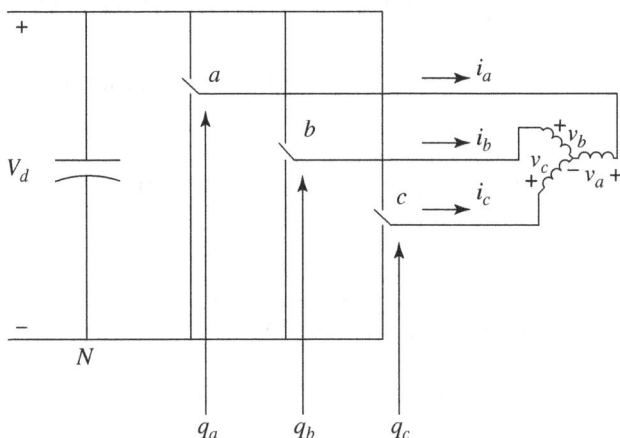

Fig. 7-1 Switch-mode inverter.

SYNTHESIS OF STATOR VOLTAGE 205

With three poles, eight switch-status combinations are possible. In Eq. (7-5), the stator voltage vector $\vec{v}_s^a(t)$ can take on one of the following seven distinct instantaneous values wherein a digital representation, the phase "a" represents the least significant digit, and phase "c" the most significant digit (for example, the resulting voltage vector due to the switch-status combination $\underbrace{011}_{(=3)}$ is represented as $\vec{v}_3$):

$$\begin{aligned}
\vec{v}_s^a(000) &= \vec{v}_0 = 0 \\
\vec{v}_s^a(001) &= \vec{v}_1 = V_d e^{j0} \\
\vec{v}_s^a(010) &= \vec{v}_2 = V_d e^{j2\pi/3} \\
\vec{v}_s^a(011) &= \vec{v}_3 = V_d e^{j\pi/3} \\
\vec{v}_s^a(100) &= \vec{v}_4 = V_d e^{j4\pi/3} \\
\vec{v}_s^a(101) &= \vec{v}_5 = V_d e^{j5\pi/3} \\
\vec{v}_s^a(110) &= \vec{v}_6 = V_d e^{j\pi} \\
\vec{v}_s^a(111) &= \vec{v}_7 = 0
\end{aligned} \qquad (7\text{-}6)$$

In Eq. (7-6), $\vec{v}_0$ and $\vec{v}_7$ are the zero vectors because of their values. The resulting instantaneous stator voltage vectors, which we will call the "basic vectors," are plotted in Fig. 7-2. The basic vectors form six sectors, as shown in Fig. 7-2.

The objective of the PWM control of the inverter switches is to synthesize the desired reference stator voltage space vector in an optimum manner with the following objectives:

- A constant switching frequency f_s.
- Smallest instantaneous deviation from its reference value.
- Maximum utilization of the available dc-bus voltages.
- Lowest ripple in the motor current.
- Minimum switching loss in the inverter.

The above conditions are generally met if the average voltage vector is synthesized by means of the two instantaneous basic nonzero

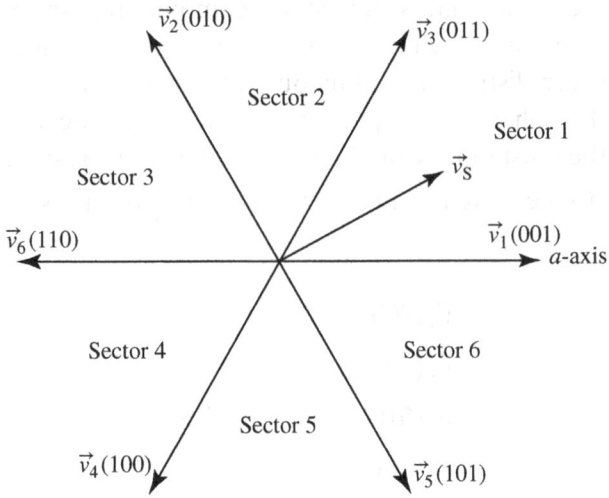

Fig. 7-2 Basic voltage vectors ($\vec{v}_0$ and $\vec{v}_7$ not shown).

voltage vectors that form the sector (in which the average voltage vector to be synthesized lies) and both the zero voltage vectors, such that each transition causes change of only one switch status to minimize the inverter switching loss.

In the following analysis, we will focus on the average voltage vector in Sector 1 with the aim of generalizing the discussion to all sectors. To synthesize an average voltage vector $\vec{v}_s^a \left(= \hat{V}_s e^{j\theta_s} \right)$ over a time period T_s in Fig. 7-3, the adjoining basic vectors $\vec{v}_1$ and $\vec{v}_3$ are applied for intervals xT_s and yT_s, respectively, and the zero vectors $\vec{v}_0$ and $\vec{v}_7$ are applied for a total duration of zT_s. In terms of the basic voltage vectors, the average voltage vector can be expressed as

$$\vec{v}_s^a = \frac{1}{T_s}\left[xT_s\vec{v}_1 + yT_s\vec{v}_3 + zT_s \cdot 0\right] \qquad (7\text{-}7)$$

or

$$\vec{v}_s^a = x\vec{v}_1 + y\vec{v}_3 \qquad (7\text{-}8)$$

where

$$x + y + z = 1 \qquad (7\text{-}9)$$

SYNTHESIS OF STATOR VOLTAGE 207

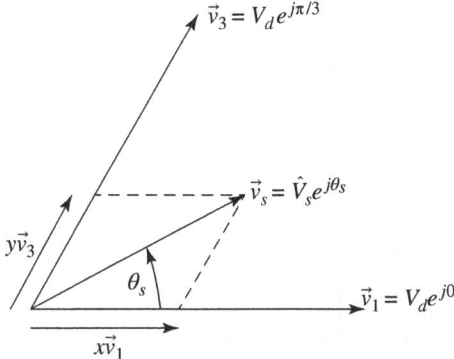

Fig. 7-3 Voltage vector in Sector 1.

In Eq. (7-8), expressing voltage vectors in terms of their amplitude and phase angles results in

$$\hat{V}_s e^{j\theta_s} = xV_d e^{j0} + yV_d e^{j\pi/3} \quad (7\text{-}10)$$

By equating real and imaginary terms on both sides of Eq. (7-10), we can solve for x and y (in terms the given values of $\hat{V}_s$, θ_s, and V_d) to synthesize the desired average space vector in sector 1 (see Problem 7-1).

Having determined the durations for the adjoining basic vectors, and the two zero vectors, the next task is to relate the above discussion to the actual poles (a, b, and c). Note in Fig. 7-2 that in any sector, the adjoining basic vectors differ in one position, for example, in Sector 1 with the basic vectors $\vec{v}_1(001)$ and $\vec{v}_3(011)$, only the pole "b" differs in the switch position. For Sector 1, the switching pattern in Fig. 7-4 shows that pole-a is in "up" position during the sum of xT_s, yT_s, and z_7T_s intervals, and hence for the longest interval of the three poles. Next in the length of duration in the "up" position is pole-b for the sum of yT_s and z_7T_s intervals. The smallest in the length of duration is pole-c for the only z_7T_s interval. Each transition requires a change in switch state in only one of the poles, as shown in Fig. 7-4. Similar switching patterns for the three poles can be generated for any other sector (see Problem 7-2).

208 SPACE VECTOR PULSE-WIDTH-MODULATED

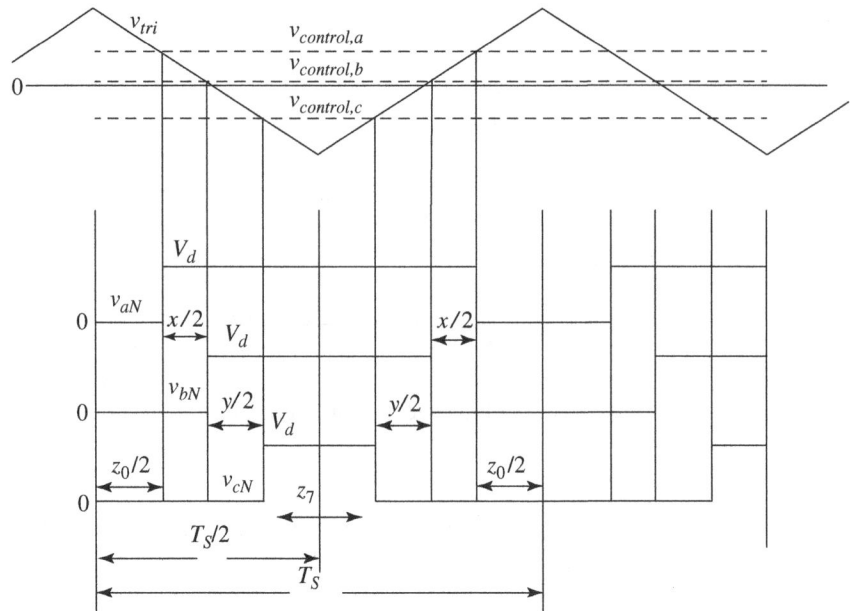

Fig. 7-4 Waveforms in Sector 1; $z = z_0 + z_7$.

7-3 COMPUTER SIMULATION OF SV-PWM INVERTER

In computer simulations, for example, using Simulink, as well as in hardware implementation using rapid prototyping tools, such as from Sciamble™, the above-described pulse-width modulation of the stator voltage space vector can be carried out by comparing control voltages with a triangular waveform signal at the switching-frequency to generate switching functions. It is similar to the sinusoidal PWM approach only to the extent of comparing control voltages with a triangular waveform signal. However, in SV-PWM, the control voltages do not have a purely sinusoidal nature as those in the sinusoidal PWM.

In an induction machine with an isolated neutral, the three-phase voltages sum to zero (see Problem 7-3),

$$v_a(t) + v_b(t) + v_c(t) = 0 \qquad (7\text{-}11)$$

COMPUTER SIMULATION OF SV-PWM INVERTER

To synthesize an average space vector $\vec{v}_s^a$ with phase components v_a, v_b, and v_c (the dc-bus voltage V_d is specified), the control voltages can be written in terms of the phase voltages as follows, expressed as a ratio of $\hat{V}_{tri}$ (the amplitude of the constant switching-frequency triangular signal v_{tri} used for comparison with these control voltages):

$$\frac{v_{control,a}}{\hat{V}_{tri}} = \frac{v_a - v_k}{V_d/2}$$

$$\frac{v_{control,b}}{\hat{V}_{tri}} = \frac{v_b - v_k}{V_d/2} \quad (7\text{-}12)$$

$$\frac{v_{control,c}}{\hat{V}_{tri}} = \frac{v_c - v_k}{V_d/2}$$

where

$$v_k = \frac{max\,(v_a, v_b, v_c) + min\,(v_a, v_b, v_c)}{2} \quad (7\text{-}13)$$

Deriving Eqs. (7-12) and (7-13) are left as a homework problem (Problem 7-5).

EXAMPLE 7-1

In a three-phase inverter, the dc-bus voltage $V_d = 700$ V. Using the space vector modulation principles, calculate and plot the control voltages in steady state to synthesize a 60-Hz output with an L-L, rms value of 460 V. Assume that $\hat{V}_{tri} = 5$ V and the switching frequency $f_s = 10$ kHz.

Solution

Figure 7-5 shows the block diagram in Simulink, which is included on the attached cd-rom, to synthesize the ac output voltages. The results are plotted in Fig. 7-6.

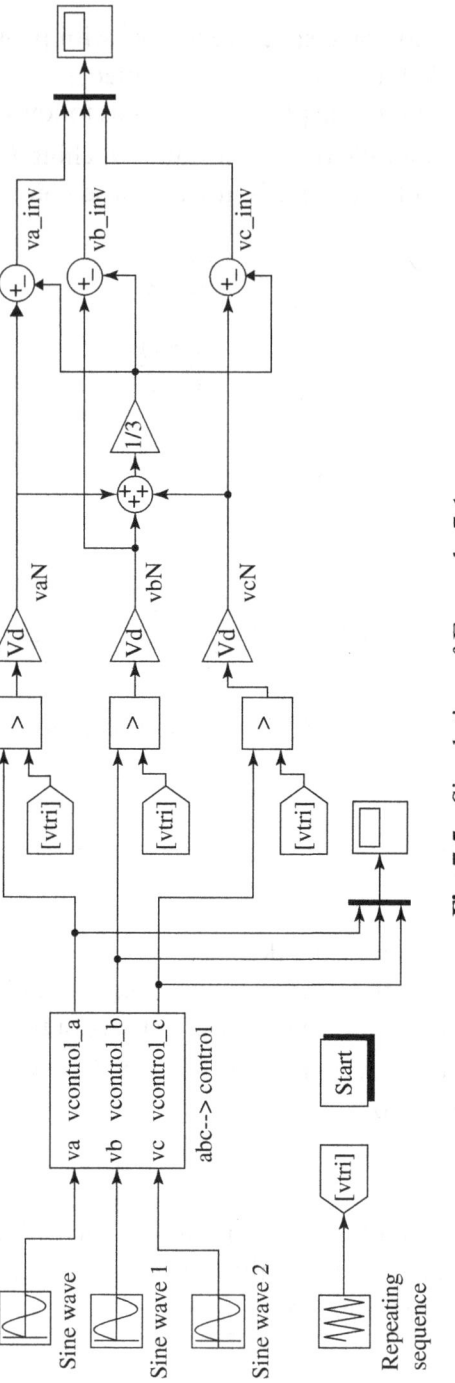

Fig. 7-5 Simulation of Example 7-1.

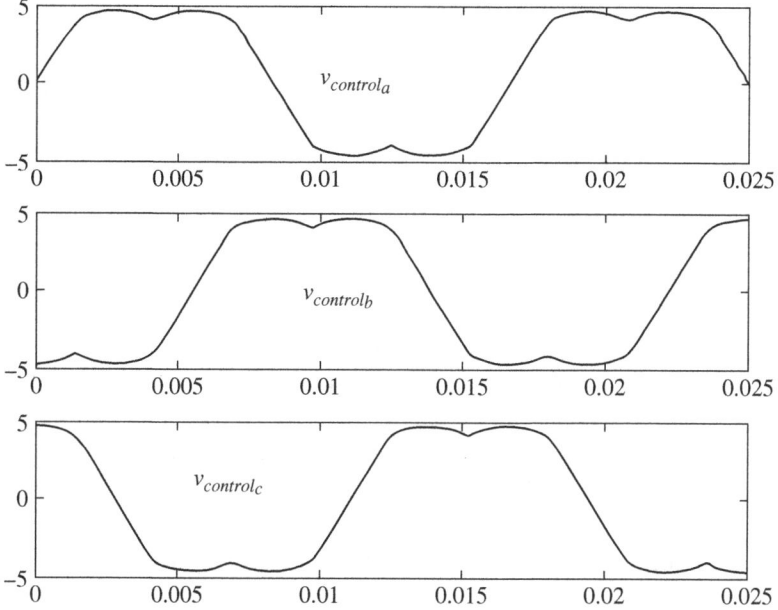

Fig. 7-6 Simulation results of Example 7-1.

7-4 LIMIT ON THE AMPLITUDE $\hat{V}_s$ OF THE STATOR VOLTAGE SPACE VECTOR $\vec{v}_s^a$

First, we will establish the absolute limit on the amplitude $\hat{V}_s$ of the average stator voltage space vector at various angles. The limit on the amplitude equals V_d (the dc-bus voltage) if the average voltage vector lies along a nonzero basic voltage vector. In between the basic vectors, the limit on the average voltage vector amplitude is that its tip can lie on the straight lines shown in Fig. 7-7 forming a hexagon (see Problem 7-6).

However, the maximum amplitude of the output voltage $\vec{v}_s^a$ should be limited to the circle within the hexagon in Fig. 7-7, to prevent distortion in the resulting currents. This can easily be concluded from the fact that, in a balanced sinusoidal steady state, the voltage vector $\vec{v}_s^a$ rotates at the synchronous speed with its constant amplitude. At its maximum amplitude,

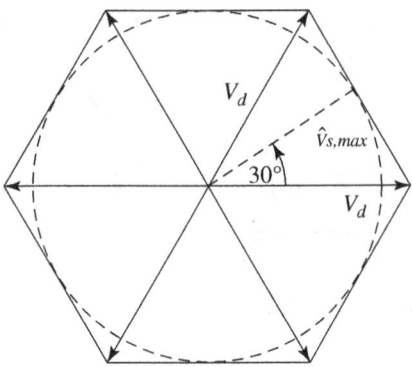

Fig. 7-7 Limit on amplitude $\hat{V}_s$.

$$\vec{v}_{s,max}^{a}(t) = \hat{V}_{s,max}\, e^{j\omega_{syn}t} \tag{7-14}$$

Therefore, the maximum value that $\hat{V}_s$ can attain is

$$\hat{V}_{s,max} = V_d \cos\left(\frac{60°}{2}\right) = \frac{\sqrt{3}}{2} V_d \tag{7-15}$$

From Eq. (7-15), the corresponding limits on the phase voltage and the L-L voltages are as follows:

$$\hat{V}_{phase,max} = \frac{2}{3}\hat{V}_{s,max} = \frac{V_d}{\sqrt{3}} \tag{7-16}$$

and

$$V_{LL,max}(rms) = \sqrt{3}\,\frac{\hat{V}_{phase,max}}{\sqrt{2}} = \frac{V_d}{\sqrt{2}} = 0.707\, V_d \tag{7-17}$$

The sinusoidal pulse-width modulation in the linear range discussed in the previous course on electric drives and power electronics results in a maximum voltage

$$V_{LL,max}(rms) = \frac{\sqrt{3}}{2\sqrt{2}} V_d = 0.612\, V_d \quad \text{(sinusoidal PWM)} \tag{7-18}$$

Comparison of Eqs. (7-17) and (7-18) shows that the SV-PWM discussed in this chapter better utilizes the dc-bus voltage and results in a higher limit on the available output voltage by a factor of $(2/\sqrt{3})$, or by approximately 15% higher, compared to the sinusoidal PWM.

7-5 HARDWARE PROTOTYPING OF SPACE VECTOR PULSE WIDTH MODULATION

In this section, the hardware implementation of both Sine-PWM, covered in Chapter 4, and SV-PWM for a three-phase two-level inverter is presented. The experiment is performed on a 42 V dc–ac inverter hardware and Workbench software as explained step-by-step in [1]. The Workbench model to perform real-time hardware implementation of Sine-PWM is shown in Fig. 7-8a, and SV-PWM is shown in Fig. 7-8b.

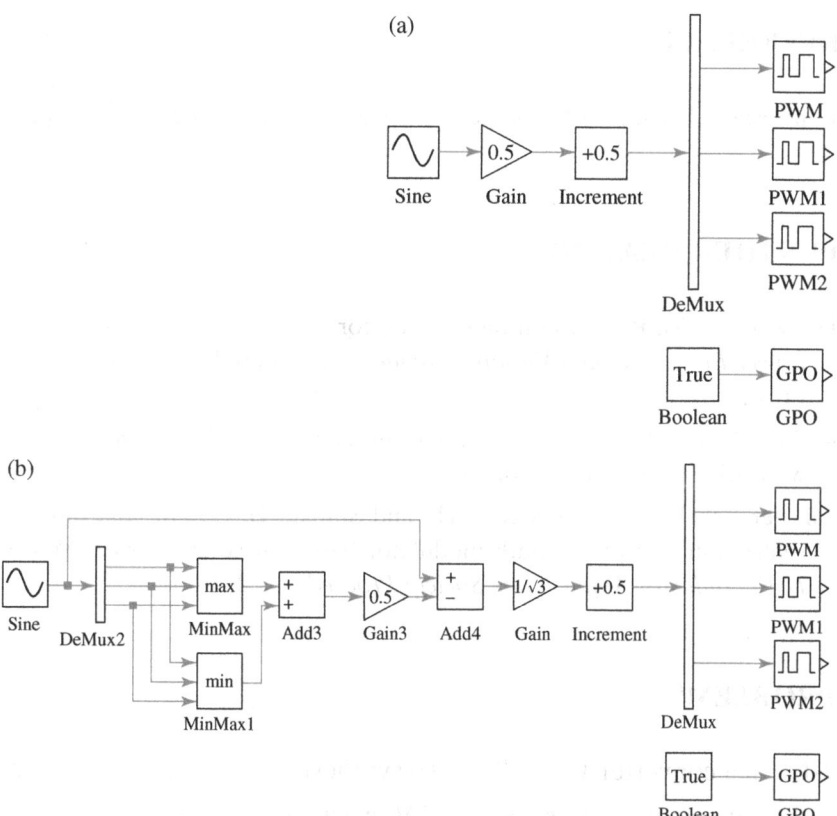

Fig. 7-8 Real-time three-phase inverter (a) Sine-PWM and (b) SV-PWM.

These two models are used in the following chapters for the hardware implementation. The step-by-step procedure to implement the above model in hardware using Workbench can be found on the accompanying website [1].

7-6 SUMMARY

In this chapter, an approach called SV-PWM is discussed, which is better than the sinusoidal PWM approach in utilizing the available dc-bus voltage. Its modeling using simulations is described. Finally, the hardware implementation of both Sine-PWM and SV-PWM is presented.

REFERENCE

1. https://sciamble.com/Resources/pe-drives-lab/advanced-drives/svpwm

FURTHER READING

Holtz, J. (1997). Pulse width modulation for electric power converters. In: *Power Electronics and Variable Frequency Drives* (ed. B.K. Bose). IEEE Press.

Mohan, N. (2011). *Electric Machines and Drives – A First Course*. Wiley www.wiley.com/college/mohan.

van der Broek, H.W., Skudelny, H., and Stanke, G. (1986). Analysis and realization of a pulse width modulator based on voltage space vectors. *IEEE Industry Applications Society Proceedings*: 244–251.

PROBLEMS

7-1 In a converter $V_d = 700$ V. To synthesize an average stator voltage vector $\vec{v}_s^a = 563.38\, e^{\,j0.44}$ V, calculate x, y, and z.

7-2 Repeat if $\vec{v}_s^a = 563.38\, e^{\,j2.53}$ V. Plot results similar to those in Fig. 7-4.

7-3 Show that in an induction machine with isolated neutral, at any instant of time, $v_a(t) + v_b(t) + v_c(t) = 0$.

7-4 Given that $\vec{v}_s^a = 563.38\, e^{\,j0.44}$ V, calculate the phase voltage components.

7-5 Derive Eqs. (7-11) and (7-12).

7-6 Derive that the maximum limit on the amplitude of the space vector forms the hexagonal trajectory shown in Fig. 7-7.

8 Sinusoidal Permanent-Magnet ac (PMAC) Drives in Steady State

8-1 INTRODUCTION

Many electric machines other than induction motors consist of permanent magnets in smaller ratings. However, the use of permanent magnets will undoubtedly extend to machines of higher power ratings, because permanent magnets provide a "free" source of flux, which otherwise has to be created by current-carrying windings that incur i^2R losses in winding resistance. The higher efficiency and the higher power density offered by permanent-magnet machines are very attractive. In recent years, significant advances have been made in Nd-Fe-B material, which has a very attractive magnetic characteristic, shown in comparison to other permanent-magnet materials in Fig. 8-1.

Nd-Fe-B magnets offer high flux-density operation, high energy densities, and a high ability to resist demagnetization. They are used in a wide range of power ratings.

Having been introduced to ac machines and their analysis using space vector theory, we will now study an important class of ac drives, namely sinusoidal-waveform, permanent-magnet ac (PMAC) drives. The motors in these drives have three-phase, sinusoidally distributed

---

(Adapted from chapter 10 of *Electric Machines and Drives: A First Course* ISBN: 978-1-118-07481-7 by Ned Mohan, January 2012)

Analysis and Control of Electric Drives: Simulations and Laboratory Implementation, First Edition. Ned Mohan and Siddharth Raju.
© 2021 John Wiley & Sons, Inc. Published 2021 by John Wiley & Sons, Inc.
Companion website: www.wiley.com/go/Mohan/Vectorcontrolinelectricdrives

218 SINUSOIDAL PERMANENT-MAGNET ac (PMAC)

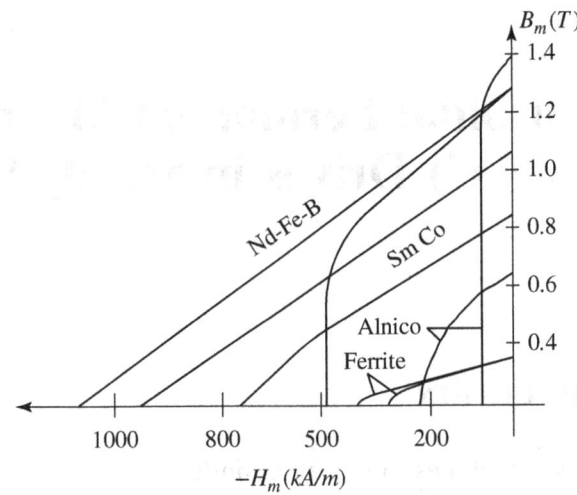

Fig. 8-1 Characteristics of various permanent-magnet materials.

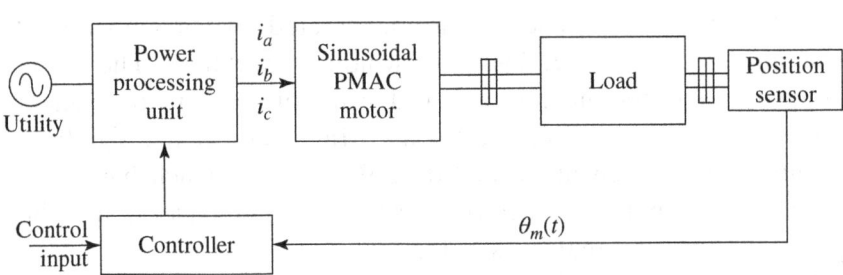

Fig. 8-2 Block diagram of the closed-loop operation of a PMAC drive.

ac stator windings, and the rotor has dc excitation in the form of permanent magnets. In such drives, the stator windings of the machine are supplied by controlled currents, which require a closed-loop operation, as shown in the block diagram of Fig. 8-2.

In these machines, the permanent magnets on the rotor are shaped to induce (in the stator windings) back-emfs that are ideally sinusoidally varying with time. PMAC drives are capable of producing a smooth torque, and thus they are used in high-performance applications. They do not suffer from the maintenance problems associated with brush-type dc machines. They are used where high efficiency and high power density are required.

8-2 THE BASIC STRUCTURE OF PMAC MACHINES

We will first consider 2-pole machines, like the one shown schematically in Fig. 8-3a, and then we will generalize our analysis to p-pole machines where $p > 2$. The stator contains three-phase, wye-connected, sinusoidally distributed windings (discussed in Chapter 9), which are shown in the cross-section of Fig. 8-3a. These sinusoidally distributed windings, produce a sinusoidally distributed mmf in the air gap.

8-3 PRINCIPLE OF OPERATION

8-3-1 Rotor-Produced Flux-Density Distribution

The permanent-magnet pole pieces mounted on the rotor surface are shaped to ideally produce a sinusoidally distributed flux density in the air gap. Without delving into detailed construction, Fig. 8-3a schematically shows a 2-pole rotor. Flux lines leave the rotor at the north pole to re-enter the air gap at the south pole. The rotor-produced flux-density distribution in the air gap (due to flux lines that completely cross the two air gaps) has its positive peak $\hat{B}_r$ directed along the north pole axis. Because this flux density is sinusoidally distributed, it can be represented, as shown in Fig. 8-3b, by a space vector of length $\hat{B}_r$, and

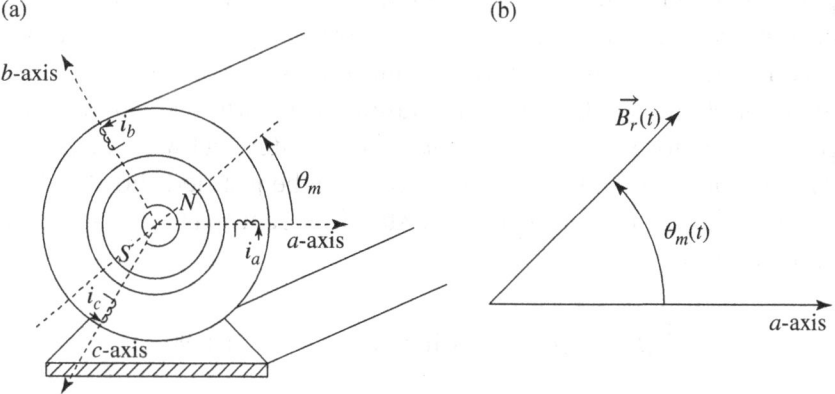

Fig. 8-3 Two-pole PMAC machine.

its orientation can be established by the location of the positive peak of the flux-density distribution. As the rotor turns, the entire rotor-produced flux-density distribution in the air gap rotates with it. Therefore, using the stationary stator phase-a axis as the reference, we can represent the rotor-produced flux-density space vector at a time t as

$$\vec{B}_r(t) = \hat{B}_r \angle \theta_m(t) \qquad (8\text{-}1)$$

where the rotor flux-density distribution axis is at an angle $\theta_m(t)$ with respect to the a-axis. In Eq. (8-1), permanent magnets produce a constant $\hat{B}_r$, but $\theta_m(t)$ is a function of time, as the rotor turns.

8-3-2 Torque Production

We would like to compute the electromagnetic torque produced by the rotor. However, the rotor consists of permanent magnets, and we have no direct way of computing this torque. Therefore, we will first calculate the torque exerted on the stator; this torque is transferred to the motor foundation. The torque exerted on the rotor is equal in magnitude to the stator torque but acts in the opposite direction.

An important characteristic of the machines under consideration is that they are supplied through the power-processing unit (PPU) shown in Fig. 8-2, which controls the currents $i_a(t)$, $i_b(t)$, and $i_c(t)$ supplied to the stator at any instant of time. At any time t, the three stator currents combine to produce a stator-current space vector $\vec{i}_s(t)$, which is controlled to be ahead of (or leading) the space vector $\vec{B}_r(t)$ by an angle of 90° in the direction of rotation, as shown in Fig. 8-4a. This produces a torque on the rotor in a counterclockwise direction. The reason for maintaining a 90° angle will be justified shortly. With the a-axis as the reference axis, the stator-current space vector can be expressed as

$$\vec{i}_s(t) = \hat{I}_s(t) \angle \theta_{i_s}(t) \text{ where } \theta_{i_s}(t) = \theta_m(t) + 90° \qquad (8\text{-}2)$$

During a steady-state operation, $\hat{I}_s$ is kept constant while $\theta_m(=\omega_m t)$ changes linearly with time.

PRINCIPLE OF OPERATION 221

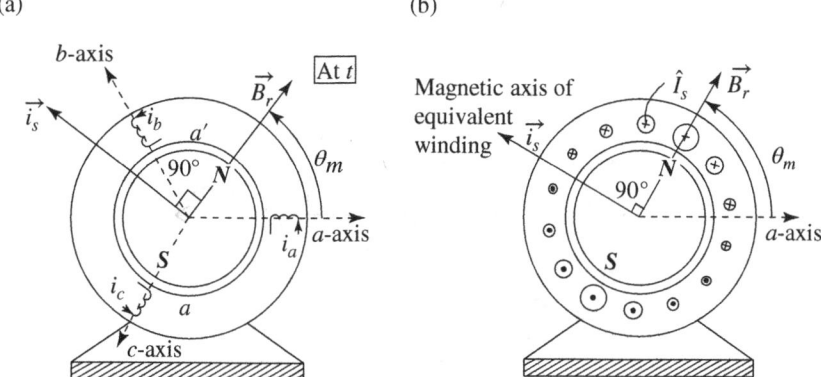

Fig. 8-4 The stator current and the rotor field space vectors in PMAC drives.

We have seen the physical interpretation of the current space vector $\vec{i}_s(t)$ before. In Fig. 8-4a at a time t, the three stator phase currents combine to produce an mmf distribution in the air gap.

This mmf distribution is the same as that produced in Fig. 8-4b by a single equivalent stator winding, which has N_s sinusoidally distributed turns supplied by a current $\hat{I}_s$ and which has its magnetic axis situated along the $\vec{i}_s(t)$ vector. As seen in Fig. 8-4b, by controlling the stator-current space vector $\vec{i}_s(t)$ to be 90° ahead of $\vec{B}_r(t)$, all of the conductors in the equivalent stator winding will experience a force acting in the same direction which, in this case, is clockwise on the stator (and hence produces a counterclockwise torque on the rotor). This justifies the choice of 90°: it results in the maximum torque per ampere of stator current because, at any other angle, some conductors will experience a force in the direction opposite to that on other conductors, a condition which will result in a smaller net torque.

As $\vec{B}_r(t)$ rotates with the rotor, the space vector $\vec{i}_s(t)$ is made to rotate at the same speed, maintaining a "lead" of 90°. Thus, the torque developed in the machine of Fig. 8-4 depends only on $\hat{B}_r$ and $\hat{I}_s$, and is independent of θ_m. Therefore, to simplify our calculation of this torque in terms of the machine parameters, we will redraw Fig. 8-4b, as in Fig. 8-5 by assuming $\theta_m = 0°$.

Using the expression for force ($f_{em} = B\ell i$), we can calculate the clockwise torque acting on the stator as follows: in the equivalent stator winding shown in Fig. 8-5, at an angle ξ, the differential angle

222 SINUSOIDAL PERMANENT-MAGNET ac (PMAC)

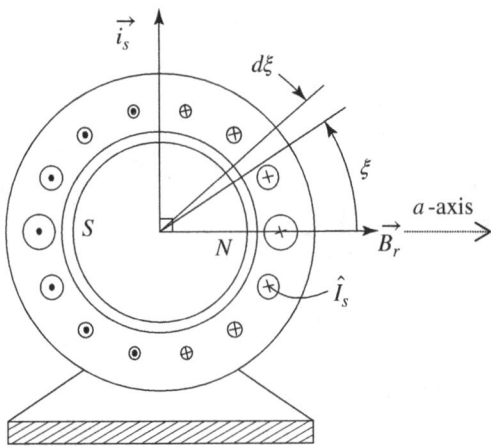

Fig. 8-5 Torque calculation on the stator.

$d\xi$ contains $n_s(\xi) \cdot d\xi$ conductors. Using Eq. (6-5) and noting that the angle ξ is measured here from the location of the peak conductor density, the conductor density $n_s(\xi) = (N_s/2) \cdot \cos \xi$. Therefore, the number of conductors in the differential angle $d\xi$ is

$$\frac{N_s}{2} \cos \xi \cdot d\xi \tag{8-3}$$

The rotor-produced flux density at an angle ξ is $\hat{B}_r \cos \xi$. Therefore, the torque $dT_{em}(\xi)$ produced by these conductors (due to the current $\hat{I}_s$ flowing through them) located at an angle ξ, at a radius r, and of length ℓ is

$$dT_{em}(\xi) = r \underbrace{\hat{B}_r \cos \xi}_{\text{flux density at } \xi} \cdot \underbrace{\ell}_{\text{cond. length}} \cdot \hat{I}_s \cdot \underbrace{\frac{N_s}{2} \cos \xi \cdot d\xi}_{\text{no. of cond. in } d\xi} \tag{8-4}$$

To account for the torque produced by all of the stator conductors, we will integrate the above expression from $\xi = -\pi/2$ to $\xi = \pi/2$, and then multiply by a factor of 2, making use of symmetry:

$$T_{em} = 2 \times \int_{\xi = -\pi/2}^{\xi = \pi/2} dT_{em}(\xi) = 2 \frac{N_s}{2} r\ell \hat{B}_r \hat{I}_s \int_{-\pi/2}^{\pi/2} \cos^2 \xi \cdot d\xi = \left(\pi \frac{N_s}{2} r\ell \hat{B}_r \right) \hat{I}_s$$

$$\tag{8-5}$$

PRINCIPLE OF OPERATION

In the above equation, all quantities within the brackets, including $\hat{B}_r$ in a machine with permanent magnets, depend on the machine design parameters and are constants. As noted earlier, the electromagnetic torque produced by the rotor is equal to that in Eq. (8-5) in the opposite direction (counterclockwise in this case). This torque in a 2-pole machine can be expressed as

$$T_{em} = k_T \hat{I}_s \text{ where } k_T = \pi \frac{N_s}{2} r\ell \hat{B}_r \quad (p = 2) \quad (8\text{-}6)$$

In the above equation, k_T is the *machine torque constant*, which has the units of Nm/A. Equation (8-6) shows that, by controlling the stator phase currents so that the corresponding stator-current space vector is ahead (in the desired direction) of the rotor-produced flux-density space vector by 90°, the torque developed is only proportional to $\hat{I}_s$.

The similarities between the brush-type dc motor drives and the PMAC motor drives are shown by means of Fig. 8-6.

In the brush-type dc motors, the flux ϕ_f produced by the stator and the armature flux ϕ_a produced by the armature winding remains directed orthogonal (at 90°) to each other, as shown in Fig. 8-6a. The stator flux ϕ_f is stationary, and so is ϕ_a (due to the commutator action), even though the rotor is turning. The torque produced is controlled by the armature current $i_a(t)$. In PMAC motor drives,

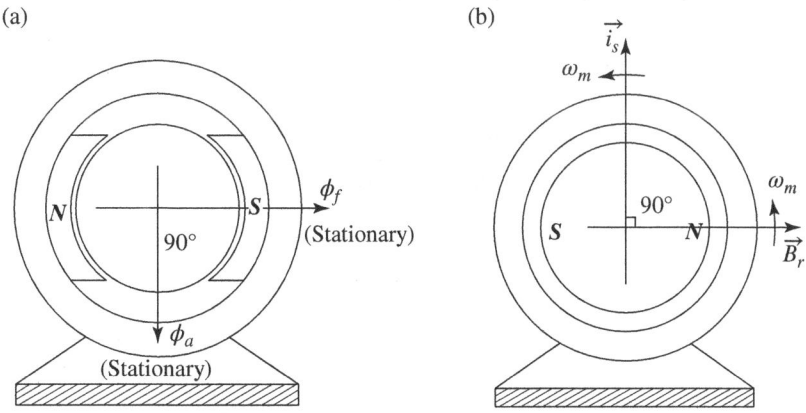

Fig. 8-6 Similarities between (a) dc motor and (b) PMAC motor drives.

the stator-produced flux-density $\vec{B}_{s,\vec{i}_s}(t)$ due to $\vec{i}_s(t)$ is controlled to be directed orthogonal (at 90° in the direction of rotation) to the rotor flux-density $\vec{B}_r(t)$, as shown in Fig. 8-6b. Both of these space vectors rotate at the speed ω_m of the rotor, maintaining the 90° angle between the two. The torque is controlled by the magnitude $\hat{I}_s(t)$ of the stator-current space vector.

In PMAC drives, this synchronism is established by a closed feedback loop in which the measured instantaneous position of the rotor directs the PPU to locate the stator mmf distribution 90° ahead of the rotor-field distribution. Therefore, there is no possibility of losing synchronism between the two, unlike in the conventional synchronous machines.

Generator Mode PMAC drives can be operated as generators. In fact, these drives are used in this mode in wind turbines. In this mode, the stator-current space vector $\vec{i}_s$ is controlled to be 90° behind the rotor flux-density vector, in the direction of rotation, in Fig. 8-4. Therefore, the resulting electromagnetic torque T_{em}, as calculated by Eq. (8-5), acts in a direction to oppose the direction of rotation.

8-3-3 Mechanical System of PMAC Drives

The electromagnetic torque acts on the mechanical system connected to the rotor, as shown in Fig. 8-7, and the resulting speed ω_m can be obtained from the following equation:

$$\frac{d\omega_m}{dt} = \frac{T_{em} - T_L}{J_{eq}} \qquad (8\text{-}7)$$

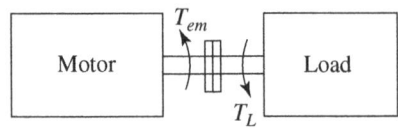

Fig. 8-7 Rotor-load mechanical system.

PRINCIPLE OF OPERATION   225

where J_{eq} is the combined motor-load inertia and T_L is the load torque, which may include friction. The rotor position $\theta_m(t)$ is

$$\theta_m(t) = \theta_m(0) + \int_o^t \omega_m(\tau) \cdot d\tau \quad (\tau = \text{variable of integration}) \quad (8\text{-}8)$$

where $\theta_m(0)$ is the rotor position at time $t = 0$.

8-3-4 Calculation of the Reference Values $i_a^*(t)$, $i_b^*(t)$, and $i_c^*(t)$ of the Stator Currents

The controller in Fig. 8-2 is responsible for controlling the torque, speed, and position of the mechanical system. It does so by calculating the instantaneous value of the desired (reference) torque $T_{em}^*(t)$ that the motor must produce. The reference torque may be generated by the cascaded controller discussed in Chapter 5. From Eq. (8-6), $\hat{I}_s^*(t)$, the reference value of the amplitude of the stator-current space vector, can be calculated as

$$\hat{I}_s^*(t) = \frac{T_{em}^*(t)}{k_T} \quad (8\text{-}9)$$

where k_T is the motor torque constant given in Eq. (8-6) (k_T is usually listed in the motor specification sheet).

The controller in Fig. 8-2 receives the instantaneous rotor position θ_m, which is measured, as shown in Fig. 8-2, by means of a mechanical sensor such as a resolver or an optical encoder (with some restrictions).

With $\theta_m(t)$ as one of the inputs and $\hat{I}_s^*(t)$ calculated from Eq. (8-9), the instantaneous reference value of the stator-current space vector becomes

$$\vec{i}_s^{\,*}(t) = \hat{I}_s^*(t) \angle \theta_{i_s}^*(t) \text{ where } \theta_{i_s}^*(t) = \theta_m(t) + \frac{\pi}{2} \quad (8\text{-}10)$$

Equation (8-10) assumes a 2-pole machine and the desired rotation to be in the counterclockwise direction. For clockwise rotation, the angle $\theta_{i_s}^*(t)$ in Eq. (8-10) will be $\theta_m(t) - \pi/2$. In a multi-pole machine with $p > 2$, the electrical angle $\theta_{i_s}^*(t)$ will be

$$\theta_{i_s}^*(t) = \frac{p}{2}\theta_m(t) \pm \frac{\pi}{2} \quad (p \geq 2) \tag{8-11}$$

where $\theta_m(t)$ is the mechanical angle. From $\vec{i_s}^*(t)$ in Eq. (8-10) (with Eq. (8-11) for $\theta_{i_s}^*(t)$ in a machine with $p > 2$), the instantaneous reference values $i_a^*(t)$, $i_b^*(t)$, and $i_c^*(t)$ of the stator phase currents can be calculated using the analysis in Chapter 6 (Eqs. (6-24a) through (6-24c)):

$$i_a^*(t) = \frac{2}{3}\text{Re}\left[\vec{i_s}^*(t)\right] = \frac{2}{3}\hat{I}_s^*(t)\cos\theta_{i_s}^*(t) \tag{8-12a}$$

$$i_b^*(t) = \frac{2}{3}\text{Re}\left[\vec{i_s}^*(t)\angle -\frac{2\pi}{3}\right] = \frac{2}{3}\hat{I}_s^*(t)\cos\left(\theta_{i_s}^*(t) - \frac{2\pi}{3}\right) \tag{8-12b}$$

and

$$i_c^*(t) = \frac{2}{3}\text{Re}\left[\vec{i_s}^*(t)\angle -\frac{4\pi}{3}\right] = \frac{2}{3}\hat{I}_s^*(t)\cos\left(\theta_{i_s}^*(t) - \frac{4\pi}{3}\right) \tag{8-12c}$$

Section 8-5, which deals with the PPU and the controller, describes how the phase currents, based on the above reference values, are supplied to the motor. Equations (8-12a) through (8-12c) show that in the balanced sinusoidal steady state, the currents have the constant amplitude of $\hat{I}_s^*$; they vary sinusoidally with time as the angle $\theta_{i_s}^*(t)$ in Eq. (8-10) or Eq. (8-11) changes continuously with time at a constant speed ω_m:

$$\theta_{i_s}^*(t) = \frac{p}{2}[\theta_m(0) + \omega_m t] \pm \frac{\pi}{2} \tag{8-13}$$

where $\theta_m(0)$ is the initial rotor angle, measured with respect to the phase-a magnetic axis.

EXAMPLE 8-1

In a three-phase, 2-pole, PMAC motor, the torque constant $k_T = 0.5$ Nm/A. Calculate the phase currents if the motor is to produce a counterclockwise holding torque of 5 Nm to keep the rotor, which is at an angle of $\theta_m = 45°$, from turning.

Solution

From Eq. (8-6), $\hat{I}_s = T_{em}/k_T = 10$ A. From Eq. (8-10), $\theta_{i_s} = \theta_m + 90° = 135°$. Therefore, $\vec{i}_s(t) = \hat{I}_s \angle \theta_{i_s} = 10 \angle 135°$ A, as shown in Fig. 8-8. From Eqs. (8-12a) through (8-12c),

$$i_a = \frac{2}{3}\hat{I}_s \cos\theta_{i_s} = -4.71 \text{ A},$$

$$i_b = \frac{2}{3}\hat{I}_s \cos(\theta_{i_s} - 120°) = 6.44 \text{ A}$$

and

$$i_c = \frac{2}{3}\hat{I}_s \cos(\theta_{i_s} - 240°) = -1.73 \text{ A}$$

Since the rotor is not turning, the phase currents in this example are dc.

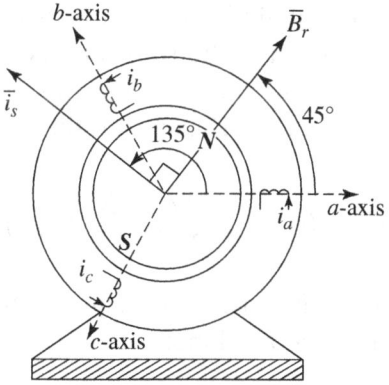

Fig. 8-8 Stator-current space vector for Example 8-1.

8-3-5 Induced EMFs in the Stator Windings During Balanced Sinusoidal Steady State

In the stator windings, emfs are induced due to two flux-density distributions:

1. As the rotor rotates with an instantaneous speed of $\omega_m(t)$, so does the space vector $\vec{B}_r(t)$ shown in Fig. 8-4a. This rotating flux-density distribution "cuts" the stator windings to induce a back-emf in them.
2. The stator phase-winding currents under a balanced sinusoidal steady state produce a rotating flux-density distribution due to the rotating $\vec{i}_s(t)$ space vector. This rotating flux-density distribution induces emfs in the stator windings, similar to those induced by the magnetizing currents in the previous chapter.

Neglecting saturation in the magnetic circuit, the emfs induced due to the two causes mentioned above can be superimposed to calculate the resultant emf in the stator windings. In the following subsections, we will assume a 2-pole machine in a balanced sinusoidal steady state, with a rotor speed of ω_m in the counterclockwise direction. We will also assume that at $t = 0$ the rotor is at $\theta_m = -90°$ for ease of drawing the space vectors.

Induced EMF in the Stator Windings Due to Rotating $\vec{B}_r(t)$

We can make use of the analysis in the previous chapter that led to Eq. (6-41). In the present case, the rotor flux-density vector $\vec{B}_r(t)$ is rotating at the instantaneous speed of ω_m with respect to the stator windings. Therefore, in Eq. (6-41), substituting $\vec{B}_r(t)$ for $\vec{B}_{ms}(t)$ and ω_m for ω_{syn},

$$\vec{e_{ms}}_{,\vec{B}_r}(t) = j\omega_m \frac{3}{2}\left(\pi r \ell \frac{N_s}{2}\right)\vec{B}_r(t) \qquad (8\text{-}14)$$

We can define a voltage constant k_E, equal to the torque constant k_T in Eq. (8-6) for a 2-pole machine:

$$k_E\left[\frac{\text{V}}{\text{rad/s}}\right] = k_T\left[\frac{\text{Nm}}{\text{A}}\right] = \pi r \ell \frac{N_s}{2}\hat{B}_r \qquad (8\text{-}15)$$

PRINCIPLE OF OPERATION 229

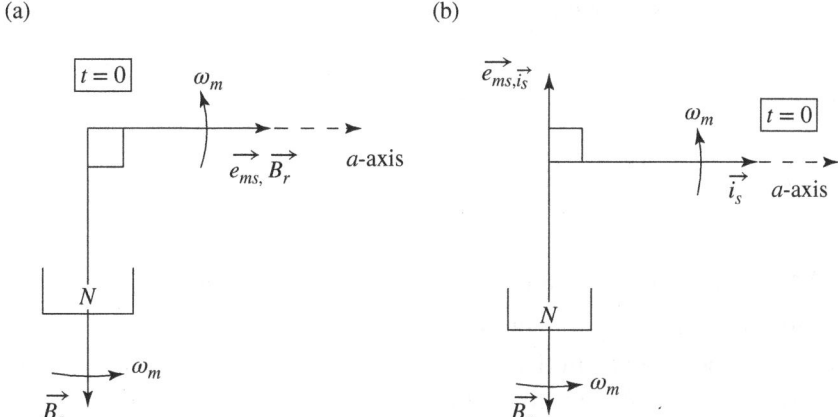

Fig. 8-9 (a) Induced emf due to rotating rotor flux-density space vector and (b) induced emf due to rotating stator-current space vector.

where $\hat{B}_r$ (the peak of the rotor-produced flux density) is a constant in permanent-magnet synchronous motors. In terms of the voltage constant k_E, the induced voltage space vector in Eq. (8-14) can be written as

$$\vec{e}_{ms,\vec{B}_r}(t) = j\frac{3}{2}k_E\omega_m\angle\theta_m(t) = \frac{3}{2}k_E\omega_m\angle\{\theta_m(t) + 90°\} \quad (8\text{-}16)$$

The rotor flux-density space vector $\vec{B}_r(t)$ and the induced-emf space vector $\vec{e}_{ms,\vec{B}_r}(t)$ are drawn for the time $t = 0$ in Fig. 8-9a.

Induced EMF in the Stator Windings Due to Rotating $\vec{i}_s(t)$: Armature Reaction

In addition to the flux-density distribution in the air gap created by the rotor magnets, another flux-density distribution is established by the stator phase currents. As shown in Fig. 8-9b, the stator-current space vector $\vec{i}_s(t)$ at the time $t = 0$ is made to lead the rotor position by 90°. Because we are operating under a balanced sinusoidal steady state, we can make use of the analysis in Chapter 6, where Eq. (6-40) showed the relationship between the induced-emf space vector and the stator-current space vector. Thus, in the present case, due to the rotation of $\vec{i}_s(t)$, the induced voltages in the stator phase windings can be represented as

$$\vec{e_{ms}}_{,\vec{i}_s}(t) = j\omega_m L_m \vec{i}_s(t) \tag{8-17}$$

Space vectors $\vec{e_{ms}}_{,\vec{i}_s}$ and $\vec{i}_s$ are shown in Fig. 8-9b at time $t = 0$.
Note that the magnetizing inductance L_m in the PMAC motor has the same meaning as in the generic ac motors discussed in Chapter 6. However, in PMAC motors, the rotor on its surface has permanent magnets (exceptions are motors with interior permanent magnets) whose permeability is effectively that of the air gap. Therefore, PMAC motors have a larger equivalent air gap, thus resulting in a smaller value of L_m (see Eq. (6-36)).

Superposition of the Induced EMFs in the Stator Windings
In PMAC motors, rotating $\vec{B}_r(t)$ and $\vec{i}_s(t)$ are present simultaneously. Therefore, the emfs induced due to each one can be superimposed (assuming no magnetic saturations) to obtain the resultant emf (excluding the leakage flux of the stator windings):

$$\vec{e_{ms}}(t) = \vec{e_{ms}}_{,\vec{B}_r}(t) + \vec{e_{ms}}_{,\vec{i}_s}(t) \tag{8-18}$$

Substituting from Eqs. (8-16) and (8-17) into Eq. (8-18), the resultant induced emf $\vec{e_{ms}}(t)$ is

$$\vec{e_{ms}}(t) = \frac{3}{2}k_E \omega_m \angle\{\theta_m(t) + 90°\} + j\omega_m L_m \vec{i}_s(t) \tag{8-19}$$

The space vector diagram is shown in Fig. 8-10a at time $t = 0$. The phase-*a* phasor equation corresponding to the space vector equation above can be written, noting that the phasor amplitudes are smaller than the space vector amplitudes by a factor of 3/2, but the phasor and the corresponding space vector have the same orientation:

PRINCIPLE OF OPERATION 231

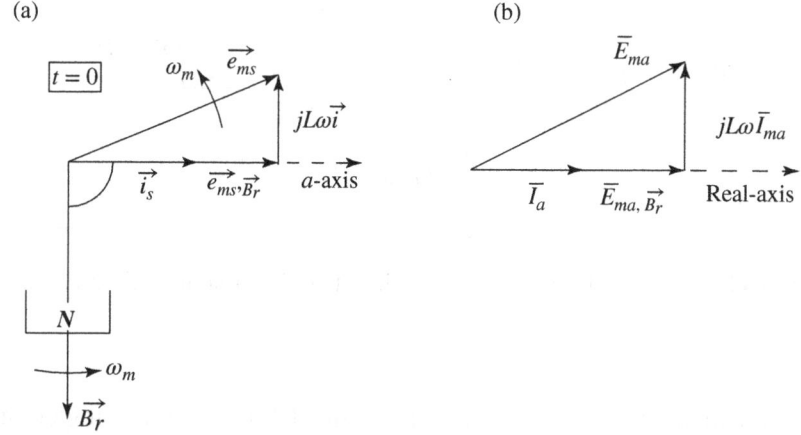

Fig. 8-10 (a) Space vector diagram of induced emfs and (b) phasor diagram for phase-a.

$$\overline{E}_{ma} = \underbrace{k_E \omega_m \angle\{\theta_m(t) + 90°\}}_{\overline{E}_{ma,\vec{B}_r}} + j\omega_m L_m \overline{I}_a \qquad (8\text{-}20)$$

The phasor diagram from Eq. (8-20) for phase-a is shown in Fig. 8-10b.

Per-Phase Equivalent Circuit Corresponding to the phasor representation in Eq. (8-20) and the phasor diagram in Fig. 8-10b, a per-phase equivalent circuit for phase-a can be drawn, as shown in Fig. 8-11a. The voltage $\overline{E}_{ma,\vec{B}_r}$, induced due to the rotation of the rotor field distribution $\vec{B}_r$, is represented as an induced back-emf. The second term on the right side of Eq. (8-20) is represented as a voltage drop across the magnetizing inductance L_m. To complete this per-phase equivalent circuit, the stator-winding leakage inductance $L_{\ell s}$ and the resistance R_s are added in series. The sum of the magnetizing inductance L_m and the leakage inductance $L_{\ell s}$ is called the synchronous inductance L_s:

$$L_s = L_{\ell s} + L_m \qquad (8\text{-}21)$$

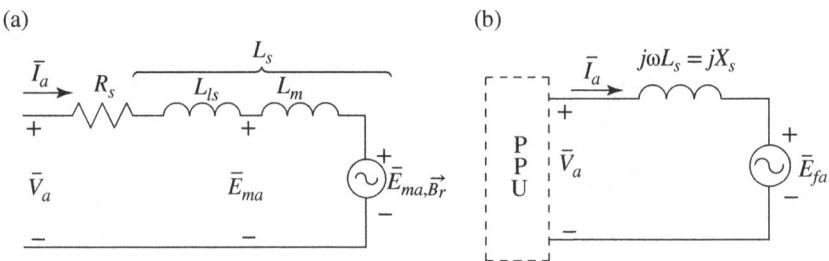

Fig. 8-11 (a) Per-phase equivalent circuit and (b) simplified equivalent circuit.

We can simplify the equivalent circuit of Fig. 8-11a by neglecting the resistance, and by representing the two inductances by their sum, L_s, as done in Fig. 8-11b. To simplify the notation, the induced back-emf is called the field-induced back-emf $\overline{E}_{fa}$ in phase-a, where, from Eq. (8-20), the peak of this voltage in each phase is

$$\hat{E}_f = k_E \omega_m \quad (8\text{-}22)$$

Notice that in PMAC drives, the PPU is a source of controlled currents such that $\overline{I}_a$ is in phase with the field-induced back-emf $\overline{E}_{fa}$, as confirmed by the phasor diagram of Fig. 8-10b. The PPU supplies this current by producing a voltage which, for phase-a in Fig. 8-11b is

$$\overline{V}_a = \overline{E}_{fa} + j\omega_m L_s \overline{I}_a \quad (8\text{-}23)$$

EXAMPLE 8-2

In a 2-pole, three-phase PMAC drive, the torque constant k_T and the voltage constant k_E are 0.5 in MKS units. The synchronous inductance is 15 mH (neglect the winding resistance). This motor is supplying a torque of 3 Nm at a speed of 3000 rpm in a balanced sinusoidal steady state. Calculate the per-phase voltage across the PPU as it supplies controlled currents to this motor.

Solution

From Eq. (8-6), $\hat{I}_s = \dfrac{3.0}{0.5} = 6\,\text{A}$ and $\hat{I}_a = \dfrac{2}{3}\hat{I}_s = 4\,\text{A}$. The speed $\omega_m = \dfrac{3000}{60}(2\pi) = 314.16\,\text{rad/s}$. From Eq. (8-22), $\hat{E}_f = k_E\omega_m = 0.5 \times 314.16 = 157.08\,\text{V}$.

Assuming $\theta_m(0) = -90°$, from Eq. (8-10), $\theta_{i_s}\big|_{t=0} = 0°$. Hence, in the per-phase equivalent circuit of Fig. 8-10b, $\bar{I}_a = 4.0\angle 0°\,\text{A}$ and $\bar{E}_{fa} = 157.08\angle 0°\,\text{V}$. Therefore, from Eq. (8-23), in the per-phase equivalent circuit of Fig. 8-10b,

$$\bar{V}_a = \bar{E}_{fa} + j\omega_m L_s \bar{I}_a = 157.08\angle 0° + j314.16 \times 15 \times 10^{-3} \times 4.0\angle 0°$$
$$= 157.08 + j18.85 = 158.2\angle 6.84°\,\text{V}$$

8-3-6 Generator-Mode of Operation of PMAC Drives

PMAC drives can operate in their generator mode simply by controlling the stator-current space vector $\vec{i}_s$ to be 90° behind the $\vec{B}_r$ space vector in the direction of rotation, as shown in Fig. 8-4a. This will result in current directions in the conductors of the hypothetical winding to be opposite of what is shown in Fig. 8-4b. Hence, the electromagnetic torque produced will be in a direction to oppose the torque supplied by the prime-mover that is causing the rotor to rotate. An analysis similar to the motoring-mode can be carried out in the generator-mode of operation.

8-4 THE CONTROLLER AND THE PPU

As shown in the block diagram of Fig. 8-2, the task of the controller is to dictate the switching in the PPU, such that the desired currents are supplied to the PMAC motors. This is further illustrated in Fig. 8-12a, where phases *b* and *c* are omitted for simplification. The reference

(a)

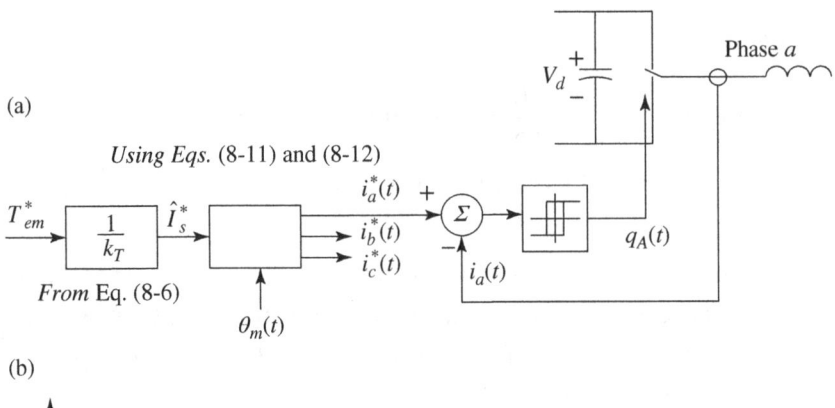

(b)

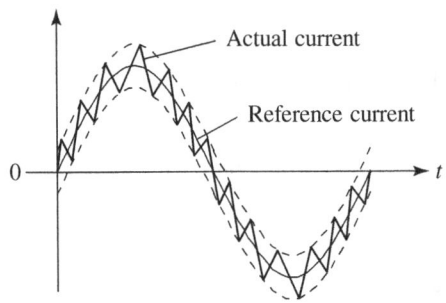

Fig. 8-12 (a) Block diagram representation of hysteresis current control and (b) current waveform.

signal T^*_{em} is generated from the outer speed and position loops discussed in Chapter 5. The rotor position θ_m is measured by the resolver connected to the shaft. Knowing the torque constant k_T, allows us to calculate the reference current $\hat{I}^*_s$ to be T^*_{em}/k_T (from Eq. (8-9)). Knowing $\hat{I}^*_s$ and θ_m allows the reference currents i^*_a, i^*_b, and i^*_c to be calculated at any instant of time from Eq. (8-11) and Eqs. (8-12a) through (8-12c).

One of the easiest ways to ensure that the motor is supplied the desired currents is to use hysteresis control. The measured phase current is compared with its reference value in the hysteresis comparator, whose output determines the switch state (up or down), resulting in current as shown in Fig. 8-11b.

In spite of the simplicity of the hysteresis control, one perceived drawback of this controller is that the switching frequency changes as a function of the back-emf waveform. For this reason, constant switching frequency controllers are used.

8-5 HARDWARE PROTOTYPING OF PMAC MOTOR HYSTERESIS CURRENT CONTROL

In this section, the hardware implementation of hysteresis current control of PMAC motor is analyzed. For the purposes of this comparison, a commercially available tabletop 36 V (L-L, rms) 3ϕ PMAC motor is modeled in Workbench, and its simulation and hardware results are compared. The exact motor parameters of the 3ϕ PMAC motor is available on the accompanying website. The Workbench file of the modeled system is shown in Fig. 8-13. The speed controller is implemented using a PI controller and has a step reference input from 0 to 100 rad/s at time $t = 1$ s. The output of the speed controller is the reference stator-current magnitude. This is synthesized using hysteresis current control described in the previous section. A step-load torque of 0.2 Nm is applied at time $t = 3$ s by means of a coupled dc generator.

The speed result from running this model in hardware is shown in Fig. 13-17. The hardware results closely match that of the reference speed, and the speed settles down rapidly to the desired speed even after a load torque is applied.

One of the major shortcomings of hysteresis current control is the changing switching frequency, which leads to significant harmonics in the stator currents as shown in Fig. 8-15. This causes torque pulsations, and hence this method is not generally used in precision motor control design. This torque pulsation can be observed as the significant band around the rotor speed in steady state as shown in Fig. 8-14. An alternative to this method, one which uses a constant switching frequency, is analyzed in Chapter 17.

The step-by-step procedure for recreating the above comparison is presented in [1].

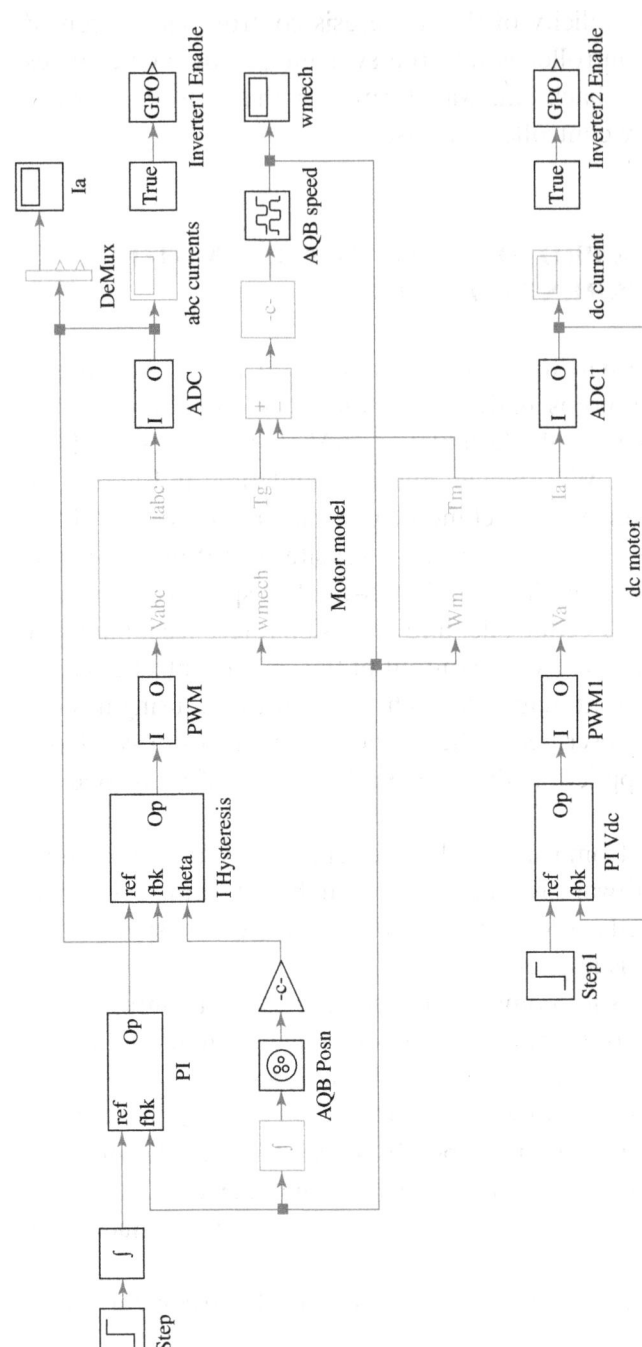

Fig. 8-13 Real-time PMAC motor hysteresis current control.

HARDWARE PROTOTYPING OF PMAC MOTOR 237

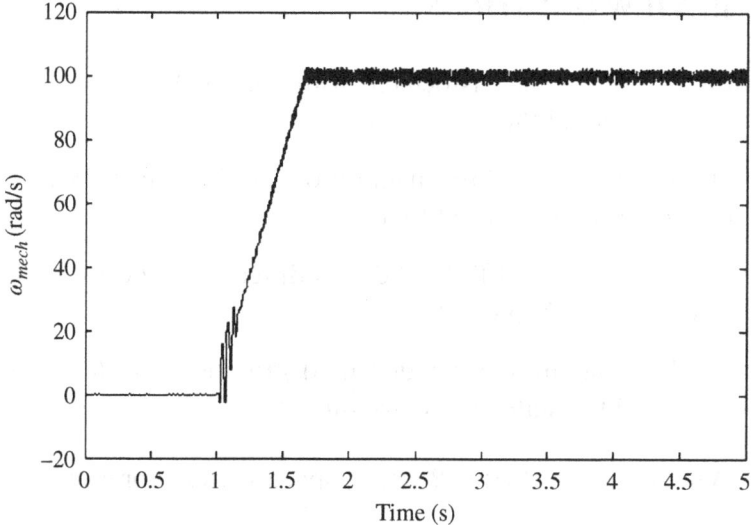

Fig. 8-14 Hardware speed result.

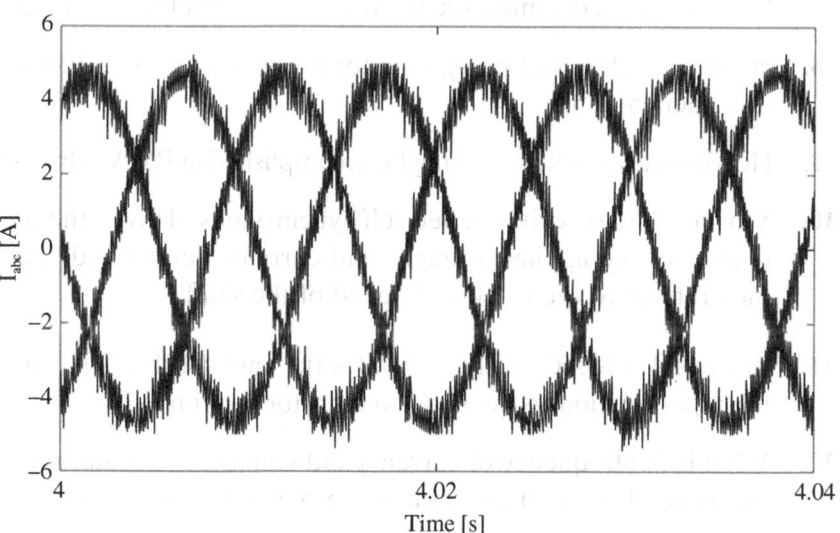

Fig. 8-15 Three phase stator current.

8-6 REVIEW QUESTIONS

1. List various names associated with the PMAC drives and the reasons behind them.
2. Draw the overall block diagram of a PMAC drive. Why must they operate in a closed loop?
3. How do sinusoidal PMAC drives differ from the ECM drives described in Chapter 7?
4. Ideally, what are the flux-density distributions produced by the rotor and the stator phase windings?
5. What does the $\vec{B}_r(t) = \hat{B}_r \angle \theta_m(t)$ space vector represent?
6. In PMAC drives, why at all times is the $\vec{i}_s(t)$ space vector placed 90° ahead of the $\vec{B}_r(t)$ space vector in the intended direction of rotation?
7. Why do we need to measure the rotor position in PMAC drives?
8. What does the electromagnetic torque produced by a PMAC drive depend on?
9. How can regenerative braking be accomplished in PMAC drives?
10. Why are PMAC drives called self-synchronous? How is the frequency of the applied voltages and currents determined? Are they related to the rotational speed of the shaft?
11. In a p-pole PMAC machine, what is the angle of the $\vec{i}_s(t)$ space vector in relation to the phase-a axis, for a given θ_m?
12. What is the frequency of currents and voltages in the stator circuit needed to produce a holding torque in a PMAC drive?
13. In calculating the voltages induced in the stator windings of a PMAC motor, what are the two components that are superimposed? Describe the procedure and the expressions.

14. Does L_m in the per-phase equivalent circuit of a PMAC machine have the same expression as in Chapter 9? Describe the differences, if any.

15. Draw the per-phase equivalent circuit and describe its various elements in PMAC drives.

16. Draw the controller block diagram and describe the hysteresis control of PMAC drives.

REFERENCE

1. https://sciamble.com/Resources/pe-drives-lab/basic-drives/pmac-torque-loadangle

FURTHER READING

Jahns, T. (1997). Variable frequency permanent magnet ac machine drives. In: *Power Electronics, and Variable Frequency Drives* (ed. B.K. Bose). IEEE Press.

Kazmierkowski, M.P. and Tunia, H. (1994). *Automatic Control of Converter-Fed Drives*. Elsevier.

Mohan, N. (2011). *Power Electronics: A First Course*. New York: Wiley.

Mohan, N., Undeland, T., and Robbins, W. (1995). *Power Electronics: Converters, Applications, and Design*, 2e. New York: Wiley.

PROBLEMS

8-1 Calculate the torque constant, similar to that in Eq. (8-6), for an 8-pole machine, where N_s equals the total number of turns per-phase.

8-2 Prove that Eq. (8-11) is correct.

8-3 Repeat Example 8-1 for $\theta_m = -45°$.

240 SINUSOIDAL PERMANENT-MAGNET ac (PMAC)

8-4 Repeat Example 8-1 for an 8-pole machine with the same value of k_T as in Example 8-1.

8-5 The PMAC machine of Example 8-2 is supplying a load torque $T_L = 5$ Nm at a speed of 5000 rpm. Draw a phasor diagram showing $\overline{V}_a$ and $\overline{I}_a$, along with their calculated values.

8-6 Repeat Problem 8-5 if the machine has $p = 4$, but has the same values of k_E, k_T, and L_s as before.

8-7 Repeat Problem 8-5, assuming that at time $t = 0$, the rotor angle $\theta_m(0) = 0°$.

8-8 The PMAC motor in Example 8-2 is driving a purely inertial load. A constant torque of 5 Nm is developed to bring the system from rest to a speed of 5000 rpm in 5 s. Neglect the stator resistance and leakage inductance. Determine and plot the voltage $v_a(t)$ and the current $i_a(t)$ as functions of time during this 8-s interval.

8-9 In Problem 8-8, the drive is expected to go into a regenerative mode at $t = 5^+$ s, with a torque $T_{em} = -5$ Nm. Assume that the rotor position at this instant is zero: $\theta_m = 0$. Calculate the three stator currents at this instant.

8-10 Redraw Fig. 8-3 if the PMAC drive is operating as a generator.

8-11 Recalculate Example 8-2 if the PMAC drive is operating as a generator, and instead of it supplying a torque of 3 Nm, it is being supplied this torque from the mechanical system connected to the machine-shaft.

8-12 The 2-pole motor in a PMAC drive has the following parameters: $R_s = 0.416\,\Omega$, $L_s = 1.365$ mH, and $k_T = 0.0957$ Nm/A. Draw the space vector and phasor diagrams for this machine if it is supplying its continuous rated torque of $T_{em} = 3.2$ Nm at its rated speed of 6000 rpm.

9 Induction Motors in Sinusoidal Steady-State

9-1 INTRODUCTION

Induction motors with squirrel-cage rotors are the workhorses of the industry because of their low cost and rugged construction. When operated directly from line-voltages (a 50- or 60-Hz utility input at essentially a constant voltage), induction motors operate at an almost constant speed. However, by means of power electronic converters, it is possible to vary their speed efficiently.

9-2 THE STRUCTURE OF THREE-PHASE, SQUIRREL-CAGE INDUCTION MOTORS

The stator of an induction motor consists of three-phase windings, sinusoidally distributed in the stator slots, as discussed in Chapter 6. These three windings are displaced by 120° in space with respect to each other, as shown by their axes in Fig. 9-1a.

The rotor, consisting of a stack of insulated laminations, has electrically conducting bars of copper or aluminum inserted (molded) through it, close to the periphery in the axial direction. These bars

---

(Adapted from chapter 11 of *Electric Machines and Drives: A First Course* ISBN: 978-1-118-07481-7 by Ned Mohan, January 2012)

Analysis and Control of Electric Drives: Simulations and Laboratory Implementation, First Edition. Ned Mohan and Siddharth Raju.
© 2021 John Wiley & Sons, Inc. Published 2021 by John Wiley & Sons, Inc.
Companion website: www.wiley.com/go/Mohan/Vectorcontrolinelectricdrives

242 INDUCTION MOTORS IN SINUSOIDAL STEADY-STATE

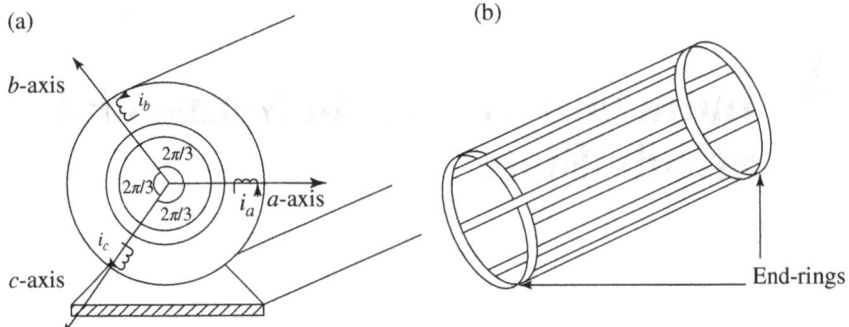

Fig. 9-1 (a) Three-phase stator winding axes and (b) squirrel-cage rotor.

are electrically shorted at each end of the rotor by electrically conducting end-rings, thus producing a cage-like structure, as shown in Fig. 9-1b. Such a rotor, called a squirrel-cage rotor, has a simple construction, low cost, and rugged nature.

9-3 THE PRINCIPLES OF INDUCTION MOTOR OPERATION

Our analysis will be under the line-fed conditions in which a balanced set of sinusoidal voltages of rated amplitude and frequency are applied to the stator windings. In the following discussion, we will assume a 2-pole structure which can be extended to a multipole machine with $p > 2$.

Figure 9-2 shows the stator windings. Under a balanced sinusoidal steady-state condition, the motor-neutral "n" is at the same potential as the source-neutral. Therefore, the source voltages v_a and so on appear across the respective phase windings, as shown in Fig. 9-2a. These phase voltages are shown in the phasor diagram of Fig. 9-2b, where

$$\overline{V}_a = \hat{V}\angle 0°, \quad \overline{V}_b = \hat{V}\angle -120°, \text{ and } \overline{V}_c = \hat{V}\angle -240° \quad (9\text{-}1)$$

and $f\left(=\dfrac{\omega}{2\pi}\right)$ is the frequency of the applied line-voltages to the motor.

THE PRINCIPLES OF INDUCTION MOTOR OPERATION 243

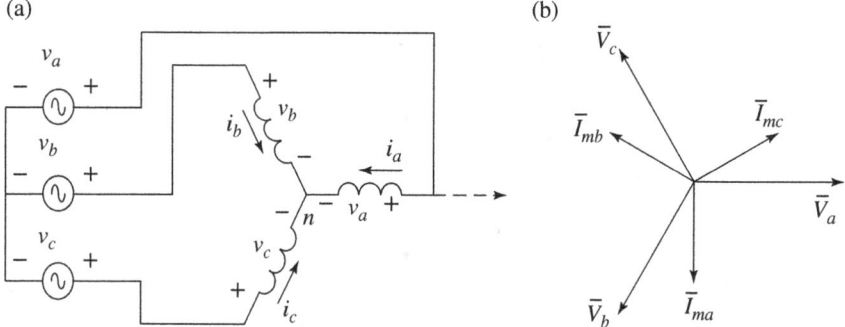

Fig. 9-2 Balanced three-phase sinusoidal voltages applied to the stator, rotor open-circuited.

To simplify our analysis, we will initially assume that the stator windings have a zero resistance ($R_s = 0$). Also, we will assume that $L_{\ell s} = 0$, implying that the leakage flux is zero; that is, all of the flux produced by each stator winding crosses the air gap and links with the other two stator windings and the rotor.

9-3-1 Electrically Open-Circuited Rotor

Initially, we will assume that the rotor is magnetically present but that its rotor bars are somehow open-circuited so that no current can flow. Therefore, we can use the analysis of this chapter, where the applied stator voltages given in Eq. (9-1) result only in the following magnetizing currents, which establish the rotating flux-density distribution in the air gap:

$$\overline{I}_{ma} = \hat{I}_m \angle -90°, \quad \overline{I}_{mb} = \hat{I}_m \angle -210°, \text{ and } \overline{I}_{mc} = \hat{I}_m \angle -330° \quad (9\text{-}2)$$

These phasors are shown in Fig. 9-2b, where, in terms of the per-phase magnetizing inductance L_m, the amplitude of the magnetizing currents is

$$\hat{I}_m = \frac{\hat{V}}{\omega L_m} \quad (9\text{-}3)$$

244 INDUCTION MOTORS IN SINUSOIDAL STEADY-STATE

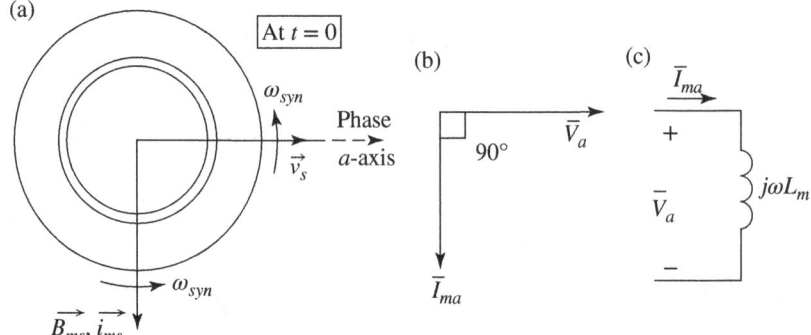

Fig. 9-3 Space vector representations at time $t = 0$, (b) voltage and current phasors for phase-a, and (c) equivalent circuit for phase-a.

The space vectors at $t = 0$ are shown in Fig. 9-3a, where, from this chapter,

$$\vec{v_s}(t) = \frac{3}{2}\hat{V}\angle\omega t \tag{9-4}$$

$$\vec{i_{ms}}(t) = \frac{3}{2}\hat{I}_m\angle\left(\omega t - \frac{\pi}{2}\right) \tag{9-5}$$

$$\vec{B_{ms}}(t) = \frac{\mu_o N_s}{2\ell_g}\hat{I}_{ms}\angle\left(\omega t - \frac{\pi}{2}\right) \tag{9-6}$$

where $\hat{B}_{ms} = \frac{3}{2}\frac{\mu_o N_s}{2\ell_g}\hat{I}_m$ and

$$\vec{v_s}(t) = \vec{e_{ms}}(t) = j\omega\left(\frac{3}{2}\pi r\ell\frac{N_s}{2}\right)\vec{B_{ms}}(t) \tag{9-7}$$

These space vectors rotate at a constant synchronous speed ω_{syn}, which in a 2-pole machine is

$$\omega_{syn} = \omega \left(\omega_{syn} = \frac{\omega}{p/2} \text{ for a } p\text{-pole machine}\right) \tag{9-8}$$

EXAMPLE 9-1

A 2-pole, three-phase induction motor has the following physical dimensions: radius $r = 7$ cm, length $\ell = 9$ cm, and the air gap length $\ell_g = 0.5$ mm. Calculate N_s, the number of turns per-phase, so that the peak of the flux-density distribution does not exceed 0.8 T when the rated voltages of 208 V (L-L, rms) are applied at the 60-Hz frequency.

Solution

From Eq. (9-7), the peak of the stator voltage and the flux-density distribution space vectors are related as follows:

$$\hat{V}_s = \frac{3}{2}\pi r \ell \frac{N_s}{2} \omega \hat{B}_{ms} \text{ where } \hat{V}_s = \frac{3}{2}\hat{V} = \frac{3}{2}\frac{208\sqrt{2}}{\sqrt{3}} = 254.75 \text{ V}$$

Substituting given values in the above expression, $N_s = 56.9$ turns. Since the number of turns must be an integer, $N_s = 57$ turns is selected.

9-3-2 The Short-Circuited Rotor

The voltages applied to the stator completely dictate the magnetizing currents (see Eqs. (9-2) and (9-3)) and the flux-density distribution, which is represented in Eq. (9-6) by $\overrightarrow{B_{ms}}(t)$ and is "cutting" the stator windings. Assuming the stator winding resistances and the leakage inductances to be zero, this flux-density distribution is unaffected by the currents in the rotor circuit, as illustrated by the transformer analogy below.

Transformer Analogy A two-winding transformer is shown in Fig. 9-4a, where two air gaps are introduced to bring the analogy closer to the case of induction machines, where flux lines must cross the air gap twice. The primary winding resistance and the leakage inductance are neglected (similar to neglecting the stator winding

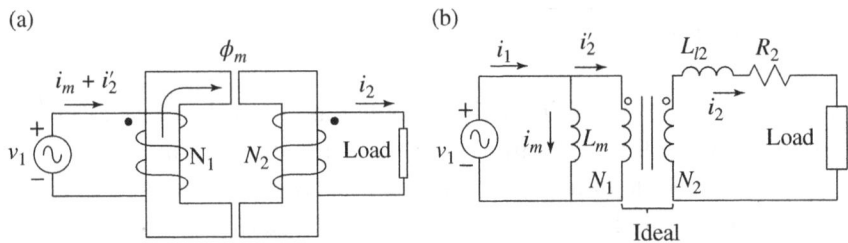

Fig. 9-4 (a) Two winding transformer and (b) equivalent circuit of the two winding transformer.

resistances and leakage inductances). The transformer equivalent circuit is shown in Fig. 9-4b. The applied voltage $v_1(t)$ and the flux $\phi_m(t)$ linking the primary winding are related by Faraday's Law:

$$v_1 = N_1 \frac{d\phi_m}{dt} \tag{9-9}$$

or, in the integral form,

$$\phi_m(t) = \frac{1}{N_1} \int v_1 \cdot dt \tag{9-10}$$

This shows that in this transformer, the flux $\phi_m(t)$ linking the primary winding is completely determined by the time-integral of $v_1(t)$, independent of the current i_2 in the secondary winding.

This observation is confirmed by the transformer equivalent circuit of Fig. 9-4b, where the magnetizing current i_m is completely dictated by the time-integral of $v_1(t)$, independent of the currents i_2 and i'_2:

$$i_m(t) = \frac{1}{L_m} \int v_1 \cdot dt \tag{9-11}$$

In the ideal transformer portion of Fig. 9-4b, the ampere-turns produced by the load current $i_2(t)$ are "nullified" by the additional current $i'_2(t)$ drawn by the primary winding, such that

$$N_1 i'_2(t) = N_2 i_2(t) \text{ or } i'_2(t) = \frac{N_2}{N_1} i_2(t) \tag{9-12}$$

THE PRINCIPLES OF INDUCTION MOTOR OPERATION   247

Thus, the total current drawn by the primary winding is

$$i_1(t) = i_m(t) + \underbrace{\frac{N_2}{N_1} i_2(t)}_{i'_2(t)} \tag{9-13}$$

Returning to our discussion of induction machines, the rotor consists of a short-circuited cage made up of rotor bars and the two end-rings. Regardless of what happens in the rotor circuit, the flux-density distribution "cutting" the stator windings must remain the same as under the assumption of an open-circuited rotor, as represented by $\overrightarrow{B_{ms}}(t)$ in Eq. (9-6).

Assume that the rotor is turning (due to the electromagnetic torque developed, as will be discussed shortly) at a speed ω_m in the same direction as the rotation of the space vectors, which represent the stator voltages and the air gap flux-density distribution. For now, we will assume that $\omega_m < \omega_{syn}$. The space vectors at time $t = 0$ are shown in the cross-section of Fig. 9-5a.

There is a relative speed between the flux-density distribution rotating at ω_{syn} and the rotor conductors rotating at ω_m. This relative

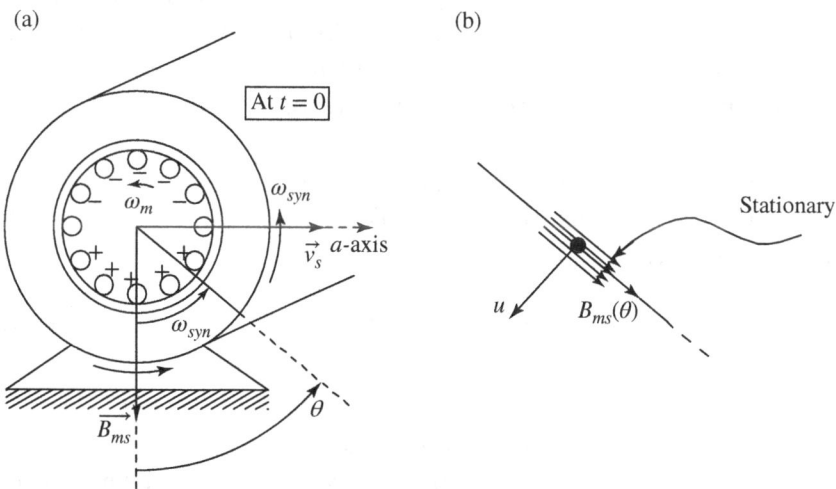

Fig. 9-5 (a) Induced voltages in the rotor bar and (b) motion of the rotor bar relative to the flux density.

248 INDUCTION MOTORS IN SINUSOIDAL STEADY-STATE

speed – that is, the speed at which the rotor is "slipping" with respect to the rotating flux-density distribution – is called the slip speed:

$$\text{slip speed} \quad \omega_{slip} = \omega_{syn} - \omega_m \quad (9\text{-}14)$$

By Faraday's Law ($e = B\ell u$), voltages are induced in the rotor bars due to the relative motion between the flux-density distribution and the rotor. At this time $t = 0$, the bar located at an angle θ from $\overrightarrow{B_{ms}}$ in Fig. 9-5a is "cutting" a flux density $B_{ms}(\theta)$. The flux-density distribution is moving ahead of the bar at a position θ at the angular speed of ω_{slip} rad/s or at the linear speed of $u = r\,\omega_{slip}$, where r is the radius. To determine the voltage induced in this rotor bar, we can consider the flux-density distribution to be stationary and the bar (at the angle θ) to be moving in the opposite direction at the speed u, as shown in Fig. 9-5b. Therefore, the voltage induced in the bar can be expressed as

$$e_{bar}(\theta) = B_{ms}(\theta)\ell\underbrace{r\,\omega_{slip}}_{u} \quad (9\text{-}15)$$

where the bar is of length ℓ and is at a radius r. The direction of the induced voltage can be established by visualizing that, on a positive charge q in the bar, the force f_q equals $u \times B$ where u and B are vectors shown in Fig. 9-5b. This force will cause the positive charge to move towards the front-end of the bar, establishing that the front-end of the bar will have a positive potential with respect to the back-end, as shown in Fig. 9-5a.

At any time, the flux-density distribution varies as the cosine of the angle θ from its positive peak. Therefore, in Eq. (9-15), $B_{ms}(\theta) = \hat{B}_{ms}\cos\theta$. Hence,

$$e_{bar}(\theta) = \ell r\,\omega_{slip}\hat{B}_{ms}\cos\theta \quad (9\text{-}16)$$

The Assumption of Rotor Leakage $L'_{\ell r} = 0$

At this point, we will make another *extremely* important simplifying assumption, to be analyzed later in more detail. The assumption is that the rotor cage has no leakage inductance; that is, $L'_{\ell r} = 0$. This assumption implies that the rotor has no leakage flux, and that all

THE PRINCIPLES OF INDUCTION MOTOR OPERATION   249

of the flux produced by the rotor-bar currents crosses the air gap and links (or "cuts") the stator windings. Another implication of this assumption is that, at any time, the current in each squirrel-cage bar, short-circuited at both ends by conducting end-rings, is inversely proportional to the bar resistance R_{bar}.

In Fig. 9-6a at $t = 0$, the induced voltages are maximum in the top and the bottom rotor bars "cutting" the peak flux density. Elsewhere, induced voltages in rotor bars depend on $\cos\theta$, as given by Eq. (9-16). The polarities of the induced voltages at the near-end of the bars are indicated in Fig. 9-6a. Figure 9-6b shows the electrical equivalent circuit, which corresponds to the cross-section of the rotor shown in Fig. 9-6a. The size of the voltage source represents the magnitude of the voltage induced. Because of the symmetry in this circuit, it is easy to visualize that the two end-rings (assumed to have negligible resistances themselves) are at the same potential. Therefore, the rotor bar at an angle θ from the positive flux-density peak location has a current which is equal to the induced voltage divided by the bar resistance:

$$i_{bar}(\theta) = \frac{e_{bar}(\theta)}{R_{bar}} = \frac{\ell\, r\, \omega_{slip} \hat{B}_{ms} \cos\theta}{R_{bar}} \quad (\text{using Eq.}(9-16)) \quad (9\text{-}17)$$

where each bar has a resistance R_{bar}. From Eq. (9-17), the currents are maximum in the top and the bottom bars at this time, indicated

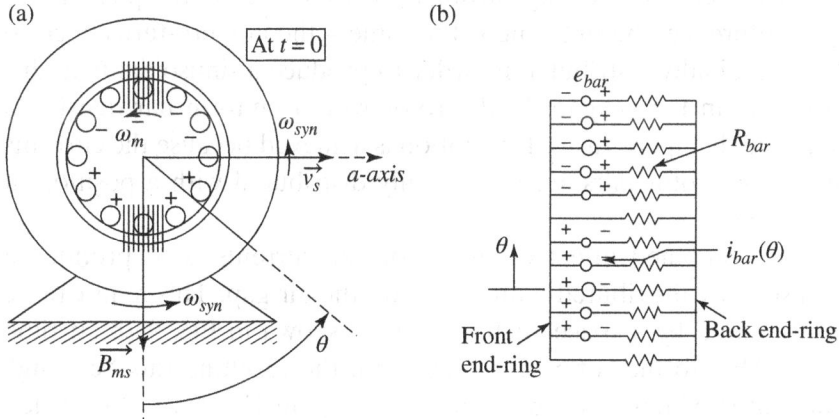

Fig. 9-6   (a) Polarities of voltages induced and (b) electrical equivalent circuit of the rotor.

250 INDUCTION MOTORS IN SINUSOIDAL STEADY-STATE

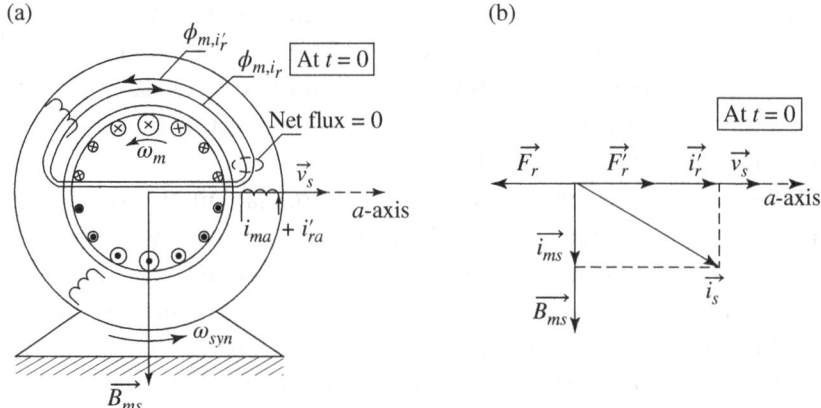

Fig. 9-7 (a) Rotor-produced flux ϕ_{m,i_r} and the flux ϕ_{m,i'_r}, and (b) space vector diagram with a short-circuited rotor ($L'_{\ell r} = 0$).

by the largest circles in Fig. 9-7a; elsewhere, the magnitude of the current depends on $\cos\theta$, where θ is the angular position of any bar as defined in Fig. 9-6a.

It is important to note that the rotor has a uniform bar density around its periphery, as shown in Fig. 9-6a. The sizes of the circles in Fig. 9-7a denote the relative current magnitudes. The sinusoidal rotor-current distribution is different than that in the stator phase winding, which has a sinusoidally distributed conductor density but the same current flowing through each conductor. In spite of this key difference, the outcome is the same – the ampere-turns need to be sinusoidally distributed in order to produce a sinusoidal field distribution in the air gap. In the rotor with a uniform bar density, a sinusoidal ampere-turns distribution is achieved because the currents in various rotor bars are sinusoidally distributed with a position at any time.

The combined effect of the rotor-bar currents is to produce a sinusoidally distributed mmf acting on the air gap. This mmf can be represented by a space vector $\vec{F}_r(t)$, as shown in Fig. 9-7b, at time $t = 0$. Due to the rotor-produced mmf, the resulting flux "cutting" the stator winding is represented by ϕ_{m,i_r} in Fig. 9-7a. As argued earlier by means of the transformer analogy, the net flux-density

THE PRINCIPLES OF INDUCTION MOTOR OPERATION 251

distribution "cutting" the voltage-supplied stator windings must remain the same as in the case of the open-circuited rotor. Therefore, in order to cancel the rotor-produced flux ϕ_{m,i_r}, the stator windings must draw the additional currents i'_{ra}, i'_{rb}, and i'_{rc} to produce the flux represented by ϕ'_{m,i_r}.

In the space vector diagram of Fig. 9-7b, the mmf produced by the rotor bars is represented by $\vec{F}_r$ at time $t = 0$. As shown in Fig. 9-7b, the stator currents i'_{ra}, i'_{rb}, and i'_{rc} (which flow in addition to the magnetizing currents) must produce an mmf $\vec{F'_r}$, which is equal in amplitude, but opposite in direction to $\vec{F}_r$, in order to neutralize its effect:

$$\vec{F'_r} = -\vec{F}_r \tag{9-18}$$

The additional currents i'_{ra}, i'_{rb}, and i'_{rc} drawn by the stator windings to produce $\vec{F'_r}$ can be expressed by a current space vector $\vec{i'_r}$, as shown in Fig. 9-7b at $t = 0$, where

$$\vec{i'_r} = \frac{\vec{F'_r}}{N_s/2} \quad \left(\hat{I}'_r = \text{the amplitude of } \vec{i'_r}\right) \tag{9-19}$$

The total stator current $\vec{i}_s$ is the vector sum of the two components: $\vec{i}_{ms}$, which sets up the magnetizing field, and $\vec{i'_r}$, which neutralizes the rotor-produced mmf:

$$\vec{i}_s = \vec{i}_{ms} + \vec{i'_r} \tag{9-20}$$

These space vectors are shown in Fig. 9-7b at $t = 0$. Equation (9-17) shows that the rotor-bar currents are proportional to the flux-density peak and the slip speed. Therefore, the "nullifying" mmf peak and the peak current $\hat{I}'_r$ must also be linearly proportional to $\hat{B}_{ms}$ and ω_{slip}. This relationship can be expressed as

$$\hat{I}'_r = k_i \hat{B}_{ms} \omega_{slip} \quad (k_i = \text{a constant}) \tag{9-21}$$

where k_i is a constant based on the design of the machine.

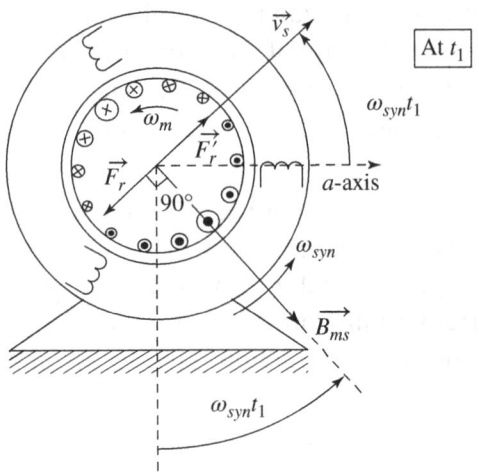

Fig. 9-8 Rotor-produced mmf and the compensating mmf at time $t = t_1$.

During the sinusoidal steady-state operating condition in Fig. 9-7b, the rotor-produced mmf distribution (represented by $\vec{F}_r$) and the compensating mmf distribution (represented by $\vec{F}'_r$) rotate at the synchronous speed ω_{syn}, and each has a constant amplitude. This can be illustrated by drawing the motor cross-section and space vectors at some arbitrary time $t_1 > 0$, as shown in Fig. 9-8, where the $\vec{B}_{ms}$ space vector has rotated by an angle $\omega_{syn}t_1$ because $\vec{v}_s$ has rotated by $\omega_{syn}t_1$. Based on the voltages and currents induced in the rotor bars, $\vec{F}_r$ is still 90° behind the $\vec{B}_{ms}$ space vector, as in Fig. 9-7a and b. This implies that the $\vec{F}_r(t)$ and $\vec{F}'_r(t)$ vectors are rotating at the same speed as $\vec{B}_{ms}(t)$ – that is, the synchronous speed ω_{syn}. At a given operating condition with constant values of ω_{slip} and $\hat{B}_{ms}$, the bar-current distribution, relative to the peak of the flux-density vector, is the same in Fig. 9-8 as it is in Fig. 9-7. Therefore, the amplitudes of $\vec{F}_r(t)$ and $\vec{F}'_r(t)$ remain constant as they rotate at the synchronous speed.

EXAMPLE 9-2

Consider an induction machine that has 2 poles and is supplied by a rated voltage of 208 V (L-L, rms) at the frequency of 60 Hz. It is operating in steady state and is loaded to its rated torque. Neglect

THE PRINCIPLES OF INDUCTION MOTOR OPERATION 253

the stator leakage impedance and the rotor leakage flux. The per-phase magnetizing current is 4.0 A (rms). The current drawn per-phase is 10 A (rms) and is at an angle of 23.56° (lagging). Calculate the per-phase current if the mechanical load decreases so that the slip speed is one-half that of the rated case.

Solution

We will consider the phase-*a* voltage to be the reference phasor. Therefore,

$$\overline{V}_a = \frac{208\sqrt{2}}{\sqrt{3}} \angle 0° = 169.8 \angle 0° \text{ V}$$

It is given that, at the rated load, as shown in Fig. 9-9a, $\overline{I}_{ma} = 4.0\sqrt{2} \angle -90°$ A and $\overline{I}_a = 10.0\sqrt{2} \angle -23.56°$ A. From the phasor diagram in Fig. 9-9a, $\overline{I}'_{ra} = 9.173\sqrt{2} \angle 0°$ A.

At one-half of the slip speed, the magnetizing current is the same, but the amplitudes of the rotor-bar currents, and hence of $\overline{I}'_{ra}$ being reduced by one-half:

$$\overline{I}_{ma} = 4.0\sqrt{2} \angle -90° \text{ A and } \overline{I}'_{ra} = 4.59\sqrt{2} \angle 0° \text{ A}$$

Therefore, $\overline{I}_a = 6.1\sqrt{2} \angle -41.16°$ A, as shown in the phasor diagram of Fig. 9-9b.

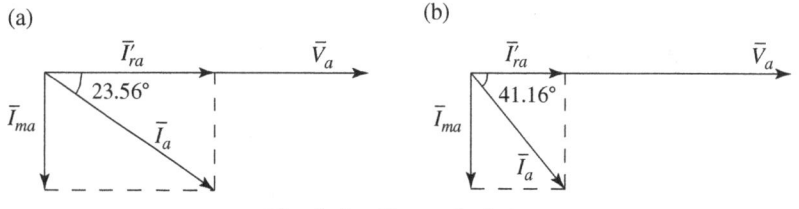

Fig. 9-9 Example 9-2.

254 INDUCTION MOTORS IN SINUSOIDAL STEADY-STATE

Revisiting the Transformer Analogy The transformer equivalent circuit in Fig. 9-4b, illustrated so that the voltage-supplied primary winding draws a compensating current to neutralize the effect of the secondary winding current, in order to ensure that the resultant flux linking the *primary winding*, remains the same as under the open-circuited condition. Similarly, in an induction motor, the stator neutralizes the rotor-produced field to ensure that the resultant flux "cutting" the *stator windings* remains the same as under the rotor open-circuited condition. In induction machines, this is how the stator "keeps track" of what is happening in the rotor. However, compared to transformers, induction machine operation is more complex where the rotor-cage quantities are at the slip frequency (discussed below) and are transformed into the stator-frequency quantities "seen" from the stator.

The Slip Frequency, f_{slip}, in the Rotor Circuit The frequency of induced voltages (and currents) in the rotor bars can be obtained by considering Fig. 9-10a. At $t = 0$, the bottom-most bar labeled "p" is being "cut" by the positive peak flux density and has a positive induced voltage at the front-end. The $\vec{B}_{ms}(t)$ space vector, which is rotating with a speed of ω_{syn}, is "pulling ahead" at the slip speed ω_{slip}

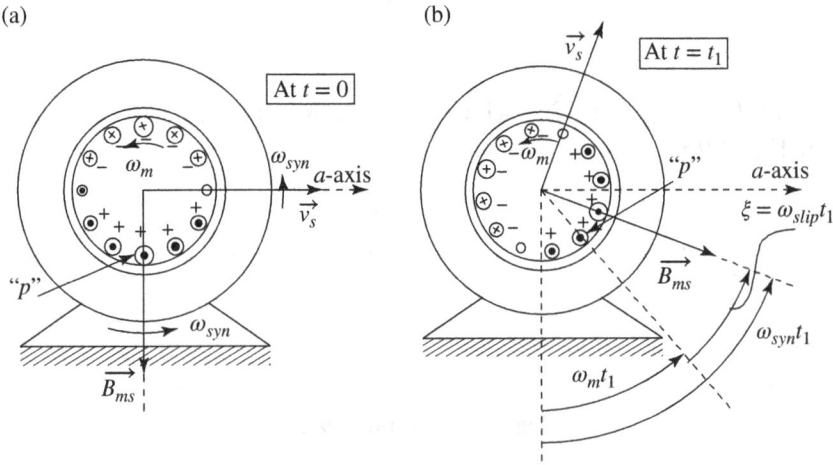

Fig. 9-10 Voltage induced in bar "p" at (a) $t = 0$ and (b) $t = t_1$.

THE PRINCIPLES OF INDUCTION MOTOR OPERATION   255

with respect to the rotor bar "p," which is rotating at ω_m. Therefore, as shown in Fig. 9-10b, at some time $t_1 > 0$, the angle between $\vec{B}_{ms}(t)$ and the rotor bar "p" is

$$\xi = \omega_{slip} \, t_1 \qquad (9\text{-}22)$$

Therefore, the first time (call it $T_{2\pi}$) when the bar "p" is again being "cut" by the positive peak flux density is when $\zeta = 2\pi$. Therefore, from Eq. (9-22),

$$T_{2\pi} = \frac{\xi(=2\pi)}{\omega_{slip}} \qquad (9\text{-}23)$$

where $T_{2\pi}$ is the time-period between the two consecutive positive peaks of the induced voltage in the rotor bar "p." Therefore, the induced voltage in the rotor bar has a frequency (which we will call the slip frequency f_{slip}), which is the inverse of $T_{2\pi}$ in Eq. (9-23):

$$f_{slip} = \frac{\omega_{slip}}{2\pi} \qquad (9\text{-}24)$$

For convenience, we will define a unit-less (dimensionless) quantity called slip, s, as the ratio of the slip speed to the synchronous speed:

$$s = \frac{\omega_{slip}}{\omega_{syn}} \qquad (9\text{-}25)$$

Substituting for ω_{slip} from Eq. (9-25) into Eq. (9-24) and noting that $\omega_{syn} = 2\pi f$ (in a 2-pole machine),

$$f_{slip} = s f \qquad (9\text{-}26)$$

In steady state, induction machines operate at ω_m, very close to their synchronous speed, with a slip s of generally less than 0.03 (or 3%). Therefore, in steady state, the frequency (f_{slip}) of voltages and currents in the rotor circuit is typically less than a few Hz.

Note that $\vec{F}_r(t)$, which is created by the slip-frequency voltages and currents in the rotor circuit, rotates at the slip speed ω_{slip}, relative to the rotor. Since the rotor itself is rotating at a speed of ω_m, the net result is that $\vec{F}_r(t)$ rotates at a total speed of $(\omega_{slip} + \omega_m)$, which is equal to the synchronous speed ω_{syn}. This confirms what we had concluded earlier about the speed of $\vec{F}_r(t)$ by comparing Figs. 9-7 and 9-8.

EXAMPLE 9-3

In Example 9-2, the rated speed (while the motor supplies its rated torque) is 3475 rpm. Calculate the slip speed ω_{slip}, the slip s, and the slip frequency f_{slip} of the currents and voltages in the rotor circuit.

Solution

This is a 2-pole motor. Therefore, at the rated frequency of 60 Hz, the rated synchronous speed, from Eq. (9-8), is $\omega_{syn} = \omega = 2\pi \times 60 = 377 \text{ rad/s}$. The rated speed is $\omega_{m,rated} = \dfrac{2\pi \times 3475}{60} = 363.9 \text{ rad/s}$

Therefore, $\omega_{slip,\,rated} = \omega_{syn,\,rated} - \omega_{m,\,rated} = 377.0 - 363.9 = 13.1$ rad/s. From Eq. (9-25),

$$\text{slip } s_{rated} = \dfrac{\omega_{slip,rated}}{\omega_{syn,rated}} = \dfrac{13.1}{377.0} = 0.0347 = 3.47\%$$ and from Eq. (9-26),

$$f_{slip,rated} = s_{rated} f = 2.08 \text{ Hz}$$

Electromagnetic Torque The electromagnetic torque on the rotor is produced by the interaction of the flux-density distribution represented by $\vec{B}_{ms}(t)$ in Fig. 9-7a, and the rotor-bar currents producing the mmf $\vec{F}_r(t)$. As in Chapter 10, it will be easier to calculate the torque produced on the rotor by first calculating the torque on the stator equivalent winding that produces the nullifying mmf $\vec{F}'_r(t)$. At $t = 0$, this equivalent stator winding, sinusoidally distributed with N_s turns, has its axis along the $\vec{F}'_r(t)$ space vector, as shown in Fig. 9-11. The winding also has a current $\hat{I}'_r$ flowing through it.

THE PRINCIPLES OF INDUCTION MOTOR OPERATION 257

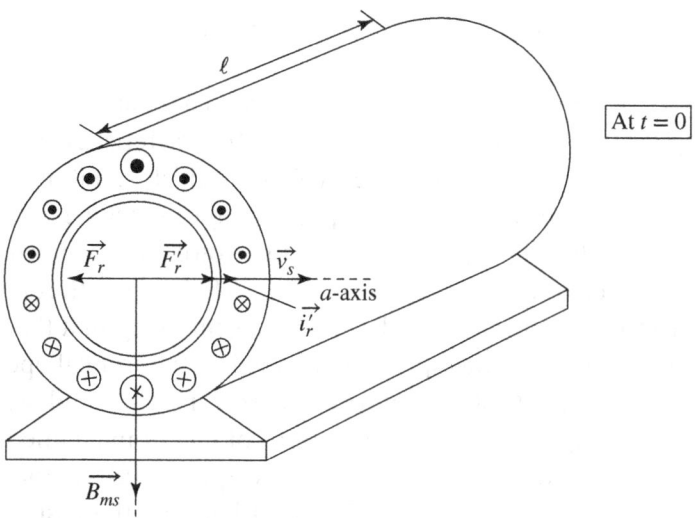

Fig. 9-11 Calculation of electromagnetic torque.

Following the derivation of the electromagnetic torque in Chapter 8, from Eq. (8-5),

$$T_{em} = \pi r \ell \frac{N_s}{2} \hat{B}_{ms} \hat{I}'_r \qquad (9\text{-}27)$$

The above equation can be written as

$$T_{em} = k_t \hat{B}_{ms} \hat{I}'_r \left(\text{where } k_t = \pi r \ell \frac{N_s}{2} \right) \qquad (9\text{-}28)$$

where k_t is a constant which depends on the machine design. The torque on the stator in Fig. 9-11 acts in a clockwise direction, and the torque on the rotor is equal in magnitude and acts in a counterclockwise direction.

The current peak $\hat{I}'_r$ depends linearly on the flux-density peak $\hat{B}_{ms}$ and the slip speed ω_{slip}, as expressed by Eq. (9-21) ($\hat{I}'_r = k_i \hat{B}_{ms} \omega_{slip}$). Therefore, substituting for $\hat{I}'_r$ in Eq. (9-28),

$$T_{em} = k_{t\omega} \hat{B}_{ms}^2 \omega_{slip} \quad (k_{t\omega} = k_t k_i) \qquad (9\text{-}29)$$

where $k_{t\omega}$ is a machine torque constant. If the flux-density peak is maintained at its rated value in Eq. (9-29),

$$T_{em} = k_{T\omega}\omega_{slip} \quad \left(k_{T\omega} = k_{t\omega}\hat{B}_{ms}^2\right) \qquad (9\text{-}30)$$

where $k_{T\omega}$ is another torque constant of the machine.

Equation (9-30) expresses the torque-speed characteristic of induction machines. For a rated set of applied voltages, which result in $\omega_{syn,rated}$ and $\hat{B}_{ms,rated}$, the torque developed by the machine increases linearly with the slip speed ω_{slip} as the rotor slows down. This torque-speed characteristic is shown in Fig. 9-12 in two different ways.

At zero torque, the slip speed ω_{slip} is zero, implying that the motor rotates at the synchronous speed. This is only a theoretical operating point, because the motor's internal bearing friction and windage losses would require that a finite amount of electromagnetic torque be generated to overcome them. The torque-speed characteristic beyond the rated torque is shown dotted because the assumptions of neglecting stator leakage impedance and the rotor leakage inductance begin to breakdown.

The torque-speed characteristic helps to explain the operating principle of induction machines, as illustrated in Fig. 9-13. In steady state, the operating speed ω_{m1} is dictated by the intersection of the electromagnetic torque and the mechanical-load torque T_{L1}. If the load torque is increased to T_{L2}, the induction motor slows down to ω_{m2}, increasing the slip speed ω_{slip}. This increased slip speed results in higher induced voltages and currents in the rotor bars, and hence a higher electromagnetic torque is produced to meet the increase in mechanical-load torque.

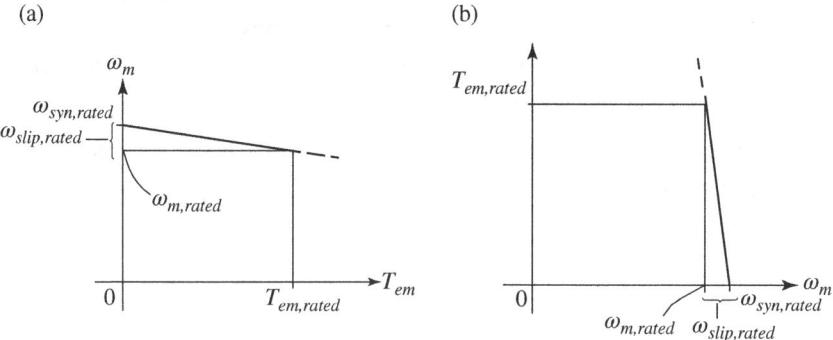

Fig. 9-12 Torque-speed characteristic of induction motors.

THE PRINCIPLES OF INDUCTION MOTOR OPERATION

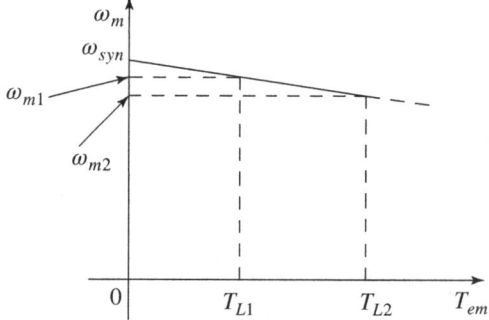

Fig. 9-13 Operation of an induction motor.

On a dynamic basis, the electromagnetic torque developed by the motor interacts with the shaft-coupled mechanical load, in accordance with the following mechanical-system equation:

$$\frac{d\omega_m}{dt} = \frac{T_{em} - T_L}{J_{eq}} \qquad (9\text{-}31)$$

where J_{eq} is the combined motor-load inertia constant and T_L (generally a function of speed) is the torque of the mechanical load opposing the rotation. The acceleration torque is $(T_{em} - T_L)$.

Note that the electromagnetic torque developed by the motor equals the load torque in steady state. Often, the torque required to overcome friction and windage (including that of the motor itself) can be included, lumped with the load torque.

EXAMPLE 9-4

In Example 9-3, the rated torque supplied by the motor is 8 Nm. Calculate the torque constant $k_{T\omega}$, which linearly relates the torque developed by the motor to the slip speed.

Solution

From Eq. (9-30), $k_{T\omega} = \dfrac{T_{em}}{\omega_{slip}}$. Therefore, using the rated conditions,

(Continued)

$$k_{T\omega} = \frac{T_{em,rated}}{\omega_{slip,rated}} = \frac{8.0}{13.1} = 0.61 \; \frac{\text{Nm}}{\text{rad/s}}$$

The torque-speed characteristic is as shown in Fig. 9-12, with the slope given above.

The Generator (Regenerative Braking) Mode of Operation
Induction machines can be used as generators, for example, many wind-electric systems use induction generators to convert wind energy to electrical output, which is fed into the utility grid. Most commonly, however, while slowing down, induction motors go into a regenerative-braking mode (which, from the machine's standpoint, is the same as the generator mode), where the kinetic energy associated with the inertia of the mechanical system is converted into electrical output. In this mode of operation, the rotor speed exceeds the synchronous speed ($\omega_m > \omega_{syn}$) where both are in the same direction. Hence, $\omega_{slip} < 0$.

Under the condition of negative slip speed shown in Fig. 9-14, the voltages and currents induced in the rotor bars are of opposite polarities and directions compared to those with positive slip speed in Fig. 9-7a. Therefore, the electromagnetic torque on the rotor acts

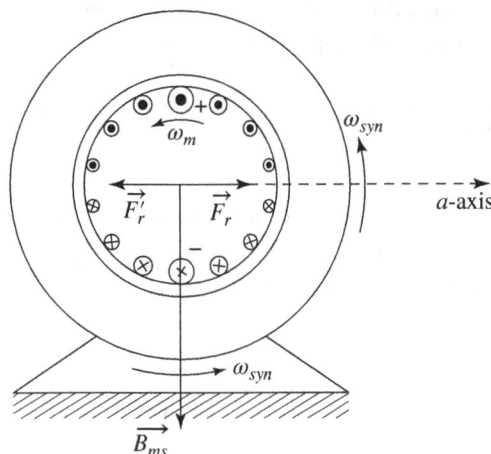

Fig. 9-14 Regenerative braking in induction motors.

THE PRINCIPLES OF INDUCTION MOTOR OPERATION

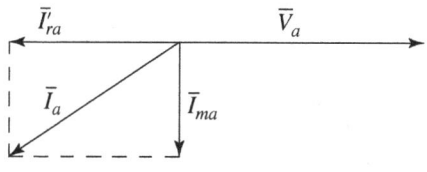

Fig. 9-15 Example 9-5.

in a clockwise direction, opposing the rotation and thus slowing down the rotor. In this regenerative-braking mode, T_{em} in Eq. (9-31) has a negative value.

EXAMPLE 9-5

The induction machine of Example 9-2 is to produce the rated torque in the regenerative-braking mode. Draw the voltage and current phasors for phase-a.

Solution

With the assumption that the stator leakage impedance can be neglected, the magnetizing current is the same as in Example 9-2, $\bar{I}_{ma} = 4.0\sqrt{2}\angle -90°$ A as shown in Fig. 9-15. However, since we are dealing with a regenerative-braking torque, $\bar{I}'_{ra} = -9.173\sqrt{2}\angle 0°$ A as shown in the phasor diagram of Fig. 9-15. Hence, $\bar{I}_a = 10.0\sqrt{2}\angle -156.44°$ A as shown in Fig. 9-15.

Reversing the Direction of Rotation Reversing the sequence of applied voltages (a-b-c to a-c-b) causes the reversal of direction, as shown in Fig. 9-16.

Including the Rotor Leakage Inductance Up to the rated torque, the slip speed and the slip frequency in the rotor circuit are small, hence, it is reasonable to neglect the effect of the rotor leakage inductance. However, loading the machine beyond the rated torque results in larger slip speeds and slip frequencies, and the effect of the rotor leakage inductance should be included in the analysis, as described below.

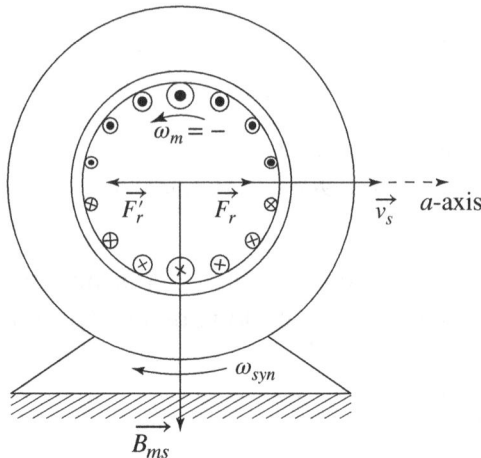

Fig. 9-16 Reversing the direction of rotation in an induction motor.

Of all the flux produced by the currents in the rotor bars, a portion (which is called the leakage flux and is responsible for the rotor leakage inductance) does not completely cross the air gap and does not "cut" the stator windings. *First, considering only the stator-established flux-density distribution* $\vec{B}_{ms}(t)$ at $t = 0$ as in Fig. 9-6a, the top and the bottom bars are "cut" by the peak $\hat{B}_{ms}$ of the flux-density distribution, and due to this flux, the voltages induced in them are the maximum. However (as shown in Fig. 9-17a), the bar currents lag, due to the inductive effect of the rotor leakage flux, and are maximum in the bars which were "cut" by $\vec{B}_{ms}(t)$ sometime earlier. Therefore, the rotor mmf space vector $\vec{F}_r(t)$ in Fig. 9-17a lags $\vec{B}_{ms}(t)$ by an angle $\frac{\pi}{2} + \theta_r$, where θ_r is called the rotor power factor angle.

At $t = 0$, the flux lines produced by the rotor currents in Fig. 9-17b can be divided into two components: ϕ_{m,i_r}, which crosses the air gap and "cuts" the stator windings, and $\phi_{\ell r}$, the rotor leakage flux, which does *not* cross the air gap to "cut" the stator windings.

The stator excited by ideal voltage sources (and assuming that R_s and L_{ls} are zero) demands that the flux-density distribution $\vec{B}_{ms}(t)$ "cutting" it be unchanged. Therefore, additional stator currents, represented by $\vec{i}_r'(t)$ in Fig. 9-17a, are drawn to produce $\phi_{m,i_r'}$ in Fig. 9-17b to compensate for ϕ_{m,i_r} (but not to compensate for $\phi_{\ell r}$,

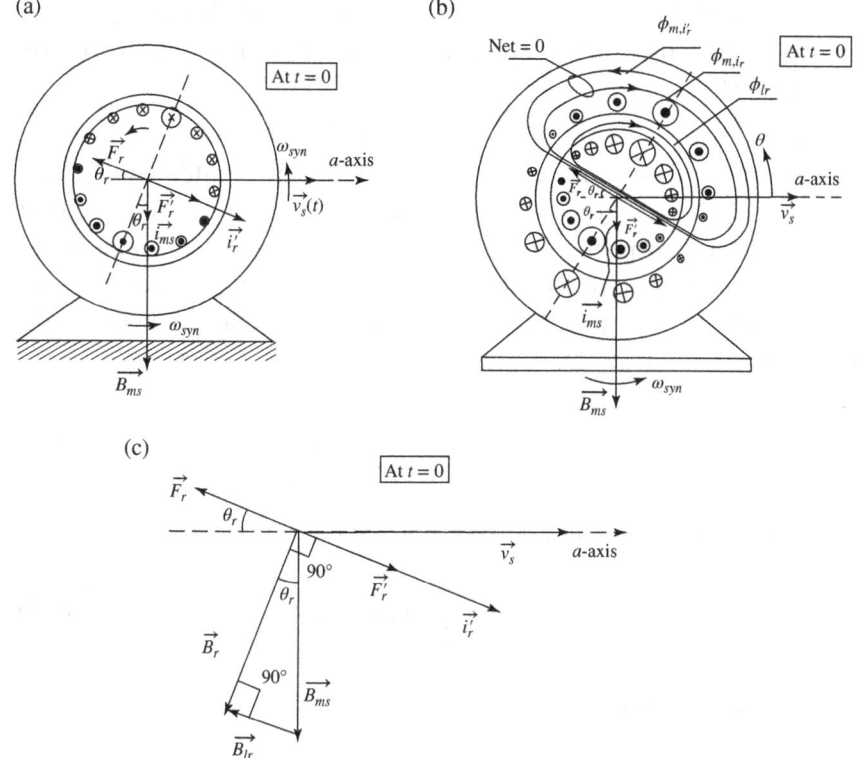

Fig. 9-17 Space vectors with the effect of rotor leakage flux included.

whose existence the stator is unaware of), such that ϕ_{m,i'_r} is equal in magnitude and opposite in direction to ϕ_{m,i_r}.

The additional currents drawn from the three stator phase windings can be represented utilizing the equivalent stator winding with N_s turns and carrying a current $\hat{i}'_r$, as shown in Fig. 9-17b. The resulting $\vec{F}_r$, $\vec{F}'_r$, and $\vec{i}'_r$ space vectors at $t = 0$ are shown in Fig. 9-17c. The rotor bars are "cut" by the net flux-density distribution represented by $\vec{B}_r(t)$, shown in Fig. 9-17c at $t = 0$, where

$$\vec{B}_r(t) = \vec{B}_{ms}(t) + \vec{B}_{\ell r}(t) \qquad (9\text{-}32)$$

$\vec{B}_{\ell r}(t)$ represents in the air gap the rotor leakage flux-density distribution (due to $\phi_{\ell r}$), which, for our purposes, is also assumed to be radial and sinusoidally distributed. Note that $\vec{B_r}$ is *not* created by the currents in the rotor bars; rather, it is the flux-density distribution "cutting" the rotor bars.

The equivalent stator winding shown in Fig. 9-17b has a current $\hat{i}'_r$ and is "cut" by the flux-density distribution represented by $\vec{B_{ms}}$. As shown in Fig. 9-17c, the $\vec{B_{ms}}$ and $\vec{i}'_r$ space vectors are at an angle of $(\pi/2 - \theta_r)$ with respect to each other. Using a procedure similar to the one which led to the torque expression in Eq. (9-28), we can show that the torque developed depends on the sine of the angle $(\pi/2 - \theta_r)$ between $\vec{B_{ms}}$ and $\vec{i}'_r$:

$$T_{em} = k_t \hat{B}_{ms} \hat{i}'_r \sin\left(\frac{\pi}{2} - \theta_r\right) \tag{9-33}$$

In the space vector diagram of Fig. 9-17c,

$$\hat{B}_{ms} \sin\left(\frac{\pi}{2} - \theta_r\right) = \hat{B}_r \tag{9-34}$$

Therefore, in Eq. (9-33),

$$T_{em} = k_t \hat{B}_r \hat{i}'_r \tag{9-35}$$

The above development suggests how we can achieve vector control of induction machines. In an induction machine, $\vec{B_r}(t)$ and $\vec{i}'_r(t)$ are naturally at right angles (90°) to each other. (Note in Fig. 9-17b that the rotor bars with the maximum current are those "cutting" the peak of the rotor flux-density distribution $\hat{B}_r$.) Therefore, if we can keep the rotor flux-density peak $\hat{B}_r$ constant, then

$$T_{em} = k_T \hat{i}'_r \text{ where } k_T = k_t \hat{B}_r \tag{9-36}$$

The torque developed by the motor can be controlled by $\hat{i}'_r$. This allows induction-motor drives to emulate the performance of dc-motor and brushless-dc motor drives.

9-3-3 Per-Phase Steady-State Equivalent Circuit (Including Rotor Leakage)

The space vector diagram at $t = 0$ is shown in Fig. 9-18a for the rated voltages applied. This results in the phasor diagram for phase-a in Fig. 9-18b.

The current $\vec{I}'_{ra}$, which is lagging behind the applied voltage $\overline{V}_a$, can be represented as flowing through an inductive branch in the equivalent circuit of Fig. 9-18c, where R_{eq} and L_{eq} are yet to be determined. For the above determination, assume that the rotor is blocked and that the voltages applied to the stator create the same conditions ($\vec{B}_{ms}$ with the same $\hat{B}_{ms}$ and at the same ω_{slip} with respect to the rotor) in the rotor circuit, as in Fig. 9-18. Therefore, in Fig. 9-19a with the blocked rotor, we will apply stator voltages at the slip frequency $f_{slip}(=\omega_{slip}/2\pi)$ from Eq. (9-8) and of amplitude $(\omega_{slip}/\omega_{syn})\hat{V}$ from Eq. (9-7), as shown in Fig. 9-19a and b.

The blocked-rotor bars, similar to those in the rotor turning at ω_m, are "cut" by an identical flux-density distribution (which has the same peak value $\hat{B}_{ms}$ and which rotates at the same slip speed ω_{slip} with respect to the rotor). The phasor diagram in the blocked-rotor case

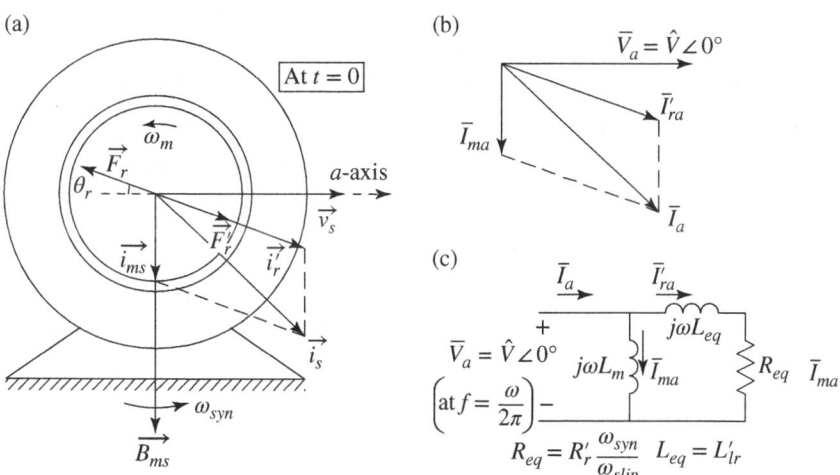

Fig. 9-18 Rated voltage applied.

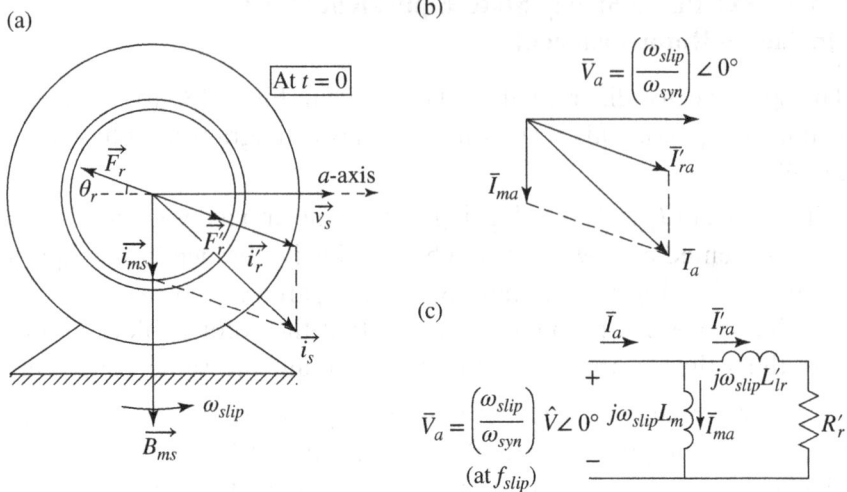

Fig. 9-19 Blocked rotor and slip-frequency voltages applied.

is shown in Fig. 9-19b and the phase equivalent circuit is shown in Fig. 9-19c. (The quantities at the stator terminals in the blocked-rotor case of Fig. 9-19 are similar to those of a transformer primary, with its secondary winding short-circuited.) The current $\overline{I}'_{ra}$ in Fig. 9-19c is at the slip frequency f_{slip} and flowing through an inductive branch, which consists of R'_r and $L'_{\ell r}$ connected in series. Note that R'_r and $L'_{\ell r}$ are the equivalent rotor resistance and the equivalent rotor leakage inductance, "seen" on a per-phase basis from the stator side. The impedance of the inductive branch with $\overline{I}'_{ra}$ in this blocked-rotor case is

$$Z_{eq,blocked} = R'_r + j\omega_{slip}L'_{\ell r} \qquad (9\text{-}37)$$

The three-phase power loss in the bar resistances of the blocked rotor is

$$P_{r,loss} = 3R'_r(I'_{ra})^2 \qquad (9\text{-}38)$$

where I'_{ra} is the rms value.

As far as the conditions "seen" by an observer sitting on the rotor are concerned, they are identical to the original case with the rotor

THE PRINCIPLES OF INDUCTION MOTOR OPERATION 267

turning at speed ω_m but slipping at speed ω_{slip} with respect to ω_{syn}. Therefore, in both cases, the current component $\bar{I}'_{ra}$ has the same amplitude, and the same phase angle, with respect to the applied voltage. Therefore, in the original case of Fig. 9-18, where the applied voltages are higher by a factor of $\dfrac{\omega_{syn}}{\omega_{slip}}$, the impedance must be higher by the same factor; that is, from Eq. (9-37),

$$\underbrace{Z_{eq}}_{\text{at } f} = \dfrac{\omega_{syn}}{\omega_{slip}} \underbrace{(R'_r + j\omega_{slip}L'_{\ell r})}_{Z_{eq,\text{blocked}} \text{ at } f_{slip}} = R'_r \dfrac{\omega_{syn}}{\omega_{slip}} + j\omega_{syn}L'_{\ell r} \qquad (9\text{-}39)$$

Therefore, in the equivalent circuit of Fig. 9-18c at a frequency f, $R_{eq} = R'_r \dfrac{\omega_{syn}}{\omega_{slip}}$ and $L_{eq} = L'_{\ell r}$. The per-phase equivalent circuit of Fig. 9-18c is repeated in Fig. 9-20a, where $\omega_{syn} = \omega$ for a 2-pole machine. The power loss $P_{r,\,loss}$ in the rotor circuit in Fig. 9-20a is the same as that given by Eq. (9-38) for the blocked-rotor case of Fig. 9-19c. Therefore, the resistance $R'_r \dfrac{\omega_{syn}}{\omega_{slip}}$ can be divided into two parts: R'_r and $R'_r \dfrac{\omega_m}{\omega_{slip}}$, as shown in Fig. 9-20b, where $P_{r,\,loss}$ is lost as heat in R'_r and the power dissipation in $R'_r \dfrac{\omega_m}{\omega_{slip}}$, on a three-phase basis, gets converted into mechanical power (which also equals T_{em} times ω_m):

$$P_{em} = 3\dfrac{\omega_m}{\omega_{slip}} R'_r (I'_{ra})^2 = T_{em}\omega_m \qquad (9\text{-}40)$$

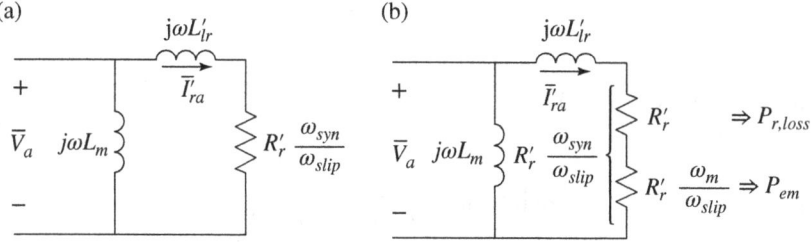

Fig. 9-20 Splitting the rotor resistance into the loss component and power output component (neglecting the stator-winding leakage impedance).

Therefore,

$$T_{em} = 3R'_r \frac{(I'_{ra})^2}{\omega_{slip}} \qquad (9\text{-}41)$$

From Eqs. (9-38) and (9-41),

$$\frac{P_{r,loss}}{T_{em}} = \omega_{slip} \qquad (9\text{-}42)$$

This is an important relationship because it shows that to produce the desired torque T_{em}, we should minimize the value of the slip speed in order to minimize the power loss in the rotor circuit.

EXAMPLE 9-6

Consider a 60-Hz induction motor with $R'_r = 0.45\,\Omega$ and $X'_{\ell r} = 0.85\,\Omega$. The rated slip speed is 4%. Ignore the stator leakage impedance. Compare the torque at the rated slip speed by (a) ignoring the rotor leakage inductance and (b) including the rotor leakage inductance.

Solution

To calculate T_{em} at the rated slip speed, we will make use of Eq. (9-41), where I'_{ra} can be calculated from the per-phase equivalent circuit of Fig. 9-20a. Ignoring the rotor leakage inductance,

$$I'_{ra}\big|_{L'_{\ell r}=0} = \frac{V_a}{R'_r \dfrac{\omega_{syn}}{\omega_{slip}}}$$

and from Eq. (9-41),

$$T_{em}\big|_{L'_{\ell r}=0} = \frac{3R'_r}{\omega_{slip}} \frac{V_a^2}{\left(R'_r \dfrac{\omega_{syn}}{\omega_{slip}}\right)^2}$$

Including the rotor leakage inductance,

$$I'_{ra} = \frac{V_a}{\sqrt{\left(R'_r \frac{\omega_{syn}}{\omega_{slip}}\right)^2 + (X'_{\ell r})^2}}$$

and from Eq. (9-41),

$$T_{em} = \frac{3R'_r}{\omega_{slip}} \frac{V_a^2}{\left(R'_r \frac{\omega_{syn}}{\omega_{slip}}\right)^2 + (X'_{\ell r})^2}$$

At the rated slip speed of 4%, $\frac{\omega_{slip}}{\omega_{syn}} = 0.04$. Therefore, comparing the above two expressions for torque by substituting the numerical values,

$$\frac{T_{em}|_{L'_{\ell r} = 0}}{T_{em}} = \frac{\left(R'_r \frac{\omega_{syn}}{\omega_{slip}}\right)^2 + (X'_{\ell r})^2}{\left(R'_r \frac{\omega_{syn}}{\omega_{slip}}\right)^2} = \frac{126.56^2 + 0.85^2}{126.56^2} = 1.0$$

The above example shows that under normal operation, when the motor is supplying a torque within its rated value, it does so at very small values of slip speed. Therefore, as shown in this example, we are justified in ignoring the effect of the rotor leakage inductance under normal operation. In high-performance applications requiring vector control, the effect of the rotor leakage inductance can be included, as discussed in Chapter 13.

Including the Stator Winding Resistance R_s and Leakage Inductance $L_{\ell s}$ Including the effect of the stator winding resistance R_s and the leakage inductance $L_{\ell s}$ is analogous to including the effect of primary winding impedance in the transformer equivalent circuit, as shown in Fig. 9-21a.

270 INDUCTION MOTORS IN SINUSOIDAL STEADY-STATE

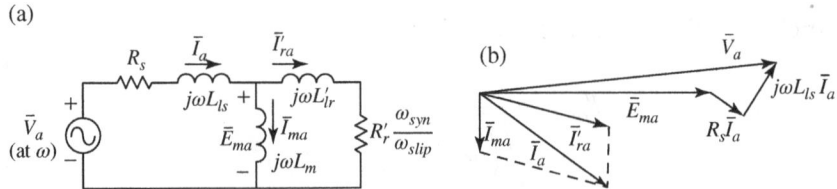

Fig. 9-21 (a) Per-phase equivalent circuit including the stator leakage and (b) phasor diagram.

In the per-phase equivalent circuit of Fig. 9-21a, the applied voltage $\overline{V}_a$ is reduced by the voltage drop across the stator-winding leakage impedance to yield $\overline{E}_{ma}$:

$$\overline{E}_{ma} = \overline{V}_a - (R_s + j\omega L_{\ell s})\overline{I}_s \qquad (9\text{-}43)$$

where $\overline{E}_{ma}$ represents the voltage induced in the stator phase-a by the rotating flux-density distribution $\overrightarrow{B_{ms}}(t)$. The phasor diagram with $\overline{E}_{ma}$ as the reference phasor is shown in Fig. 9-21b.

9-4 TESTS TO OBTAIN THE PARAMETERS OF THE PER-PHASE EQUIVALENT CIRCUIT

The parameters of the per-phase equivalent circuit of Fig. 9-21a are usually not supplied by the motor manufacturers. The three tests described below can be performed to estimate these parameters.

9-4-1 dc-Resistance Test to Estimate R_s

The stator resistance R_s can best be estimated by the dc measurement of the resistance between the two phases:

$$R_s(dc) = \frac{R_{phase\text{-}phase}}{2} \qquad (9\text{-}44)$$

This dc resistance value, measured by passing a dc current through two of the phases, can be modified by a skin-effect factor [1] to help estimate its line-frequency value more closely.

9-4-2 The No-Load Test to Estimate L_m

The magnetizing inductance L_m can be calculated from the no-load test. In this test, the motor is applied its rated stator voltages in steady state and no mechanical load is connected to the rotor shaft. Therefore, the rotor turns almost at the synchronous speed with $\omega_{slip} \cong 0$. Hence, the resistance $R'_r \dfrac{\omega_{syn}}{\omega_{slip}}$ in the equivalent circuit of Fig. 9-21a becomes very large, allowing us to assume that $\bar{I}'_{ra} \cong 0$, as shown in Fig. 9-22a.

The following quantities are measured: the per-phase rms voltage $V_a(=V_{LL}/\sqrt{3})$, the per-phase rms current I_a, and the three-phase power $P_{3-\phi}$ drawn by the motor. Subtracting the calculated power dissipation in R_s from the measured power, the remaining power $P_{FW,core}$ (the sum of the core losses, the stray losses, and the power to overcome friction and windage) is

$$P_{FW,core} = P_{3-\phi} - 3R_s I_a^2 \qquad (9\text{-}45)$$

With rated voltages applied to the motor, the above loss can be assumed to be a constant value which is independent of the motor loading.

Assuming that $L_m \gg L_{\ell s}$, the magnetizing inductance L_m can be calculated based on the per-phase reactive power Q from the following equation:

$$Q = \sqrt{(V_a I_a)^2 - \left(\dfrac{P_{3-\phi}}{3}\right)^2} = (\omega L_m) I_a^2 \qquad (9\text{-}46)$$

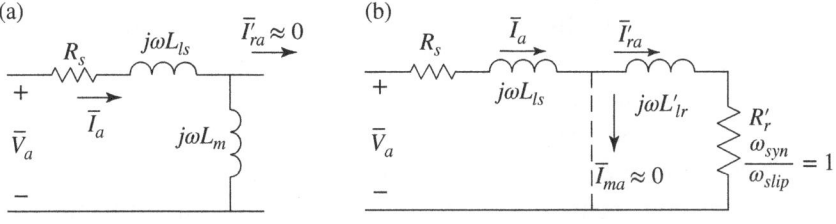

Fig. 9-22 (a) No-load test and (b) blocked-rotor test.

9-4-3 Blocked-Rotor Test to Estimate R'_r and the Leakage Inductances

The blocked-rotor (or locked-rotor) test is conducted to determine both R'_r, the rotor resistance "seen" from the stator on a per-phase basis, and the leakage inductances in the equivalent circuit of Fig. 9-21a. Note that the rotor is blocked from turning, and the stator is applied line-frequency, three-phase voltages with a small magnitude such that the stator currents equal their rated value. With the rotor blocked, $\omega_m = 0$ and hence $\dfrac{\omega_{syn}}{\omega_{slip}} = 1$. The resulting equivalent impedance $(R'_r + j\omega L'_{\ell r})$ in Fig. 9-22b, can be assumed to be much smaller than the magnetizing reactance $(j\omega L_m)$, which can be considered to be infinite. Therefore, by measuring V_a, I_a, and the three-phase power into the motor, we can calculate R'_r (having already estimated R_s previously) and $(L_{\ell s} + L'_{\ell r})$. In order to determine these two leakage inductances explicitly, we need to know their ratio, which depends on the design of the machine. As an approximation for general-purpose motors, we can assume that

$$L_{\ell s} \cong \frac{2}{3} L'_{\ell r} \tag{9-47}$$

This allows both leakage inductances to be calculated explicitly.

9-5 INDUCTION MOTOR CHARACTERISTICS AT RATED VOLTAGES IN MAGNITUDE AND FREQUENCY

The typical torque-speed characteristic for general-purpose induction motors, with name-plate (rated) values of applied voltages, is shown in Fig. 9-23a, where the normalized torque (as a ratio of its rated value) is plotted as a function of the rotor speed ω_m/ω_{syn}.

With no load connected to the shaft, the torque T_{em} demanded from the motor is very low (only enough to overcome the internal bearing friction and windage), and the rotor turns at speed very close in value to the synchronous speed ω_{syn}. Up to the rated torque, the torque developed by the motor is linear with respect to ω_{slip}, a

INDUCTION MOTOR CHARACTERISTICS 273

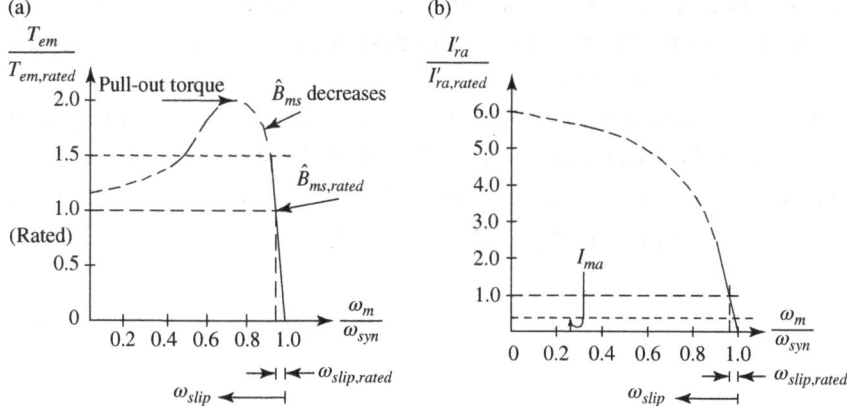

Fig. 9-23 (a) Torque-speed characteristic and (b) current-speed characteristic.

relationship given by Eq. (9-30). Far beyond the rated condition, for which the machine is designed to operate in steady state, T_{em} no longer increases linearly with ω_{slip} for the following reasons:

1. The effect of leakage inductance in the rotor circuit at a higher frequency can no longer be ignored, and, from Eq. (9-33), the torque is less due to the declining value of $\sin(\pi/2 - \theta_r)$.
2. Large values of I'_{ra}, and hence of I_a, cause a significant voltage drop across the stator-winding leakage impedance $(R_s + j\omega L_{\ell s})$. This voltage drop causes E_{ma} to decrease, which in turn decreases $\hat{B}_{ms}$.

The above effects take place simultaneously, and the resulting torque characteristic for large values of ω_{slip} (which are avoided in the induction-motor drives discussed in the next chapter) is shown dotted in Fig. 9-23a. The rated value of the slip-speed ω_{slip} at which the motor develops its rated torque is typically in a range of 0.03–0.05 times the synchronous speed ω_{syn}.

In the torque-speed characteristic of Fig. 9-23a, the maximum torque that the motor can produce is called the pull-out (breakdown) torque. The torque when the rotor speed is zero is called the starting

274 INDUCTION MOTORS IN SINUSOIDAL STEADY-STATE

torque. The values of the pull-out and the starting torques, as a ratio of the rated torque, depend on the design class of the motor, as discussed in the next section.

Figure 9-23b shows the plot of the normalized rms current I'_{ra} as a function of the rotor speed. Up to the rated slip speed (up to the rated torque), I'_{ra} is linear with respect to the slip speed. This can be seen from Eq. (9-21) (with $\hat{B}_{ms} = \hat{B}_{ms,rated}$):

$$I'_r = \left(k_i \hat{B}_{ms,rated} \right) \omega_{slip} \qquad (9\text{-}48)$$

Hence,

$$I'_{ra} = k_I \omega_{slip} \quad \left(k_I = \frac{1}{\sqrt{2}} \frac{2}{3} k_i \hat{B}_{ms,rated} \right) \qquad (9\text{-}49)$$

where k_I is a constant, which linearly relates the slip speed to the rms current I'_{ra}. Notice that this plot is linear up to the rated slip speed, beyond which, the effects of the stator and the rotor leakage inductances come into effect. At the rated operating point, the value of the rms magnetizing current I_{ma} is typically in a range of 20–40% of the per-phase stator rms current I_a. The magnetizing current I_{ma} remains relatively constant with speed, decreasing slightly at very large values of ω_{slip}. At or below the rated torque, the per-phase stator current magnitude I_a can be calculated by assuming the $\vec{I}'_{ra}$ and $\vec{I}_{ma}$ phasors to be at 90° with respect to each other; thus

$$I_a \cong \sqrt{I'^2_{ra} + I^2_{ma}} \text{ (below the rated torque)} \qquad (9\text{-}50)$$

While delivering a torque higher than the rated torque, $\vec{I}'_{ra}$ is much larger in magnitude than the magnetizing current $\vec{I}_{ma}$ (also considering a large phase shift between the two). This allows the stator current to be approximated as follows:

$$I_a \cong I'_{ra} \text{ (above the rated torque)} \qquad (9\text{-}51)$$

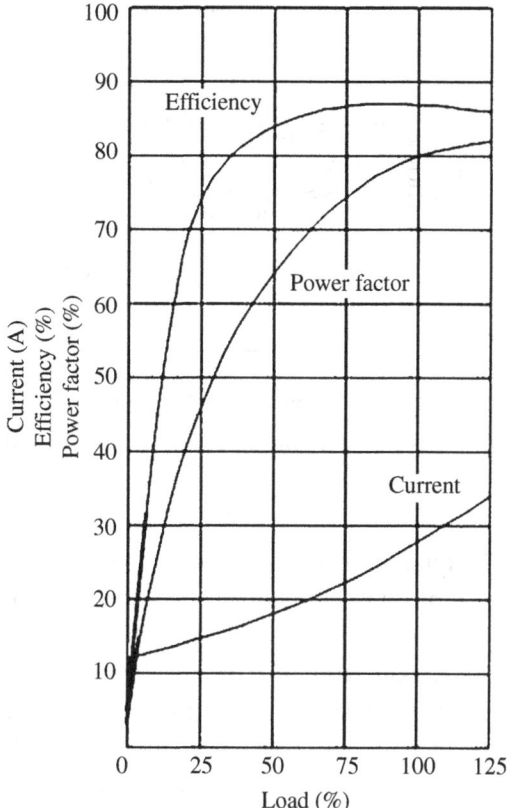

Fig. 9-24 Typical performance curves for Design B 10 kW, 4-pole, three-phase induction motor.

Figure 9-24 shows the typical variations of the power factor and the motor efficiency as a function of motor loading. These curves depend on the class and size of the motor.

9-6 INDUCTION MOTORS OF NEMA DESIGN A, B, C, AND D

Three-phase induction machines are classified in the American Standards (NEMA) under five design letters: A, B, C, D, and F. Each design class of motors has different torque and current specifications. Figure 9-25 illustrates typical torque-speed curves for design A, B, C,

276 INDUCTION MOTORS IN SINUSOIDAL STEADY-STATE

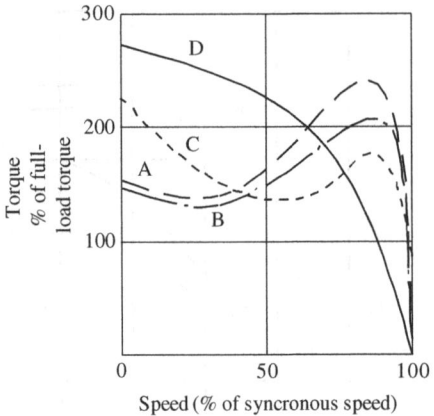

Fig. 9-25 Typical torque-speed characteristics of NEMA design A, B, C, and D motors.

and D motors; Design F motors have low pull-out and starting torques and thus are very limited in applications. As a ratio of the rated quantities, each design class specifies minimum values of pull-out and starting torques, and a maximum value of the starting current.

As noted previously, Design Class B motors are used most widely for general-purpose applications. These motors must have a minimum of a 200% pull-out torque.

Design A motors are similar to the general-purpose Design B motors, except that they have a somewhat higher pull-out (breakdown) torque and a smaller full-load slip. Design A motors are used when unusually low values of winding losses are required – in totally enclosed motors, for example.

Design C motors are high starting-torque, low starting-current machines. They also have a somewhat lower pull-out (breakdown) torque than Design A and B machines. Design C motors are almost always designed with double-cage rotor windings to enhance the rotor-winding skin effect.

Finally, Design D motors are high starting-torque, high-slip machines. The minimum starting torque is 275% of the rated torque. The starting torque in these motors can be assumed to be the same as the pull-out torque.

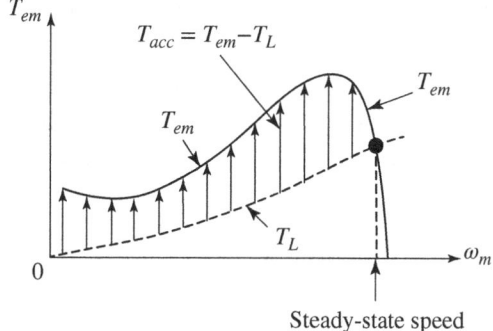

Fig. 9-26 Available acceleration torque during start-up.

9-7 LINE START

It should be noted that the induction-motor drives, discussed in detail in the next chapter, are operated so as to keep ω_{slip} at low values. Hence, the dotted portions of the characteristics shown in Fig. 9-23 are of no significance. However, if an induction motor is started from the line-voltage supply without an electronic power converter, it would at first draw 6 to 8 times its rated current, as shown in Fig. 9-23b, limited mainly by the leakage inductances. Figure 9-26 shows that the available acceleration torque $T_{acc}(=T_{em} - T_L)$ causes the motor to accelerate from a standstill, in accordance with Eq. (9-31). In Fig. 9-26, an arbitrary torque-speed characteristic of the load is assumed, and the intersection of the motor and the load characteristics determines the steady-state point of operation.

9-8 HARDWARE PROTOTYPING OF INDUCTION MOTOR PARAMETER ESTIMATION

In this chapter, the steady-state operation of an induction motor was presented. The procedure to estimate the motor parameters was discussed in Section 9-4. Accurate estimation of motor parameters is critical for proper controller design, to ensure a stable motor operation, as will be seen in the following chapters.

278 INDUCTION MOTORS IN SINUSOIDAL STEADY-STATE

The step-by-step procedure to obtain the motor parameters of a 24 V (L-L, rms) 3ϕ ac induction motor using Workbench in hardware is provided in the accompanying website [2]. The estimated motor electrical and mechanical parameters, which will be used in the controller design for hardware implementation, are discussed in the following chapters.

9-9 REVIEW QUESTIONS

1. Describe the construction of squirrel-cage induction machines.

2. With the rated voltages applied, what does the magnetizing current depend on? Does this current, to a significant extent, depend on the mechanical load on the motor? How large is it in relation to the rated motor current?

3. Draw the space vector diagram at $t = 0$, and the corresponding phasor diagram, assuming the rotor to be open-circuited.

4. Under a balanced, three-phase, sinusoidal steady-state excitation, what is the speed of the rotating flux-density distribution called? How is this speed related to the angular frequency of the electrical excitation in a p-pole machine?

5. In our analysis, why did we initially assume the stator leakage impedance to be zero? How does the transformer analogy, with the primary winding leakage impedance assumed to be zero, help? Under the assumption that the stator leakage impedance is zero, is $\vec{B}_{ms}(t)$ completely independent of the motor loading?

6. What is the definition of the slip speed ω_{slip}? Does ω_{slip} depend on the number of poles? How large is the rated slip speed, compared to the rated synchronous speed?

7. Write the expressions for the voltage and the current (assuming the rotor leakage inductance to be zero) in a rotor bar located at an angle θ from the peak of $\vec{B}_{ms}$.

REVIEW QUESTIONS 279

8. The rotor bars located around the periphery of the rotor are of uniform cross-section. In spite of this, what allows us to represent the mmf produced by the rotor bar currents by a space vector $\vec{F}_r(t)$ at any time t?

9. Assuming the stator leakage impedance and the rotor inductance to be zero, draw the space vector diagram, the phasor diagram, and the per-phase equivalent circuit of a loaded induction motor.

10. In the equivalent circuit of Problem 9-9, what quantities does the rotor-bar current peak, represented by $\hat{I}'_{ra}$, depend on?

11. What is the frequency of voltages and currents in the rotor circuit called? How is it related to the slip speed? Does it depend on the number of poles?

12. What is the definition of slip s, and how does it relate the frequency of voltages and currents in the stator circuit to that in the rotor circuit?

13. What is the speed of rotation of the mmf distribution produced by the rotor bar currents: (a) with respect to the rotor? (b) in the air gap with respect to a stationary observer?

14. Assuming $L'_{\ell r}$ to be zero, what is the expression for the torque T_{em} produced? How and why does it depend on ω_{slip} and $\hat{B}_{ms}$? Draw the torque-speed characteristic.

15. Assuming $L'_{\ell r}$ to be zero, explain how induction motors meet load-torque demand.

16. What makes an induction machine go into the regenerative-braking mode? Draw the space vectors and the corresponding phasors under the regenerative-braking condition.

17. Can an induction machine be operated as a generator that feeds into a passive load, for example, a bank of three-phase resistors?

18. How is it possible to reverse the direction of rotation of an induction machine?

19. Explain the effect of including the rotor leakage flux by means of a space vector diagram.

20. How do we derive the torque expression, including the effect of $L'_{\ell r}$?

21. What is $\vec{B_r}(t)$ and how does it differ from $\vec{B_{ms}}(t)$? Is $\vec{B_r}(t)$ perpendicular to the $\vec{F_r}(t)$ space vector?

22. Including the rotor leakage flux, which rotor bars have the highest currents at any instant of time?

23. Describe how to obtain the per-phase equivalent circuit, including the effect of the rotor leakage flux.

24. What is the difference between $\vec{I}'_{ra}$ in Fig. 9-18c and in Fig. 9-19c, in terms of its frequency, magnitude, and phase angle?

25. Is the torque expression in Eq. (9-41) valid in the presence of the rotor leakage inductance and the stator leakage impedance?

26. When producing a desired torque T_{em}, what is the power loss in the rotor circuit proportional to?

27. Draw the per-phase equivalent circuit, including the stator leakage impedance.

28. Describe the tests and the procedure to obtain the parameters of the per-phase equivalent circuit.

29. In steady state, how is the mechanical torque at the shaft different than the electromechanical torque T_{em} developed by the machine?

30. Do induction machines have voltage and torque constants similar to other machines that we have studied so far? If so, write their expressions.

31. Plot the torque-speed characteristic of an induction motor for applied rated voltages. Describe various portions of this characteristic.

32. What are the various classes of induction machines? Briefly describe their differences.

33. What are the problems associated with the line-starting of induction motors? Why is the starting current so high?

34. Why is reduced-voltage starting used? Show the circuit implementation and discuss the pros and cons of using it to save energy.

REFERENCES

1. Mohan, N., Undeland, T., and Robbins, W. (1995). *Power Electronics: Converters, Applications, and Design*, 2e. New York: Wiley.
2. https://sciamble.com/Resources/pe-drives-lab/basic-drives/im-characterization

FURTHER READING

Fitzgerald, K. and Umans, S.D. (1990). *Electric Machinery*, 5e. McGraw Hill.

Slemon, G.R. (1992). *Electric Machines, and Drives*. Addison-Wesley, Inc.

PROBLEMS

9-1 Consider a three-phase, 2-pole induction machine. Neglect the stator winding resistance and the leakage inductance. The rated voltage is 208 V (L-L, rms) at 60 Hz. L_m = 60 mH, and the peak flux density in the air gap is 0.85 T. Consider that the phase-a voltage reaches its positive peak at $\omega t = 0$. Assuming that the rotor circuit is somehow open-circuited, calculate and draw the following space vectors at $\omega t = 0$ and at $\omega t = 60°$: $\vec{v_s}$, $\vec{i_{ms}}$, and $\vec{B_{ms}}$. Draw the phasor diagram with $\overline{V}_a$ and $\overline{I}_{ma}$. What is the relationship between $\hat{B}_{ms}$, $\hat{I}_{ms}$, and $\hat{I}_m$?

282 INDUCTION MOTORS IN SINUSOIDAL STEADY-STATE

9-2 Calculate the synchronous speed in machines with a rated frequency of 60 Hz and with the following number of poles p: 2, 4, 6, 8, and 12.

9-3 The machines in Problem 9-2 produce the rated torque at a slip $s = 4\%$ when supplied with rated voltages. Under the rated torque condition, calculate in each case the slip speed ω_{slip} in rad/s and the frequency f_{slip} (in Hz) of the currents and voltages in the rotor circuit.

9-4 In the transformer of Fig. 9-4a, each air gap has a length $\ell_g = 1.0$ mm. The core iron can be assumed to have an infinite permeability. $N_1 = 100$ turns and $N_2 = 50$ turns. In the air gap, $\hat{B}_g = 1.1$ T and $v_1(t) = 100\sqrt{2}\cos\omega t$ at a frequency of 60 Hz. The leakage impedance of the primary winding can be neglected. With the secondary winding open-circuited, calculate and plot $i_m(t)$, $\phi_m(t)$, and the induced voltage $e_2(t)$ in the secondary winding due to $\phi_m(t)$ and $v_1(t)$.

9-5 In Example 9-1, calculate the magnetizing inductance L_m.

9-6 In an induction machine, the torque constant $k_{T\omega}$ (in Eq. (9-30)) and the rotor resistance R'_r are specified. Calculate $\hat{I}'_r$ as a function of ω_{slip}, in terms of $k_{T\omega}$ and R'_r, for torques below the rated value. Assume that the flux density in the air gap is at its rated value. Hint: use Eq. (9-41).

9-7 An induction motor produces rated torque at a slip speed of 100 rpm. If a new machine is built with bars of a material that has twice the resistivity of the old machine (and nothing else is changed), calculate the slip speed in the new machine when it is loaded to the rated torque.

9-8 In the transformer circuit of Fig. 9-4b, the load on the secondary winding is a pure resistance R_L. Show that the emf induced in the secondary winding (due to the time-derivative of the combination of ϕ_m and the secondary-winding leakage flux) is in phase with the secondary current i_2. Note: this is analogous to the induction-motor case, where the rotor leakage flux is included, and the current is maximum in the bar which is "cut" by $\hat{B}_r$, the peak of the rotor flux-density distribution (represented by $\overrightarrow{B_r}$).

9-9 In a 60-Hz, 208 V (L-L, rms), 5-kW motor, $R'_r = 0.45\,\Omega$ and $X'_{\ell r} = 0.83\,\Omega$. The rated torque is developed at the slip $s = 0.04$. Assuming that the motor is supplied with rated voltages and is delivering the rated torque, calculate the rotor power factor angle. What is $\hat{B}_r/\hat{B}_{ms}$?

9-10 In a 2-pole, 208 V (L-L, rms), 60-Hz motor, $R_s = 0.5\,\Omega$, $R'_r = 0.45\,\Omega$, $X_{\ell s} = 0.6\,\Omega$, and $X'_{\ell r} = 0.83\,\Omega$. The magnetizing reactance $X_m = 28.5\,\Omega$. This motor is supplied by its rated voltages. The rated torque is developed at the slip $s = 0.04$. At the rated torque, calculate the rotor power loss, the input current, and the input power factor of operation.

9-11 In a 208-V (L-L, rms), 60-Hz, 5-kW motor, tests are carried out with the following results: $R_{phase\text{-}phase} = 1.1\,\Omega$. No-Load Test: applied voltages of 208 V (L-L, rms), $I_a = 6.5$ A, and $P_{no\text{-}load,\,3\text{-}phase} = 175$ W. Blocked-Rotor Test: applied voltages of 53 V (L-L, rms), $I_a = 18.2$ A, and $P_{blocked,3\text{-}phase} = 900$ W. Estimate the per-phase equivalent circuit parameters.

10 Induction-Motor Drives: Speed Control

10-1 INTRODUCTION

Induction-motor drives are used in the process-control industry to adjust the speeds of fans, compressors, pumps, and the like. In many applications, the capability to vary speed efficiently can lead to large savings in energy. Adjustable-speed induction-motor drives are also used for electric traction, and for motion control to automate factories.

Figure 10-1 shows the block diagram of an adjustable-speed induction-motor drive. The utility input can be either single-phase or three-phase. It is converted by the power-processing unit (PPU) into three-phase voltages of appropriate magnitude and frequency, based on the controller input. In most general-purpose adjustable-speed drives (ASDs), the speed is not sensed, and hence the speed-sensor block and its input to the controller are shown dotted.

It is possible to adjust the induction-motor speed by controlling only the magnitude of the line-frequency voltages applied to the motor. Though simple and inexpensive to implement, this method is extremely energy-*in*efficient if the speed is to be varied over a wide range. Also, there are various other methods of speed control, but

(Adapted from chapter 12 of *Electric Machines and Drives: A First Course* ISBN: 978-1-118-07481-7 by Ned Mohan, January 2012)

Analysis and Control of Electric Drives: Simulations and Laboratory Implementation, First Edition. Ned Mohan and Siddharth Raju.
© 2021 John Wiley & Sons, Inc. Published 2021 by John Wiley & Sons, Inc.
Companion website: www.wiley.com/go/Mohan/Vectorcontrolinelectricdrives

286 INDUCTION-MOTOR DRIVES

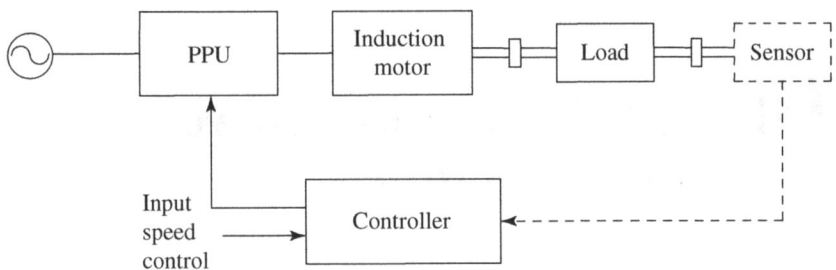

Fig. 10-1 Block diagram of an induction-motor drive.

they require wound-rotor induction motors. Their description can be found in references listed at the end of Chapter 9. Our focus in this chapter is on examining energy-efficient speed control of squirrel-cage induction motors over a wide range.

The emphasis is on general-purpose speed control rather than precise control of position using vector control, which is discussed in Chapter 13.

10-2 CONDITIONS FOR EFFICIENT SPEED CONTROL OVER A WIDE RANGE

In the block diagram of the induction-motor drive shown in Fig. 10-1, we find that an energy-efficient system requires that both the PPU and the induction motor maintain high energy efficiency over a wide range of speed and torque conditions. In Chapter 4, it was shown that the switch-mode techniques result in very high efficiencies of the PPUs. Therefore, the focus in this section will be on achieving high efficiency of induction motors over a wide range of speed and torque.

We will begin this discussion by first considering the case in which an induction motor is applied the rated voltages (line-frequency sinusoidal voltages of the rated amplitude $\hat{V}_{rated}$ and the rated frequency f_{rated}, which are the same as the name-plate values). In Chapter 9, we derived the following expressions for a line-fed induction motor:

$$\frac{P_{r,loss}}{T_{em}} = \omega_{slip} \quad \text{(Eq.(9-42), repeated)} \qquad (10\text{-}1)$$

CONDITIONS FOR EFFICIENT SPEED CONTROL

and

$$T_{em} = k_{t\omega} \hat{B}_{ms}^2 \omega_{slip} \quad \text{(Eq.(9-29), repeated)} \quad (10\text{-}2)$$

Equation (10-1) shows that to meet the load-torque demand ($T_{em} = T_L$), the motor should be operated with as small a slip speed ω_{slip} as possible in order to minimize power loss in the rotor circuit (this also minimizes the loss in the stator resistance). Equation (10-2) can be written as

$$\omega_{slip} = \frac{T_{em}}{k_{t\omega} \hat{B}_{ms}^2} \quad (10\text{-}3)$$

This shows that to minimize ω_{slip} at the required torque, the peak flux density $\hat{B}_{ms}$ should be kept as high as possible, the highest value being $\hat{B}_{ms,rated}$, for which the motor is designed and beyond which the iron in the motor will become saturated. (For additional discussion, please see Section 8-9.) Therefore, keeping $\hat{B}_{ms}$ constant at its rated value, the electromagnetic torque developed by the motor depends linearly on the slip speed ω_{slip}:

$$T_{em} = k_{T\omega} \omega_{slip} \left(k_{T\omega} = k_{t\omega} \hat{B}_{ms,rated}^2 \right) \quad (10\text{-}4)$$

This is similar to Eq. (9-30).

Applying rated voltages (of amplitude $\hat{V}_{rated}$ and frequency f_{rated}), the resulting torque-speed characteristic based on Eq. (10-4) is shown in Fig. 10-2a, repeated from Fig. 9-12a.

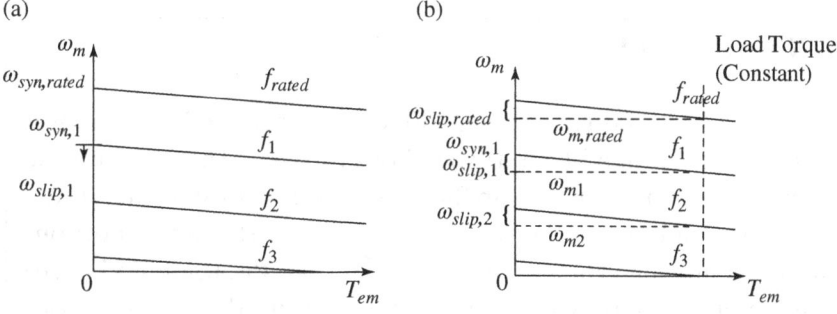

Fig. 10-2 Operation characteristics with constant $\hat{B}_{ms} = \hat{B}_{ms,rated}$.

The synchronous speed is $\omega_{syn,rated}$. This characteristic is a straight line based on the assumption that the flux-density peak is maintained at its rated value $\hat{B}_{ms,rated}$ throughout the torque range up to $T_{em,rated}$. As shown in Fig. 10-2a, a family of such characteristics corresponding to various frequencies $f_3 < f_2 < f_1 < f_{rated}$ can be achieved (assuming that the flux-density peak is maintained throughout at its rated value $\hat{B}_{ms,rated}$, as discussed in the next section). Focusing on the frequency f_1 corresponding to one of the characteristics in Fig. 10-2a, the synchronous speed at which the flux-density distribution in the air gap rotates is given by

$$\omega_{syn,1} = \frac{2\pi f_1}{p/2} \qquad (10\text{-}5)$$

Therefore, at a rotor speed $\omega_m (<\omega_{syn,1})$, the slip speed, measured with respect to the synchronous speed $\omega_{syn,1}$, is

$$\omega_{slip,1} = \omega_{syn,1} - \omega_m \qquad (10\text{-}6)$$

Using the above $\omega_{slip,1}$ in Eq. (10-4), the torque-speed characteristic at f_1 has the same slope as at f_{rated}. This shows that the characteristics at various frequencies are parallel to each other, as shown in Fig. 10-2a. Considering a load whose torque requirement remains independent of speed, as shown by the dotted line in Fig. 10-2b, the speed can be adjusted by controlling the frequency of the applied voltages; for example, the speed is $\omega_{m,1}(=\omega_{syn,1} - \omega_{slip,1})$ at a frequency of f_1, and $\omega_{m,2}(=\omega_{syn,2} - \omega_{slip,2})$ at f_2.

EXAMPLE 10-1

A three-phase, 60-Hz, 4-pole, 440-V (L-L, rms) induction-motor drive has a full-load (rated) speed of 1746 rpm. The rated torque is 40 Nm. Keeping the air gap flux-density peak constant at its rated value, (a) plot the torque-speed characteristics (the linear portion) for the following values of the frequency f: 60, 45, 30, and 15 Hz. (b) This motor is supplying a load whose torque demand increases linearly with speed such that it equals the rated motor torque at the rated motor speed. Calculate the speeds of operation at the four values of frequency in part (a).

Solution

(a) In this example, it is easier to make use of speed (denoted by the symbol "n") in rpm. At the rated frequency of 60 Hz, the synchronous speed in a 4-pole motor can be calculated as follows: from Eq. (10-5),

$$\omega_{syn,rated} = \frac{2\pi f_{rated}}{p/2}$$

Therefore,

$$n_{syn,rated} = \underbrace{\frac{\omega_{syn,rated}}{2\pi}}_{rev.\,per\,sec.} \times 60 \text{ rpm} = \frac{f_{rated}}{p/2} \times 60 \text{ rpm} = 1800 \text{ rpm}$$

Therefore,

$$n_{slip,rated} = 1800 - 1746 = 54 \text{ rpm}$$

The synchronous speeds corresponding to the other three frequency values are: 1350 rpm at 45 Hz, 900 rpm at 30 Hz, and 450 rpm at 15 Hz. The torque-speed characteristics are parallel, as shown in Fig. 10-3, for the four frequency values, keeping $\hat{B}_{ms} = \hat{B}_{ms,rated}$.

(b) The torque-speed characteristic in Fig. 10-3 can be described for each frequency by the equation below, where n_{syn} is the synchronous speed corresponding to that frequency:

$$T_{em} = k_{Tn}(n_{syn} - n_m) \qquad (10\text{-}7)$$

In this example, $k_{Tn} = \dfrac{40 \text{ Nm}}{(1800 - 1746) \text{ rpm}} = 0.74 \, \dfrac{\text{Nm}}{\text{rpm}}$.

(Continued)

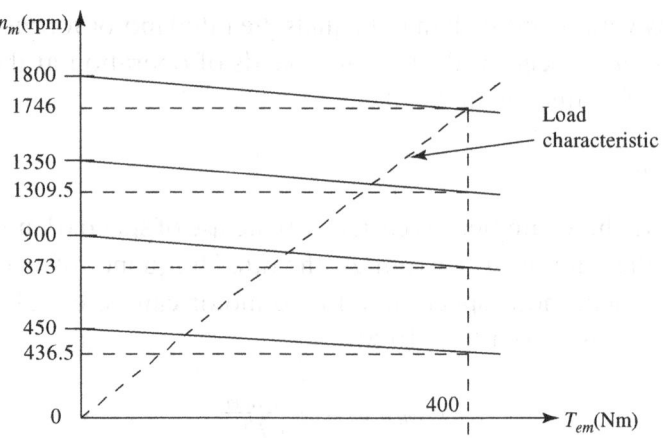

Fig. 10-3 Example 10-1.

The linear load torque-speed characteristic can be described as

$$T_L = c_n n_m \tag{10-8}$$

where, in this example, $c_n = \dfrac{40\,\text{Nm}}{1746\,\text{rpm}} = 0.023\,\dfrac{\text{Nm}}{\text{rpm}}$.

In steady state, the electromagnetic torque developed by the motor equals the load torque. Therefore, equating the right sides of Eqs. (10-7) and (10-8),

$$k_{Tn}(n_{syn} - n_m) = c_n n_m \tag{10-9}$$

Hence,

$$n_m = \dfrac{k_{Tn}}{k_{Tn} + c_n} n_{syn} = 0.97 n_{syn} \text{(in this example)} \tag{10-10}$$

Therefore, we have the following speeds and slip speeds at various values of f:

f (Hz)	n_{syn} (rpm)	n_m (rpm)	n_{slip} (rpm)
60	1800	1746	54
45	1350	1309.5	40.5
30	900	873	27
15	450	436.5	13.5

10-3 APPLIED VOLTAGE AMPLITUDES TO KEEP $\hat{B}_{ms} = \hat{B}_{ms,rated}$

Maintaining $\hat{B}_{ms}$ at its rated value minimizes power loss in the rotor circuit. To maintain $\hat{B}_{ms,rated}$ at various frequencies and torque loading, the applied voltages should be of the appropriate amplitude, as discussed in this section.

The per-phase equivalent circuit of an induction motor under the balanced sinusoidal steady state is shown in Fig. 10-4a. With the rated voltages at $\hat{V}_{a,rated}$ and f_{rated} applied to the stator, loading the motor by its rated (full-load) torque $T_{em,rated}$ establishes the rated operating point. At the rated operating point, all quantities related to the motor are at their rated values: the synchronous speed $\omega_{syn,rated}$, the motor speed $\omega_{m,rated}$, the slip speed $\omega_{slip,rated}$, the flux-density peak $\hat{B}_{ms,rated}$, the internal voltage $\hat{E}_{ma,rated}$, the magnetizing current $\hat{I}_{ma,rated}$, the rotor-branch current $\hat{I}'_{ra,rated}$, and the stator current $\hat{I}_{a,rated}$.

The objective of maintaining the flux density at $\hat{B}_{ms,rated}$, implies that in the equivalent circuit of Fig. 10-4a, the magnetizing current should be maintained at $\hat{I}_{ma,rated}$:

$$\hat{I}_{ma} = \hat{I}_{ma,rated} \text{(a constant)} \tag{10-11}$$

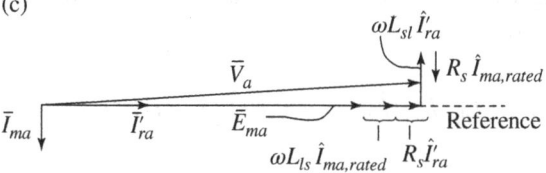

Fig. 10-4 (a) Per-phase equivalent circuit in balanced steady state, (b) equivalent circuit with the rotor-leakage neglected, and (c) phasor diagram in steady state at the rated flux-density.

With this magnetizing current, the internal voltage $\overline{E}_{ma}$ in Fig. 10-4a has the following amplitude:

$$\hat{E}_{ma} = \omega L_m \hat{I}_{ma,rated} = \underbrace{2\pi L_m \hat{I}_{ma,rated}}_{constant} f \qquad (10\text{-}12)$$

This shows that $\hat{E}_{ma}$ is linearly proportional to the frequency f of the applied voltages.

For torques below the rated value, the leakage inductance of the rotor can be neglected (see Example 9-6), as shown in the equivalent circuit of Fig. 10-4b. With this assumption, the rotor-branch current $\overline{I}'_{ra}$ is in phase with the internal voltage $\overline{E}_{ma}$, and its amplitude $\hat{I}'_{ra}$ depends linearly on the electromagnetic torque developed by the motor (as in Eq. (9-28)) to provide the load torque. Therefore, in terms of the rated values,

$$\overline{I}'_{ra} = \left(\frac{T_{em}}{T_{em,rated}}\right)\hat{I}'_{ra,rated} \qquad (10\text{-}13)$$

At some frequency and torque, the phasor diagram corresponding to the equivalent circuit in Fig. 10-4b is shown in Fig. 10-4c. If the internal emf is the reference phasor $\overline{E}_{ma} = \hat{E}_{ma} \angle 0°$, then $\overline{I}'_{ra} = \hat{I}'_{ra} \angle 0°$ and the applied voltage is

$$\overline{V}_a = \hat{E}_{ma} \angle 0° + (R_s + j2\pi f L_{\ell s})\overline{I}_s \qquad (10\text{-}14)$$

where

$$\overline{I}_s = \hat{I}'_{ra} \angle 0° - j\hat{I}_{ma,rated} \qquad (10\text{-}15)$$

Substituting Eq. (10-15) into Eq. (10-14) and separating the real and imaginary parts,

$$\overline{V}_a = \left[\hat{E}_{ma} + (2\pi f L_{\ell s})\hat{I}_{ma,rated} + R_s \hat{I}'_{ra}\right] + j\left[(2\pi f L_{\ell s})\hat{I}'_{ra} - R_s \hat{I}_{ma,rated}\right]$$
$$(10\text{-}16)$$

These phasors are plotted in Fig. 10-4c near the rated operating condition, using reasonable parameter values. This phasor diagram shows that in determining the magnitude $\hat{V}_a$ of the applied voltage phasor $\overline{V}_a$, the perpendicular component in Eq. (10-16) can be neglected, yielding

$$\hat{V}_a \cong \hat{E}_{ma} + (2\pi f L_{\ell s})\hat{I}_{ma,rated} + R_s \hat{I}'_{ra} \qquad (10\text{-}17)$$

Substituting for $\hat{E}_{ma}$ from Eq. (10-12) into Eq. (10-17) and rearranging terms,

$$\hat{V}_a = \underbrace{2\pi(L_m + L_{\ell s})\hat{I}_{ma,rated}}_{constant\ slope} f + R_s \hat{I}'_{ra} \text{ or } \hat{V}_a = (slope)f + R_s \hat{I}'_{ra} \qquad (10\text{-}18)$$

This shows that to maintain flux density at its rated value, the applied voltage amplitude $\hat{V}_a$ depends linearly on the frequency f of the applied voltages, except for the offset due to the resistance R_s of the stator windings. At a constant torque value, the relationship in Eq. (10-18) between $\hat{V}_a$ and f is a straight line, as shown in Fig. 10-5. This line has a constant slope equal to $2\pi(L_m + L_{\ell s})\hat{I}_{ma,rated}$. This slope can be obtained by using the values at the rated operating point of the motor in Eq. (10-18):

$$slope = \frac{\hat{V}_{a,rated} - R_s \hat{I}'_{ra,rated}}{f_{rated}} \qquad (10\text{-}19)$$

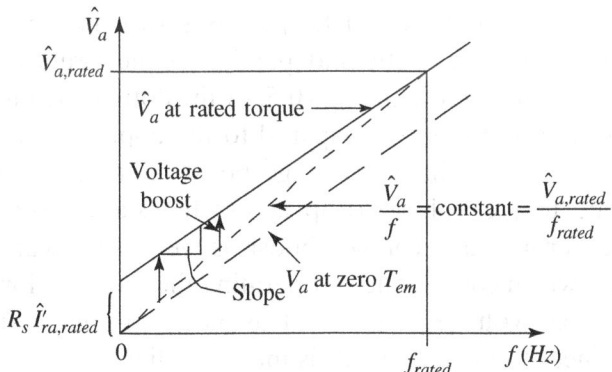

Fig. 10-5 Relation of applied voltage and frequency at the rated flux density.

294 INDUCTION-MOTOR DRIVES

Therefore, in terms of the slope in Eq. (10-19), the relationship in Eq. (10-18) can be expressed as

$$\hat{V}_a = \left(\frac{\hat{V}_{a,rated} - R_s \hat{I}'_{ra,rated}}{f_{rated}}\right) f + R_s \hat{I}'_{ra} \qquad (10\text{-}20)$$

At the rated torque, in Eq. (10-20), $\hat{V}_a$, $\hat{I}'_{ra}$, and f are all at their rated values. This establishes the rated point in Fig. 10-5. Continuing to provide the rated torque, as the frequency f is reduced to nearly zero at very low speeds, from Eq. (10-20),

$$\hat{V}_a\big|_{T_{em,rated}, f \simeq 0} = R_s \hat{I}'_{ra,rated} \qquad (10\text{-}21)$$

This is shown by the offset above the origin in Fig. 10-5. Between this offset point (at $f \simeq 0$) and the rated point, the voltage-frequency characteristic is linear, as shown, while the motor is loaded to deliver its rated torque. We will consider another case of no-load connected to the motor, where $\hat{I}'_{ra} \simeq 0$ in Eq. (10-20), and hence at nearly zero frequency

$$\hat{V}_a\big|_{T_{em} = 0, f \simeq 0} = 0 \qquad (10\text{-}22)$$

This condition shifts the entire characteristic at no-load downwards compared to that at the rated torque, as shown in Fig. 10-5. An approximate V/f characteristic (independent of the torque developed by the motor) is also shown in Fig. 10-5, by the dotted line through the origin and the rated point. Compared to the approximate relationship, Fig. 10-5 shows that a "voltage boost" is required at higher torques, due to the voltage drop across the stator resistance. In percentage terms, this voltage boost is very significant at low frequencies, which correspond to operating the motor at low speeds; the percentage voltage boost that is necessary near the rated frequency (near the rated speed) is much smaller.

EXAMPLE 10-2

In the motor drive of Example 10-1, the induction motor is such that while applied the rated voltages and loaded to the rated torque, it draws 10.39 A (rms) per-phase at a power factor of 0.866 (lagging). $R_s = 1.5\,\Omega$. Calculate the voltages corresponding to the four values of the frequency f to maintain $\hat{B}_{ms} = \hat{B}_{ms,rated}$.

Solution

Neglecting the rotor leakage inductance, as shown in the phasor diagram of Fig. 10-6, the rated value of the rotor-branch current can be calculated as

$$\hat{I}'_{ra,rated} = 10.39\sqrt{2}(0.866) = 9.0\sqrt{2}\text{ A}$$

Using Eq. (10-20) and the rated values, the slope of the characteristic can be calculated as

$$slope = \frac{\hat{V}_{a,rated} - R_s \hat{I}'_{ra,rated}}{f_{rated}} = \frac{\frac{440\sqrt{2}}{\sqrt{3}} - 1.5 \times 9.0\sqrt{2}}{60} = 5.67\text{ V/Hz}$$

In Eq. (10-20), $\hat{I}'_{ra}$ depends on the torque that the motor is supplying. Therefore, substituting for $\hat{I}'_{ra}$ from Eq. (10-13) into Eq. (10-20),

$$\hat{V}_a = \left(\frac{\hat{V}_{a,rated} - R_s \hat{I}'_{ra,rated}}{f_{rated}}\right) f + R_s \left(\frac{T_{em}}{T_{em,rated}} \hat{I}'_{ra,rated}\right) \quad (10\text{-}23)$$

While the drive is supplying a load whose torque depends linearly on speed (and demands the rated torque at the rated speed as in Example 10-1), the torque ratio in Eq. (10-23) is

$$\frac{T_{em}}{T_{em,rated}} = \frac{n_m}{n_{m,rated}}$$

(Continued)

296 INDUCTION-MOTOR DRIVES

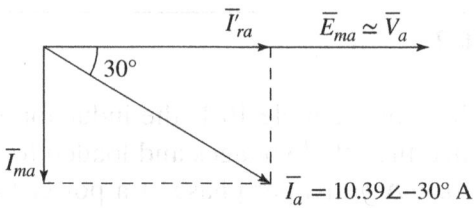

Fig. 10-6 Example 10-2.

Therefore, Eq. (10-23) can be written as

$$\hat{V}_a = \left(\frac{\hat{V}_{a,rated} - R_s\hat{I}'_{ra,rated}}{f_{rated}}\right)f + R_s\left(\frac{n_m}{n_{m,rated}}\hat{I}'_{ra,rated}\right) \quad (10\text{-}24)$$

Substituting the four values of the frequency f and their corresponding speeds from Example 10-1, the voltages can be tabulated as below. The values obtained by using the approximate dotted characteristic plotted in Fig. 10-5 (which assumes a linear V/f relationship) are nearly identical to the values in the table below because at low values of frequency (hence, at low speeds) the torque is also reduced in this example – therefore, no voltage boost is necessary.

f	60 Hz	45 Hz	30 Hz	15 Hz
$\hat{V}_a$	359.3 V	269.5 V	179.6 V	89.8 V

10-4 STARTING CONSIDERATIONS IN DRIVES

Starting currents are primarily limited by the leakage inductances of the stator and the rotor, and can be six to eight times the rated current of the motor, as shown in the plot of Fig. 9-23b in Chapter 9. In the motor drives of Fig. 10-1, if large currents are drawn even for a short

STARTING CONSIDERATIONS IN DRIVES 297

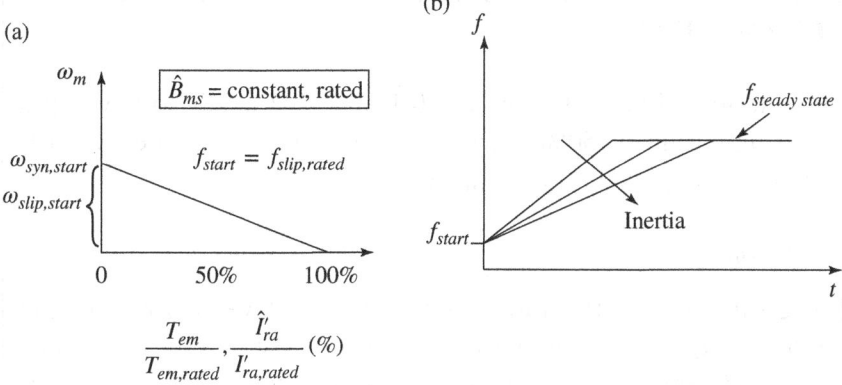

Fig. 10-7 Start-up considerations in induction-motor drives.

time, the current rating required of the PPU will become unacceptably large.

At starting, the rotor speed ω_m is zero, and hence the slip speed ω_{slip} equals the synchronous speed ω_{syn}. Therefore, at start-up, we must apply voltages of a low frequency in order to keep ω_{slip} low, and hence avoid large starting currents. Figure 10-7a shows the torque-speed characteristic at a frequency $f_{start}(=f_{slip,rated})$ of the applied voltages, such that the starting torque (at $\omega_m = 0$) is equal to the rated value. The same is true of the rotor-branch current. It is assumed that the applied voltage magnitudes are appropriately adjusted to maintain $\hat{B}_{ms}$ constant at its rated value.

As shown in Fig. 10-7b, as the rotor speed builds up, the frequency f of the applied voltages is increased continuously at a preset rate until the final desired speed is reached in steady state. The rate at which the frequency is increased should not let the motor current exceed a specific limit (usually 150% of the rated). The rate should be decreased for higher inertia loads to allow the rotor speed to catch up. Note that the voltage amplitude is adjusted, as a function of the frequency f, as discussed in the previous section, to keep $\hat{B}_{ms}$ constant at its rated value.

EXAMPLE 10-3

The motor drive in Examples 10-1 and 10-2 needs to develop a starting torque of 150% of the rated in order to overcome the starting friction. Calculate f_{start} and $\hat{V}_{a,start}$.

Solution

The rated slip of this motor is 54 rpm. To develop 150% of the rated torque, the slip speed at start-up should be $1.5 \times n_{slip,rated} = 81$ rpm. Note that at start-up, the synchronous speed is the same as the slip speed. Therefore, $n_{syn,start} = 81$ rpm. Hence, from Eq. (10-5) for this 4-pole motor,

$$f_{start} = \underbrace{\left(\frac{n_{syn,start}}{60}\right)}_{rev.\ per\ second} \frac{p}{2} = 2.7\,\text{Hz}$$

At 150% of the rated torque, from Eq. (10-13),

$$\hat{I}'_{ra,start} = 1.5 \times \hat{I}'_{ra,rated} = 1.5 \times 9.0\sqrt{2}\,\text{A}$$

Substituting various values at start-up into Eq. (10-20), $\hat{V}_{a,start} = 43.9\,\text{V}$.

10-5 CAPABILITY TO OPERATE BELOW AND ABOVE THE RATED SPEED

Due to the rugged construction of the squirrel-cage rotor, induction-motor drives can be operated at speeds in the range of zero to almost twice the rated speed. The following constraints on the drive operation should be noted:

1. The magnitude of applied voltages is limited to their rated value. Otherwise, the motor insulation may be stressed, and the rating of the PPU will have to be larger.

2. The motor currents are also limited to their rated values. This is because the rotor-branch current $\hat{I}'_{ra}$ is limited to its rated value, in order to limit the loss $P_{r,loss}$ in the rotor bar resistances. This loss, dissipated as heat, is difficult to remove; beyond its rated value, it will cause the motor temperature to exceed its design limit, thus shortening the motor life.

The torque-capability regions below and above the rated speed are shown in Fig. 10-8 and discussed in the following sections.

10-5-1 Rated Torque Capability Below the Rated Speed (With $\hat{B}_{ms,rated}$)

This region of operation has already been discussed in Section 10-3, where the motor is operated at the rated flux density $\hat{B}_{ms,rated}$. Therefore, at any speed below the rated speed, a motor in steady state can deliver its rated torque while $\hat{I}'_{ra}$ stays equal to its rated value. This capability region is shown in Fig. 10-8 as the rated-torque capability region. At low speeds, due to poor cooling, the steady-state torque capability may have to be reduced, as shown by the dotted curve.

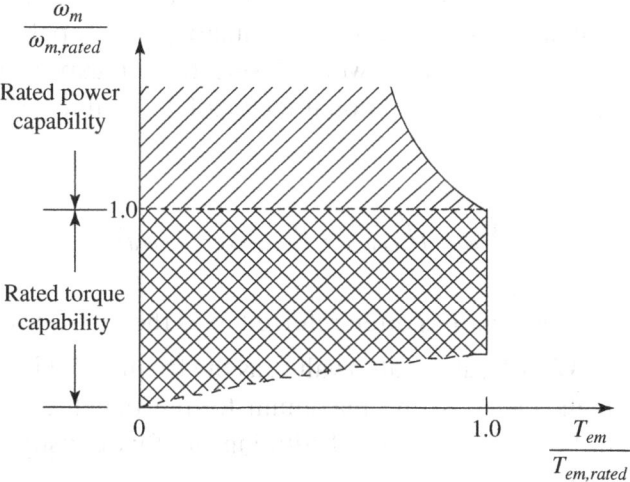

Fig. 10-8 Capability below and above the rated speed.

10-5-2 Rated Power Capability Above the Rated Speed by Flux-Weakening

Speeds above the rated value are obtained by increasing the frequency f of the applied voltages above the rated frequency, thus increasing the synchronous speed at which the flux-density distribution rotates in the air gap:

$$\omega_{syn} > \omega_{syn,rated} \qquad (10\text{-}25)$$

The amplitude of the applied voltages is limited to its rated value $\hat{V}_{a,rated}$, as discussed earlier. Neglecting the voltage drop across the stator winding leakage inductance and resistance, in terms of the rated values, the peak flux density $\hat{B}_{ms}$ declines below its rated value, such that it is inversely proportional to the increasing frequency f (per Eq. (9-7)):

$$\hat{B}_{ms} = \hat{B}_{ms,rated} \frac{f_{rated}}{f} \quad (f > f_{rated}) \qquad (10\text{-}26)$$

In the equivalent circuit of Fig. 10-4b, the rotor-branch current should not exceed its rated value $\hat{I}'_{ra,rated}$ in steady state; otherwise, the power loss in the rotor will exceed its rated value. Neglecting the rotor leakage inductance when estimating the capability limit, the maximum three-phase power crossing the air gap, in terms of the peak quantities (the additional factor of 1/2 is due to the peak quantities) is

$$P_{max} = \frac{3}{2} \hat{V}_{a,rated} \hat{I}'_{a,rated} = P_{rated} (f > f_{rated}) \qquad (10\text{-}27)$$

Therefore, this region is often referred to as the rated-power capability region. With $\hat{I}'_{ra}$ at its rated value, as the frequency f is increased to obtain higher speeds, the maximum torque that the motor can develop can be calculated by substituting the flux density given by Eq. (10-26) into Eq. (9-28):

$$T_{em}|_{\hat{I}'_{r,rated}} = k_t \hat{I}'_{r,rated} \hat{B}_{ms} = \underbrace{k_t \hat{I}'_{r,rated} \hat{B}_{ms,rated}}_{T_{em,rated}} \frac{f}{f_{rated}}$$

$$= T_{em,rated} \frac{f_{rated}}{f} \qquad (f > f_{rated}) \quad (10\text{-}28)$$

This shows that the maximum torque, plotted in Fig. 10-8, is inversely proportional to the frequency.

10-6 INDUCTION-GENERATOR DRIVES

Induction machines can operate as generators, as discussed in Section 9-3-2. For an induction machine to operate as a generator, the applied voltages must be at a frequency at which the synchronous speed is less than the rotor speed, resulting in a negative slip speed:

$$\omega_{slip} = (\omega_{syn} - \omega_m) < 0 \quad \omega_{syn} < \omega_m \qquad (10\text{-}29)$$

Maintaining the flux density at $\hat{B}_{ms,rated}$ by controlling the voltage amplitudes, the torque developed, according to Eq. (10-4), is negative (in a direction opposite that of rotation) for negative values of slip speed. Figure 10-9 shows the motor torque-speed characteristics at two frequencies, assuming a constant $\hat{B}_{ms} = \hat{B}_{ms,rated}$. These characteristics are extended into the negative torque region (generator region) for the rotor speeds above the corresponding synchronous speeds. Consider that the induction machine is initially operating as a motor with a stator frequency f_0 and at the rotor speed of ω_{m_0}, which is less than ω_{syn_0}. If the stator frequency is decreased to f_1, the new synchronous speed is ω_{syn_1}. This makes the slip speed negative, and thus T_{em} becomes negative, as shown in Fig. 10-9. This torque acts in the direction opposite to the direction of rotation.

Therefore, in the wind-turbine application, if a squirrel-cage induction machine is to operate as a generator, then the synchronous-speed ω_{syn} (corresponding to the frequency of the applied voltages to the motor terminals by the power electronics interface) must be less than

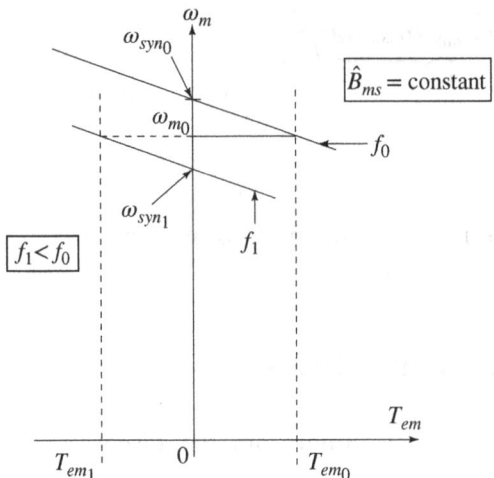

Fig. 10-9 Induction-generator drives.

the rotor-speed ω_m, so that the slip-speed ω_{slip} is negative, and the machine operates as a generator.

10-7 SPEED CONTROL OF INDUCTION-MOTOR DRIVES

The focus of this section is to discuss speed control of induction-motor drives in general-purpose applications where very precise speed control is not necessary, and therefore, as shown in Fig. 10-10, the speed is not measured (rather, it is estimated).

The reference speed $\omega_{m,ref}$ is set either manually or by a slow-acting control loop of the process where the drive is used. The use of induction-motor drives in high-performance servo-drive applications is discussed in the next chapter.

In addition to the reference speed, the other two inputs to the controller are the measured dc-link voltage V_d and the input current i_d of the inverter. This dc-link current represents the instantaneous three-phase currents of the motor. Some of the salient points of control in Fig. 10-10 are described below.

SPEED CONTROL OF INDUCTION-MOTOR DRIVES   303

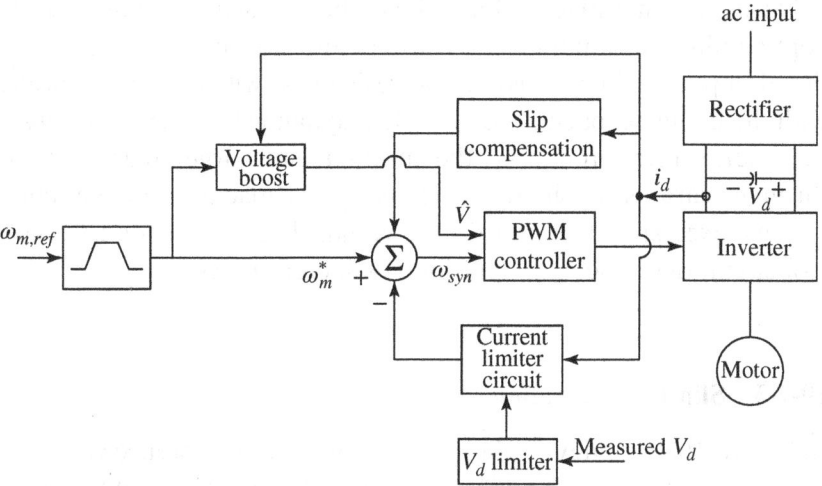

Fig. 10-10 Speed control of induction-motor drives.

10-7-1 Limiting of Acceleration/Deceleration

During acceleration and deceleration, it is necessary to keep the motor currents and the dc-link voltage V_d within their design limits. Therefore, in Fig. 10-10, the maximum acceleration and deceleration are usually set by the user, resulting in a dynamically modified reference speed signal ω_m^*.

10-7-2 Current-Limiting

In the motoring mode, if ω_{syn} increases too quickly compared to the motor speed, then ω_{slip} and the motor currents may exceed their limits. To limit acceleration, so that the motor currents stay within their limits, i_d (representing the actual motor current) is compared with the current limit and the error though the controller acts on the speed control circuit by reducing acceleration (i.e. by reducing ω_{syn}).

In the regenerative-braking mode, if ω_{syn} is reduced too quickly, the negative slip will become too large in magnitude and will result in a large current through the motor and the inverter of the PPU. To restrict this current within the limit, i_d is compared with the current limit, and the error is fed through a controller to decrease deceleration (i.e. by increasing ω_{syn}).

During regenerative-braking, the dc-bus capacitor voltage must be kept within a maximum limit. If the rectifier of the PPU is unidirectional in power flow, a dissipation resistor is switched on, in parallel with the dc-link capacitor, to provide a dynamic braking capability. If the energy recovered from the motor is still larger than that lost through various dissipation means, the capacitor voltage could become excessive. Therefore, if the voltage limit is exceeded, the control circuit decreases deceleration (by increasing ω_{syn}).

10-7-3 Slip Compensation

In Fig. 10-10, to achieve a rotor speed equal to its reference value, the machine should be applied voltages at a frequency f, with a corresponding synchronous speed ω_{syn}, such that it is the sum of ω_m^* and the slip speed:

$$\omega_{syn} = \omega_m^* + \underbrace{T_{em}/k_{T\omega}}_{\omega_{slip}} \quad (10\text{-}30)$$

where the required slip speed, in accordance with Eq. (10-4), depends on the torque to be developed. The slip speed is calculated by the slip-compensation block of Fig. 10-10. Here, T_{em} is estimated as follows: the dc power input to the inverter is measured as a product of V_d and the average of i_d. From this, the estimated losses in the inverter of the PPU and the stator resistance are subtracted to estimate the total power P_{ag} crossing the air gap into the rotor. We can show, by adding Eqs. (9-40) and (9-42), that $T_{em} = P_{ag}/\omega_{syn}$.

10-7-4 Voltage Boost

To keep the air gap flux density $\hat{B}_{ms}$ constant at its rated value, the motor voltage must be controlled per Eq. (10-18), where $\hat{I}'_{ra}$ is linearly proportional to T_{em} estimated earlier.

10-8 PULSE-WIDTH-MODULATED PPU

In the block diagram of Fig. 10-10, the inputs $\hat{V}$ and ω_{syn} generate the three control voltages that are compared with a switching-frequency triangular waveform v_{tri} of constant amplitude. The PPU of Fig. 10-11a, as described in Chapter 4, supplies the desired voltages to the stator windings. By averaging, each pole is represented by an ideal transformer in Fig. 10-11b whose turns-ratio is continuously controlled, proportional to the control voltage.

10-9 HARMONICS IN THE PPU OUTPUT VOLTAGES

The instantaneous voltage waveforms corresponding to the logic signals are shown in Fig. 10-12a. These are best discussed by means of computer simulations. The harmonic spectrum of the L-L output voltage waveform, shows the presence of harmonic voltages as the sidebands of the switching frequency f_s and its multiples. The PPU output voltages, for example $v_a(t)$, can be decomposed into the fundamental-frequency component (designated by the subscript "1") and the ripple voltage

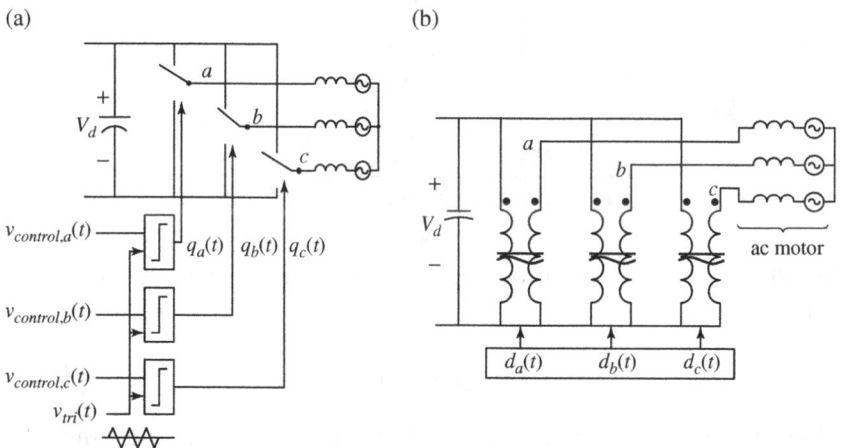

Fig. 10-11 PPU of induction-motor drives.

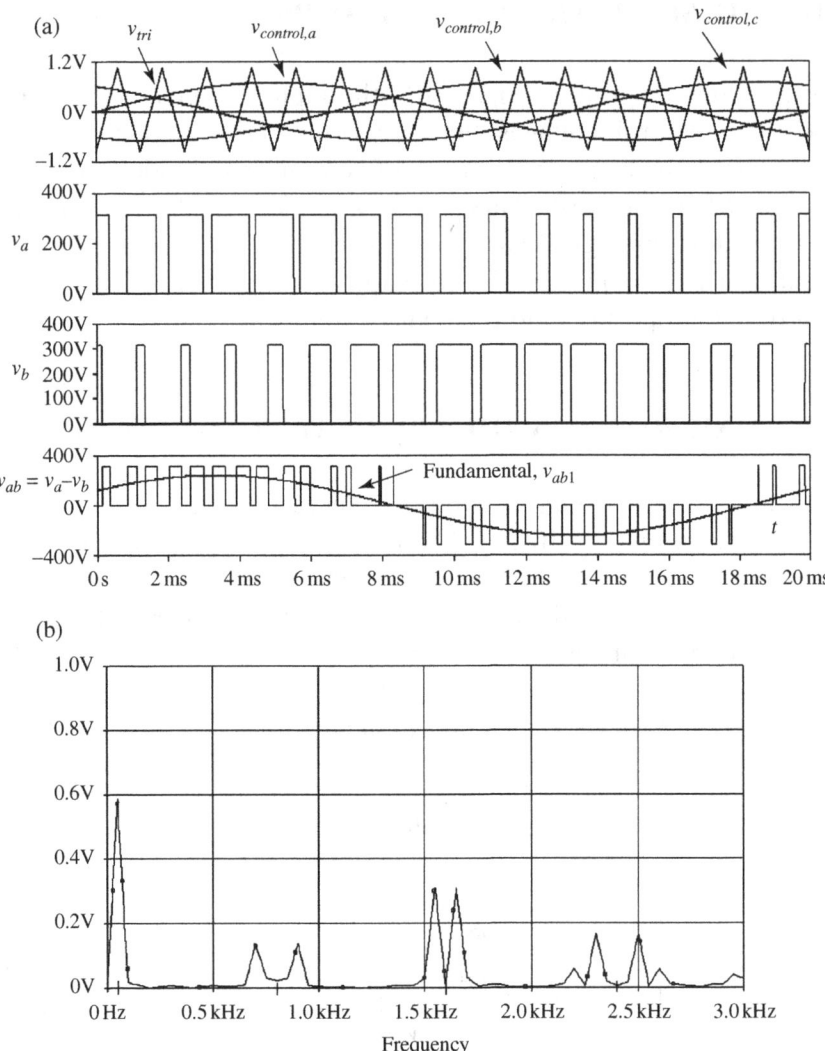

Fig. 10-12 (a) PPU output voltage waveforms and (b) harmonic spectrum of L-L voltages.

$$v_a(t) = v_{a1}(t) + v_{a,ripple}(t) \qquad (10\text{-}31)$$

where the ripple voltage consists of the components in the range of, and higher than, the switching frequency f_s, as shown in Fig. 10-12b. With the availability of higher switching-speed power devices such as modern IGBTs, the switching frequency in low- and medium-power motor drives approach, and in some cases exceed, 20 kHz.

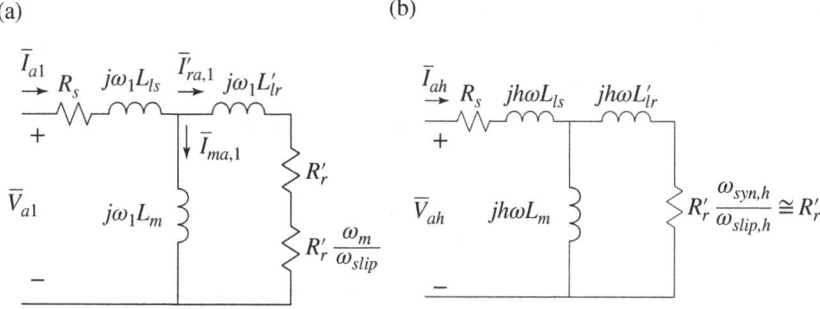

Fig. 10-13 Per-phase equivalent circuit (a) at the fundamental frequency and (b) at harmonic frequencies.

The motivation for selecting a high switching frequency f_s, if the switching losses in the PPU can be kept manageable, is to reduce the ripple in the motor currents, thus reducing the electromagnetic torque ripple and the power losses in the motor resistances.

To analyze the motor's response to the applied voltages with ripple, we will make use of superposition. The motor's dominant response is determined by the fundamental-frequency voltages, which establish the synchronous speed ω_{syn} and the rotor speed ω_m. The per-phase equivalent circuit at the fundamental frequency is shown in Fig. 10-13a.

In the PPU output voltages, the voltage components at a harmonic frequency $f_h \gg f$ produce rotating flux distribution in the air gap at a synchronous speed $\omega_{syn,\,h}$, where

$$\omega_{syn,h}(=h \times \omega_{syn}) \gg \omega_{syn},\ \omega_m \qquad (10\text{-}32)$$

The flux-density distribution at a harmonic frequency may be rotating in the same or opposite direction as the rotor. In any case, because it is rotating at a much faster speed compared to the rotor speed ω_m, the slip speed for the harmonic frequencies is

$$\omega_{slip,h} = \omega_{syn,h} \pm \omega_m \cong \omega_{syn,h} \qquad (10\text{-}33)$$

Therefore, in the per-phase equivalent circuit at harmonic frequencies,

$$R'_r \frac{\omega_{syn,h}}{\omega_{slip,h}} \cong R'_r \qquad (10\text{-}34)$$

which is shown in Fig. 10-13b. At high switching frequencies, the magnetizing reactance is very large and can be neglected in the circuit of Fig. 10-13b, and the harmonic frequency current is determined primarily by the leakage reactances (which dominate over R'_r):

$$\hat{I}_{ah} \cong \frac{\hat{V}_{ah}}{X_{\ell s,h} + X'_{\ell r,h}} \qquad (10\text{-}35)$$

The additional power loss, due to these harmonic frequency currents in the stator and the rotor resistances, on a three-phase basis, can be expressed as

$$\Delta P_{loss,R} = 3 \sum_h \frac{1}{2} (R_s + R'_r) \hat{I}_{ah}^2 \qquad (10\text{-}36)$$

In addition to these losses, there are additional losses in the stator and the rotor iron due to eddy currents and hysteresis at harmonic frequencies. These are further discussed in Chapter 15 dealing with efficiencies in drives.

10-9-1 Modeling the PPU-Supplied Induction Motors in Steady State

In steady state, an induction motor supplied by voltages from the PPU should be modeled such that it allows the fundamental-frequency currents in Fig. 10-13a, and the harmonic-frequency currents in Fig. 10-13b, to be superimposed. This can be done if the per-phase equivalent is drawn, as shown in Fig. 10-14a, where the voltage drop across the resistance $R'_r \dfrac{\omega_m}{\omega_{slip}}$ in Fig. 10-13a at the fundamental frequency is represented by a fundamental-frequency voltage $R'_r \dfrac{\omega_m}{\omega_{slip}} i'_{ra,1}(t)$. All three phases are shown in Fig. 10-14b.

10-10 REDUCTION OF $\hat{B}_{ms}$ AT LIGHT LOADS

In Section 10-2, no attention was paid to core losses (only to the copper losses) in justifying that the machine should be operated at its rated flux density at any torque while operating at speeds below

HARDWARE PROTOTYPING OF CLOSED-LOOP 309

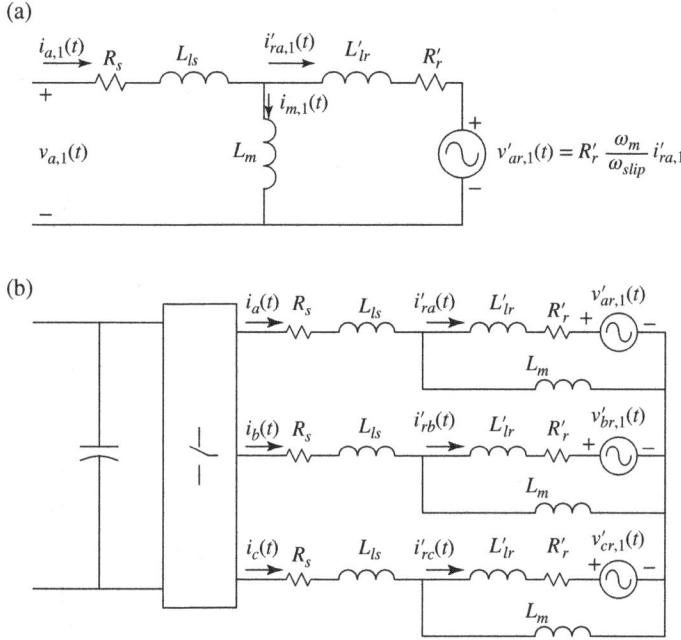

Fig. 10-14 (a) Equivalent circuit for fundamental and harmonic frequencies in steady state and (b) three-phase equivalent circuit.

the rated value. As illustrated by the discussion in Section 9-9, it is possible to improve the overall efficiency under lightly loaded conditions by reducing $\hat{B}_{ms}$ below its rated value.

10-11 HARDWARE PROTOTYPING OF CLOSED-LOOP SPEED CONTROL OF INDUCTION MOTOR

In the previous chapter, the procedure to obtain the induction motor parameters was briefly discussed. In this section, the torque-speed characteristic of an actual induction motor is obtained using Workbench as shown in Fig. 10-15.

Results of running the above experiment have shown that the speed of the induction motor does not remain as constant as load torque changes. To overcome this, closed-loop V/f speed control is used as shown in Fig. 10-10. The Workbench model of the same to run the closed-loop speed control in hardware is shown in Fig. 10-16.

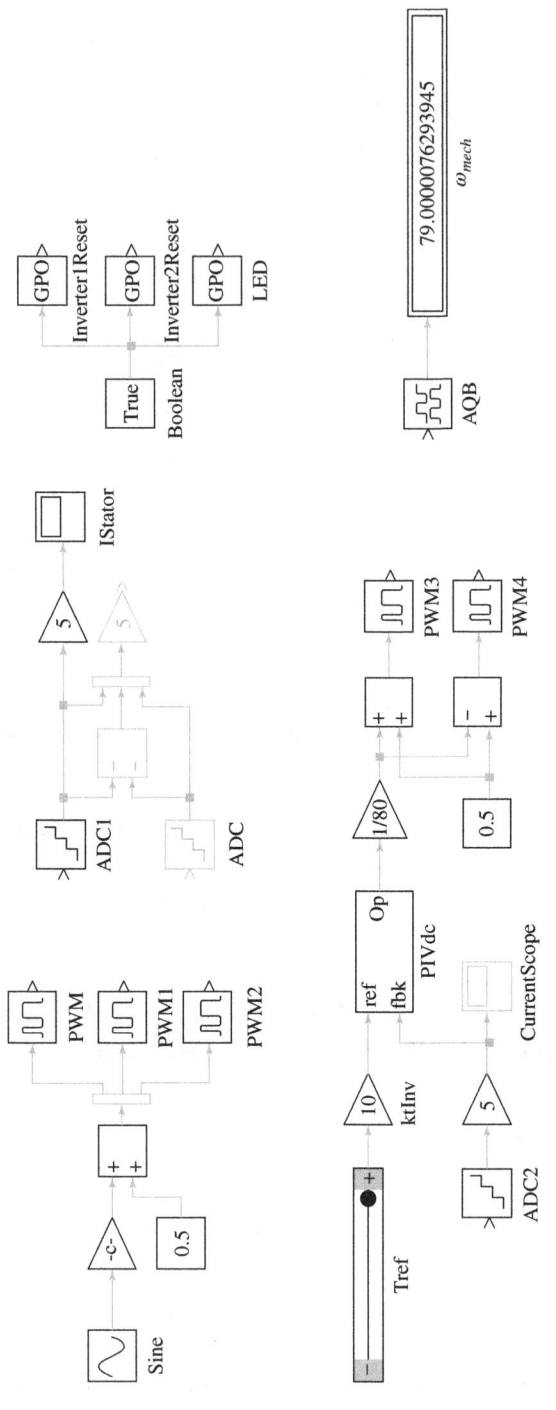

Fig. 10-15 Real-time model to obtain induction motor torque-speed characteristic.

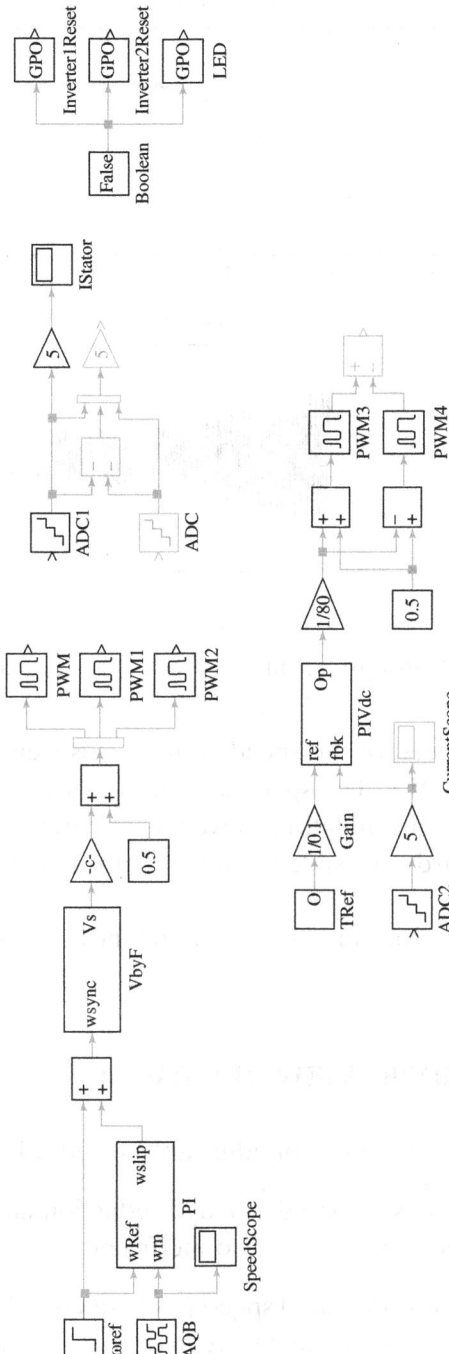

Fig. 10-16 Real-time induction motor closed-loop speed control.

312 INDUCTION-MOTOR DRIVES

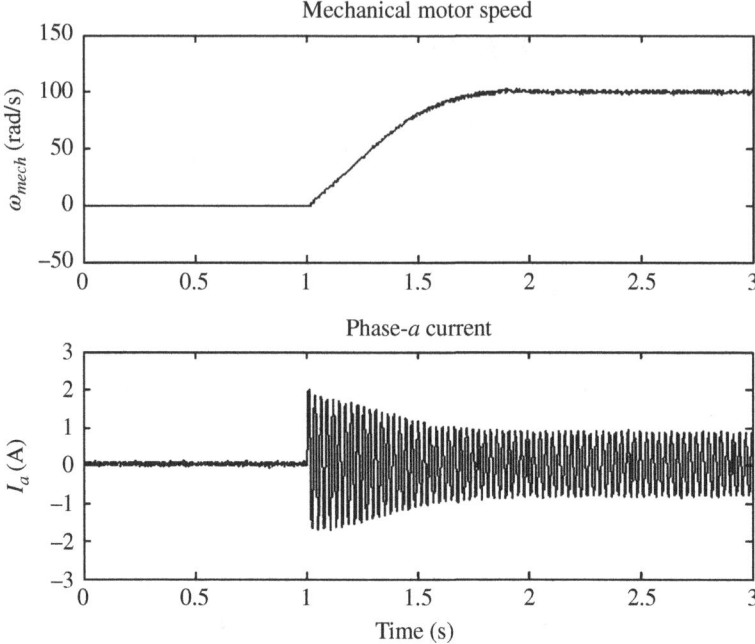

Fig. 10-17 Hardware result of closed-loop speed control.

The reference speed of the speed controller is stepped from 0 to 100 rad/s at time $t = 1$ s. The speed and phase-a current result from running this model in simulation as well as in hardware is shown in Fig. 10-17. The hardware speed result rapidly settles down to the desired speed.

The step-by-step procedure for recreating the above comparison is presented in [1].

10-12 SUMMARY/REVIEW QUESTIONS

1. What are the applications of adjustable-speed drives?

2. Why are the thyristor-based, voltage reduction circuits for controlling induction-motor speed so inefficient?

3. In operating below the rated speed (and not considering the core losses), why is it most efficient to keep the flux-density peak in the air gap at the rated value?

4. Since an induction motor is operated at different values of frequency, hence different values of synchronous speed, how is the slip speed defined?

5. Supplying a load that demands a constant torque independent of speed, what is the slip speed at various values of the frequency f of the applied voltages?

6. To keep the flux-density peak in the air gap at the rated value, why do the voltage magnitudes, at a given frequency of operation, depend on the torque being supplied by the motor?

7. At start-up, why should small-frequency voltages be applied initially? What determines the rate at which the frequency can be ramped up?

8. At speeds below the rated value, what is the limit on the torque that can be delivered, and why?

9. At speeds above the rated value, what is the limit on the power that can be delivered, and why? What does it mean for the torque that can be delivered above the rated speed?

REFERENCE

1. https://sciamble.com/Resources/pe-drives-lab/basic-drives/im-torque-speed

FURTHER READING

N. Mohan, T. Undeland, and W. P. Robbins, *Power Electronics: Converters, Applications, and Design*, 2nd edition, 1995, Wiley, New York.

Bose, B.K. (1986). *Power Electronics and AC Drives*. Prentice-Hall.

Kazmierkowski, M., Krishnan, R., and Blaabjerg, F. (2002). *Control of Power Electronics*, 518. Academic Press.

PROBLEMS

10-1 Repeat Example 10-1 as if the load is a centrifugal load that demands a torque, proportional to the speed squared, such that it equals the rated torque of the motor at the motor rated speed.

10-2 Repeat Example 10-2 as if the load is a centrifugal load that demands a torque, proportional to the speed squared, such that it equals the rated torque of the motor at the motor rated speed.

10-3 Repeat Example 10-3 as if the starting torque is to be equal to the rated torque.

10-4 Consider the drive in Examples 10-1 and 10-2, operating at the rated frequency of 60 Hz and supplying the rated torque. At the rated operating speed, calculate the voltages (in frequency and amplitude) needed to produce a regenerative-braking torque that equals the rated torque in magnitude.

10-5 A 6-pole, three-phase induction machine used for wind turbines has the following specifications: V_{LL} = 600 V(rms) at 60 Hz, the rated output power P_{out} = 1.5 MW, the rated slip s_{rated} = 1%. Assuming the machine efficiency to be approximately 95% while operating close to the rated power, calculate the frequency and the amplitude of the voltages to be applied to this machine by the power electronics converter if the rotational speed is 1100 rpm. Estimate the power output from this generator.

SIMULATION PROBLEM

10-1 Using the average representation of the PWM inverter, simulate the drive in Examples 10-1 and 10-2, while operating in steady state, at a frequency of 60 Hz. The dc-bus voltage is 800 V, and the stator and the rotor leakage inductances are 2.2 Ω each. Estimate the rotor resistance R'_r from the data given in Examples 10-1 and 10-2.

Part III

Vector Control of ac Machines

11 Induction Machine Equations in Phase Quantities: Assisted by Space Vectors

11-1 INTRODUCTION

In ac machines, the stator windings are intended to have a sinusoidally distributed conductor density in order to produce a sinusoidally distributed field distribution in the air gap. In the squirrel-cage rotor of induction machines, the bar density is uniform. Yet, the currents in the rotor bars produce an mmf that is sinusoidally distributed. Therefore, it is possible to replace the squirrel-cage with an equivalent wound rotor with three sinusoidally distributed windings.

In this chapter, we will briefly review the sinusoidally distributed windings and then calculate their inductances for developing equations for induction machines in phase (a-b-c) quantities. The development of these equations is assisted by space vectors, which are briefly reviewed. The analysis in this chapter establishes the framework for the dq-winding-based analysis of induction machines under dynamic conditions carried out in Chapter 12.

(Adapted from chapter 2 of *Advanced Electric Drives: Analysis, Control, and Modeling Using MATLAB/Simulink* ISBN: 978-1-118-48548-4 by Ned Mohan, August 2014)

Analysis and Control of Electric Drives: Simulations and Laboratory Implementation, First Edition. Ned Mohan and Siddharth Raju.
© 2021 John Wiley & Sons, Inc. Published 2021 by John Wiley & Sons, Inc.
Companion website: www.wiley.com/go/Mohan/Vectorcontrolinelectricdrives

11-2 SINUSOIDALLY DISTRIBUTED STATOR WINDINGS

In the following analysis, we will also assume that the magnetic material in the stator and the rotor is operated in its linear region and has an infinite permeability.

In the ac machines of Fig. 11-1a, windings for each phase should ideally produce a sinusoidally distributed radial field (F, H, and B) in the air gap. Theoretically, this requires a sinusoidally distributed winding in each phase. If each phase winding has a total of N_s turns (that is, $2N_s$ conductors), the conductor density $n_s(\theta)$ in phase-a of Fig. 11-1b can be defined as

$$n_s(\theta) = \frac{N_s}{2} \sin\theta \quad 0 \leq \theta \leq \pi \tag{11-1}$$

The angle θ is measured in the counterclockwise direction, with respect to the phase-a magnetic axis. Rather than restricting the conductor density expression to a region $0 < \theta < \pi$, we can interpret the negative of the conductor density in the region $\pi < \theta < 2\pi$ in Eq. (11-1) as being associated with carrying the current in the opposite direction, as indicated in Fig. 11-1b.

In a multipole machine (with $p > 2$), the peak conductor density remains $N_s/2$, as in Eq. (11-1), for an 11-pole machine but the angle θ is expressed in electrical radians. Therefore, we will always express

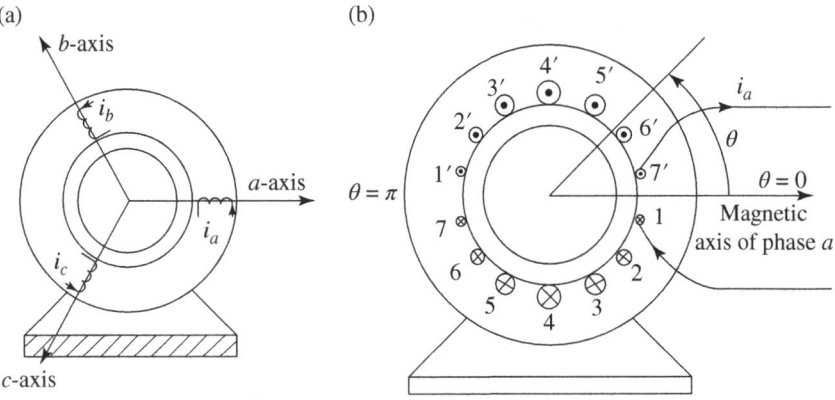

Fig. 11-1 Stator windings.

angles in all equations throughout this book by θ in electrical radians, thus making the expressions for field distributions and space vectors applicable to 2-pole, as well as multipole machines.

The current i_a through this sinusoidally distributed winding, results in the air gap a magnetic field (mmf, flux density, and field intensity) that is co-sinusoidally distributed, with respect to the position θ away from the magnetic axis of the phase:

$$H_a(\theta) = \frac{N_s}{p\ell_g} i_a \cos\theta \qquad (11\text{-}2)$$

$$B_a(\theta) = \mu_o H_a(\theta) = \left(\frac{\mu_o N_s}{p\ell_g}\right) i_a \cos\theta \qquad (11\text{-}3)$$

and

$$F_a(\theta) = \ell_g H_a(\theta) = \frac{N_s}{p} i_a \cos\theta \qquad (11\text{-}4)$$

The radial field distribution in the air gap peaks along the phase-a magnetic axis, and at any instant of time, the amplitude is linearly proportional to the value of i_a at that time. Notice that regardless of the positive or negative current in phase-a, the flux-density distribution produced by it in the air gap always has its peak (positive or negative) along the phase-a magnetic axis.

11-2-1 Three-Phase, Sinusoidally Distributed Stator Windings

In the previous section, we focused only on phase-a, which has its magnetic axis along $\theta = 0°$. There are two more identical sinusoidally distributed windings for phases b and c, with magnetic axes along $\theta = 120°$ and $\theta = 240°$, respectively, as represented in Fig. 11-2a. These three windings are generally connected in a wye-arrangement by joining terminals a', b', and c' together, as shown in Fig. 11-2b. A positive current into a winding terminal is assumed to produce flux in the radially outward direction. Field distributions in the air gap due to currents i_b and i_c are identical in sinusoidal shape to those due to i_a, but they peak along their respective phase-b and phase-c magnetic axes.

320 INDUCTION MACHINE EQUATIONS IN PHASE QUANTITIES

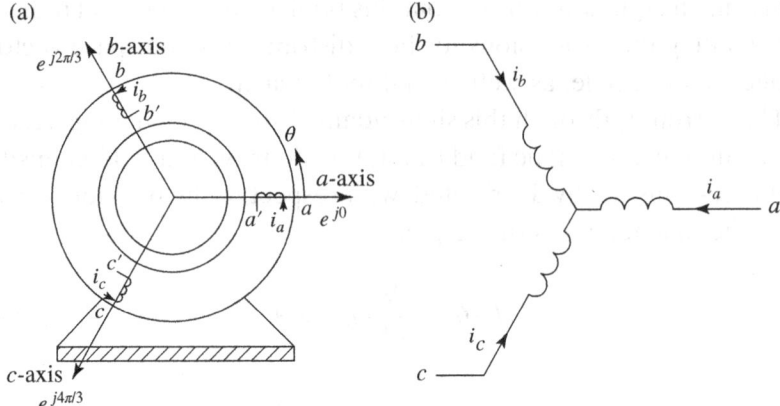

Fig. 11-2 Three-phase windings.

11-3 STATOR INDUCTANCES (ROTOR OPEN-CIRCUITED)

The stator windings are assumed to be wye-connected, as shown in Fig. 11-2b where the neutral is not accessible. Therefore, at any time

$$i_a(t) + i_b(t) + i_c(t) = 0 \qquad (11\text{-}5)$$

For defining stator-winding inductances, we will assume that the rotor is present, but it is electrically inert, that is "somehow" hypothetically of-course, it is electrically open-circuited.

11-3-1 Stator Single-Phase Magnetizing Inductance $L_{m,one\text{-}phase}$

As shown in Fig. 11-3a, hypothetically exciting only phase-a (made possible only if the neutral is accessible) by a current i_a results in two equivalent flux components represented in Fig. 11-3b: (i) magnetizing flux, which crosses the air gap and links with other stator phases and the rotor, and (ii) the leakage flux, which links phase-a only. Therefore, the self-inductance of a stator phase winding is

$$L_{s,self} = \left.\frac{\lambda_a}{i_a}\right|_{i_a\ only} = \underbrace{\frac{\lambda_{a,leakage}}{i_a}}_{L_{\ell s}} + \underbrace{\frac{\lambda_{a,magentizing}}{i_a}}_{L_{m,one\text{-}phase}} \qquad (11\text{-}6a)$$

STATOR INDUCTANCES (ROTOR OPEN-CIRCUITED)

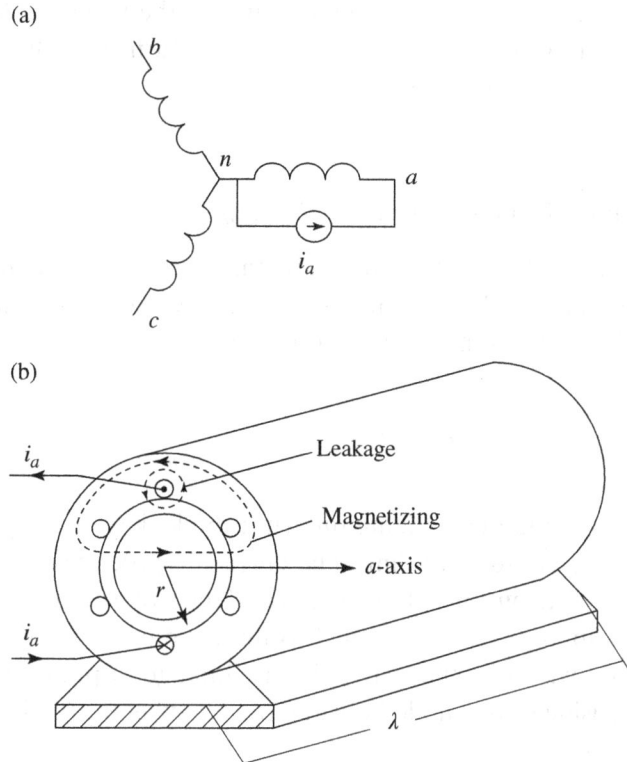

Fig. 11-3 Single-phase magnetizing inductance $L_{m,one\text{-}phase}$ and leakage inductance $L_{\ell s}$.

Therefore,

$$L_{s,self} = L_{\ell s} + L_{m,one\text{-}phase} \tag{11-6b}$$

It requires no-load and blocked-rotor tests to estimate the leakage inductance $L_{\ell s}$, but the single-phase magnetizing inductance $L_{m,one\text{-}phase}$ can be easily calculated by equating the energy storage in the air gap (integrating the energy density $\frac{1}{2}\frac{B_a^2(\theta)}{\mu_o}$ over the air gap volume, where the flux-density $B_a(\theta)$ is given by Eq. (11-3)) to $\frac{1}{2}Li^2$ (see Problem 11-1):

$$L_{m,one\text{-}phase} = \frac{\pi\mu_o r\ell}{\ell_g}\left(\frac{N_s}{p}\right)^2 \tag{11-7}$$

where r is the mean radius at the air gap, ℓ is the length of the rotor along its shaft axis, N_s equals the number of turns per phase, and p equals the number of poles.

11-3-2 Stator Mutual-Inductance L_{mutual}

As shown in Fig. 11-4, the mutual-inductance L_{mutual} between two stator phases can be calculated by hypothetically exciting phase-a by i_a and calculating the flux linkage with phase-b

$$L_{mutual} = \left.\frac{\lambda_b}{i_a}\right|_{i_b, i_c = 0, rotor\ open} \tag{11-8}$$

Note that only the magnetizing flux (not the leakage flux) produced by i_a links the phase-b winding. The current i_a produces a sinusoidal flux-density distribution in the air gap, and the two windings are sinusoidally distributed. Therefore, the flux linking phase-b winding due to i_a can be shown to be the magnetic flux linkage of phase-a winding times the cosine of the angle between the two windings (which in this case is 120°):

$$\lambda_{b, due\ to\ i_a} = \cos(120°)\lambda_{a, magnetizing\ due\ to\ i_a} \tag{11-9a}$$

$$= -\frac{1}{2}\lambda_{a, magnetizing\ due\ to\ i_a} \tag{11-9b}$$

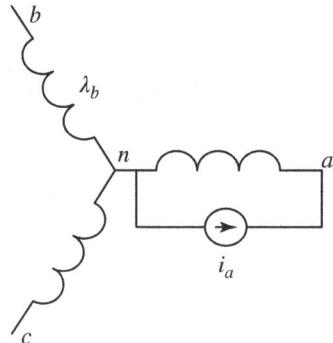

Fig. 11-4 Mutual inductance L_{mutual}.

STATOR INDUCTANCES (ROTOR OPEN-CIRCUITED) 323

Therefore, in Eq. (11-8), using Eqs. (11-6a) and (11-9b),

$$L_{mutual} = -\frac{1}{2} L_{m,one-phase} \quad (11\text{-}10)$$

The same mutual inductance exists between phase-a and phase-c, and between phase-b and phase-c.

The expression for the mutual inductance can also be derived from energy storage considerations (see Problem 11-2).

11-3-3 Per-Phase Magnetizing-Inductance L_m

Under the condition that the rotor is open-circuited, and all three phases are excited in Fig. 11-2b, such that the sum of the three-phase currents is zero as given by Eq. (11-5),

$$\lambda_{a,magnetizing} \Big|_{\substack{(rotor\ open\text{-}circuited) \\ i_a + i_b + i_c = 0}} = L_{m,one-phase} i_a + L_{mutual} i_b + L_{mutual} i_c$$

(11-11)

Using Eq. (11-10) for L_{mutual}, and from Eq. (11-5) replacing $(-i_b - i_c)$ by i_a in Eq. (11-11),

$$L_m = \frac{\lambda_{a,magnetizing}}{i_a} \Big|_{i_a + i_b + i_c = 0,\ rotor\ open} = \frac{3}{2} L_{m,one-phase} \quad (11\text{-}12)$$

Using Eq. (11-7),

$$L_m = \frac{3}{2} \frac{\pi \mu_o r \ell}{\ell_g} \left(\frac{N_s}{p}\right)^2 \quad (11\text{-}13)$$

Note that the single-phase magnetizing inductance $L_{m,one-phase}$ does not include the effect of mutual coupling from the other two phases, whereas the per-phase magnetizing-inductance L_m in Eq. (11-13) does. Hence, L_m is 3/2 times $L_{m,one-phase}$.

11-3-4 Stator-Inductance L_s

Due to all three stator currents (not including the flux linkage due to the rotor currents), the total flux linkage of phase-a can be expressed as

$$\lambda_a|_{rotor\text{-}open} = \lambda_{a,leakage} + \lambda_{a,magnetizing}$$
$$= L_{\ell s} i_a + L_m i_a \qquad (11\text{-}14)$$
$$= L_s i_a$$

where the stator-inductance L_s is

$$L_s = L_{\ell s} + L_m \qquad (11\text{-}15)$$

11-4 EQUIVALENT WINDINGS IN A SQUIRREL-CAGE ROTOR

For developing equations for dynamic analysis, we will replace the squirrel-cage on the rotor by a set of three sinusoidally distributed phase windings. The number of turns in each phase of these equivalent rotor windings can be selected arbitrarily. However, the simplest, hence an obvious choice, is to assume that each rotor phase has N_s turns (similar to the stator windings), as shown in Fig. 11-5a. The voltages and currents in these windings are defined in Fig. 11-5b, where the dotted connection to the rotor-neutral is redundant for the following reason: in a balanced rotor, all the bar currents sum to zero at any instant of time (equal currents in either direction). Therefore, in Fig. 11-5b, the three-rotor phase currents add up to zero at any instant of time

$$i_A(t) + i_B(t) + i_C(t) = 0 \qquad (11\text{-}16)$$

Note that, similar to the stator windings, a positive current into a rotor winding produces flux lines in the radially outward direction along its magnetic axis.

EQUIVALENT WINDINGS IN A SQUIRREL-CAGE ROTOR 325

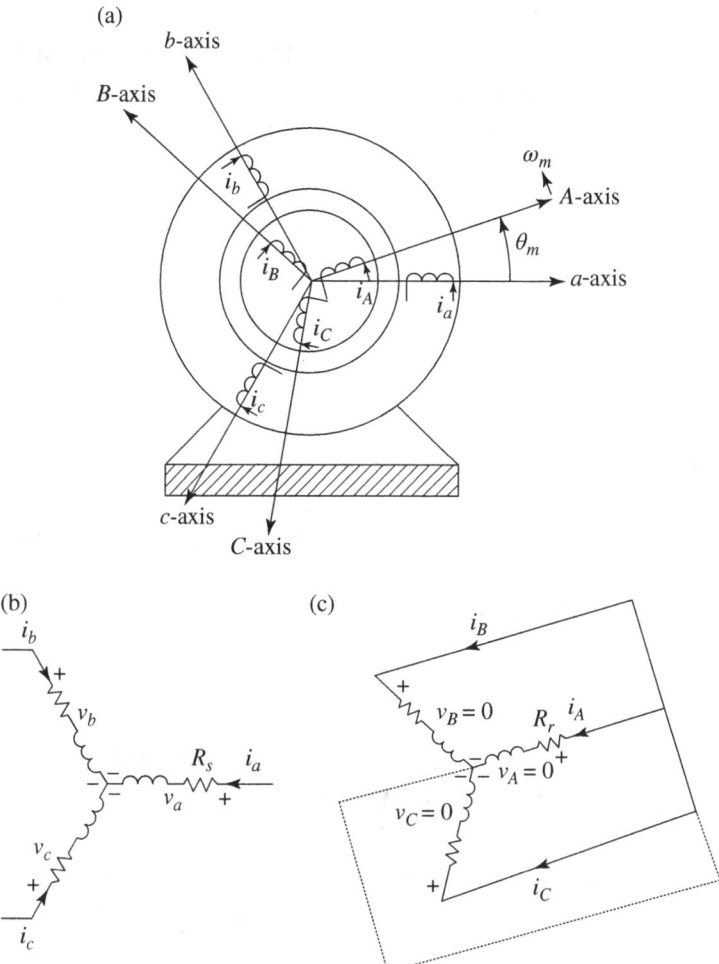

Fig. 11-5 Rotor circuit represented by three-phase windings.

11-4-1 Rotor-Winding Inductances (Stator Open-Circuited)

The magnetizing flux produced by each rotor equivalent winding, has the same magnetic path in crossing the air gap and the same number of turns as the stator phase windings. Hence, each rotor phase has the same magnetizing inductance $L_{m,one\text{-}phase}$ as the magnetic flux produced by the stator phase winding, although its leakage inductance $L_{\ell r}$ may be different than $L_{\ell s}$. Similarly, L_{mutual} between the two rotor phases would be the same as that between two stator phases. The above equalities also imply that the per-phase magnetizing-inductance L_m in the rotor circuit (under the condition that at any time, $i_A + i_B + i_C = 0$) is the same as that in the stator

$$L_m = \frac{3}{2} L_{m,one\text{-}phase} \quad (11\text{-}17)$$

and

$$L_r = L_{\ell r} + L_m \quad (11\text{-}18)$$

Note that with the choice of the same number of turns in the equivalent three-phase rotor windings as in the stator windings, the rotor leakage inductance $L_{\ell r}$ in Eq. (11-18) is the same as $L'_{\ell r}$ in the per-phase, steady-state equivalent circuit of an induction motor. The same applies to the resistances of these equivalent rotor windings, i.e. $R_r = R'_r$.

11-5 MUTUAL INDUCTANCES BETWEEN THE STATOR AND THE ROTOR PHASE WINDINGS

If $\theta_m = 0$ in Fig. 11-5a, so that the magnetic axis of stator phase-a is aligned with the rotor phase-A, the mutual inductance between the two is at its positive peak and equals $L_{m,one\text{-}phase}$. At any other position of the rotor (including $\theta_m = 0$), this mutual inductance between the stator phase-a and the rotor phase-A can be expressed as

$$L_{aA} = L_{m,one\text{-}phase} \cdot \cos \theta_m \quad (11\text{-}19)$$

Similar expressions can be written for mutual inductances between any of the stator phases and any of the rotor phases (see Problem 11-3). Equation (11-19) shows that the mutual inductance, and hence the flux linkages between the stator and the rotor phases, vary with the position θ_m as the rotor turns.

11-6 REVIEW OF SPACE VECTORS

At any instant of time, each phase winding produces a sinusoidal flux-density distribution (or mmf) in the air gap, which can be represented by a space vector (of the appropriate length) along the magnetic axis of that phase (or opposite to, if the phase current at that instant is negative). These mmf space vectors are $\vec{F}_a^a(t)$, $\vec{F}_b^a(t)$, and $\vec{F}_c^a(t)$, as shown in Fig. 11-6a with an "$\rightarrow$" on top of an instantaneous quantity where the superscript "a" indicates that the space vectors are expressed as complex numbers, with the stator a-axis chosen as the reference axis with an angle of 0°. Assuming that there is no magnetic saturation, the resultant mmf distribution in the air gap due to all three phases at that instant can be represented, using vector addition, by the resultant space vector shown in Fig. 11-6b, where the subscript "s" represents the combined stator quantities:

$$\vec{F}_s^a(t) = \vec{F}_a^a(t) + \vec{F}_b^a(t) + \vec{F}_c^a(t) \qquad (11\text{-}20)$$

The above explanation provides a physical basis for understanding space vectors. We should note that unlike phasors, space vectors are also applicable under dynamic conditions.

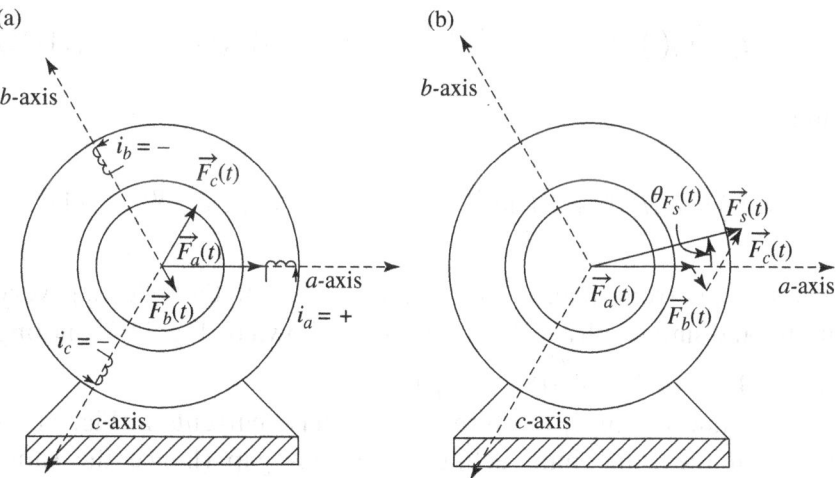

Fig. 11-6 Space vector representation of various mmf quantities.

328 INDUCTION MACHINE EQUATIONS IN PHASE QUANTITIES

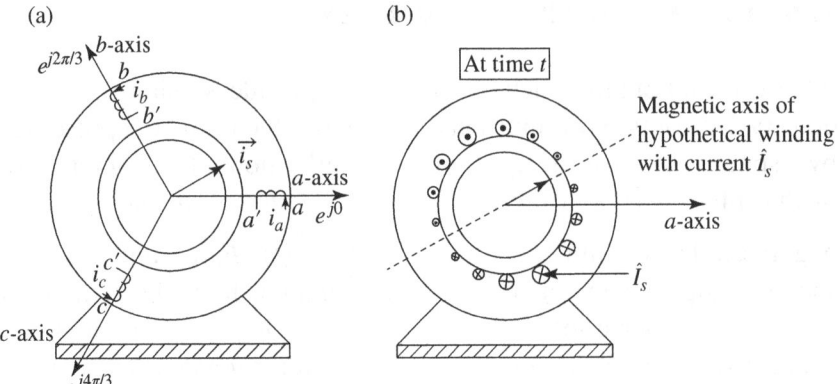

Fig. 11-7 Physical interpretation of stator current space vector.

It is easy to visualize the use of space vectors to represent field distributions (F, B, H), which are distributed sinusoidally in the air gap at any instant of time. However, unlike the field quantities, the currents, the voltages, and the flux linkages of phase windings are treated as terminal quantities. The resultant current, voltage, and flux linkage space vectors for the stator are calculated by multiplying instantaneous phase values by the stator-winding orientations shown in Fig. 11-7a:

$$\vec{i}_s^a(t) = i_a(t)e^{j0} + i_b(t)e^{j2\pi/3} + i_c(t)e^{j4\pi/3} = \hat{I}_s(t)e^{j\theta_{i_s}(t)} \quad (11\text{-}21)$$

$$\vec{v}_s^a(t) = v_a(t)e^{j0} + v_b(t)e^{j2\pi/3} + v_c(t)e^{j4\pi/3} = \hat{V}_s(t)e^{j\theta_{v_s}(t)} \quad (11\text{-}22)$$

and

$$\vec{\lambda}_s^a(t) = \lambda_a(t)e^{j0} + \lambda_b(t)e^{j2\pi/3} + \lambda_c(t)e^{j4\pi/3} = \hat{\lambda}_s(t)e^{j\theta_{\lambda_s}(t)} \quad (11\text{-}23)$$

The stator current space vector $\vec{i}_s^a(t)$ lends itself to the following, very useful and simple, physical interpretation shown by Fig. 11-7b, noting that in Eq. (11-20), $\vec{F}_s^a(t) = (N_s/p)\vec{i}_s^a(t)$.

At a time instant t, the three stator phase currents in Fig. 11-7a result in the same mmf acting on the air gap (hence the same flux-density distribution) as that produced by $\vec{i}_s^a\left(=\hat{I}_s e^{j\theta_{i_s}}\right)$, that is by a current equal to its peak value $\hat{I}_s$ flowing through a hypothetical sinusoidally distributed winding shown in Fig. 11-7b, with its magnetic axis oriented at $\theta_{i_s}(=\theta_{F_s})$. This hypothetical winding has the same number of turns N_s sinusoidally distributed as any of the phase windings.

The above physical explanation not only permits the stator current space vector to be visualized, but it also simplifies the derivation of the electromagnetic torque, which can now be calculated on just this single hypothetic winding, rather than having to calculate torques separately on each of the phase windings and then summing them. Similar space vector equations can be written in the rotor circuit with the rotor axis-A as the reference.

11-6-1 Relationship Between Phasors and Space Vectors in Sinusoidal Steady State

Under a balanced sinusoidal steady-state condition, the voltage and current phasors in phase-a have the same orientation as the stator voltage and current space vectors at the time $t = 0$, as shown for the current in Fig. 11-8; the amplitudes are related by a factor of 3/2:

$$\vec{i}_s^a \bigg|_{t=0} = \frac{3}{2}\bar{I}_a \left(\hat{I}_s = \frac{3}{2}\hat{I}_a \right) \qquad (11\text{-}24)$$

This relationship is very useful because in our dynamic analysis, we often begin with the induction machine initially operating in a balanced, sinusoidal steady state (see Problem 11-5).

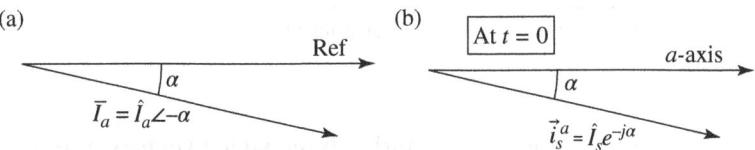

Fig. 11-8 Relationship between space vector and phasor in sinusoidal steady state.

11-7 FLUX LINKAGES

In this section, we will develop equations for stator and rotor flux linkages in terms of currents. We will begin by assuming the stator and the rotor to be open-circuited, one-at-a-time. Then, by superposition, based on the assumption of magnetic material in its linear range, we will be able to obtain flux linkages when the stator and the rotor currents are simultaneously present.

11-7-1 Stator Flux Linkage (Rotor Open-Circuited)

In accordance with the Kirchhoff's current law, the currents in the stator windings sum to zero. Initially, we will assume that the rotor is "somehow" open-circuited. Using Eqs. (11-14) and (11-15), writing the flux-linkage equation for each phase and multiplying each equation with its winding orientation,

$$[\lambda_{a,i_s}(t) = L_{\ell s} i_a(t) + L_m i_a(t)] \times e^{j0} \quad (11\text{-}25a)$$

$$[\lambda_{b,i_s}(t) = L_{\ell s} i_b(t) + L_m i_b(t)] \times e^{j2\pi/3} \quad (11\text{-}25b)$$

and

$$[\lambda_{c,i_s}(t) = L_{\ell s} i_c(t) + L_m i_c(t)] \times e^{j4\pi/3} \quad (11\text{-}25c)$$

Using Eqs. (11-25a) through (11-25c) into Eq. (11-23) (where the stator flux linkage due to the rotor currents is not included), the stator flux linkage space vector is

$$\vec{\lambda}^a_{s,i_s}(t) = \underbrace{L_{\ell s} \vec{i}^a_s(t)}_{\text{due to leakage flux}} + \underbrace{L_m \vec{i}^a_s(t)}_{\text{due to magnetizing flux}} = L_s \vec{i}^a_s(t) \text{ (rotor open)}$$

$$(11\text{-}26)$$

As in the case of stator current and voltage space vectors, the projection of the stator flux-linkage space vector along a phase axis, multiplied by a factor of 2/3, equals the flux linkage of that phase.

FLUX LINKAGES

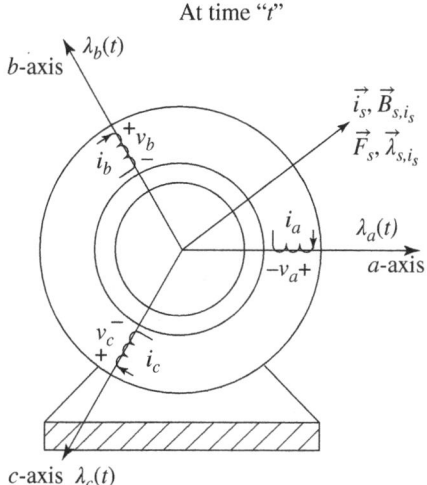

Fig. 11-9 All stator space vectors are collinear (rotor open-circuited).

We have seen earlier that $\vec{i}_s^a(t)$ and $\vec{F}_s^a(t)$ space vectors are collinear, as shown in Fig. 11-9; they are related by a constant. Collinear with $\vec{F}_s^a(t)$, related by a constant μ_0/ℓ_g, is the $\vec{B}_{s,i_s}^a(t)$ space vector, which represents the flux-density distribution due to the stator currents only, "cutting" the stator conductors. Similarly, the stator flux linkage $\vec{\lambda}_{s,i_s}^a(t)$ in Fig. 11-9 (not including the flux linkage due to the rotor currents) is related to $\vec{i}_s^a(t)$ by a constant L_s, as shown by Eq. (11-26). Therefore, all the field quantities, with the rotor open-circuited, are collinear, as shown in Fig. 11-9. Note that the superscript "a" is not used while drawing the various space vectors; it needs to be used only while expressing them mathematically, as defined with respect to a reference axis, which here is the phase-a magnetic axis.

11-7-2 Rotor Flux Linkage (Stator Open-Circuited)

The currents in the rotor equivalent windings sum to zero, as expressed by Eq. (11-16). Assuming that the rotor "somehow" has

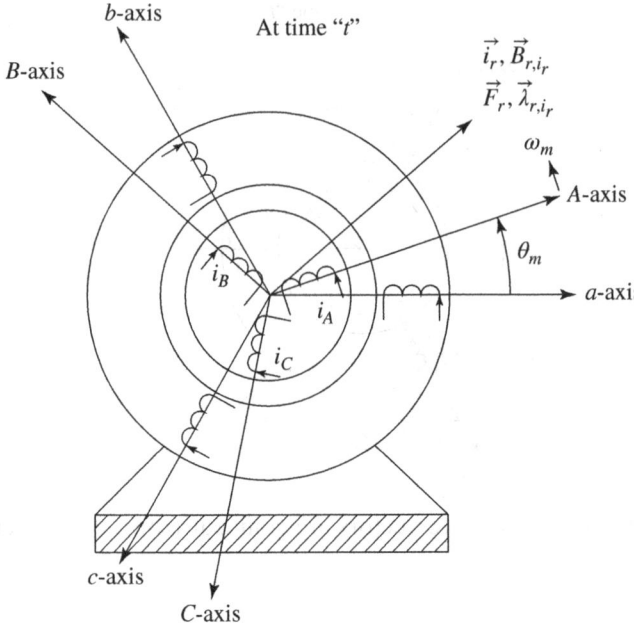

Fig. 11-10 All rotor space vectors are collinear (stator open-circuited).

currents while the stator is open-circuited, by analogy, we can write the expression for the rotor flux linkage space vector as

$$\vec{\lambda}_{r,i_r}^A(t) = \underbrace{L_{\ell r}\vec{i}_r^A(t)}_{\text{due to leakage flux}} + \underbrace{L_m\vec{i}_r^A(t)}_{\text{due to magnetizing flux}} = L_r\vec{i}_r^A(t) \text{ (stator open)}$$

(11-27)

where the superscript "A" indicates that the rotor phase-A axis is chosen as the reference axis with an angle of $0°$, and $L_r = L_{\ell r} + L_m$. Similar to the stator case, all the field quantities with the stator open-circuited are collinear, as shown in Fig. 11-10.

11-7-3 Stator and Rotor Flux Linkages (Simultaneous Stator and Rotor Currents)

When the stator and the rotor currents are present simultaneously, the flux linking any of the stator phases is due to the stator currents

STATOR AND ROTOR VOLTAGE EQUATIONS 333

as well as the mutual magnetizing flux due to the rotor currents. The magnetizing flux-density space vectors in the air gap due to the stator and the rotor currents add up as vectors when these currents are simultaneously present. Therefore, the stator flux linkage, including the leakage flux due to the stator currents, can be obtained using Eqs. (11-26) and (11-27) as

$$\vec{\lambda}_s^a(t) = L_s \vec{i}_s^a(t) + L_m \vec{i}_r^a(t) \tag{11-28}$$

where the rotor current space vector is also defined with respect to the stator phase-a axis.

Similarly, in the rotor circuit, we can write

$$\vec{\lambda}_r^A(t) = L_m \vec{i}_s^A(t) + L_r \vec{i}_r^A(t) \tag{11-29}$$

where the stator current space vector is also defined with respect to the rotor phase-A axis.

11-8 STATOR AND ROTOR VOLTAGE EQUATIONS IN TERMS OF SPACE VECTORS

The individual phase equations can be combined to obtain the space vector equation as follows:

$$\left[v_a(t) = R_s i_a(t) + \frac{d}{dt}\lambda_a(t)\right] \times e^{j0} \tag{11-30a}$$

$$\left[v_b(t) = R_s i_b(t) + \frac{d}{dt}\lambda_b(t)\right] \times e^{j2\pi/3} \tag{11-30b}$$

and

$$\left[v_c(t) = R_s i_c(t) + \frac{d}{dt}\lambda_c(t)\right] \times e^{j4\pi/3} \tag{11-30c}$$

Adding the above three equations and applying the definitions of space vectors, the stator equation can be written as

$$\vec{v}_s^a(t) = R_s \vec{i}_s^a(t) + \frac{d}{dt}\vec{\lambda}_s^a(t) \qquad (11\text{-}31)$$

Similar to the development in the stator circuit, in the rotor circuit,

$$\underbrace{\vec{v}_r^A(t)}_{=0} = R_r \vec{i}_r^A(t) + \frac{d}{dt}\vec{\lambda}_r^A(t) \qquad (11\text{-}32)$$

where in a squirrel-cage rotor, all the equivalent phase voltages are individually zero and $\vec{v}_r^A(t) = 0$.

11-9 MAKING A CASE FOR A dq-WINDING ANALYSIS

At this point, we should assess how far we have come. The use of space vectors has very quickly allowed us to express the stator and the rotor flux linkages (Eqs. (11-28) and (11-29)), which in a compact form include mutual coupling between the six windings: three on the stator and three on the equivalent rotor. In terms of phase quantities of an induction machine, we have developed voltage equations for the rotor and the stator, expressed in a compact space vector form (Eqs. (11-31) and (11-32)). These voltage equations include the time-derivatives of flux linkages that depend on the rotor position. This dependence can be seen if we examine the flux linkage equations by expressing them with current space vectors defined with respect to their own reference axes in Fig. 11-5 as

$$\vec{i}_r^a(t) = \vec{i}_r^A(t)e^{j\theta_m} \qquad (11\text{-}33)$$

and

$$\vec{i}_s^A(t) = \vec{i}_s^a(t)e^{-j\theta_m} \qquad (11\text{-}34)$$

Using the above two equations in the flux linkage equations (Eqs. (11-28) and (11-29)),

$$\vec{\lambda}_s^a(t) = L_s \vec{i}_s^a(t) + L_m \vec{i}_r^A(t) e^{j\theta_m} \quad (11\text{-}35)$$

and

$$\vec{\lambda}_r^A(t) = L_m \vec{i}_s^a(t) e^{-j\theta_m} + L_r \vec{i}_r^A(t) \quad (11\text{-}36)$$

The flux linkage equations in the above form, clearly show their dependence on the rotor position θ_m for given values of the stator and the rotor currents at any instant of time. For this reason, the voltage equations in phase quantities, expressed in a space vector form by Eqs. (11-31) and (11-32), which include the time-derivatives of flux linkages, are complicated to solve. It is possible to make these equations simpler by using a transformation called dq transformation; which is the topic of the next chapter.

The above argument alone, on the basis of simplifying the equations, especially in the age of fast (and faster!) computers, is not sufficient to search for an alternative, such as the dq-winding analysis. The power of the dq-winding analysis lies in the fact that it allows the torque and the flux in the machine to be controlled independently under dynamic conditions, which is not clear in our foregoing analysis based on the phase (a-b-c) quantities. An obvious question at this point is if the analysis in this chapter has been a waste. The answer is a resounding "no." We will use every bit of the analysis in this chapter to carry out the dq-analysis in Chapter 12.

Before we embark on the dq-analysis in the next chapter, we will further look at the analysis of an induction machine in phase quantities by means of the following examples.

EXAMPLE 11-1

First, take a two coupled-coil system, one on the stator and the other on the rotor. Derive the electromagnetic torque expression

(*Continued*)

336 INDUCTION MACHINE EQUATIONS IN PHASE QUANTITIES

by energy considerations and then generalize it in terms of three-phase stator and rotor currents.

Solution

Neglecting losses, the differential electrical input energy, mechanical energy output, and the stored field energy can be written as follows:

$$dW_{in} = dW_{mech} + dW_{mag} \tag{11-37}$$

For coupled two-coil system of coils 1 and 2, the differential electrical input energy is as follows:

$$\begin{aligned} dW_{in} &= v_1 i_1 dt + v_2 i_2 dt \\ &= i_1 d\lambda_1 + i_2 d\lambda_2 \\ &= i_1 d(L_{11} i_1 + L_{12} i_2) + i_2 d(L_{12} i_1 + L_{22} i_2) \\ &= L_{11} i_1 di_1 + L_{12} i_1 di_2 + i_1^2 dL_{11} + i_1 i_2 dL_{12} \\ &\quad + L_{22} i_2 di_2 + L_{12} i_2 di_1 + i_2^2 dL_{22} + i_1 i_2 dL_{12} \end{aligned} \tag{11-38}$$

The stored magnetic energy is

$$W_{mag} = \frac{1}{2} L_{11} i_1^2 + \frac{1}{2} L_{22} i_2^2 + L_{12} i_1 i_2 \tag{11-39}$$

Therefore, the differential increase in the stored magnetic energy is

$$\begin{aligned} dW_{mag} &= v_1 i_1 dt + v_2 i_2 dt \\ &= i_1 d\lambda_1 + i_2 d\lambda_2 \\ &= i_1 d(L_{11} i_1 + L_{12} i_2) + i_2 d(L_{12} i_1 + L_{22} i_2) \\ &= L_{11} i_1 di_1 + L_{12} i_1 di_2 + i_1^2 dL_{11} + i_1 i_2 dL_{12} \\ &\quad + L_{22} i_2 di_2 + L_{12} i_2 di_1 + i_2^2 dL_{22} + i_1 i_2 dL_{12} \end{aligned} \tag{11-40}$$

MAKING A CASE FOR A dq-WINDING ANALYSIS 337

From Eqs. (11-37), (11-38), and (11-40),

$$T_{em} = \frac{1}{2}i_1^2 \frac{dL_{11}}{d\theta_m} + \frac{1}{2}i_2^2 \frac{dL_{22}}{d\theta_m} + i_1 i_2 \frac{dL_{12}}{d\theta_m} \qquad (11\text{-}41)$$

In terms of a matrix equation, Eq. (11-41) can be written as

$$T_{em} = \frac{1}{2} [i_1 \; i_2] \frac{d}{d\theta_m} \begin{bmatrix} L_{11} & L_{12} \\ L_{12} & L_{22} \end{bmatrix} \begin{bmatrix} i_1 \\ i_2 \end{bmatrix}$$

$$= \frac{1}{2} [i]^t \frac{d}{d\theta_m} [L][i] \qquad (11\text{-}42)$$

Equation (11-42) can be generalized to six windings, a-b-c on the stator and A-B-C on the rotor, for a p-pole machine as follows:

$$T_{em} = \frac{p}{2} \frac{1}{2} [i]^t \frac{d}{d\theta_m} [L][i] \qquad (11\text{-}43)$$

where $\theta_m = \frac{p}{2} \theta_{mech}$.

EXAMPLE 11-2

For the machine parameters given below, simulate the induction machine start-up in MATLAB using the equations derived in this chapter, from a completely powered-down state with no external load connected to it. Plot rotor speed, electromagnetic torque, stator, and rotor currents.

The parameters for a 1.5 MW wind-turbine induction machine are as follows:

Power:	1.5 MW
Voltage:	575 V (L-L, rms)
Frequency:	60 Hz
Phases:	3

(*Continued*)

Number of poles: 6
Full-load slip: 1%
Moment of inertia: 75 kg·m²

Per-Phase Circuit Parameters:

$R_s = 0.0014 \, \Omega$

$R_r = 0.000\,991 \, \Omega$

$X_{\ell s} = 0.034 \, \Omega$

$X_{\ell r} = 0.031 \, \Omega$

$X_m = 0.0576 \, \Omega$

Solution

See the complete results, including the computer files, in Appendix 11-A on the accompanying website.

EXAMPLE 11-3

Verify the results in Example 11-2 by simulations.

Solution

See the complete results, including the simulation files, on the accompanying website.

11-10 SUMMARY

In this chapter, we have briefly reviewed the sinusoidally distributed windings and then calculated their inductances for developing equations for induction machines in phase (*a-b-c*) quantities. The development of these equations is assisted by space vectors, which are briefly reviewed. The analysis in this chapter establishes the framework and the rationale for the dq-windings-based analysis of induction machines under dynamic conditions carried out in Chapter 12.

PROBLEMS

11-1 Derive Eq. (11-7) for $L_{m,one\text{-}phase}$.

11-2 Derive the expression for L_{mutual} in Eq. (11-8) by energy storage considerations. Hint: Assume that only the stator phases a and b are excited so that $i_b = -i_a$. To keep this derivation general, begin by assuming an arbitrary angle θ between the magnetic axes of the two windings.

11-3 Write the expressions for L_{kJ} as functions of θ_m, where $k \equiv a, b, c$ and $J \equiv A, B, C$.

11-4 Calculate L_s, L_r, and L_m for the motor in Example 11-2.

11-5 A motor with the following nameplate data is operating in a balanced sinusoidal steady state under its rated condition (with rated voltages applied to it and it is loaded to its rated torque). Assume that the voltage across phase-a is at its positive peak at $t = 0$. (a) Obtain at the time $t = 0$, $\vec{v}_s(0)$, $\vec{i}_s(0)$, and $\vec{i}_r'(0)$, and (b) express phase voltages and currents as functions of time.
Nameplate Data:

Power	3 HP/2.4 kW
Voltage	460 V (L-L, rms)
Frequency	60 Hz
Phases	3
Full-load current	4 A
Full-load speed	1750 RPM
Full-load efficiency	88.5%
Power factor	80.0%
Number of poles	4

340 INDUCTION MACHINE EQUATIONS IN PHASE QUANTITIES

Per-Phase Motor Circuit Parameters:

$$R_s = 1.77 \, \Omega$$

$$R_r = 1.34 \, \Omega$$

$$X_{\ell s} = 5.25 \, \Omega \text{ (at 60 Hz)}$$

$$X_{\ell r} = 4.57 \, \Omega \text{ (at 60 Hz)}$$

$$X_m = 139.0 \, \Omega \text{ (at 60 Hz)}$$

Full-load slip = 1.72%

The iron losses are specified as 78 W, and the mechanical (friction and windage) losses are specified as 24 W. The inertia of the machine is given. Assuming that the reflected load inertia is approximately the same as the motor inertia, the total equivalent inertia of the system is $J_{eq} = 0.025 \text{ kg} \cdot \text{m}^2$.

11-6 At an instant of time in an induction machine, hypothetically assume that the stator currents $i_a = 10 \text{ A}$, $i_b = -3 \text{ A}$, $i_c = -7 \text{ A}$ and the rotor currents $i_A = 3 \text{ A}$, $i_B = -1 \text{ A}$, $i_C = -2 \text{ A}$. Calculate $\vec{\lambda}_s^a \big|_{\vec{i}_s}(t), \vec{\lambda}_r^A \big|_{\vec{i}_r}(t), \vec{\lambda}_s^a(t)$, and $\vec{\lambda}_r^A(t)$ in terms of machine inductances L_m, L_s, and L_r, if the rotor angle θ_m has the following values: (a) 0° and (b) 30°.

11-7 Write the expression for the stator phase-a flux linkage, in terms of three stator and three rotor phase currents and the appropriate inductances, for a rotor position of θ_m. Repeat this for the other stator and rotor phases.

11-8 Show that Eqs. (11-28) and (11-29) can be written with respect to any arbitrary axis, rather than a-axis or A-axis.

11-9 Show the intermediate steps in generalizing Eq. (11-42) to Eq. (11-43).

12 Dynamic Analysis of Induction Machines in Terms of *dq*-Windings

12-1 INTRODUCTION

In this chapter, we will develop equations to analyze induction machine operation under dynamic conditions. We will make use of space vectors as intermediary in transforming *a-b-c* phase winding quantities into equivalent *dq*-winding quantities that we will use for dynamic (non-steady state) analysis. We will see in later chapters the benefits of *d*- and *q*-axis analysis in controlling ac machines.

12-2 *dq*-WINDING REPRESENTATION

We studied in the previous chapter, that the stator and the rotor flux linkages, $\vec{\lambda}_s^a(t)$ and $\vec{\lambda}_r^a(t)$, depend on the rotor angle θ_m because the mutual inductances between the stator and the rotor windings are position-dependent. The main reason for the *d*- and *q*-axis analysis in machines like the induction machines is to control them properly, for example, using vector control principles. In most textbooks, this analysis is discussed as a mathematical transformation called Park's

---

(Adapted from chapter 3 of *Advanced Electric Drives: Analysis, Control, and Modeling Using MATLAB/Simulink* ISBN: 978-1-118-48548-4 by Ned Mohan, August 2014)

Analysis and Control of Electric Drives: Simulations and Laboratory Implementation, First Edition. Ned Mohan and Siddharth Raju.
© 2021 John Wiley & Sons, Inc. Published 2021 by John Wiley & Sons, Inc.
Companion website: www.wiley.com/go/Mohan/Vectorcontrolinelectricdrives

Transformation. In this chapter, we will take a physical approach to this transformation, which is much easier to visualize and arrive at identical results.

12-2-1 Stator dq-Winding Representation

In Fig. 12-1a at a time t, phase currents $i_a(t)$, $i_b(t)$, and $i_c(t)$ are represented by a stator current space vector $\vec{i}_s(t)$. A collinear mmf space vector $\vec{F}_s(t)$ is related to $\vec{i}_s(t)$ by a factor of (N_s/p), where N_s equals the number of turns per phase and p equals the number of poles:

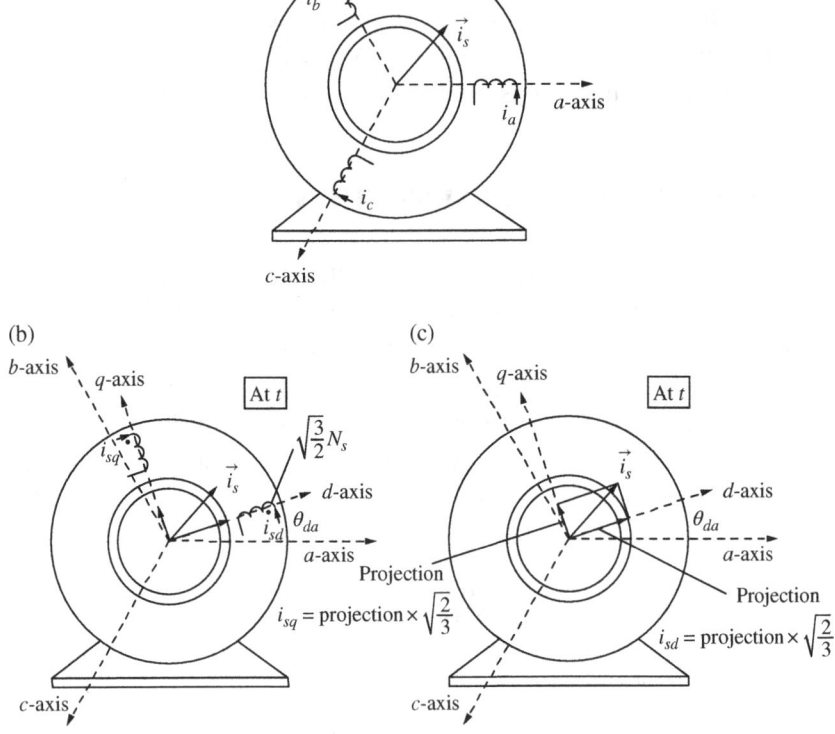

Fig. 12-1 Representation of stator mmf by equivalent dq-windings.

dq-WINDING REPRESENTATION

$$\vec{i}_s^{\,a}(t) = i_a(t) + i_b(t)\, e^{j2\pi/3} + i_c(t)\, e^{j4\pi/3} \tag{12-1}$$

and

$$\vec{F}_s^{\,a}(t) = \frac{N_s}{p}\, \vec{i}_s^{\,a}(t) \tag{12-2}$$

We should note that the space vector $\vec{i}_s(t)$ in Fig. 12-1 is written without a superscript "a." The reason is that a reference axis is needed *only* to express it mathematically utilizing complex numbers. However, $\vec{i}_s(t)$ in Fig. 12-1 depends on the instantaneous values of phase currents and is independent of the choice of the reference axis to draw it.

In the previous chapters on analyzing ac machines under balanced sinusoidal steady-state conditions, we replaced the three windings by a single hypothetical equivalent winding that produced the same mmf distribution in the air gap. This single winding was sinusoidally distributed with the same number of turns N_s (as any phase winding), with its magnetic axis along the stator current space vector and a current $\hat{I}_s$ (peak value of $\vec{i}_s$) flowing through it.

However, for dynamic analysis and control of ac machines, we need two orthogonal windings such that the torque and the flux within the machine can be controlled independently. At any instant of time, the air gap mmf distribution by three-phase windings can also be produced by a set of two orthogonal windings shown in Fig. 12-1b, each sinusoidally distributed with $\sqrt{3/2}\,N_s$ turns: one winding along the d-axis, and the other along the q-axis. The reason for choosing $\sqrt{3/2}\,N_s$ turns will be explained shortly. This dq-winding set may be at any arbitrary angle θ_{da}, with respect to the phase-a axis. However, the currents i_{sd} and i_{sq} in these two windings must have specific values that can be obtained by equating the mmf produced by the dq-windings to that produced by the three-phase windings, and represented by a single winding with N_s turns in Eq. (12-2),

$$\frac{\sqrt{3/2}\,N_s}{p}\left(i_{sd} + j i_{sq}\right) = \frac{N_s}{p}\, \vec{i}_s^{\,d} \tag{12-3}$$

where the stator current space vector is expressed using the d-axis as the reference axis, hence the superscript "d." Equation (12-3) results in

$$\left(i_{sd} + ji_{sq}\right) = \sqrt{\frac{2}{3}}\, \vec{i}_s^{\,d} \tag{12-4}$$

which shows that the dq-winding currents are $\sqrt{2/3}$ times the projections of $\vec{i}_s(t)$ vector along the d- and q-axis, as shown in Fig. 12-1c,

$$i_{sd} = \sqrt{2/3} \times \text{projection of } \vec{i}_s(t) \text{ along the } d\text{-axis} \tag{12-5}$$

and

$$i_{sq} = \sqrt{2/3} \times \text{projection of } \vec{i}_s(t) \text{ along the } q\text{-axis} \tag{12-6}$$

The factor $\sqrt{2/3}$, reciprocal of the factor $\sqrt{3/2}$ used in choosing the number of turns for the dq-windings, ensures that the dq-winding currents produce the same mmf distribution as the three-phase winding currents.

In Fig. 12-1b, the d and the q windings are mutually decoupled magnetically due to their orthogonal orientation. Choosing $\sqrt{3/2}\, N_s$ turns for each of these windings, results in their magnetizing inductance to be L_m (same as the per-phase magnetizing inductance in Chapter 2 for three-phase windings with $i_a + i_b + i_c = 0$) for the following reason: the inductance of a winding is proportional to the square of the number of turns, and therefore, the magnetizing inductance of any dq-winding (noting that there is no mutual inductance between the two orthogonal windings) is

$$\begin{aligned} dq \text{ winding magnetizing inductance} &= \left(\sqrt{3/2}\right)^2 L_{m,one\text{-}phase} \\ &= (3/2) L_{m,one\text{-}phase} \\ &= L_m \text{ (using Eq.}(2-12)) \end{aligned} \tag{12-7}$$

Each of these equivalent windings has a resistance R_s and a leakage inductance $L_{\ell s}$, similar to the a-b-c phase windings (see Problem 12-1).

dq-WINDING REPRESENTATION   345

In fact, if a 12-phase machine were to be converted to a two-phase machine using the same stator shell (but the windings could be different) to deliver the same power output and speed, we will choose the number of turns in each of the two-phase windings to be $\sqrt{3/2}\, N_s$.

12-2-2 Rotor dq-Windings (Along the Same dq-Axes as in the Stator)

The rotor mmf space vector $\vec{F}_r(t)$ is produced by the combined effect of the rotor bar currents, or by the three equivalent phase windings, each with N_s turns as shown in Fig. 12-2 (short-circuited in a squirrel-cage rotor). The phase currents in these equivalent rotor phase windings can be represented by a rotor current space vector, where

$$\vec{i}_r^{\,A}(t) = i_A(t) + i_B(t)\, e^{j2\pi/3} + i_C(t)\, e^{j4\pi/3} \qquad (12\text{-}8)$$

where

$$\vec{i}_r^{\,A}(t) = \frac{\vec{F}_r^{\,A}(t)}{N_s/p} \qquad (12\text{-}9)$$

The mmf $\vec{F}_r(t)$ and the rotor current $\vec{i}_r(t)$ in Fig. 12-2, can also be produced by the components $i_{rd}(t)$ and $i_{rq}(t)$ flowing through their respective windings, as shown. (Note that the d- and the q-axis are the same as those chosen for the stator in Fig. 12-1. Otherwise, all benefits of

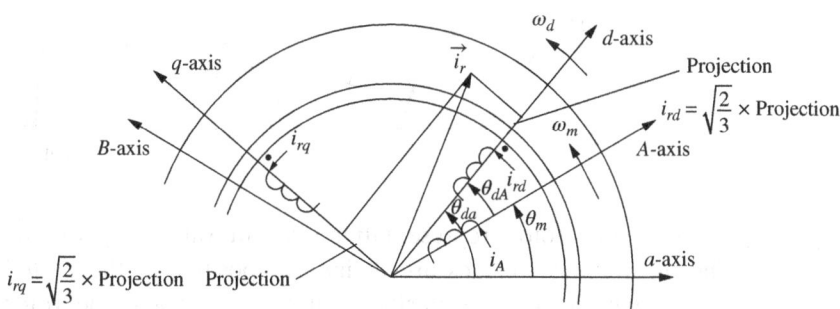

Fig. 12-2 Representation of rotor mmf by equivalent dq-winding currents.

the dq-analysis will be lost.) Similar to the stator case, each of the dq-windings on the rotor has $\sqrt{3/2}N_s$ turns, and a magnetizing inductance of L_m, which is the same as that for the stator dq-windings because of the same number of turns (by choice) and the same magnetic path for flux lines. Each of these rotor equivalent windings has a resistance R_r and a leakage inductance $L_{\ell r}$. The mutual inductance between these two orthogonal windings is zero.

12-2-3 Mutual Inductance Between dq-Windings on the Stator and the Rotor

The equivalent dq-windings for the stator and the rotor are shown in Fig. 12-3. The mutual inductance between the stator and the rotor d-axis windings is equal to L_m due to the magnetizing flux crossing the air gap. Similarly, the mutual inductance between the stator and the rotor q-axis windings equals L_m. Out of four dq-windings,

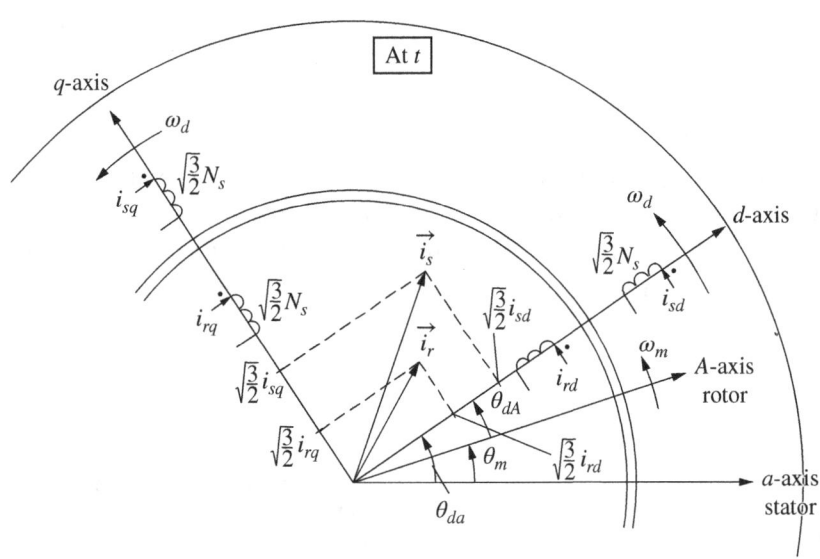

Fig. 12-3 Stator and rotor representation by equivalent dq-winding currents. The dq-winding voltages are defined as positive at the dotted terminals. Note that the relative positions of the stator and the rotor current space vectors are not actual, rather only for definition purposes.

MATHEMATICAL RELATIONSHIPS OF THE dq-WINDINGS 347

the mutual inductance between any d-axis winding with any q-axis winding is zero because of their orthogonal orientation, which results in zero mutual magnetic coupling of flux.

12-3 MATHEMATICAL RELATIONSHIPS OF THE dq-WINDINGS (AT AN ARBITRARY SPEED ω_d)

Next, we will describe relationships between the stator and the rotor quantities, and their equivalent dq-winding components in Fig. 12-3, which in combination produce the same mmf as the actual three-phase windings.

It is worth repeating that the space vectors at some arbitrary time t in Fig. 12-3 are expressed without a superscript "a" or "A." The reason is that a reference axis is needed *only* to express them mathematically, by means of complex numbers. In other words, these space vectors in Fig. 12-3 would be in the same position, independent of the choice of the reference axis to express them. We should note that the relative position of $\vec{i}_s$ and $\vec{i}_r$ is shown arbitrarily here just for definition purposes (in an induction machine, the angle between $\vec{i}_s$ and $\vec{i}_r$ is very large – more than 145°).

Hereafter, we will drop the superscript to any space vector expressed using d-axis as the reference.

From Fig. 12-3, we note that at a time t, the d-axis is shown at an angle θ_{da} with respect to the stator a-axis. Therefore,

$$\vec{i}_s(t) = \vec{i}_s^{\,a}(t) e^{-j\theta_{da}(t)} \qquad (12\text{-}10)$$

Substituting for $\vec{i}_s^{\,a}$ from Eq. (12-1),

$$\vec{i}_s(t) = i_a e^{-j\theta_{da}} + i_b e^{-j(\theta_{da} - 2\pi/3)} + i_c e^{-j(\theta_{da} - 4\pi/3)} \qquad (12\text{-}11)$$

Equating the real and imaginary components on the right side of Eq. (12-11) to i_{sd} and i_{sq} in Eq. (12-4),

$$\begin{bmatrix} i_{sd}(t) \\ i_{sq}(t) \end{bmatrix} =$$

$$\sqrt{\frac{2}{3}} \underbrace{\begin{bmatrix} \cos(\theta_{da}) & \cos\left(\theta_{da} - \frac{2\pi}{3}\right) & \cos\left(\theta_{da} - \frac{4\pi}{3}\right) \\ -\sin(\theta_{da}) & -\sin\left(\theta_{da} - \frac{2\pi}{3}\right) & -\sin\left(\theta_{da} - \frac{4\pi}{3}\right) \end{bmatrix}}_{[T_s]_{abc \to dq}} \begin{bmatrix} i_a(t) \\ i_b(t) \\ i_c(t) \end{bmatrix}$$

(12-12)

where $[T_s]_{abc \to dq}$ is the transformation matrix to transform stator *a-b-c* phase winding currents to the corresponding *dq*-winding currents. This transformation procedure is illustrated by the block diagram in Fig. 12-4a. The same transformation matrix relates the stator flux linkages, and the stator voltages, in phase windings to those in the equivalent stator *dq*-windings.

A similar procedure to that in the stator case is followed for the rotor where, in terms of the phase currents, the rotor current space vector is

$$\vec{i}_r^A(t) = i_A(t) + i_B(t)\, e^{j2\pi/3} + i_C(t)\, e^{j4\pi/3} \qquad (12\text{-}13)$$

From Fig. 12-3, we note that at a time *t*, *d*-axis is at an angle θ_{dA} with respect to the rotor *A*-axis. Therefore,

$$\vec{i}_r(t) = \vec{i}_r^A(t)\, e^{-j\theta_{dA}(t)} \qquad (12\text{-}14)$$

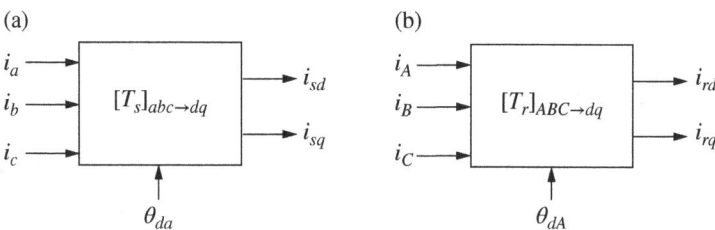

Fig. 12-4 Transformation of phase quantities into *dq*-winding quantities: (a) stator and (b) rotor.

MATHEMATICAL RELATIONSHIPS OF THE dq-WINDINGS 349

The currents in the dq rotor windings must be i_{rd} and i_{rq} where these two current components are $\sqrt{2/3}$ times the projections of $\vec{i}_r(t)$ vector along the d- and q-axis, as shown in Fig. 12-3,

$$i_{rd} = \sqrt{2/3} \times \text{projection of } \vec{i}_r(t) \text{ along the } d\text{-axis} \quad (12\text{-}15)$$

and

$$i_{rq} = \sqrt{2/3} \times \text{projection of } \vec{i}_r(t) \text{ along the } q\text{-axis} \quad (12\text{-}16)$$

Similar to Eq. (12-12), replacing θ_{da} by θ_{dA},

$$\begin{bmatrix} i_{rd}(t) \\ i_{rq}(t) \end{bmatrix} =$$

$$\underbrace{\sqrt{\frac{2}{3}} \begin{bmatrix} \cos(\theta_{dA}) & \cos\left(\theta_{dA} - \frac{2\pi}{3}\right) & \cos\left(\theta_{dA} - \frac{4\pi}{3}\right) \\ -\sin(\theta_{dA}) & -\sin\left(\theta_{dA} - \frac{2\pi}{3}\right) & -\sin\left(\theta_{dA} - \frac{4\pi}{3}\right) \end{bmatrix}}_{[T_r]_{ABC \to dq}} \begin{bmatrix} i_A(t) \\ i_B(t) \\ i_C(t) \end{bmatrix}$$

$$(12\text{-}17)$$

where $[T_r]_{ABC \to dq}$ is the transformation matrix for the rotor. This transformation procedure is illustrated by the block diagram in Fig. 12-4b, similar to that in Fig. 12-4a. The same transformation matrix relates the rotor flux linkages and the rotor voltages in the equivalent A-B-C windings to those in the equivalent rotor dq-windings. Same relationships apply to voltages and flux linkages.

12-3-1 Relating dq-Winding Variables to Phase Winding Variables

In case of an isolated neutral, where all three-phase currents add up to zero at any time, the variables in a-b-c phase windings can be calculated in terms of the dq-winding variables. In Eq. (12-12), we can add a row at the bottom to represent the condition that all three-phase currents

350 DYNAMIC ANALYSIS OF INDUCTION MACHINES

sum to zero. Inverting the resulting matrix and discarding the last column whose contribution is zero, we obtain the desired relationship

$$\begin{bmatrix} i_a(t) \\ i_b(t) \\ i_c(t) \end{bmatrix} = \sqrt{\frac{2}{3}} \underbrace{\begin{bmatrix} \cos(\theta_{da}) & -\sin(\theta_{da}) \\ \cos\left(\theta_{da} + \frac{4\pi}{3}\right) & -\sin\left(\theta_{da} + \frac{4\pi}{3}\right) \\ \cos\left(\theta_{da} + \frac{2\pi}{3}\right) & -\sin\left(\theta_{da} + \frac{2\pi}{3}\right) \end{bmatrix}}_{[T_s]_{dq \to abc}} \begin{bmatrix} i_{sd} \\ i_{sq} \end{bmatrix}$$

(12-18)

where $[T_s]_{dq \to abc}$ is the transformation matrix in the reverse direction (dq to abc). A similar transformation matrix $[T_r]_{dq \to ABC}$ for the rotor can be written by replacing θ_{da} in Eq. (12-18) by θ_{dA}.

12-3-2 Flux Linkages of dq-Windings in Terms of Their Currents

We have a set of four dq-windings, as shown in Fig. 12-3. There is no mutual coupling between the windings on the d-axis and those on the q-axis. The flux linking any winding is due to its own current, and that due to the other winding on the same axis. Let us select the stator d-winding as an example. Due to i_{sd}, both the magnetizing flux as well as the leakage flux link this winding. However, due to i_{rd}, only the magnetizing flux (leakage flux does not cross the air gap) links this stator winding. Using this logic, we can write the following flux expressions for all four windings.

Stator Windings

$$\lambda_{sd} = L_s i_{sd} + L_m i_{rd} \qquad (12\text{-}19)$$

and

$$\lambda_{sq} = L_s i_{sq} + L_m i_{rq} \qquad (12\text{-}20)$$

where in Eqs. (12-19) and (12-20), $L_s = L_{\ell s} + L_m$.

Rotor Windings

$$\lambda_{rd} = L_r i_{rd} + L_m i_{sd} \qquad (12\text{-}21)$$

and

$$\lambda_{rq} = L_r i_{rq} + L_m i_{sq} \qquad (12\text{-}22)$$

where in Eqs. (12-21) and (12-22), $L_r = L_{\ell r} + L_m$.

12-3-3 dq-Winding Voltage Equations

Stator Windings To derive the dq-winding voltages, we will first consider a set of orthogonal $\alpha\beta$ windings affixed to the stator, as shown in Fig. 12-5, where the α-axis is aligned with the stator a-axis. In all windings, the voltage polarity is defined to be positive at the dotted terminal. In $\alpha\beta$ windings in terms of their variables,

$$v_{s\alpha} = R_s i_{s\alpha} + \frac{d}{dt}\lambda_{s\alpha} \qquad (12\text{-}23)$$

and

$$v_{s\beta} = R_s i_{s\beta} + \frac{d}{dt}\lambda_{s\beta} \qquad (12\text{-}24)$$

The above two equations can be combined by multiplying both sides of Eq. (12-24) by the operator (j) and then adding to Eq. (12-23). In terms of resulting space vectors,

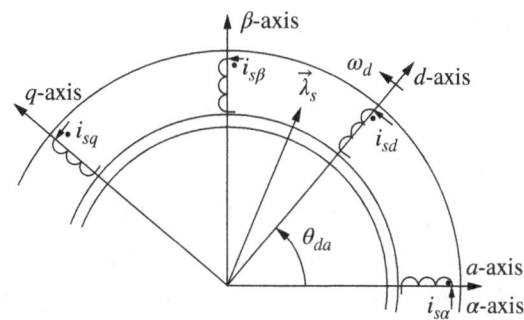

Fig. 12-5 Stator $\alpha\beta$ and dq equivalent windings.

352 DYNAMIC ANALYSIS OF INDUCTION MACHINES

$$\vec{v}_{s\_\alpha\beta}^{\alpha} = R_s \vec{i}_{s\_\alpha\beta}^{\alpha} + \frac{d}{dt}\vec{\lambda}_{s\_\alpha\beta}^{\alpha} \qquad (12\text{-}25)$$

where $\vec{v}_{s\_\alpha\beta}^{\alpha} = v_{s\alpha} + jv_{s\beta}$ and so on. As can be seen from Fig. 12-5, the current, voltage, and flux linkage space vectors with respect to the α-axis are related to those with respect to the d-axis as follows:

$$\vec{v}_{s\_\alpha\beta}^{\alpha} = \vec{v}_{s\_dq} \cdot e^{j\theta_{da}} \qquad (12\text{-}26a)$$

$$\vec{i}_{s\_\alpha\beta}^{\alpha} = \vec{i}_{s\_dq} \cdot e^{j\theta_{da}} \qquad (12\text{-}26b)$$

and

$$\vec{\lambda}_{s\_\alpha\beta}^{\alpha} = \vec{\lambda}_{s\_dq} \cdot e^{j\theta_{da}} \qquad (12\text{-}26c)$$

where $\vec{v}_{s\_dq} = v_{sd} + jv_{sq}$ and so on. Substituting expressions from Eqs. (12-26a) through (12-26c) into Eq. (12-25),

$$\vec{v}_{s\_dq} \cdot e^{j\theta_{da}} = R_s \vec{i}_{s\_dq} \cdot e^{j\theta_{da}} + \frac{d}{dt}\left(\vec{\lambda}_{s\_dq} \cdot e^{j\theta_{da}}\right)$$

or

$$\vec{v}_{s\_dq} \cdot e^{j\theta_{da}} = R_s \vec{i}_{s\_dq} \cdot e^{j\theta_{da}} + \frac{d\vec{\lambda}_{s\_dq}}{dt} \cdot e^{j\theta_{da}} + j\underbrace{\frac{d\theta_{da}}{dt}}_{\omega_d} \cdot \vec{\lambda}_{s\_dq} \cdot e^{j\theta_{da}}$$

Hence,

$$\vec{v}_{s\_dq} = R_s \vec{i}_{s\_dq} + \frac{d}{dt}\vec{\lambda}_{s\_dq} + j\omega_d \vec{\lambda}_{s\_dq} \qquad (12\text{-}27)$$

where $\frac{d}{dt}\theta_{da} = \omega_d$ is the instantaneous speed (in electrical radians per second) of the dq-winding set in the air gap, as shown in Figs. 12-3 and 12-5. Separating the real and imaginary components in Eq. (12-27), we obtain

MATHEMATICAL RELATIONSHIPS OF THE dq-WINDINGS 353

$$v_{sd} = R_s i_{sd} + \frac{d}{dt}\lambda_{sd} - \omega_d \lambda_{sq} \qquad (12\text{-}28)$$

and

$$v_{sq} = R_s i_{sq} + \frac{d}{dt}\lambda_{sq} + \omega_d \lambda_{sd} \qquad (12\text{-}29)$$

In Eqs. (12-28) and (12-29), the speed terms are the components that are proportional to ω_d (the speed of the dq reference frame relative to the actual physical stator winding speed) and to the flux linkage of the orthogonal winding.

Equations (12-28) and (12-29) can be written as follows in a vector form, where each vector contains a pair of variables – the first entry corresponds to the d-winding and the second to the q-winding:

$$\begin{bmatrix} v_{sd} \\ v_{sq} \end{bmatrix} = R_s \begin{bmatrix} i_{sd} \\ i_{sq} \end{bmatrix} + \frac{d}{dt}\begin{bmatrix} \lambda_{sd} \\ \lambda_{sq} \end{bmatrix} + \omega_d \underbrace{\begin{bmatrix} 0 & -1 \\ 1 & 0 \end{bmatrix}}_{[M_{rotate}]} \begin{bmatrix} \lambda_{sd} \\ \lambda_{sq} \end{bmatrix} \qquad (12\text{-}30)$$

Note that the 2×2 matrix $[M_{rotate}]$ in Eq. (12-30) in the vector form corresponds to the operator (j) in Eq. (12-27), where $j(=e^{j\pi/2})$ has the role of rotating the space vector $\vec{\lambda}_{s\_dq}$ by an angle of $\pi/2$.

Rotor Windings An analysis similar to the stator case is carried out for the rotor, where the $\alpha\beta$ windings affixed to the rotor are shown in Fig. 12-6 with the α-axis aligned with the rotor A-axis. The d-axis (same as the d-axis for the stator) in this case is at an angle θ_{dA} with respect to the A-axis. Following the procedure for the stator case by replacing θ_{da} with θ_{dA}, results in the following equations for the rotor winding voltages:

$$v_{rd} = R_r i_{rd} + \frac{d}{dt}\lambda_{rd} - \omega_{dA}\lambda_{rq} \qquad (12\text{-}31)$$

and

354 DYNAMIC ANALYSIS OF INDUCTION MACHINES

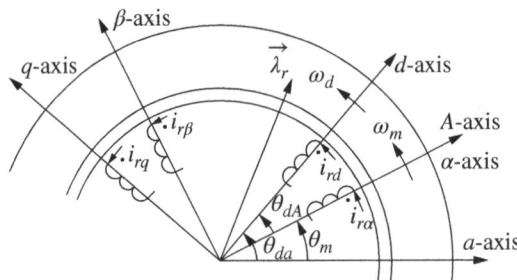

Fig. 12-6 Rotor $\alpha\beta$ and dq equivalent windings.

$$v_{rq} = R_r i_{rq} + \frac{d}{dt}\lambda_{rq} + \omega_{dA}\lambda_{rd} \quad (12\text{-}32)$$

where $\frac{d}{dt}\theta_{dA} = \omega_{dA}$ is the instantaneous speed (in electrical radians per second) of the dq-winding set in the air gap with respect to the rotor A-axis speed (rotor speed), that is,

$$\omega_{dA} = \omega_d - \omega_m \quad (12\text{-}33)$$

In Eq. (12-33), ω_m is the rotor speed in electrical radians per second. It is related to ω_{mech}, the rotor speed in actual radians per second, by the pole-pairs as follows:

$$\omega_m = (p/2)\,\omega_{mech} \quad (12\text{-}34)$$

In Eqs. (12-31) and (12-32), the speed terms are the components that are proportional to ω_{dA} (the speed of the dq reference frame relative to the actual physical rotor winding speed) and to the flux linkage of the orthogonal winding.

Equations (12-31) and (12-32) can be written as follows in a vector form, where each vector contains a pair of numbers – the first entry corresponds to the d-axis and the second to the q-axis:

$$\begin{bmatrix} v_{rd} \\ v_{rq} \end{bmatrix} = R_r \begin{bmatrix} i_{rd} \\ i_{rq} \end{bmatrix} + \frac{d}{dt}\begin{bmatrix} \lambda_{rd} \\ \lambda_{rq} \end{bmatrix} + \omega_{dA}\underbrace{\begin{bmatrix} 0 & -1 \\ 1 & 0 \end{bmatrix}}_{[M_{rotate}]}\begin{bmatrix} \lambda_{rd} \\ \lambda_{rq} \end{bmatrix} \quad (12\text{-}35)$$

12-3-4 Obtaining Fluxes and Currents with Voltages as Inputs

We can write Eqs. (12-30) and (12-35) in a state space form as follows:

$$\frac{d}{dt}\begin{bmatrix}\lambda_{sd}\\ \lambda_{sq}\end{bmatrix} = \begin{bmatrix}v_{sd}\\ v_{sq}\end{bmatrix} - R_s\begin{bmatrix}i_{sd}\\ i_{sq}\end{bmatrix} - \omega_d\underbrace{\begin{bmatrix}0 & -1\\ 1 & 0\end{bmatrix}}_{[M_{rotate}]}\begin{bmatrix}\lambda_{sd}\\ \lambda_{sq}\end{bmatrix} \quad (12\text{-}36)$$

and

$$\frac{d}{dt}\begin{bmatrix}\lambda_{rd}\\ \lambda_{rq}\end{bmatrix} = \begin{bmatrix}v_{rd}\\ v_{rq}\end{bmatrix} - R_r\begin{bmatrix}i_{rd}\\ i_{rq}\end{bmatrix} - \omega_{dA}\underbrace{\begin{bmatrix}0 & -1\\ 1 & 0\end{bmatrix}}_{[M_{rotate}]}\begin{bmatrix}\lambda_{rd}\\ \lambda_{rq}\end{bmatrix} \quad (12\text{-}37)$$

Assigning $[\lambda_{s\_dq}]$, $[v_{s\_dq}]$, and so on to represent these vectors, Eqs. (12-36) and (12-37) can be written as

$$\frac{d}{dt}[\lambda_{s\_dq}] = [v_{s\_dq}] - R_s[i_{s\_dq}] - \omega_d[M_{rotate}][\lambda_{s\_dq}] \quad (12\text{-}38)$$

and

$$\frac{d}{dt}[\lambda_{r\_dq}] = [v_{r\_dq}] - R_r[i_{r\_dq}] - \omega_{dA}[M_{rotate}][\lambda_{r\_dq}] \quad (12\text{-}39)$$

Equations (12-38) and (12-39) are represented by a block diagram in Fig. 12-7, where the calculation of dq-winding currents from flux linkages is formalized in Section 12.9.

12-4 CHOICE OF THE dq-WINDING SPEED ω_d

It is possible to assume any arbitrary value for the dq-winding speed ω_d. However, there is one value (out of three) that usually makes sense: $\omega_d = \omega_{syn}$, 0 or ω_m, where ω_{syn} is the synchronous speed in electrical radians per second. The corresponding values for ω_{dA} equal $\omega_{slip}, -\omega_m$ or 0, respectively, where $\omega_{slip} = \omega_{syn} - \omega_m$ in electrical radians per second.

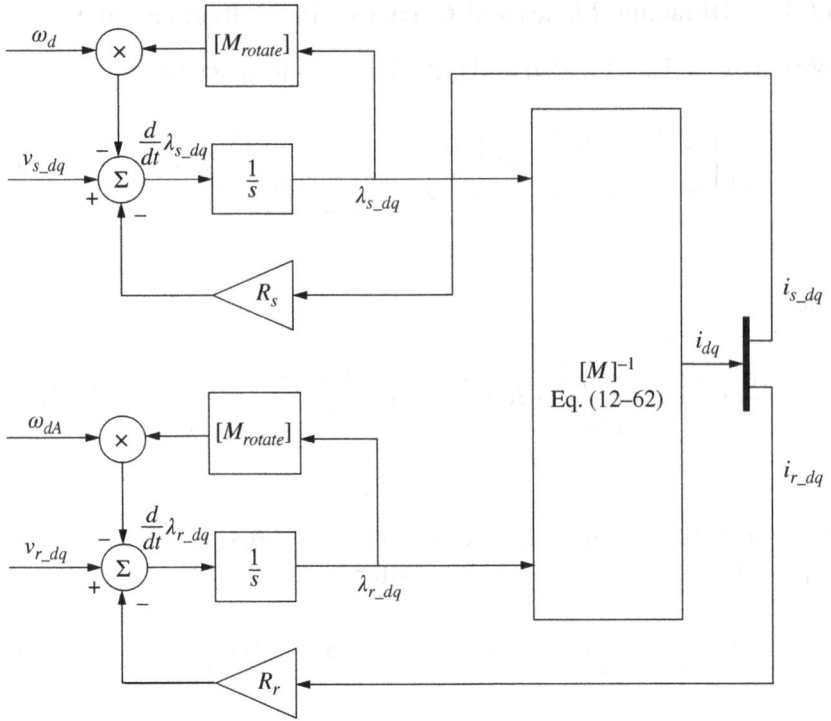

Fig. 12-7 Calculating dq-winding flux linkages and currents.

Under a balanced sinusoidal steady state, the choice of $\omega_d = \omega_{syn}$ (hence $\omega_{dA} = \omega_{slip}$) results in the hypothetical dq-windings rotating at the same speed as the field distribution in the air gap. Therefore, all the currents, voltages, and flux linkages associated with the stator, and the rotor dq-windings, are dc in a balanced sinusoidal steady state. It is easy to design *PI* controllers for dc quantities, hence ω_{syn} is often the choice for ω_d.

In contrast, choosing $\omega_d = 0$, that is, a stationary d-axis (often chosen to be aligned with the a-axis of the stator with $\theta_{da} = 0$) leads to the rotor and the stator dq-winding voltages and currents oscillating at the synchronous frequency in a balanced sinusoidal steady state. The choice of $\omega_d = \omega_m$ results in dq-winding voltages and currents in the stator and the rotor varying at the slip frequency; this choice is made for analyzing synchronous machines, as we will discuss in Chapter 17.

12-5 ELECTROMAGNETIC TORQUE

12-5-1 Torque on the Rotor d-Axis Winding

On the rotor d-axis winding, the torque produced is due to the flux density produced by the q-axis windings in Fig. 12-8. The peak of the flux density distribution "cutting" the rotor d winding due to i_{sq} and i_{rq}, each flowing through $\sqrt{3/2}N_s$ turns of the q-axis windings (using Eq. (2-3)) is

$$\hat{B}_{rq} = \frac{\mu_0}{\ell_g} \underbrace{\left(\frac{\sqrt{3/2}N_s}{p}\right)\left(i_{sq} + \frac{L_r}{L_m}i_{rq}\right)}_{mmf} \qquad (12\text{-}40)$$

where the factor L_r/L_m allows us to include both the magnetizing and the leakage flux produced by i_{rq}. Using the torque expression in Eq. (9-27), and noting that the current i_{rd} in the rotor d-axis winding

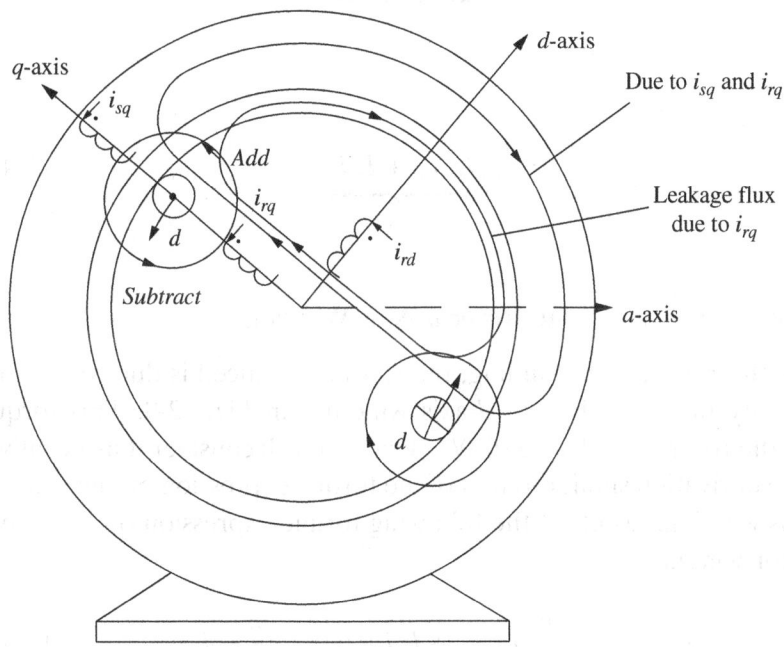

Fig. 12-8 Torque on the rotor d-axis.

358 DYNAMIC ANALYSIS OF INDUCTION MACHINES

flows through $\sqrt{3/2}N_s$ turns, the instantaneous torque on the d-axis rotor winding is

$$T_{d,rotor} = \frac{p}{2}\left(\pi\frac{\sqrt{3/2}N_s}{p}r\ell\hat{B}_{rq}\right)i_{rd} \qquad (12\text{-}41)$$

As shown in Fig. 12-8, this torque on the rotor is counterclockwise (CCW), hence we will consider it as positive. Substituting for $\hat{B}_{rq}$ from Eq. (12-40) into Eq. (12-41),

$$T_{d,rotor} = \frac{p}{2}\left(\pi\frac{\mu_0}{\ell_g}r\ell\right)\left(\frac{\sqrt{3/2}N_s}{p}\right)^2\left(i_{sq} + \frac{L_r}{L_m}i_{rq}\right)i_{rd} \qquad (12\text{-}42)$$

Rewriting Eq. (12-42) below, we can recognize L_m from Eq. (2-13),

$$T_{d,rotor} = \frac{p}{2}\underbrace{\left(\frac{3}{2}\pi\frac{\mu_0}{\ell_g}r\ell\left(\frac{N_s}{p}\right)^2\right)}_{L_m}\left(i_{sq} + \frac{L_r}{L_m}i_{rq}\right)i_{rd}$$

Hence,

$$T_{d,rotor} = \frac{p}{2}\underbrace{(L_m i_{sq} + L_r i_{rq})}_{\lambda_{rq}}i_{rd} = \frac{p}{2}\lambda_{rq}i_{rd} \qquad (12\text{-}43)$$

12-5-2 Torque on the Rotor q-Axis Winding

On the rotor q-axis winding, the torque produced is due to the flux density produced by the d-axis windings in Fig. 12-9. This torque on the rotor is clockwise (CW), hence we will consider it as negative. The derivation similar to that of the torque expression on the rotor d-axis winding, results in the following torque expression on the q-axis rotor winding:

$$T_{q,rotor} = -\frac{p}{2}\underbrace{(L_m i_{sd} + L_r i_{rd})}_{\lambda_{rd}}i_{rq} = -\frac{p}{2}\lambda_{rd}i_{rq} \qquad (12\text{-}44)$$

ELECTROMAGNETIC TORQUE 359

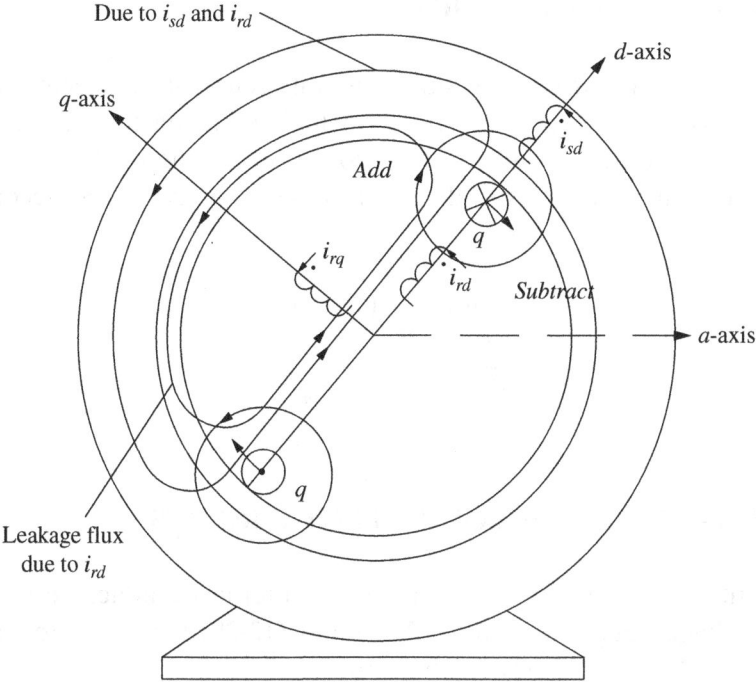

Fig. 12-9 Torque on the rotor q-axis.

12-5-3 Net Electromagnetic Torque T_{em} on the Rotor

By superposition, adding the torques acting on the d-axis and the q-axis of the rotor windings, the instantaneous torque is

$$T_{em} = T_{d,rotor} + T_{q,rotor} \quad (12\text{-}45)$$

which, using Eqs. (12-43) and (12-44) results in

$$T_{em} = \frac{p}{2}\left(\lambda_{rq}i_{rd} - \lambda_{rd}i_{rq}\right) \quad (12\text{-}46)$$

Substituting for flux linkages in Eq. (12-46), the electromagnetic torque can be expressed in terms of inductances as

$$T_{em} = \frac{p}{2}L_m\left(i_{sq}i_{rd} - i_{sd}i_{rq}\right) \quad (12\text{-}47)$$

12-6 ELECTRODYNAMICS

The acceleration is determined by the difference of the electromagnetic torque and the load torque (including friction torque) acting on J_{eq}, the combined inertia of the load and the motor. In terms of the actual (mechanical) speed of the rotor ω_{mech} in radians per second, where

$$\omega_{mech} = (2/p)\omega_m,$$
$$\frac{d}{dt}\omega_{mech} = \frac{T_{em} - T_L}{J_{eq}} \qquad (12\text{-}48)$$

12-7 d- AND q-AXIS EQUIVALENT CIRCUITS

Substituting for flux-linkage derivatives in terms of inductances into the voltage equations (Eqs. (12-28) and (12-29) for the stator and Eqs. (12-31) and (12-32) for the rotor),

$$v_{sd} = R_s i_{sd} - \omega_d \lambda_{sq} + L_{\ell s}\frac{d}{dt}i_{sd} + L_m\frac{d}{dt}(i_{sd} + i_{rd}) \qquad (12\text{-}49)$$

$$v_{sq} = R_s i_{sq} + \omega_d \lambda_{sd} + L_{\ell s}\frac{d}{dt}i_{sq} + L_m\frac{d}{dt}(i_{sq} + i_{rq}) \qquad (12\text{-}50)$$

and

$$\underbrace{v_{rd}}_{=\,0} = R_r i_{rd} - \omega_{dA}\lambda_{rq} + L_{\ell r}\frac{d}{dt}i_{rd} + L_m\frac{d}{dt}(i_{sd} + i_{rd}) \qquad (12\text{-}51)$$

$$\underbrace{v_{rq}}_{=\,0} = R_r i_{rq} + \omega_{dA}\lambda_{rd} + L_{\ell r}\frac{d}{dt}i_{rq} + L_m\frac{d}{dt}(i_{sq} + i_{rq}) \qquad (12\text{-}52)$$

For each axis, the stator and the rotor winding equations are combined to result in the dq equivalent circuits shown in Fig. 12-10a and b. Using Eq. (12-28), we can label the terminals across which the voltage is $d\lambda_{sd}/dt$ in Fig. 12-10a. Similarly, using Eqs. (12-29),

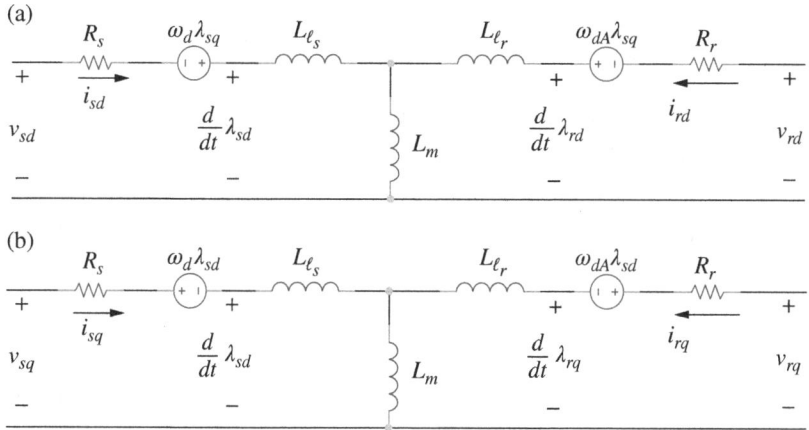

Fig. 12-10 dq-winding equivalent circuits: (a) d-axis and (b) q-axis.

(12-31), and (12-32), respectively, we can label terminals in Fig. 12-10a and b with $d\lambda_{sq}/dt$, $d\lambda_{rd}/dt$, and $d\lambda_{rq}/dt$.

12-8 RELATIONSHIP BETWEEN THE dq-WINDINGS AND THE PER-PHASE PHASOR-DOMAIN EQUIVALENT CIRCUIT IN BALANCED SINUSOIDAL STEADY STATE

In this section, we will see that under a balanced sinusoidal steady-state condition, the dq-winding equations combine to result in the per-phase equivalent circuit of an induction machine that we have derived in Chapter 9. It will be easiest to choose $\omega_d = \omega_{syn}$ (although any other choice of reference speed would lead to the same results; see Problem 12-8) so that the dq-winding quantities are dc and their time derivatives are zero, under a balanced sinusoidal steady-state condition. Therefore, in the stator voltage equation Eq. (12-27) in steady state,

$$\vec{v}_{s\_dq} = R_s \vec{i}_{s\_dq} + j\omega_{syn} \vec{\lambda}_{s\_dq} \quad \text{(steady state)} \quad (12\text{-}53)$$

Similarly, the voltage equation for the rotor dq-windings, under a balanced sinusoidal steady state with $\omega_d = \omega_{syn}$ (thus $\omega_{dA} = \omega_{slip} = s\,\omega_{syn}$), results in

$$0 = \frac{R_r}{s}\vec{i}_{r\_dq} + j\omega_{syn}\vec{\lambda}_{r\_dq} \quad \text{(steady state)} \quad (12\text{-}54)$$

where s is the slip. Substituting for flux linkage space vectors in Eqs. (12-53) and (12-54) results in

$$\vec{v}_{s\_dq} = R_s \vec{i}_{s\_dq} + j\omega_{syn}L_{\ell s}\vec{i}_{s\_dq} + j\omega_{syn}L_m\left(\vec{i}_{s\_dq} + \vec{i}_{r\_dq}\right) \quad (12\text{-}55)$$

and

$$0 = \frac{R_r}{s}\vec{i}_{r\_dq} + j\omega_{syn}L_{\ell r}\vec{i}_{r\_dq} + j\omega_{syn}L_m\left(\vec{i}_{s\_dq} + \vec{i}_{r\_dq}\right) \quad (12\text{-}56)$$

The above space vector equations, in a balanced sinusoidal steady state, correspond to the following phasor equations for phase a:

$$\overline{V}_a = R_s\overline{I}_a + j\omega_{syn}L_{\ell s}\overline{I}_a + j\omega_{syn}L_m(\overline{I}_a + \overline{I}_A) \quad (12\text{-}57)$$

and

$$0 = \frac{R_r}{s}\overline{I}_A + j\omega_{syn}L_{\ell r}\overline{I}_A + j\omega_{syn}L_m(\overline{I}_a + \overline{I}_A) \quad (12\text{-}58)$$

The above two equations combined correspond to the per-phase equivalent circuit of Fig. 12-11, under a balanced sinusoidal steady-state condition. Note that in Fig. 12-11, $\overline{I}_A = -\overline{I}'_{ra}$.

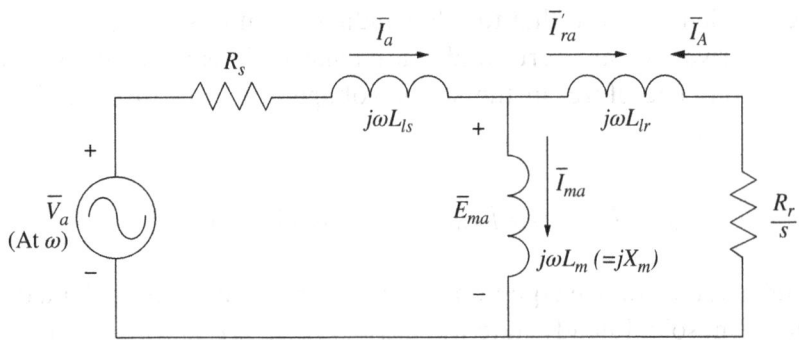

Fig. 12-11 Per-phase equivalent circuit in steady state.

12-9 COMPUTER SIMULATION

In dq-windings, the flux linkages and voltage equations are derived earlier. We will use λ_{sd}, λ_{sq}, λ_{rd}, and λ_{rq} as state variables, and express i_{sd}, i_{sq}, i_{rd}, and i_{rq} in terms of these state variables. The reason for choosing flux linkages as state variables, have to do with the fact these quantities change slowly compared to currents which can change *almost* instantaneously.

We can calculate dq-winding currents from the stator and the rotor flux linkages of the respective windings as follows: referring to Fig. 12-3, the stator and the rotor d-winding flux linkages are related to their winding currents (rewriting Eqs. (12-19) and (12-21) in a matrix form) as

$$\begin{bmatrix} \lambda_{sd} \\ \lambda_{rd} \end{bmatrix} = \underbrace{\begin{bmatrix} L_s & L_m \\ L_m & L_r \end{bmatrix}}_{[L]} \begin{bmatrix} i_{sd} \\ i_{rd} \end{bmatrix} \qquad (12\text{-}59)$$

Similarly, in the q-axis windings, from Eqs. (12-20) and (12-22), the matrix $[L]$ of the above equation relates flux linkages to respective currents:

$$\begin{bmatrix} \lambda_{sq} \\ \lambda_{rq} \end{bmatrix} = \underbrace{\begin{bmatrix} L_s & L_m \\ L_m & L_r \end{bmatrix}}_{[L]} \begin{bmatrix} i_{sq} \\ i_{rq} \end{bmatrix} \qquad (12\text{-}60)$$

Combining matrix Eqs. (12-59) and (12-60), we can relate fluxes to currents as follows:

$$\begin{bmatrix} \lambda_{sd} \\ \lambda_{sq} \\ \lambda_{rd} \\ \lambda_{rq} \end{bmatrix} = \underbrace{\begin{bmatrix} L_s & 0 & L_m & 0 \\ 0 & L_s & 0 & L_m \\ L_m & 0 & L_r & 0 \\ 0 & L_m & 0 & L_r \end{bmatrix}}_{[M]} \begin{bmatrix} i_{sd} \\ i_{sq} \\ i_{rd} \\ i_{rq} \end{bmatrix} \qquad (12\text{-}61)$$

364 DYNAMIC ANALYSIS OF INDUCTION MACHINES

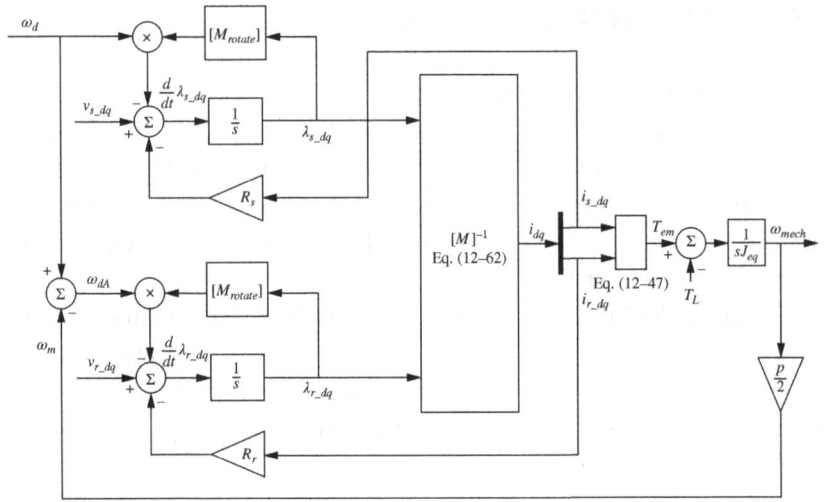

Fig. 12-12 Induction motor model in terms of dq-windings.

From Eq. (12-61), currents can be calculated by using the inverse of matrix $[M]$:

$$\begin{bmatrix} i_{sd} \\ i_{sq} \\ i_{rd} \\ i_{rq} \end{bmatrix} = [M]^{-1} \begin{bmatrix} \lambda_{sd} \\ \lambda_{sq} \\ \lambda_{rd} \\ \lambda_{rq} \end{bmatrix} \quad (12\text{-}62)$$

With voltages as input and choosing the speed ω_d of the dq-windings, the flux linkages are calculated, as shown in Fig. 12-7 using Eqs. (12-38) and (12-39) derived earlier. The currents are calculated using Eq. (12-62) derived above. Combining these with the torque equation Eq. (12-47), and the electrodynamics equation (12-48), we can draw the overall block diagram for computer modeling in Fig. 12-12.

12-9-1 Calculation of Initial Conditions

In order to carry out computer simulations, we need to calculate the initial values of the state variables, that is, of the flux linkages of the dq-windings. These can be calculated in terms of the initial values of the dq-winding currents. These currents allow us to compute the electromagnetic torque in steady state, thus the initial loading of the induction machine. To accomplish this, we will make use of the phasor analysis in the initial steady state as follows.

12-10 PHASOR ANALYSIS

In the sinusoidal steady state, we can calculate current phasors $\overline{I}_a$ and $\overline{I}'_{ra}(=-\overline{I}_A)$ in Fig. 12-11 for a given $\overline{V}_a$. All the space vectors and the dq-winding variables at $t = 0$ can be calculated. The phasor current for phase-a allows the stator current space vector at $t = 0$ to be calculated as follows:

$$\overline{I}_a = \hat{I}_a \angle \theta_i \quad \Rightarrow \quad \vec{i}_s(0) = \underbrace{\frac{3}{2}\hat{I}_a}_{\hat{i}_s} e^{j\theta_i} \quad (12\text{-}63)$$

Assuming the initial value of θ_{da} to be zero (that is, the d-axis along the stator a-axis), and using Fig. 12-3 and Eqs. (12-5) and (12-6),

$$i_{sd}(0) = \sqrt{\frac{2}{3}} \times \text{projection of } \vec{i}_s(0) \text{ on } d\text{-axis} = \sqrt{\frac{2}{3}}\underbrace{\left(\frac{3}{2}\hat{I}\right)}_{\hat{i}_s} \cos(\theta_i)$$

$$(12\text{-}64)$$

and

$$i_{sq}(0) = \sqrt{\frac{2}{3}} \times \text{projection of } \vec{i}_s(0) \text{ on } q\text{-axis} = \sqrt{\frac{2}{3}}\underbrace{\left(\frac{3}{2}\hat{I}\right)}_{\hat{i}_s} \sin(\theta_i)$$

$$(12\text{-}65)$$

Similarly, we can calculate $v_{sd}(0)$ and $v_{sq}(0)$. The phasor $\overline{I}_A(=-\overline{I}'_{ra})$ allows $i_{rd}(0)$ and $i_{rq}(0)$ to be calculated. Knowing the currents in the dq-windings at $t = 0$ allows the initial values of the flux linkages to be calculated from Eq. (12-61). Also, these currents allow the computation of the electromagnetic torque in steady state by Eq. (12-47) to calculate the initial loading of the induction machine.

EXAMPLE 12-1

An induction machine with the following nameplate data is initially operating under its rated condition in steady state, supplying its rated torque. Calculate initial values of flux linkages and the load torque. Assume that the phase-a voltage has its positive peak at the time $t = 0$.
Nameplate Data:

Power	3 HP/2.4 kW
Voltage	460 V (L-L, rms)
Frequency	60 Hz
Phases	3
Full-load current	4 A
Full-load speed	1750 RPM
Full-load efficiency	88.5%
Power factor	80.0%
Number of poles	4

Per-phase motor circuit parameters:

$$R_s = 1.77\,\Omega$$

$$R_r = 1.34\,\Omega$$

$$X_{\ell s} = 5.25\,\Omega \text{ (at 60 Hz)}$$

$$X_{\ell r} = 4.57\,\Omega \text{ (at 60 Hz)}$$

$$X_m = 139.0\,\Omega \text{ (at 60 Hz)}$$

Full-load slip = 1.72%
The iron losses are specified as 78 W, and the mechanical (friction and windage) losses are specified as 24 W. The inertia of the

machine is given. Assuming that the reflected load inertia is approximately the same as the motor inertia, the total equivalent inertia of the system is $J_{eq} = 0.025 \text{ kg} \cdot \text{m}^2$.

Solution

A MATLAB file EX12_1.m on the accompanying website is based on the following steps:

Step 1. Calculate by phasor analysis $\overline{V}_a, \overline{I}_a,$ and $\overline{I}_A \left(= -\overline{I}'_{ra} \right)$, given that the phase-*a* voltage has a positive peak at the time $t = 0$.

Step 2. Calculate the current space vectors $\vec{i}_s^{\,a}$ and $\vec{i}_r^{\,a}$ at the time $t = 0$ from the phasors for phase-*a*.

Step 3. In the *dq* analysis, choose $\omega_d = \omega_{syn}$ and $\theta_{da}(0) = 0$. Calculate $i_{s\alpha}, i_{s\beta}, i_{r\alpha},$ and $i_{r\beta}$ from the space vectors in step 2, using equations similar to Eqs. (12-64) and (12-65).

Step 4. Calculate flux linkages of *dq*-windings using Eq. (12-61).

Step 5. Calculate torque $T_L(0)$, which equals T_{em} in steady state, from Eq. (12-46) or Eq. (12-47).

The results from EX12_1.m are listed below:

$\lambda_{sd}(0) = 0.0174$ Wb-turns
$\lambda_{rd}(0) = -0.1237$ Wb-turns
$\lambda_{sq}(0) = -1.1951$ Wb-turns
$\lambda_{rq}(0) = -1.1363$ Wb-turns
$T_{em}(0) = T_L(0) = 12.644$ Nm.

EXAMPLE 12-2

Calculate the initial conditions of the induction machine operating in steady state in Example 12-1 using the voltage equations

(*Continued*)

Eqs. (12-28), (12-29), (12-31), and (12-32). Also, calculate the load torque.

Solution

In a balanced steady state with ω_d chosen as the synchronous speed ω_{syn}, all dq-winding variables are dc quantities and $\omega_{dA} = s\omega_{syn}$. Therefore, their time derivatives are zero in Eqs. (12-28), (12-29), (12-31), and (12-32), resulting in the following equations:

$$v_{sd} = R_s i_{sd} - \omega_{syn} \lambda_{sq}$$

$$v_{sq} = R_s i_{sq} + \omega_{syn} \lambda_{sd}$$

$$0 = R_r i_{rd} - s\omega_{syn} \lambda_{rq}$$

$$0 = R_r i_{rq} + s\omega_{syn} \lambda_{rd}$$

Substituting in the above equations for flux linkages from Eqs. (12-19) through (12-22),

$$\begin{bmatrix} v_{sd} \\ v_{sq} \\ 0 \\ 0 \end{bmatrix} = \underbrace{\begin{bmatrix} R_s & -\omega_{syn} L_s & 0 & -\omega_{syn} L_m \\ \omega_{syn} L_s & R_s & \omega_{syn} L_m & 0 \\ 0 & -s\omega_{syn} L_m & R_r & -s\omega_{syn} L_r \\ s\omega_{syn} L_m & 0 & s\omega_{syn} L_r & R_r \end{bmatrix}}_{[A]} \begin{bmatrix} i_{sd} \\ i_{sq} \\ i_{rd} \\ i_{rq} \end{bmatrix}$$

The machine currents can be calculated from the equation above by inverting matrix $[A]$:

$$\begin{bmatrix} i_{sd} \\ i_{sq} \\ i_{rd} \\ i_{rq} \end{bmatrix} = [A]^{-1} \begin{bmatrix} v_{sd} \\ v_{sq} \\ 0 \\ 0 \end{bmatrix}$$

Once the dq-winding currents are calculated, the flux linkages can be calculated from Eq. (12-61). The results from a MATLAB file EX12_2.m on the accompanying website are as follows for currents, with flux linkages and the load torque exactly as in Example 12-1:

$i_{sd}(0) = 5.34$ A
$i_{sq}(0) = -3.7$ A
$i_{rd}(0) = -5.5$ A
$i_{rq}(0) = 0.60$ A

EXAMPLE 12-3

In Simulink, develop a simulation of the induction machine described in Example 12-1, operating in steady state as specified in Example 12-1. At $t = 0.1$ s, the load torque T_L suddenly goes to one-half of its initial value and stays at that level. Assume $\theta_{da}(0) = 0$ and a synchronously rotating dq reference frame.

Plot the electromagnetic torque developed by the motor and the rotor speed as functions of time.

Solution

The Simulink file EX12_3.mdl included on the accompanying website follows the block diagram in Fig. 12-12. Before its execution, initial conditions for the flux linkages, the rotor speed, and the load torque must be calculated either by executing the file for Example 12-1 (EX12_1.m) or for Example 12-2 (EX12_2.m), by double-clicking on the start icon shown in the schematic of Fig. 12-13. The resulting waveforms are plotted in Fig. 12-14.

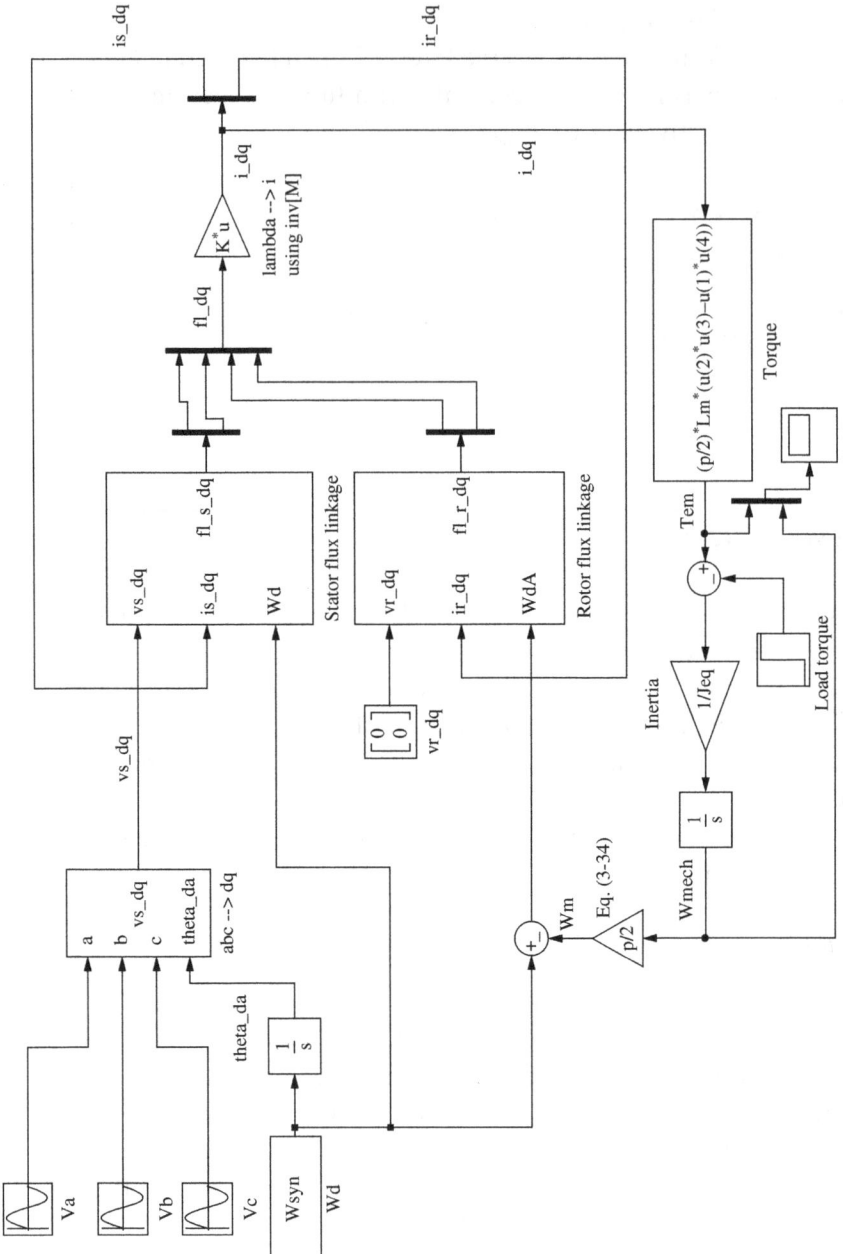

Fig. 12-13 Simulation of Example 12-3.

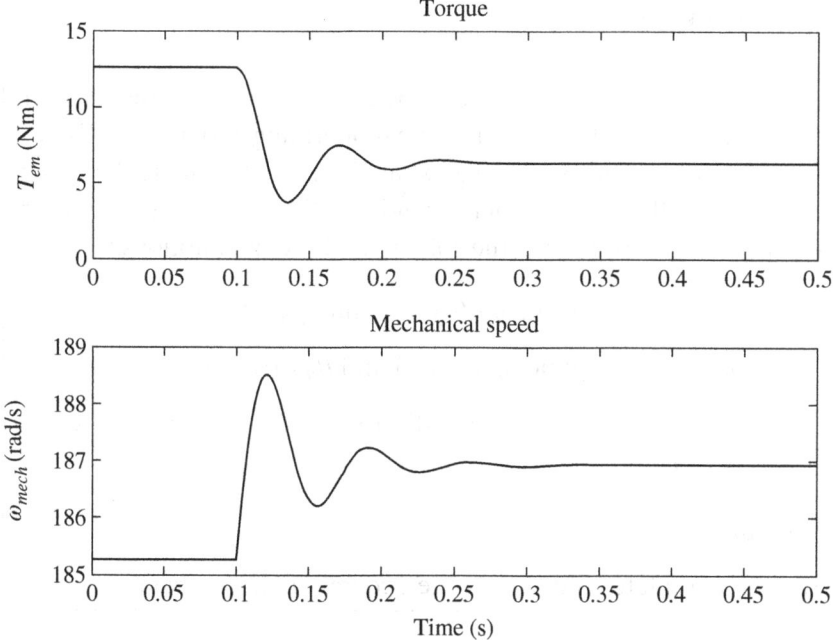

Fig. 12-14 Simulation results of Example 12-3.

EXAMPLE 12-4

Consider the "test" machine described after the back-of-the-chapter problems. Simulate the machine starting from a completely powered down state with no external load connected to it. Simulate for three different assumptions regarding the d-axis:

(a) $\omega_d = \omega_{syn}$

(b) $\omega_d = \omega_m$

(c) The d-axis is aligned with the rotor flux-linkage space vector.

Solution

See the complete solution on the accompanying website.

EXAMPLE 12-5

Consider the "test" machine described after the back-of-the-chapter problems. Assume that this machine is operating under its rated condition in steady state, supplying its rated torque. Calculate the initial values of the flux linkages of dq-windings, the dq currents, the rotor speed, and the torque for the following three assumptions:

(a) $\omega_d = \omega_{syn}$; Assume $\theta_{da}(0) = 0$ and $\theta_{dA}(0) = 0$.

(b) $\omega_d = \omega_m$; Assume $\theta_{da}(0) = 0$ and $\theta_{dA}(0) = 0$.

(c) The d-axis is aligned with the rotor flux-linkage space vector.

Solution

See the complete solution on the accompanying website.

EXAMPLE 12-6

Use the initial conditions calculated in Example 12-5 to simulate a sudden change in load torque to one-half of its initial value. Plot various quantities as a function of time.

Solution

See the complete solution on the accompanying website.

EXAMPLE 12-7

The machine in Example 12-5 is made to go into the generator mode. Make the load torque change linearly from its initial rated positive value to the rated negative value in 0.2 seconds. Plot various quantities as a function of time.

Solution

See the complete solution on the accompanying website.

12-11 SUMMARY

In this chapter, the actual stator phase windings and the equivalent rotor phase windings are represented by an equivalent set of dq-windings, which produce the same air gap mmf. There are many advantages of doing so: there is zero magnetic coupling between the windings on the d-axis and those on the q-axis due to their orthogonal orientation. This procedure results in much simpler expressions, and allows air gap flux and electromagnetic torque to be controlled independently, which will be discussed in the following chapters.

FURTHER READING

Krause, P.C. and Thomas, C.H. (1965). Simulation of symmetrical induction machinery. *IEEE Transactions on Power Apparatus and Systems* PAS-84 (11): 1038–1053.

PROBLEMS

12-1 Derive that each of the two windings with $\sqrt{3/2}\,N_s$ turns in an equivalent two-phase machine has the magnetizing inductance of L_m, the leakage inductance of $L_{\ell s}$, and the resistance R_s, where these quantities correspond to those of an equivalent two-phase machine.

12-2 Show that the instantaneous power loss in the stator resistances is the same in the 12-phase machine as in an equivalent two-phase machine.

12-3 Show that the instantaneous total input power is the same in a-b-c and in the dq circuits.

12-4 Re-derive all the equations used in obtaining the torque expressions of Eqs. (12-46) and (12-47) for a p-pole machine from energy considerations.

12-5 Draw dynamic equivalent circuits of Fig. 12-9 for the following values of the dq-winding speed ω_d: 0 and ω_m. What is the frequency of dq-winding variables in a balanced sinusoidal steady state, including the condition that $\omega_d = \omega_{syn}$.

12-6 The "test" machine described after the back-of-the-chapter problems is operating at its rated conditions. Calculate $v_{sd}(t)$ and $v_{sq}(t)$ as functions of time, (a) if $\omega_d = \omega_{syn}$, and (b) if $\omega_d = 0$.

12-7 Under a balanced sinusoidal steady state, calculate the input power factor of operation based on the d- and the q-axis equivalent circuits of Fig. 12-10 in the "test" machine described after the back-of-the-chapter problems.

12-8 Show that the equations for the dq-windings in a balanced sinusoidal steady-state result in the per-phase equivalent circuit of Fig. 12-11 for $\omega_d = 0$ and $\omega_d = \omega_m$.

12-9 Using Eqs. (12-46) and (12-47), derive the torque expressions in terms of (a) stator dq-winding flux linkages and currents, and (b) stator dq-winding currents and rotor dq-winding flux linkages.

12-10 Show that for the transformation matrix in Eq. (12-12),

$$\underbrace{[T]_{abc \to dq}}_{2 \times 3} \underbrace{[T]_{dq \to abc}}_{3 \times 2} = \underbrace{[I]}_{2 \times 2}$$

12-11 Derive the voltage equations in the dq stator windings (Eqs. (12-28) and (12-29)) using the transformation matrix of Eqs. (12-12) (also for the rotor dq-windings).

12-12 In Examples 12-1 and 12-2, plot $\vec{v}_s$, $\vec{\lambda}_s$, $\vec{\lambda}_r$, $\vec{i}_s$, and $\vec{i}_r$ at time $t = 0$.

12-13 In the simulation of Example 12-3, plot the dq-winding currents and the a-b-c phase currents. Also, plot the slip speed.

12-14 Repeat Example 12-3 assuming $\omega_d = 0$. Plot the dq-winding currents, the a-b-c phase currents, and the slip speed.

12-15 Repeat Example 12-3 assuming $\omega_d = \omega_m$. Plot the dq-winding currents, the a-b-c phase currents, and the slip speed.

12-16 Modify the simulation file of Example 12-3 to simulate line start with the rated load connected to the motor, without the load disturbance at $t = 0.1$ s.

12-17 The "test" machine described after the back-of-the-chapter problems is made to go into the generator mode. Modify the file of Example 12-3, by making the load torque change linearly from its initial rated positive value to the rated negative value in 0.2 s, starting at $t = 0.1$ s. Plot the same variables as in that example, as well as the phase voltages and currents.

TEST MACHINE

The parameters for a 1.5 MW wind-turbine induction machine are as follows:

Power	1.5 MW
Voltage	575 V (L-L, rms)
Frequency	60 Hz
Phases	3
Number of poles	6
Full-load slip	1%
Moment of Inertia	75 kg·m^2

Per-Phase Circuit Parameters:

$$R_s = 0.0014 \ \Omega$$

$$R_r = 0.000991 \ \Omega$$

$$X_{\ell s} = 0.034 \ \Omega$$

$$X_{\ell r} = 0.031 \ \Omega$$

$$X_m = 0.0576 \ \Omega$$

13 Mathematical Description of Vector Control in Induction Machines

13-1 INTRODUCTION

In vector control, the d-axis is aligned with the rotor flux-linkage space vector, such that the rotor flux linkage in the q-axis is zero. With this as the motivation, we will first develop a model of the induction machine where the above condition is always met. Such a model of the machine would be valid, regardless if the machine is vector controlled, or if the voltages and currents are applied as under a general-purpose operation (line-fed or in adjustable speed drives described in the previous course).

After developing the above-mentioned motor model, we will study vector control of induction motor drives, assuming that the exact motor parameters are known. We will first use an idealized current-regulated pulse-width-modulated (PWM) inverter to supply motor currents calculated by the controller. As the last step in this chapter, we will use an idealized space vector PWM inverter (discussed in detail in Chapter 7) to supply motor voltages that result in the desired currents calculated by the controller.

(Adapted from chapter 5 of *Advanced Electric Drives: Analysis, Control, and Modeling Using MATLAB/Simulink* ISBN: 978-1-118-48548-4 by Ned Mohan, August 2014)

Analysis and Control of Electric Drives: Simulations and Laboratory Implementation, First Edition. Ned Mohan and Siddharth Raju.
© 2021 John Wiley & Sons, Inc. Published 2021 by John Wiley & Sons, Inc.
Companion website: www.wiley.com/go/Mohan/Vectorcontrolinelectricdrives

13-2 MOTOR MODEL WITH THE d-AXIS ALIGNED ALONG THE ROTOR FLUX LINKAGE $\vec{\lambda}_r$-AXIS

As noted in the qualitative description of vector control, we will align the d-axis (common to both the stator and the rotor) to be along with the rotor flux linkage $\vec{\lambda}_r (= \hat{\lambda}_r e^{j0})$, as shown in Fig. 13-1. Therefore,

$$\lambda_{rq}(t) = 0 \qquad (13\text{-}1)$$

Equating λ_{rq} in Eq. (3-22) to zero,

$$i_{rq} = -\frac{L_m}{L_r} i_{sq} \qquad (13\text{-}2)$$

The condition that the d-axis is always aligned with $\vec{\lambda}_r$ such that $\lambda_{rq} = 0$ also results in $d\lambda_{rq}/dt$ to be zero. Using $\lambda_{rq} = 0$ and $d\lambda_{rq}/dt = 0$ in the d- and the q-axis dynamic circuits, we can obtain the simplified circuits shown in Fig. 13-2a and b. Note that Eq. (13-2) is consistent with the equivalent circuit of Fig. 13-2b, where $L_r = L_{\ell r} + L_m$.

Next, we will calculate the slip speed ω_{dA} and the electromagnetic torque T_{em} in this new motor model in terms of the rotor flux λ_{rd}

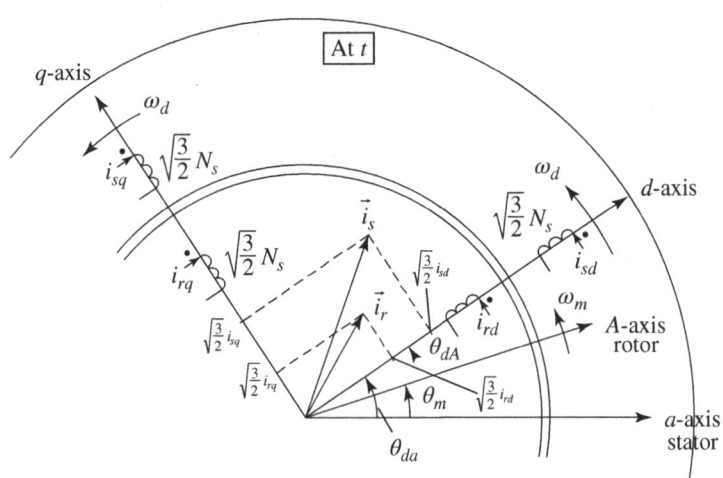

Fig. 13-1 Stator and the rotor mmf representations by equivalent dq winding current. The d-axis is aligned with $\vec{\lambda}_r$.

MOTOR MODEL WITH THE d-AXIS 379

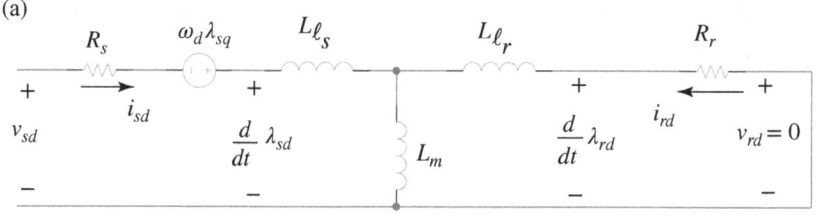

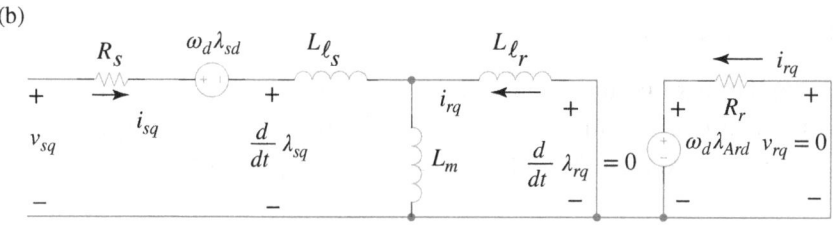

Fig. 13-2 Dynamic circuits with the d-axis aligned with $\vec{\lambda}_r$.

and the stator current component i_{sq} in the q-winding (under vector control conditions, λ_{rd} would be kept constant except in the field-weakening mode and the torque production will be controlled by i_{sq}).

We will also establish the dynamics of the rotor flux λ_{rd} in the rotor d-winding (λ_{rd} varies during flux buildup at startup and when the motor is made to go into the flux-weakening mode of operation).

13-2-1 Calculation of ω_{dA}

As discussed above, under the condition that the d-axis is always aligned with the rotor flux, the q-axis rotor flux linkage is zero, as well as $d\lambda_{rq}/dt = 0$. Therefore, in a squirrel-cage rotor with $v_{rq} = 0$, Eq. (3-32) results in

$$\omega_{dA} = -R_r \frac{i_{rq}}{\lambda_{rd}} \qquad (13\text{-}3)$$

which is consistent with the equivalent circuit of Fig. 13-2b. In the rotor circuit, the time-constant τ_r, called the rotor time-constant, is

$$\tau_r = \frac{L_r}{R_r} \qquad (13\text{-}4)$$

Substituting for i_{rq} from Eq. (13-2), in terms of τ_r, the slip speed can be expressed as

$$\omega_{dA} = \frac{L_m}{\tau_r \lambda_{rd}} i_{sq} \tag{13-5}$$

13-2-2 Calculation of T_{em}

Since the flux linkage in the q-axis of the rotor is zero, the electromagnetic torque is produced only by the d-axis flux in the rotor acting on the rotor q-axis winding. Therefore, from Eq. (3-46),

$$T_{em} = -\frac{p}{2} \lambda_{rd} i_{rq} \tag{13-6}$$

In Eq. (13-6), substituting for i_{rq} from Eq. (13-2),

$$T_{em} = \frac{p}{2} \lambda_{rd} \left(\frac{L_m}{L_r} i_{sq} \right) \tag{13-7}$$

13-2-3 d-Axis Rotor Flux-Linkage Dynamics

To obtain the dependence of λ_{rd} on i_{sd}, we will make use of the equivalent circuit in Fig. 13-2a, and redraw it as in Fig. 13-3 with a current excitation by i_{sd}. From Fig. 13-3, in terms of Laplace-domain variables,

$$i_{rd}(s) = -\frac{sL_m}{R_r + sL_r} i_{sd}(s) \tag{13-8}$$

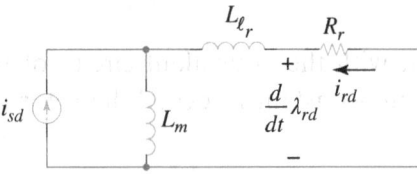

Fig. 13-3 The d-axis circuit simplified with a current excitation.

In the rotor d-axis winding, from Eq. (3-21),

$$\lambda_{rd} = L_r i_{rd} + L_m i_{sd} \qquad (13\text{-}9)$$

Substituting for i_{rd} from Eq. (13-8) into Eq. (13-9), and using τ_r from Eq. (13-4),

$$\lambda_{rd}(s) = \frac{L_m}{(1 + s\tau_r)} i_{sd}(s) \qquad (13\text{-}10)$$

In time-domain, the rotor flux-linkage dynamics expressed by Eq. (13-10) is as follows:

$$\frac{d}{dt}\lambda_{rd} + \frac{\lambda_{rd}}{\tau_r} = \frac{L_m}{\tau_r} i_{sd} \qquad (13\text{-}11)$$

13-2-4 Motor Model

Based on the above equations, a block diagram of an induction-motor model, where the d-axis is aligned with the rotor flux linkage, is shown in Fig. 13-4. The currents i_{sd} and i_{sq} are the inputs and λ_{rd}, θ_{da}, and T_{em}

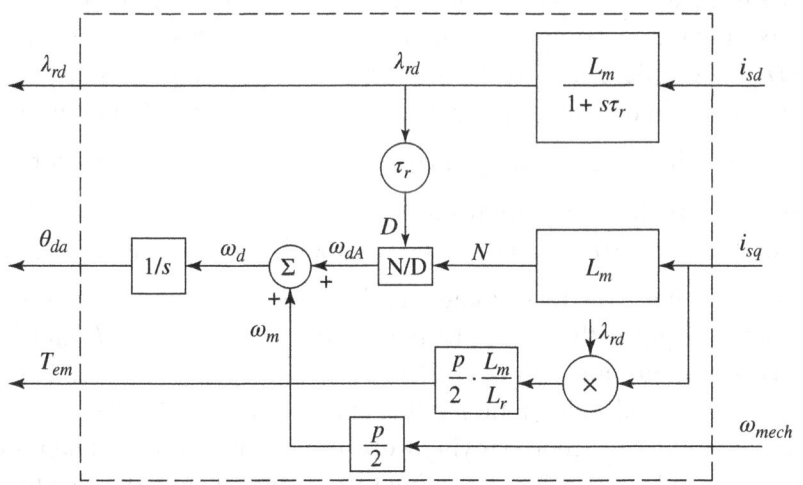

Fig. 13-4 Motor model with d-axis aligned with $\vec{\lambda}_r$.

382 MATHEMATICAL DESCRIPTION OF VECTOR CONTROL

are the outputs. Note that $\omega_d(=\omega_{dA}+\omega_m)$ is the speed of the rotor field and therefore, the rotor-field angle with respect to the stator a-axis (see Fig. 13-1) is

$$\theta_{da}(t) = 0 + \int_0^t \omega_d(\tau)\, d\tau \tag{13-12}$$

where τ is the variable of integration, and the initial value of θ_{da} is assumed to be zero at $t = 0$.

EXAMPLE 13-1

The motor model developed above, with the d-axis aligned with $\vec{\lambda}_r$, can be used to model induction machines where vector control is not the objective. To illustrate this, we will repeat the simulation of Example 12-3 of a line-fed motor using this new motor model (which is much simpler) and compare simulation results of these two examples.

Solution

We need to recalculate initial flux values, because now the rotor flux linkage is completely along the d-axis. This is done in a MATLAB file EX13_1calc.m on the accompanying website. The initial part in this file is the same as in EX12_1.m (used in Example 12-3) in which the initial values of the angles *thetar* and *thetas* for $\vec{\lambda}_r$ and $\vec{\lambda}_s$ are calculated with respect to the d-axis aligned to the stator a-axis with $\theta_{da}(0) = 0$. In the present model, with the d-axis aligned with $\vec{\lambda}_r$, the rotor flux-linkage angle is zero, and the stator flux-linkage angle with respect to the d-axis equals (*thetas–thetar*) in terms of their values in EX12.1m.

The Simulink schematic for this example is called EX13_1.mdl (included on the accompanying website), and its top-level diagram is shown in Fig. 13-5. The resulting torque and speed plots, due to a

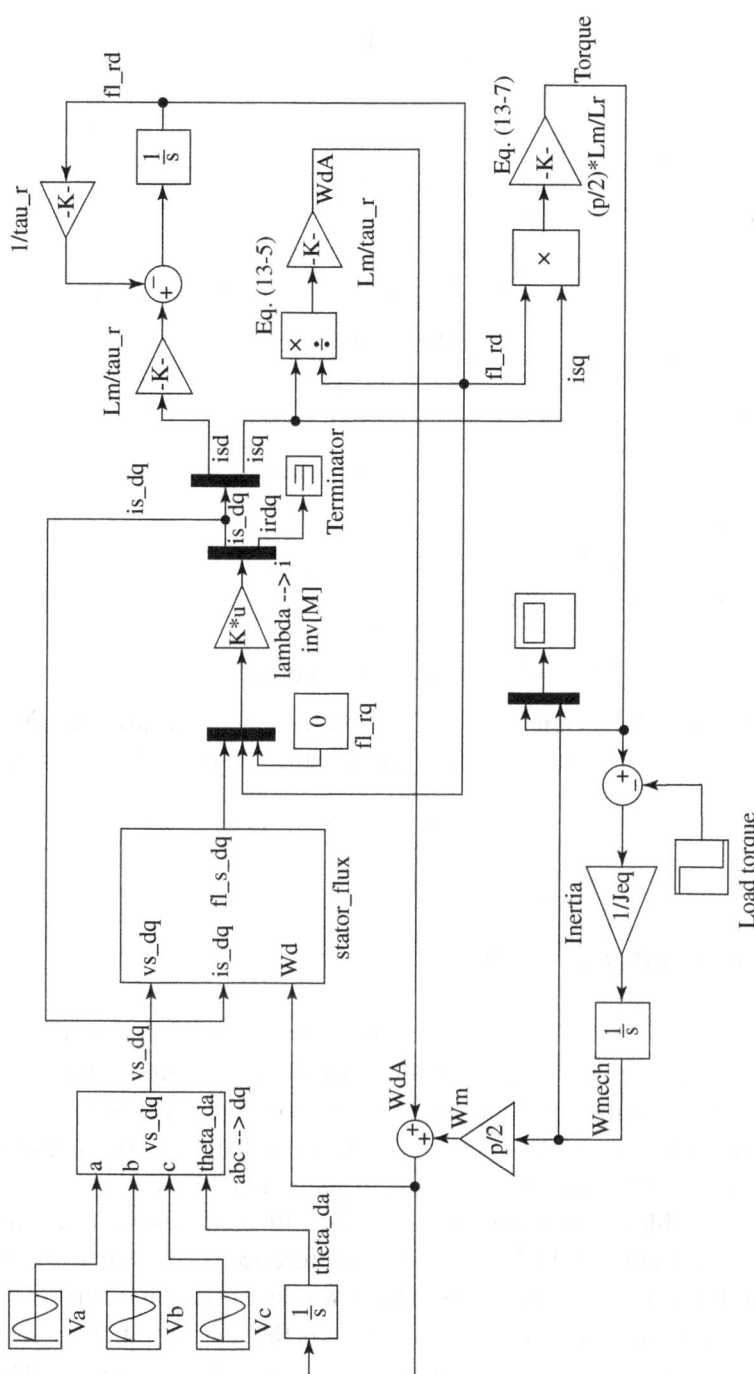

Fig. 13-5 Simulations of Example 13-1.

384 MATHEMATICAL DESCRIPTION OF VECTOR CONTROL

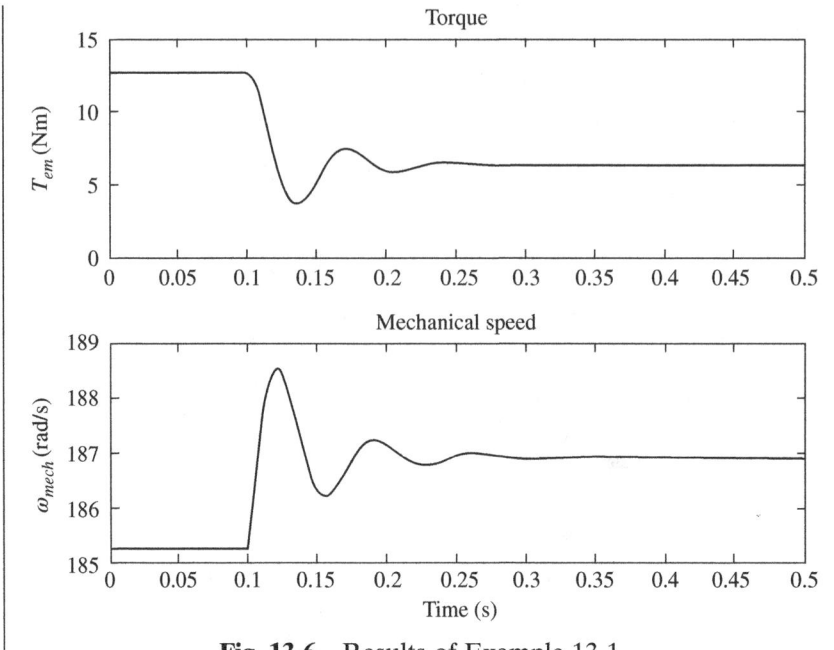

Fig. 13-6 Results of Example 13-1.

load-torque disturbance in this line-fed machine, are plotted in Fig. 13-6, which are identical to the results obtained in Example 12-3.

13-3 VECTOR CONTROL

One of the vector control methods is discussed in this section. It is called indirect vector control in the rotor flux reference frame. For many other possible methods, and their pros and cons, readers are urged to look at several books on vector control and the IEEE transactions and conference proceedings of its various societies.

A partial block diagram of a vector-controlled induction motor drive is shown in Fig. 13-7 with the two reference (or command) currents indicated by "*" as inputs. The d-winding reference current i_{sd}^* controls the rotor flux linkage λ_{rd}, whereas the q-winding current i_{sq}^* controls the electromagnetic torque T_{em} developed by the motor. The

VECTOR CONTROL 385

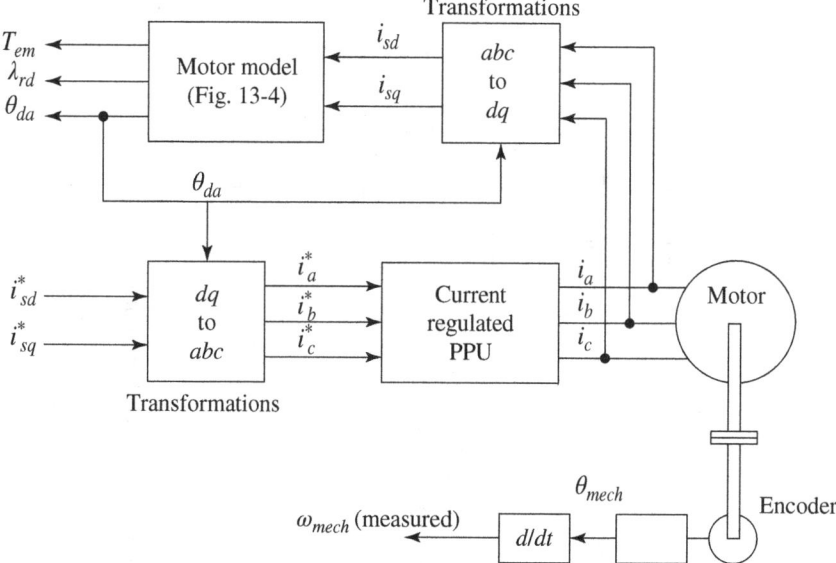

Fig. 13-7 Vector-controlled induction motor with a CR-PWM inverter.

reference dq winding currents (the outputs of the Proportional-Integral (PI) controllers described in the next section) are converted into the reference phase currents $i_a^*(t)$, $i_b^*(t)$, and $i_c^*(t)$. A current-regulated switch-mode converter (the power processing unit, PPU) can deliver the desired currents to the induction motor, using a tolerance-band control, described in Chapter 12. However, in such a current-regulation scheme, the switching frequency within the PPU does not remain constant. If it is important to keep this switching frequency constant, then an alternative is described in this section using a space-vector pulse-width-modulation scheme, which is discussed in detail in Chapter 7.

13-3-1 Speed and Position Control Loops

The current references i_{sd}^* and i_{sq}^* (inputs in the block diagram of Fig. 13-7) are generated by the cascaded torque, speed, and position control loops shown in the block diagram of Fig. 13-8, where θ_{mech}^* is the position reference input. The actual position θ_{mech} and the rotor

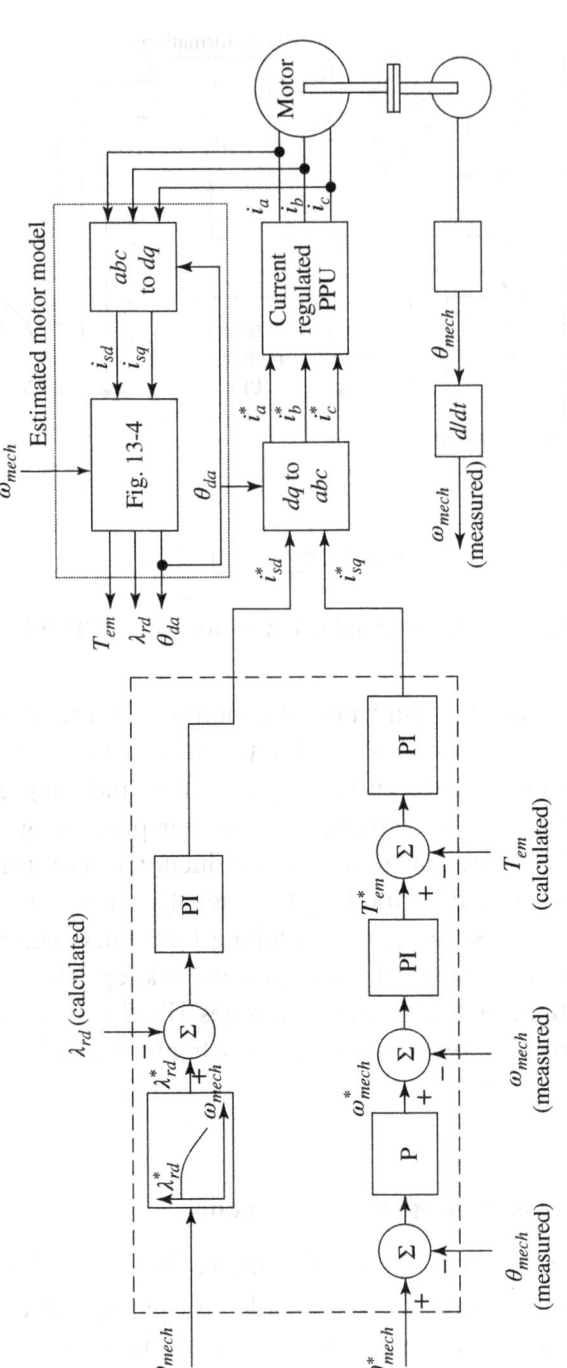

Fig. 13-8 Vector-controlled induction motor drive with a current-regulated PPU.

speed ω_{mech} (where $\omega_m = p/2 \times \omega_{mech}$) are measured, and the rotor flux linkage λ_{rd} is calculated as shown in the block diagram of Fig. 13-8. For operation in an extended speed range beyond the rated speed, the flux weakening is implemented as a function of rotor speed in computing the reference for the rotor flux linkage.

EXAMPLE 13-2

In this example, we will consider the drive system of Example 13-1 under vector control described above. The initial conditions in the motor are identical to those in the previous example. We will neglect the torque loop in this example, where all the motor parameter estimates are assumed to be perfect. (We will see the effect of estimate errors in the motor parameters in the next chapter.) The objective of the speed loop is to keep the speed at its initial value, in spite of the load-torque disturbance at $t = 0.1$ s. We will design the speed loop with a bandwidth of 25 rad/s and a phase margin of 60°.

Solution

Initial flux values are the same as in Example 13-1. These calculations are repeated in a MATLAB file EX13_2calc.m on the accompanying website. To design the speed loop (without the torque loop), the torque expression is derived as follows at the rated value of i^*_{sd}: in steady state under vector control, $i_{rd} = 0$ in Fig. 13-3. Therefore, in Eq. (13-9),

$$\lambda_{rd} = L_m i_{sd} \text{ (under vector control in steady state)} \quad (13\text{-}13)$$

Substituting for λ_{rd} from Eq. (13-13) and for i_{rq} from Eq. (13-2) into the torque expression of Eq. (13-7) at the rated i^*_{sd},

$$T_{em} = \underbrace{\frac{p}{2}\frac{L_m^2}{L_r}}_{k} i^*_{sd} i_{sq} \text{ (under vector control in steady state)} \quad (13\text{-}14)$$

(Continued)

388 MATHEMATICAL DESCRIPTION OF VECTOR CONTROL

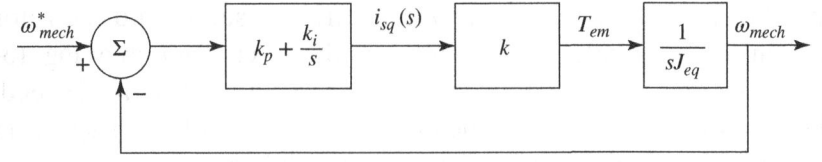

Fig. 13-9 Design of the speed-loop controller.

where k is a constant. The speed loop diagram is shown in Fig. 13-9, where the PI controller constants are calculated in EX13_2calc.m on the basis that the crossover frequency of the open loop is 25 rad/s, and the phase margin is 60°.

The simulation diagram of the file EX13_2.mdl (included on the accompanying website) is shown in Fig. 13-10, and the torque and speed are plotted in Fig. 13-11.

13-3-2 Initial Startup

Unlike the example above, where the system was operating in steady state initially, the system must be started from standstill conditions. Initially, the flux is built up to its rated value, keeping the torque to be zero. Therefore, initially i_{sq}^* is zero. The reference value λ_{rd}^* of the rotor flux at zero speed is calculated in the block diagram of Fig. 13-8. The value of the rotor-field angle θ_{da} is assumed to be zero. The division by zero in the block diagram of Fig. 13-4 is prevented until λ_{rd} takes on some finite (nonzero) value. This way, three stator currents build up to their steady-state dc magnetizing values. The rotor flux builds up entirely along the a-axis. Once the dynamics of the flux buildup is completed, the drive is ready to follow the torque, speed, and position commands.

13-3-3 Calculating the Stator Voltages to be Applied

It is usually desirable to keep the switching frequency within the switch-mode converter (PPU) constant. Therefore, it is a common practice to calculate the required stator voltages that the PPU must

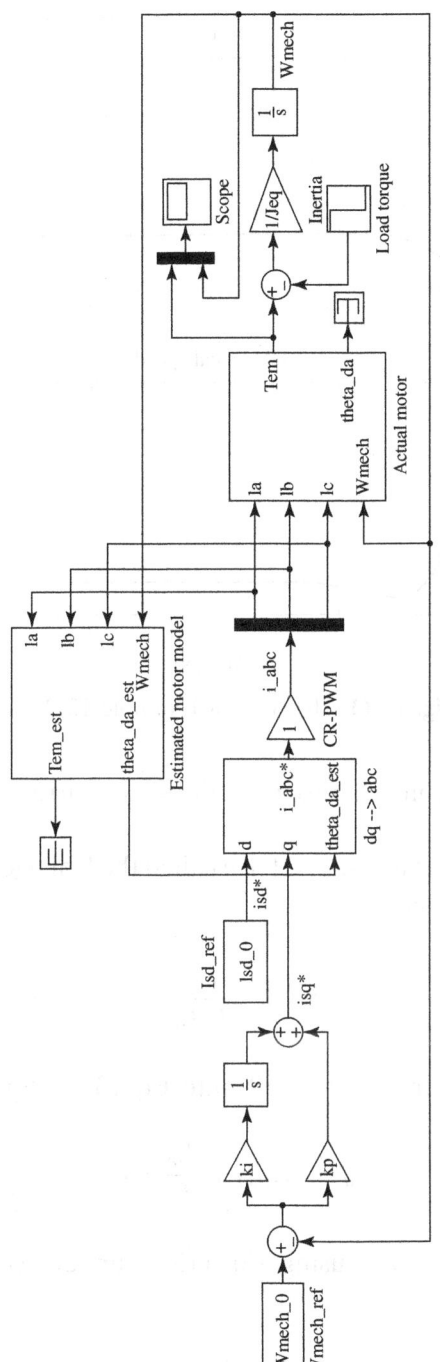

Fig. 13-10 Simulation of Example 13-2.

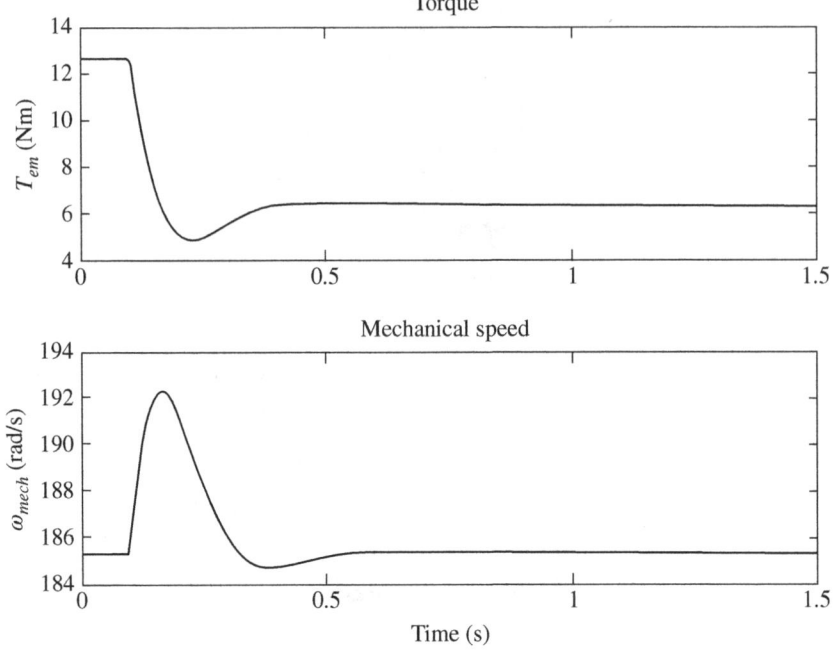

Fig. 13-11 Results of Example 13-2.

supply to the motor in order to make the stator currents equal to their reference values.

We will first define a unit-less term called the leakage factor σ of the induction machine as

$$\sigma = 1 - \frac{L_m^2}{L_s L_r} \tag{13-15}$$

Substituting for i_{rd} from Eq. (13-9) into Eq. (3-19) for λ_{sd},

$$\lambda_{sd} = \sigma L_s i_{sd} + \frac{L_m}{L_r} \lambda_{rd} \tag{13-16}$$

From Eq. (3-20) for λ_{sq}, using Eq. (13-2) under vector-controlled conditions,

$$\lambda_{sq} = \sigma L_s i_{sq} \tag{13-17}$$

Substituting these into Eqs. (3-28) and (3-29) for v_{sd} and v_{sq},

$$v_{sd} = \underbrace{R_s i_{sd} + \sigma L_s \frac{d}{dt} i_{sd}}_{v'_{sd}} + \underbrace{\frac{L_m}{L_r}\frac{d}{dt}\lambda_{rd} - \omega_d \sigma L_s i_{sq}}_{v_{sd,comp}} \quad (13\text{-}18)$$

and

$$v_{sq} = \underbrace{R_s i_{sq} + \sigma L_s \frac{d}{dt} i_{sq}}_{v'_{sq}} + \underbrace{\omega_d \frac{L_m}{L_r}\lambda_{rd} + \omega_d \sigma L_s i_{sd}}_{v_{sq,comp}} \quad (13\text{-}19)$$

13-3-4 Designing the PI Controllers

In the d-axis voltage equation of Eq. (13-18), on the right side, only the first two terms are due to the d-axis current i_{sd} and di_{sd}/dt. The other terms, due to λ_{rd} and i_{sq}, can be considered as disturbances. Similarly, in Eq. (13-19), the terms due to λ_{rd} and i_{sd} can be considered as disturbances. Therefore, we can rewrite these equations as

$$v'_{sd} = R_s i_{sd} + \sigma L_s \frac{d}{dt} i_{sd} \quad (13\text{-}20)$$

and

$$v'_{sq} = R_s i_{sq} + \sigma L_s \frac{d}{dt} i_{sq} \quad (13\text{-}21)$$

where the compensation terms are

$$v_{sd,comp} = \frac{L_m}{L_r}\frac{d}{dt}\lambda_{rd} - \omega_d \sigma L_s i_{sq} \quad (13\text{-}22)$$

and

$$v_{sq,comp} = \omega_d \left(\frac{L_m}{L_r}\lambda_{rd} + \sigma L_s i_{sd}\right) \quad (13\text{-}23)$$

392 MATHEMATICAL DESCRIPTION OF VECTOR CONTROL

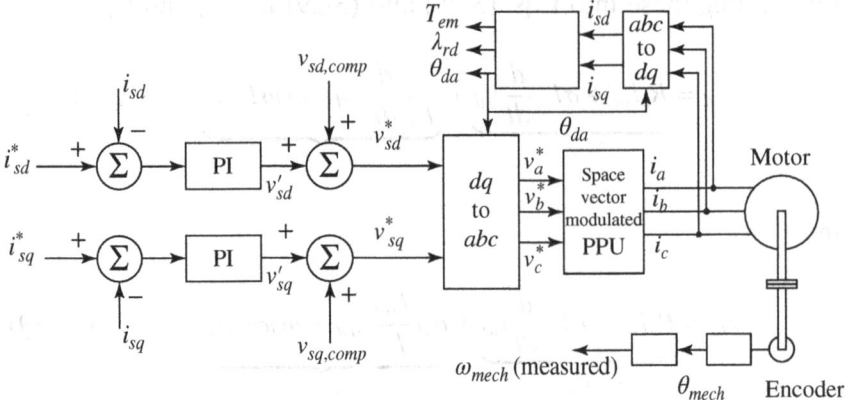

Fig. 13-12 Vector control with applied voltages.

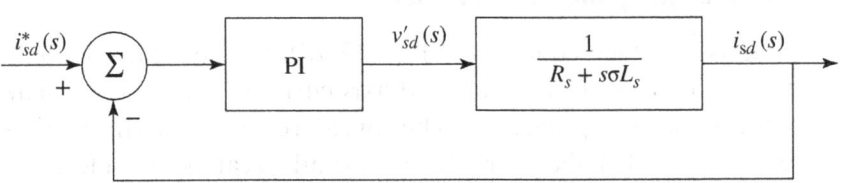

Fig. 13-13 Design of the current-loop controller.

As shown in the block diagram of Fig. 13-12, we can generate the reference voltages v_{sd}^* and v_{sq}^* from given i_{sd}^* and i_{sq}^*, and using the calculated values of λ_{rd}, i_{sd}, and i_{sq}, and the chosen value of ω_d. Using the calculated value of θ_{da} in the block diagram of Fig. 13-12, the reference values v_a^*, v_b^*, and v_c^* for the phase voltages are calculated. The actual stator voltages v_a, v_b, and v_c are supplied by the power electronics converter, using the stator voltage space vector modulation technique discussed in Chapter 7.

To obtain v'_{sd} and v'_{sq} signals in Fig. 13-12, we will employ PI controllers in the current loops. To compute the gains of the proportional and the integral portions of the PI controllers, we will assume that the compensation is perfect. Hence, each channel results in a block diagram of Fig. 13-13 (shown for d-axis), where the "motor-load plant" can be represented by the transfer functions below, based on Eqs. (13-20) and (13-21):

HARDWARE PROTOTYPING OF VECTOR CONTROL 393

$$i_{sd}(s) = \frac{1}{R_s + s\sigma L_s} v'_{sd}(s) \qquad (13\text{-}24)$$

and

$$i_{sq}(s) = \frac{1}{R_s + s\sigma L_s} v'_{sq}(s) \qquad (13\text{-}25)$$

Now, the gain constants of the PI controller in Fig. 13-13 (same in the q-winding) can be calculated using the procedure illustrated in the following example.

EXAMPLE 13-3

Repeat the vector control of Example 13-2 by replacing the current-regulated PWM inverter by a space vector PWM inverter, which is assumed to be ideal. The speed-loop specifications are the same as in Example 13-2. The current (torque) loop to generate reference voltage has ten times the bandwidth of the speed loop and the same phase margin of 60°.

Solution

Calculations for the initial conditions are repeated in the MATLAB file EX13_3calc.m, which is included on the accompanying website. It also shows the procedure for calculating the gain constants of the PI controller of the current loop. The simulation diagram of the SIMULINK file EX13_3.mdl (included on the accompanying website) is shown in Fig. 13-14, and the simulation results are plotted in Fig. 13-15.

13-4 HARDWARE PROTOTYPING OF VECTOR CONTROL OF INDUCTION MOTOR

In this chapter, vector control of induction motor drives was explained in detail, along with accompanying simulation results. In this section, the hardware implementation of the same is analyzed.

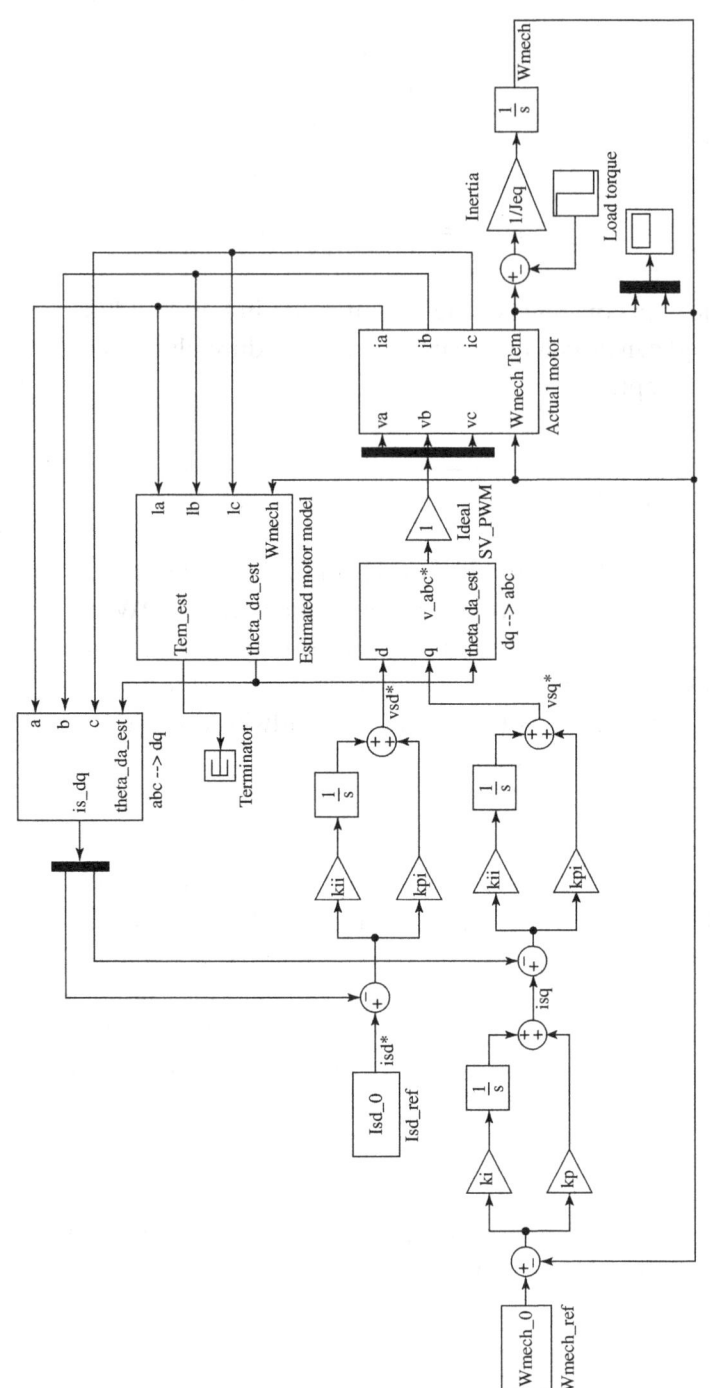

Fig. 13-14 Simulation of Example 13-3.

HARDWARE PROTOTYPING OF VECTOR CONTROL 395

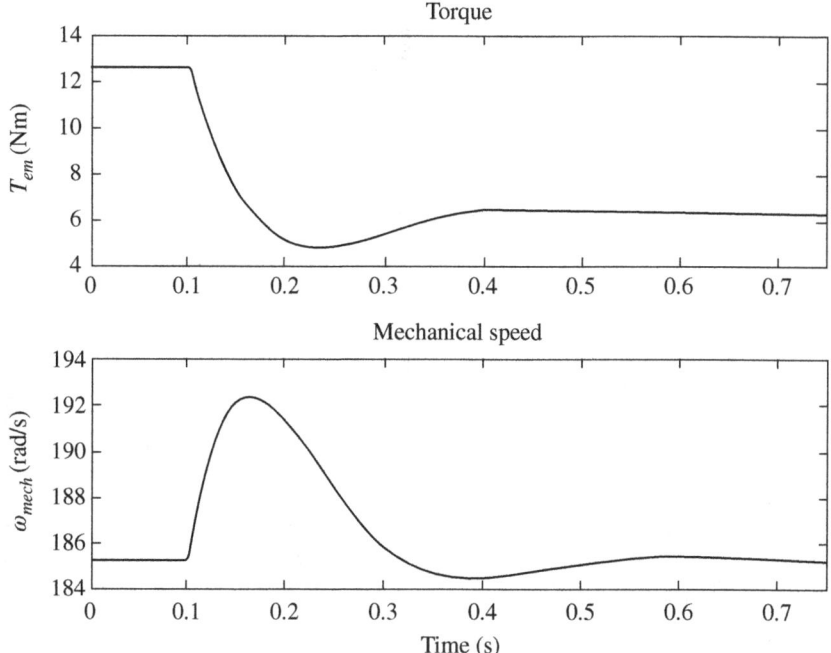

Fig. 13-15 Simulation results of Example 13-3.

For the purposes of this comparison, a commercially available tabletop 24 V (L-L, rms) 3ϕ ac induction motor is modeled in Workbench, and its simulation and hardware results are compared. Although the exact motor parameters of the 3ϕ ac induction motor are not necessary, this is available on the accompanying website. The Workbench file of the modeled system is shown in Fig. 13-16. The speed controller has a step reference input from 0 to 100 rad/s at time $t = 1$ s. A step-load torque of 0.075 Nm is applied at time $t = 3$ s by means of a coupled dc generator.

The speed result from running this model in simulation as well as in hardware is shown in Fig. 13-17. The hardware results closely match that of the simulation, and the speed settles down rapidly to the desired speed even after a load torque is applied.

The estimated electromagnetic torque, as outputted by the induction motor estimator model, and the load torque applied by the dc motor are shown in Fig. 13-18. These two torques do not converge in the

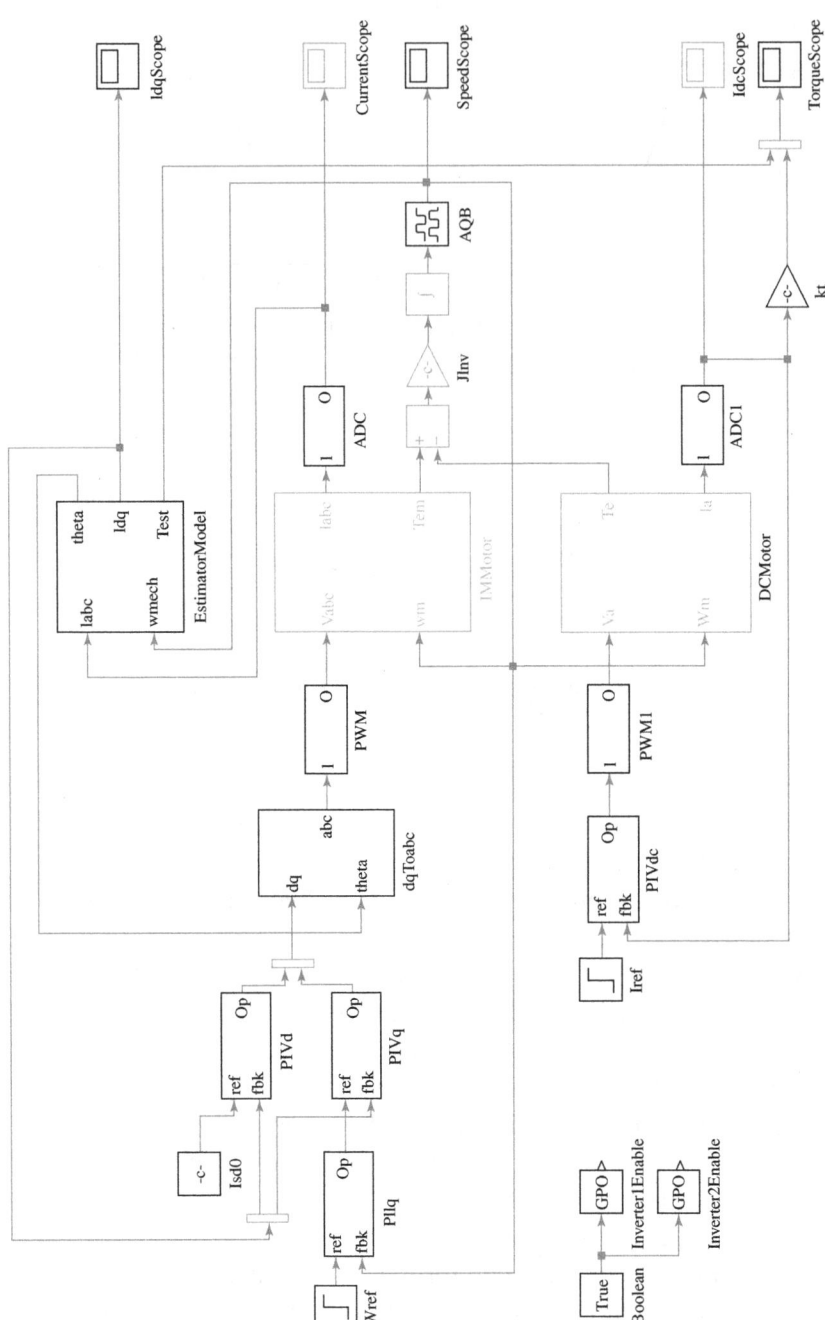

Fig. 13-16 Real-time induction motor vector control.

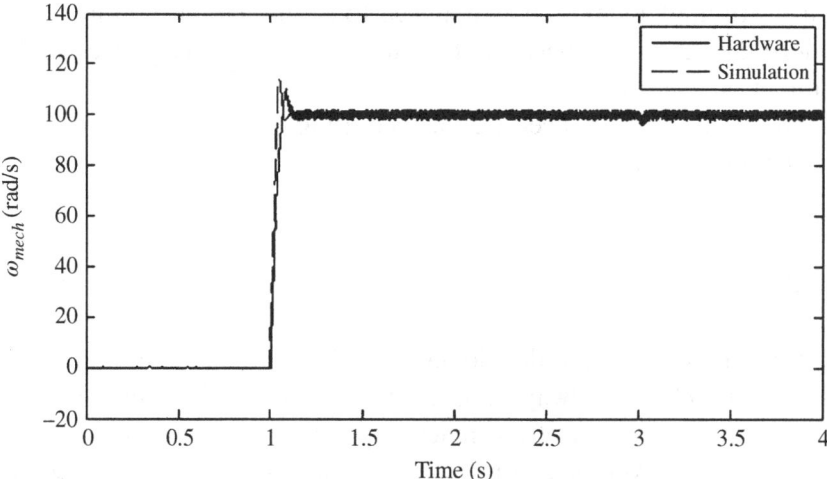

Fig. 13-17 Simulation versus hardware speed result.

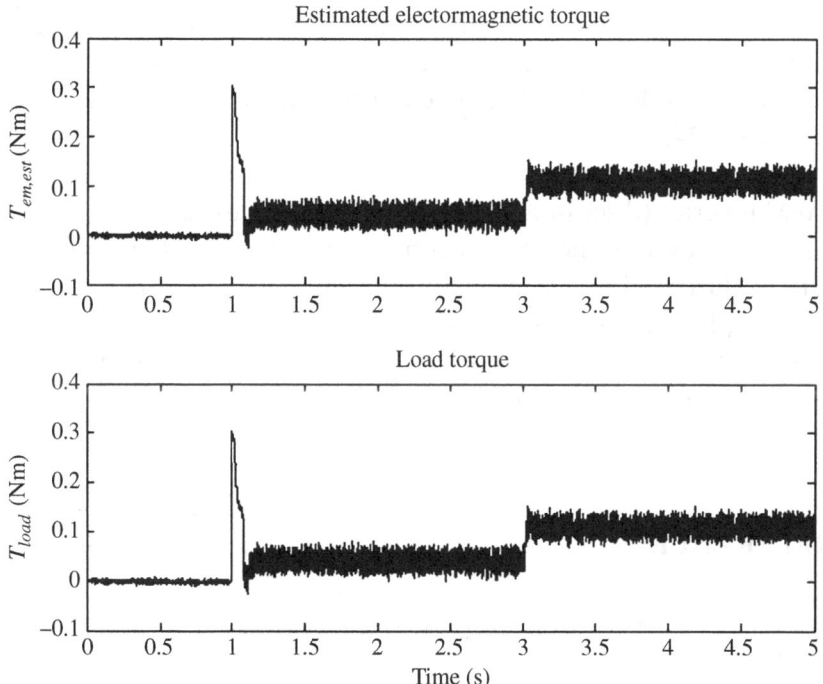

Fig. 13-18 Estimated torque versus load torque in hardware.

hardware implementation as opposed to in simulation Example 13-2, where they converge because the effect of friction was ignored in the simulation example.

The step-by-step procedure for recreating the above comparison is presented in [1].

13-5 SUMMARY

In this chapter, we first developed a model of the induction machine where the d-axis is always aligned with the rotor flux-linkage space vector. Such a model of the machine is valid regardless of whether the machine is vector controlled, or if the voltages and currents to it are applied as under a general-purpose operation, as discussed in Chapter 3. This is illustrated by Example 13-1.

After developing the above-mentioned motor model, we studied vector control of induction motor drives, assuming that the exact motor parameters are known. We first used an idealized current-regulated PWM inverter to supply motor currents calculated by the controller. This vector control is illustrated by means of Example 13-2.

As the last step in this chapter, we used an idealized space vector PWM inverter (discussed in detail in Chapter 7) to supply motor voltages that result in the desired currents calculated by the controller. This is illustrated by means of Example 13-3.

As the last step in this chapter, we implemented a real-time hardware vector control of a laboratory induction motor – dc generator coupled system.

REFERENCE

1. https://sciamble.com/Resources/pe-drives-lab/advanced-drives/im-vector-control

PROBLEMS

13-1 In Example 13-1, comment how i_{sd}, i_{sq}, and $\hat{\lambda}_r$ vary under the dynamic condition caused by the change in load torque. Plot and comment on ω_d under steady state as well as under dynamic conditions.

13-2 Modify the simulation of Example 13-1 for a line-start from a standstill at $t = 0$, with the rated load torque.

13-3 In Example 13-2, plot the stator dq-winding currents, the phase currents, ω_d and ω_{dA}.

13-4 Add the blocks necessary in the simulation of Example 13-2 to plot phase voltages.

13-5 In the simulation of Example 13-2, include the torque loop, assuming its bandwidth to be ten times larger than the speed-loop bandwidth of 25 rad/s (keeping the phase margin in both loops at 60°). Compare results with those in Example 13-2.

13-6 Include flux weakening in the simulation of Example 13-2 by modifying the simulation as follows: initially, in the steady-state operating condition, the load torque is one-half the rated torque of the motor. Instead of the load disturbance at $t = 0.1$ s, the speed reference is ramped linearly to reach 1.5 times the full-load motor speed in 2 s.

13-7 Plot phase voltages in Example 13-3.

13-8 Add the compensation terms in the simulation of Example 13-3. Compare results with those of Example 13-3.

13-9 Repeat Problem 13-6 in the simulation of Example 13-3 by including flux weakening.

14 Speed-Sensorless Vector Control of Induction Motor

14-1 INTRODUCTION

In the vector control of induction motor described in the previous chapter, the dq-frame was chosen to rotate such that the d-axis was aligned with the rotor flux-linkage space vector. The position of the rotor flux-linkage space vector was determined from the measured stator currents and the mechanical rotor speed. Due to the not insignificant additional cost associated with the need for a speed sensor, as well as reduced system reliability because of the delicate nature of optical encoders, there is significant interest in performing vector control using estimated motor speed. This would only require the stator current measurement, which can be carried out by current sense transformers, hall sensors, sense resistors, etc., with the last option being extremely inexpensive.

In this chapter, two different methodologies to estimate the motor speed based on stator current and voltage measurements are described. Following this, the modifications necessary to make these estimator models stable in a real-world application are described. Finally, the shortcomings of these systems, especially at very low speed and alternative methods to overcome this, are briefly mentioned.

Analysis and Control of Electric Drives: Simulations and Laboratory Implementation, First Edition. Ned Mohan and Siddharth Raju.
© 2021 John Wiley & Sons, Inc. Published 2021 by John Wiley & Sons, Inc.
Companion website: www.wiley.com/go/Mohan/Vectorcontrolinelectricdrives

14-2 OPEN-LOOP SPEED ESTIMATOR

By considering a static dq-frame, i.e. $\omega_d = 0$, from stator voltage Eq. (12-27), the stator flux-linkage is obtained from the measured stator currents, and the measured or computed stator terminal voltages as given by:

$$\vec{\lambda}_{s\_dq} = \int \left(\vec{v}_{s\_dq} - R_s \vec{i}_{s\_dq} \right) \cdot dt \qquad (14\text{-}1)$$

The rotor current space vector is determined from $\vec{i}_{s\_dq}$ and $\vec{\lambda}_{s\_dq}$ by substituting Eq. (14-1) in stator flux-linkage Eqs. (12-19) and (12-20):

$$\vec{i}_{r\_dq} = \frac{\vec{\lambda}_{s\_dq} - L_s \vec{i}_{s\_dq}}{L_m} \qquad (14\text{-}2)$$

Substituting Eq. (14-1) in flux-linkage Eqs. (12-19) through (12-22), yields the rotor flux-linkage space vector:

$$\vec{\lambda}_{r\_dq} = \frac{L_r}{L_m} \left(\vec{\lambda}_{s\_dq} - \sigma L_s \vec{i}_{s\_dq} \right) \qquad (14\text{-}3)$$

where σ is the leakage factor given by Eq. (13-15).

Using the solution of $\vec{i}_{r\_dq}$ from Eq. (14-2) and $\vec{\lambda}_{r\_dq}$ Eq. (14-3), ω_{dA} can be estimated from rotor voltage Eqs. (12-31) and (12-32) by setting $v_{rd} = 0$ and $v_{rq} = 0$ (since it is a squirrel-cage rotor):

$$\omega_{dA} = -\left. \frac{R_r i_{rq} + \frac{d}{dt} \lambda_{rq}}{\lambda_{rd}} \right|_{\lambda_{rd} \neq 0} \qquad (14\text{-}4a)$$

$$\omega_{dA} = \left. \frac{R_r i_{rd} + \frac{d}{dt} \lambda_{rd}}{\lambda_{rq}} \right|_{\lambda_{rq} \neq 0} \qquad (14\text{-}4b)$$

The above equations consist of two solutions that produce the same result if the estimated motor parameters match the actual motor parameters. It must be noted that the solution does not exist for Eq. (14-4a) as $\lambda_{rd} \to 0$ and does not exist for Eq. (14-4b) as $\lambda_{rq} \to 0$. These two solutions are combined to yield a solution across all operating points:

$$\omega_{dA} = \frac{\lambda_{rq}\left(R_r i_{rd} + \frac{d}{dt}\lambda_{rd}\right) - \lambda_{rd}\left(R_r i_{rq} + \frac{d}{dt}\lambda_{rq}\right)}{\lambda_{rd}^2 + \lambda_{rq}^2} \quad (14\text{-}5)$$

The rotor mechanical speed is obtained from Eqs. (12-33) and (12-34), by substituting the initial assumption of a static dq-frame, i.e. $\omega_d = 0$,

$$\omega_{mech} = -\frac{2}{P}\omega_{dA} \quad (14\text{-}6)$$

The $\vec{\lambda}_{r\_dq}$ position for performing vector control can be obtained from the λ_{rd} and λ_{rq}, acquired in Eq. (14-3):

$$\theta_{da} = \tan^{-1}\left(\frac{\lambda_{rq}}{\lambda_{rd}}\right) \quad (14\text{-}7)$$

The overall model of the open-loop speed estimator based on the above equations is shown in Fig. 14-1.

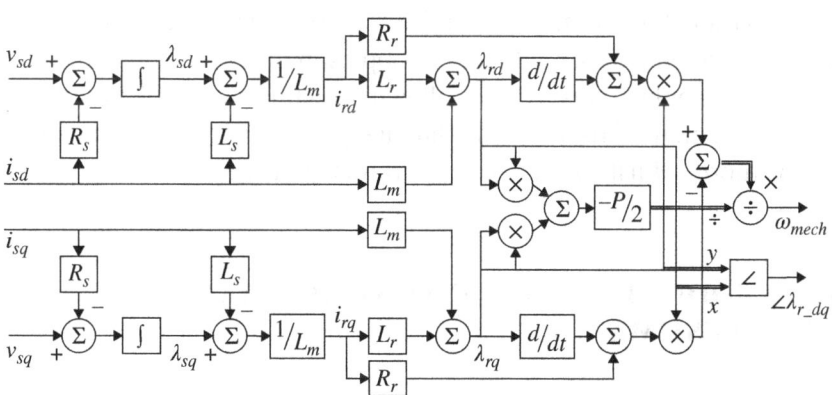

Fig. 14-1 Open-loop ω_{mech} and $\vec{\lambda}_{r\_dq}$ position estimator.

EXAMPLE 14-1

The estimator model in Example 13-3 computed $\vec{\lambda}_{r\_dq}$ position, to perform vector control, based on the measured stator currents and the rotor speed. Modify the estimator model to estimate the speed based on stator current and voltage measurements. The motor starts from zero initial conditions, and a step in $\omega_{mech} = 100$ rad/s is applied at $t = 1$ s followed by a step in load torque at $t = 2$ s.

Solution

The reference i_{sd} and the inner current and outer speed loop gains are maintained, the same as in Example 13-3, and are computed in EX14_1.m. The Simulink schematic for this model is in EX14_1.mdl. The initial conditions of all the integrator blocks are set to 0 to reflect the stated zero initial condition. The top-level diagram is shown in Fig. 14-2a. The estimator model shown in Fig. 14-1, is implemented in the Estimated Motor Model subsystem in the Simulink model, as shown in Fig. 14-2b.

The resulting actual motor mechanical speed and the estimated motor mechanical speed are plotted in Fig. 14-3. As can be seen, the estimated speed matches the actual speed.

The major drawback of open-loop estimator is its high susceptibility to motor parameter variation, and noise and offset in the measured signal. This susceptibility makes the open-loop estimator unsuitable for most real-world applications. In the following section, we shall explore an alternative method for rotor flux-linkage position and speed estimation that is more robust to changes in parameters and system noise.

14-3 MODEL-REFERENCE ADAPTIVE SYSTEM (MRAS) ESTIMATOR

Equations (14-4a) and (14-4b) reflect the inherent redundancy within the motor model equations that lead to two different sets of equations producing the same result. This redundancy can be leveraged

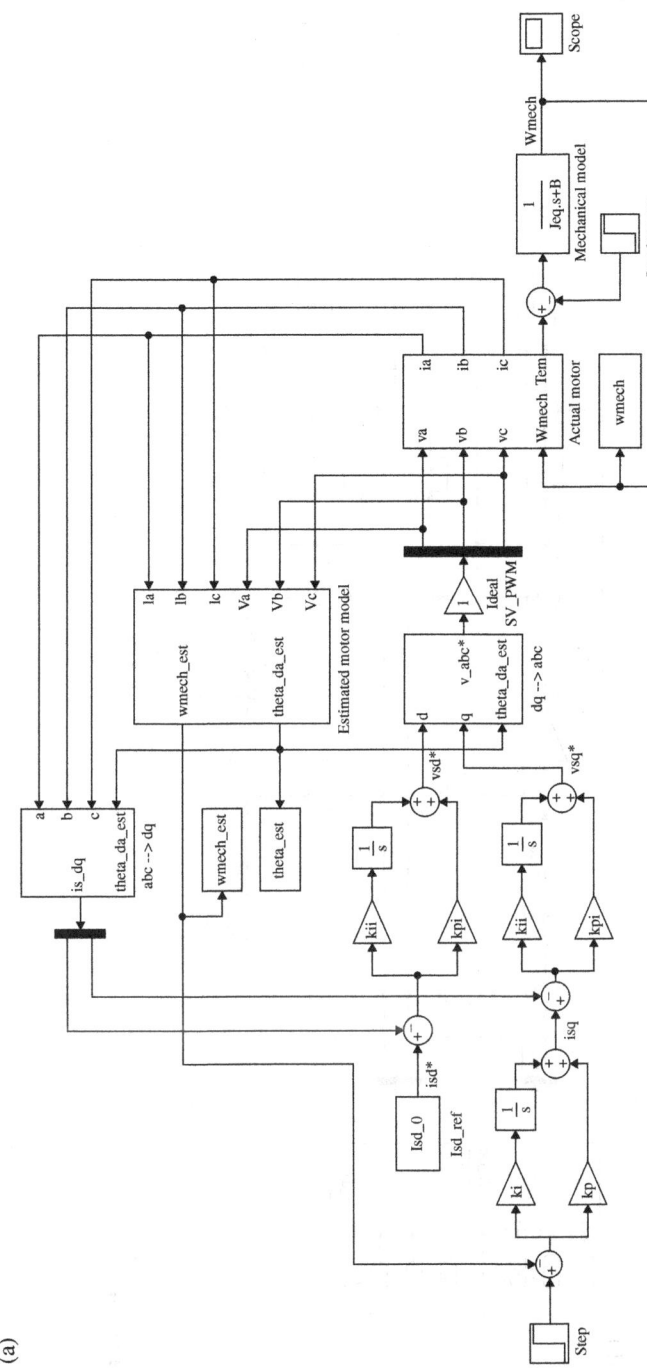

Fig. 14.2 Simulation of Example 14-1. (a) Overall Simulink model and (b) open-loop estimator subsystem.

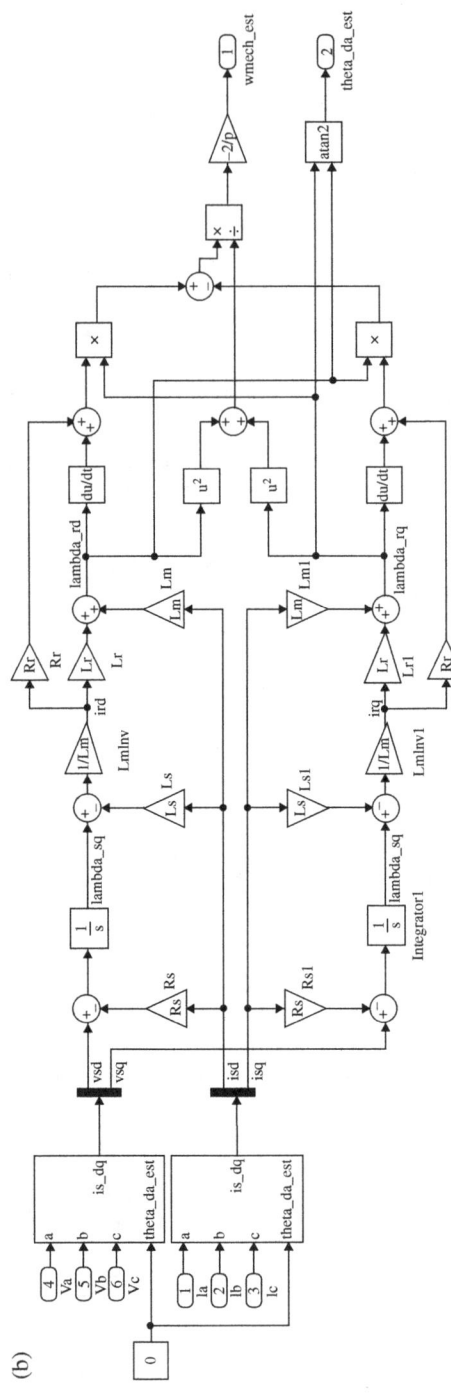

Fig. 14.2 (*Continued*)

MODEL-REFERENCE ADAPTIVE SYSTEM (MRAS) ESTIMATOR

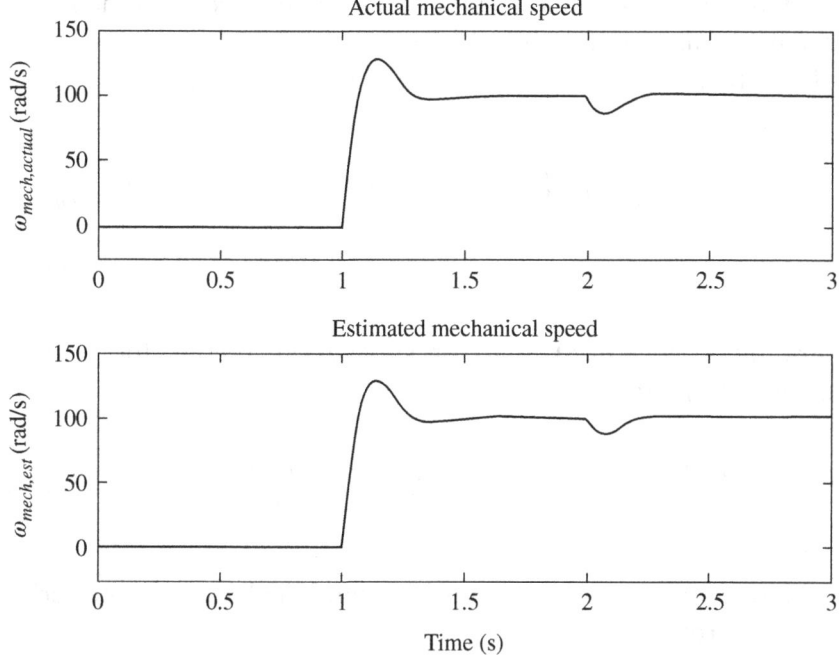

Fig. 14-3 Results of Example 14-1.

to reduce the effect of the motor parameter variation of the estimated $\vec{\lambda}_{r\_dq}$ position. This is done in the adaptive speed identification control strategy, as detailed in [1].

14-3-1 Rotor Speed Estimation

In this method, the rotor flux is estimated independently using the stator model from stator voltage (Eq. (12-27)), and flux-linkage (Eqs. (12-19) through (12-22)), and from the rotor model using rotor voltage (Eqs. (12-31) and (12-32)), rotor flux-linkage (Eq. (12-21) and (12-22)), and rotor speed (Eq. (12-33)). The results of both these models are fed to an adaptive mechanism to estimate ω_{mech}, such that the result produced by the rotor model closely tracks that produced by the stator model. The rotor model is referred to as the adaptive model as its output depends on ω_{mech}, which is the output of the adaptation mechanism. The stator model is referred to as the reference model as its output is independent of ω_{mech}.

From flux-linkage Eqs. (12-19) through (12-22),

$$\vec{\lambda}_{s\_dq} = L_s \vec{i}_{s\_dq} + L_m \left(\frac{\vec{\lambda}_{r\_dq} - L_m \vec{i}_{s\_dq}}{L_r} \right) = \frac{L_m}{L_r} \vec{\lambda}_{r\_dq} + \sigma L_s \vec{i}_{s\_dq}$$

(14-8)

Substituting Eq. (14-8) in stator voltage Eq. (12-27) for a static dq-frame ($\omega_d = 0$),

$$\frac{d}{dt} \vec{\lambda}^s_{r\_dq} = \frac{L_r}{L_m} \left(\vec{v}_{s\_dq} - R_s \vec{i}_{s\_dq} - \sigma L_s \frac{d}{dt} \vec{i}_{s\_dq} \right)$$

(14-9)

In this section, a superscript s indicates stator model, and a superscript r refers to rotor model.

Equation (14-9) represents the stator model for rotor flux-linkage estimation. Similarly, the rotor model for rotor flux-linkage estimation can be obtained by substituting rotor flux-linkage Eqs. (12-21) and (12-22), and rotor speed Eq. (12-33), in rotor voltage Eqs. (12-31) and (12-32),

$$\left(\frac{1}{\tau_r} + \frac{d}{dt} \right) \vec{\lambda}^r_{r\_dq} - j\omega_m \vec{\lambda}^r_{r\_dq} = \frac{L_m}{L_r} R_r \vec{i}_{s\_dq}$$

(14-10)

where τ_r is the rotor time-constant defined in Eq. (13-4). The objective of the MRAS adaptive mechanism, is to find the optimal ω_m that minimize the error between the rotor flux-linkage estimated using the stator model (Eq. (14-9)), $\vec{\lambda}^s_{r\_dq}$ and using the rotor model (Eq. (14-10)), $\vec{\lambda}^r_{r\_dq}$. This is achieved when the error function e defined below is minimized,

$$e = \hat{\lambda}^s_{r\_dq} \hat{\lambda}^r_{r\_dq} \sin \alpha$$

(14-11)

where $\hat{\lambda}^s_{r\_dq}$ is the magnitude of $\vec{\lambda}^s_{r\_dq}$, $\hat{\lambda}^s_{r\_dq}$ is the magnitude of $\vec{\lambda}^r_{r\_dq}$, and α is the angle between the two estimated rotor flux-linkage space vectors.

Equation (14-11) can alternatively be expressed as in Eq. (14-12), proof of which is left as a homework problem.

$$e = \lambda^r_{r\_d}\lambda^s_{r\_q} - \lambda^r_{r\_q}\lambda^s_{r\_d} \qquad (14\text{-}12)$$

This error is minimized using a PI controller whose output is the electrical rotor speed ω_m. The overall model is shown in Fig. 14-4. Note that the L_r/L_m factor in Eq. (14-9), and L_m/L_r factor in Eq. (14-10), are ignored since they cancel each other in the error function equation.

The response speed of this estimator in tracking ω_m is controlled by the PI loop bandwidth, and only limited by noise considerations in practical implementation. The procedure to compute the PI controller gains is detailed in the following section.

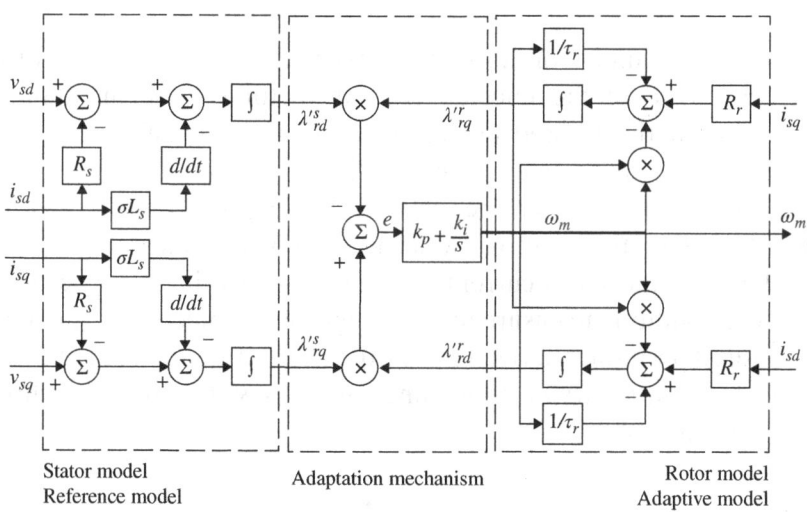

Fig. 14-4 MRAS ω_m and $\vec{\lambda}_{r\_dq}$ position estimator.

14-3-2 Stator d- and q-Axis Current Reference

Under steady-state operating condition, with the d-axis aligned along the rotor flux-linkage space vector (i.e. $\lambda_{rq} = 0$), the reference d-axis current is given by Eq. (13-13),

$$i_{sd}^* = \frac{\lambda_{rd}^*}{L_m} \qquad (14\text{-}13)$$

where λ_{rd}^* is the reference rotor flux-linkage, which is maintained at rated value to ensure fast dynamic response.

Similarly, the reference q-axis current is given by Eq. (13-5),

$$i_{sq}^* = \frac{\lambda_{rd}^*}{L_m} \omega_{dA} \tau_r \qquad (14\text{-}14)$$

The procedure to obtain the electrical slip speed ω_{dA} is described in the next subsection. From the above two reference current equations, the angle of the injected stator current space vector is

$$\angle \vec{i}_{s\_dq}^* = \tan^{-1}\left(\frac{i_{sq}^*}{i_{sd}^*}\right) = \tan^{-1}(\omega_{dA}\tau_r) \qquad (14\text{-}15)$$

The major advantage of this method over the open-loop estimator is that the angle of the injected current vector can be maintained accurately at the desired angle, even when the actual rotor time-constant $\tau_{r,actual}$ deviates significantly from the estimate rotor time-constant $\tau_{r,est}$. When the error of the adaptation mechanism in Fig. 14-4 is zero, the adaptive model accurately replicates the relationship between the stator current and the rotor flux-linkage space vectors. The condition to ensure that the angle of the injected current is at the desired position is derived from Eq. (14-10). Equation (14-10) was derived for stationary dq-frame and can be rewritten for a synchronous dq-frame as

$$\tau_r \frac{d}{dt}\vec{\lambda}_{r\_dq} = -(1 + j\tau_r\omega_{dA})\vec{\lambda}_{r\_dq} + L_m \vec{i}_{s\_dq} \qquad (14\text{-}16)$$

MODEL-REFERENCE ADAPTIVE SYSTEM (MRAS) ESTIMATOR 411

From the above equation, for the above adaptive model to replicate the steady-state interaction between $\vec{i}_{s\_dq}$ and $\vec{\lambda}_{r\_dq}$, as ensured by MRAS,

$$\tau_{r,est}\omega_{dA,est} = \tau_{r,actual}\omega_{dA,actual} \qquad (14\text{-}17)$$

Thus, even when $\tau_{r,est} \neq \tau_{r,actual}$, the product of the estimated motor slip speed and rotor time-constant is the same as that of the product of actual motor slip speed and time constant. From Eqs. (14-15) and (14-17), since the angle of the injected stator currents are only dependent on the product of rotor time-constant and the motor slip speed, the current space vector is injected at an accurate angle.

14-3-3 Estimation of ω_{dA} and θ_{da}

The rotor electrical slip speed ω_{dA} is required to compute both the desired q-axis current in Eq. (14-14) as well the θ_{da}. The relationship between ω_{dA} and the electromagnetic torque T_{em} is obtained from Eqs. (13-14) and (14-14),

$$T_{em} = \underbrace{\frac{p}{2}\frac{L_m^2}{R_r}i_{sd}^{*2}}_{k}\omega_{dA} \qquad (14\text{-}18)$$

Using the above expression, ω_{dA} can be obtained from the desired motor speed ω_{mech}^*, and the estimated motor speed in Fig. 14-4, where $\omega_{mech} = \frac{2}{p}\omega_m$, using a PI controller as done in Example 13-2.

The rotor flux-linkage space vector angle is obtained by

$$\theta_{da} = \int(\omega_m + \omega_{dA})\cdot dt \qquad (14\text{-}19)$$

The overall system of θ_{da} and i_{sq}^* determination based on ω_m estimation is shown in Fig. 14-5.

The reference currents obtained from the MRAS-based estimator can be injected into the motor using a current-regulated PWM inverter, as done in Example 13-2 or a space vector modulated PWM inverter as shown in Example 13-3.

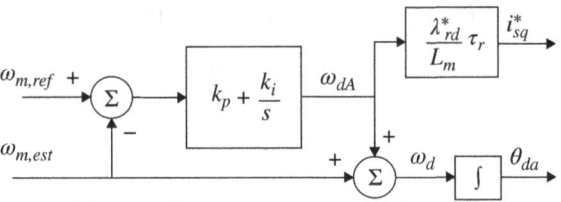

Fig. 14-5 MRAS θ_{da} and i^*_{sq} determination.

The estimated ω_m, at steady state, settles down to the value that makes the rotor flux-linkage, estimated by the rotor model, closely track that which is estimated by the stator model. This, in addition to improving stability, allows operating the motor under maximum torque, as any mismatch between the estimated and actual $\vec{\lambda}_{r\_dq}$ position will have the same effect as that of field-weakening. It must be noted that, for the estimated rotor position to be aligned with the actual rotor position, even in the presence of error in the estimated τ_r, from Eq. (14-17), the estimated slip speed and in turn the estimated motor speed will be offset from the actual motor speed.

EXAMPLE 14-2

Modify the estimator model in Example 14-1, which uses the open-loop speed estimator to MRAS-based estimator. Use the following PI controller gain values for the estimator:

$$k_p = 65 \text{ and } k_i = 4000$$

Solution

The reference i_{sd}, and the inner current and outer speed loop gains are maintained the same as in Example 14-1 and are computed in EX14_2.m. The Simulink schematic for this model is in EX14_2.mdl.

The estimator model shown in Fig. 14-4 is implemented in the Estimated Motor Model subsystem in the Simulink model, as shown in Fig. 14-6. The resulting actual motor mechanical

(Continued)

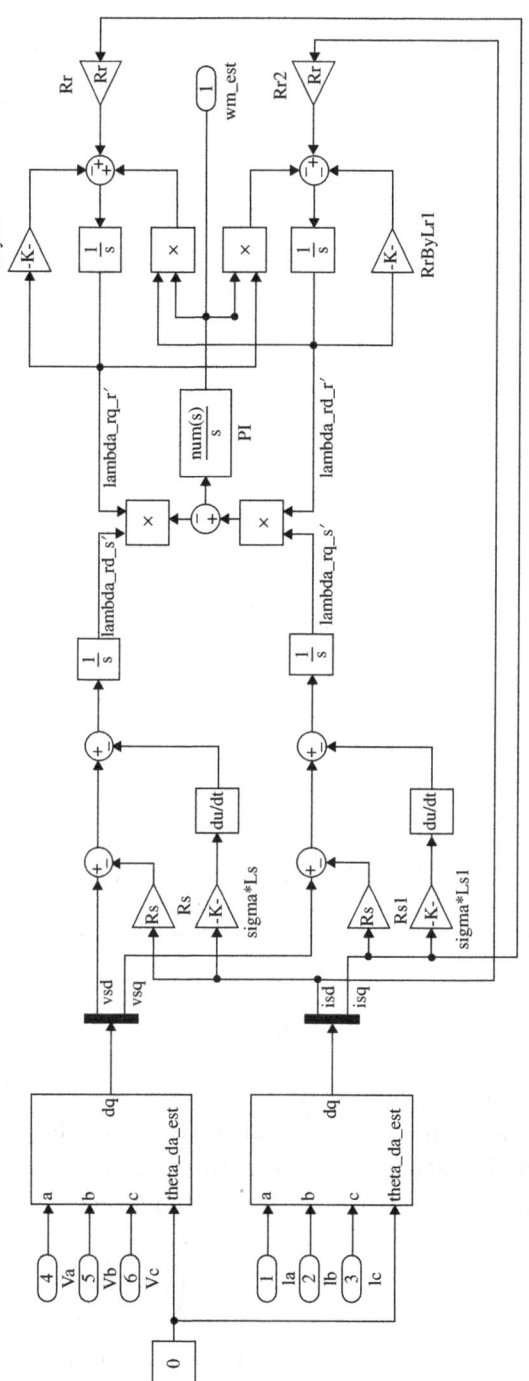

Fig. 14-6 Simulation of Example 14-2 MRAS estimator.

speed and the estimated motor mechanical speed are plotted in Fig. 14-7. As can be seen, the estimated speed closely tracks the actual speed.

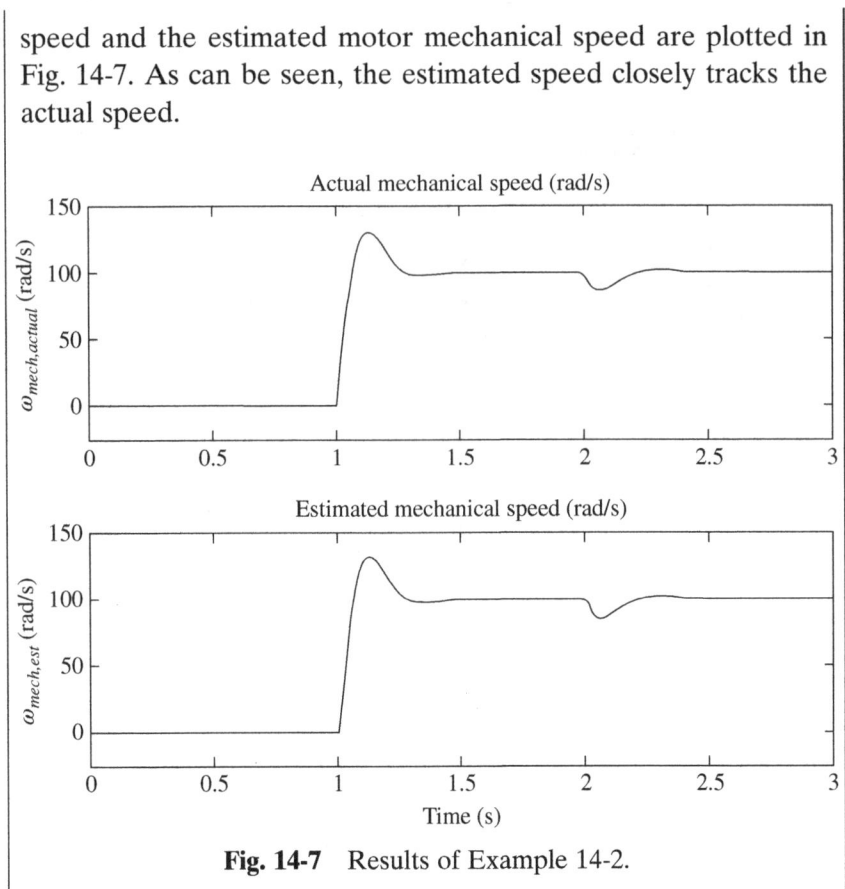

Fig. 14-7 Results of Example 14-2.

14-3-4 Designing the PI controller

Since the error function defined in Eq. (14-12) is nonlinear, the controller is tuned for a linearized model around a steady-state operating point. For a small perturbation in the estimated rotor speed $\Delta\omega_m$, the dynamic relationship between the error function Δe and $\Delta\omega_m$ is

$$\frac{\Delta e}{\Delta\omega_m} = G(s) = \hat{\lambda}_{r\_dq,ss}^2 \frac{\left(s + \frac{1}{\tau_r}\right)}{\left(s + \frac{1}{\tau_r}\right)^2 + \omega_{slip,ss}^2} \quad (14\text{-}20)$$

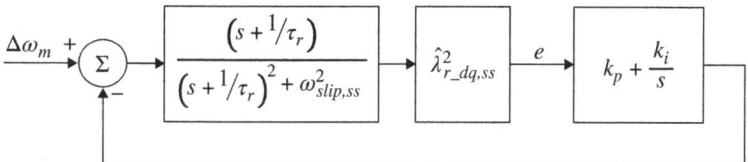

Fig. 14-8 MRAS estimator linearized system transfer function.

where $\hat{\lambda}_{r\_dq,ss}$ is the magnitude of $\vec{\lambda}_{r\_dq}$ and $\omega_{slip,ss}$ is the electrical rotor slip speed under steady-state operation around which the plant model was linearized. The derivation for the above equation is presented in Appendix 14.A. The overall linearized dynamic response model of the MRAS estimator is shown in Fig. 14-8.

EXAMPLE 14-3

The motor in Example 14-2 has the following motor parameters, and steady-state operating condition under rated voltage and load condition:

$$R_r = 1.34\,\Omega, \quad L_r = 0.3808\,\text{H}$$

$$\hat{\lambda}_{r\_dq} = 1.143\,\text{Wb}, \quad \omega_{slip} = 6.4842\,\text{rad/s}$$

Design a PI controller for MRAS estimator with a crossover frequency of 100 rad/s and a phase margin of $\pi/3$.

Solution

The rotor time-constant $\tau_r = L_r/R_r = 0.3808/1.34 = 0.2842\,\text{s}$. From Eq. (14-18), the system transfer function in the frequency domain is

$$G(j\omega)PI(j\omega) = 1.143^2 \frac{j\omega + 3.5189}{(j\omega + 3.5189)^2 + 6.4842^2} \frac{j\omega k_p + k_i}{j\omega}$$

$$\Rightarrow G(j\omega_c)PI(j\omega_c)$$

$$= (4.65 \times 10^{-6} - j1.31 \times 10^{-4})(\omega_c k_p - jk_i)$$

(Continued)

At the desired crossover frequency $\omega_c = 100$ rad/s, and phase margin $\phi_c = \pi/3$,

$$|G(j\omega_c)PI(j\omega_c)| = 1 \Rightarrow \left((100k_p)^2 + k_i^2\right) = (1/_{1.311})^2 = 0.5818$$

and

$$\angle G(j\omega_c)PI(j\omega_c) = -\pi + \phi_c \Rightarrow \frac{k_i}{100k_p}$$

$$= -\tan\left(-\pi + \frac{\pi}{3} - \tan^{-1}\left(\frac{-1.31 \times 10^{-4}}{4.65 \times 10^{-6}}\right)\right) = 0.6257$$

Solving the above two equation yields $k_p = 64.65$ and $k_i = 4045.13$.

14-4 PARAMETER SENSITIVITY OF OPEN-LOOP ESTIMATOR AND MRAS ESTIMATOR

The performance of the estimator model depends on the accuracy of the measured current and voltage signals and the estimated motor parameters. The effects of noise and offset in the measured signal significantly affect the stable operating region of the estimator model, and ways to alleviate these effects in the practical implementation of this estimator are discussed in the next section. In this section, the effect of motor parameter variation is explored through an example.

EXAMPLE 14-4

Compare the error in the estimated and actual ω_{mech} of open-loop and MRAS estimator, designed in Examples 14-1 and 14-2, for the following five cases, where the estimated parameter is different from the actual motor parameters

(a) $R_{s,est} = 0.7 \times R_s$

(b) $R_{r,est} = 0.7 \times R_r$

(c) $L_{m,est} = 0.7 \times L_m$

(d) $L_{ls,est} = 0.7 \times L_{ls}$

(e) $L_{lr,est} = 0.7 \times L_{lr}$

Solution

The combined open-loop estimator and MRAS estimator is in Simulink schematic EX14_4.mdl. The estimator parameters are modified in Ex14_4.m for each of the cases, and the error in $\vec{\lambda}_{r\_dq}$ position and ω_{mech} is plotted in Figs. 14-9 and 14-11 for open-loop and MRAS estimator, respectively.

As seen in Figs. 14-9b and 14-11b, the open-loop estimator has a steady-state error in the estimated $\vec{\lambda}_{r\_dq}$ position. This leads to a reduction in the maximum available electromagnetic torque. In comparison, the MRAS estimator has a significantly lower steady-state error in the estimated $\vec{\lambda}_{r\_dq}$ position. In the case of parameter error in the rotor time-constant, reflected by error in estimated rotor resistance, case b, the MRAS estimator has zero steady-state error in the estimated $\vec{\lambda}_{r\_dq}$ position as opposed to open-loop estimator. This significantly improves the stability of the system under wide operating conditions. The MRAS estimator tends to have more error in the estimated ω_{mech}, in comparison to the open-loop estimator as seen in Figs. 14-8a and 14-9a, and this difference is even higher at lower speeds.

14-5 PRACTICAL IMPLEMENTATION

The open-loop estimator and the MRAS estimator system both operate on measured voltage and currents, which suffers from steady-state offset and tends to be noisy. The presence of an open integrator in

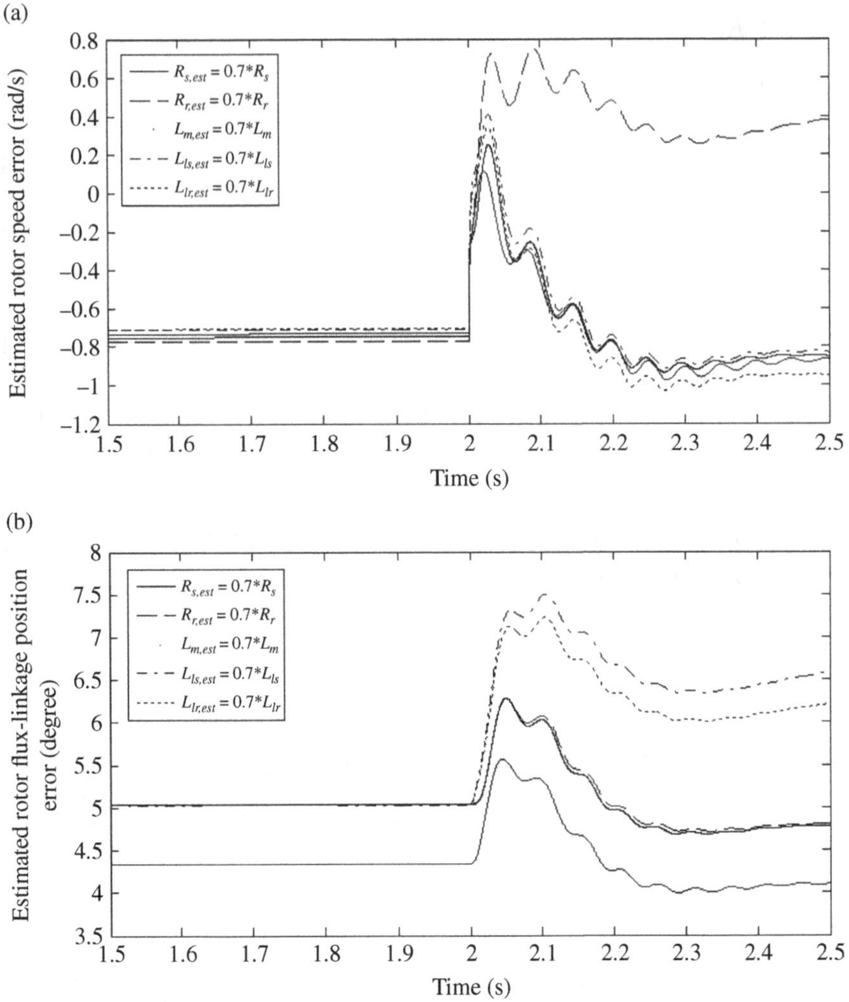

Fig. 14-9 Simulation of Example 14-4: open-speed estimator effect of parameter variation (a) ω_{mech} error and (b) $\vec{\lambda}_{r\_dq}$ position error.

both the systems, as seen in Figs. 14-1 and 14-4, leads to a gradual accumulation of any offset present in the system and could lead to a non-converging result.

To overcome this, the ideal integrator is replaced by a low-pass filter so that any steady-state offset in the current and voltage sensors

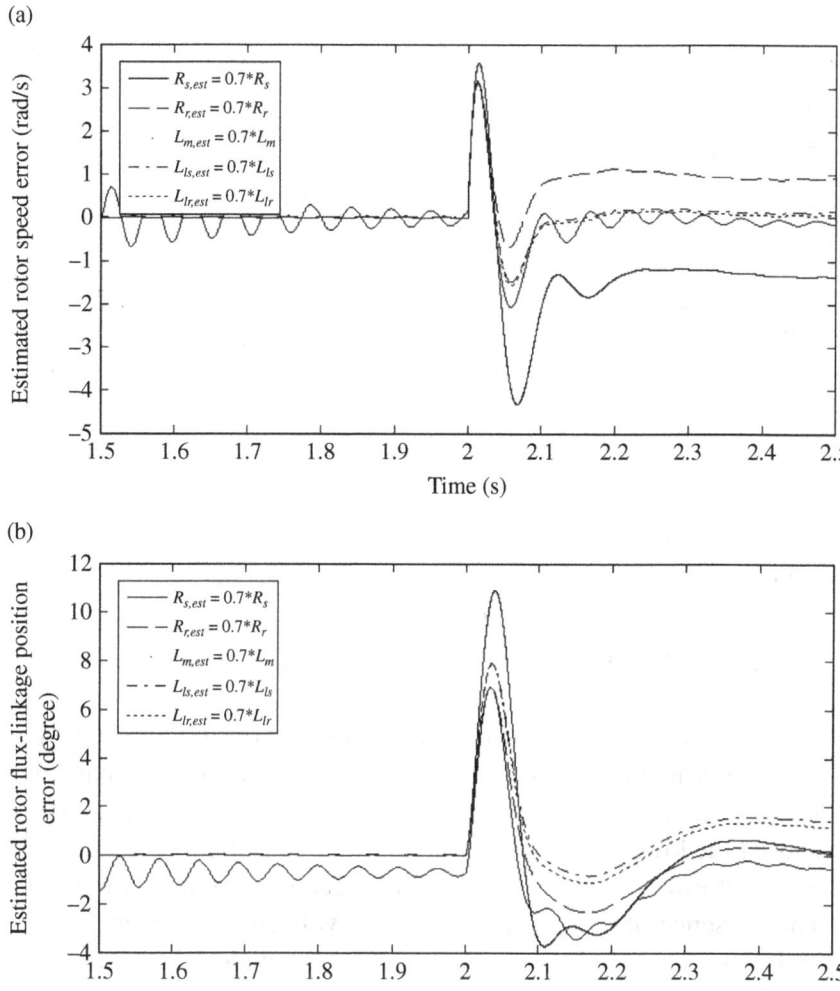

Fig. 14-10 Simulation of Example 14-4: MRAS estimator effect of parameter variation (a) ω_{mech} error and (b) $\vec{\lambda}_{r\_dq}$ position error.

do not get accumulated by the integrator. The transfer function of this modified integrator is given by

$$I(s) = \frac{1}{s + \dfrac{1}{\tau_i}} \qquad (14\text{-}21)$$

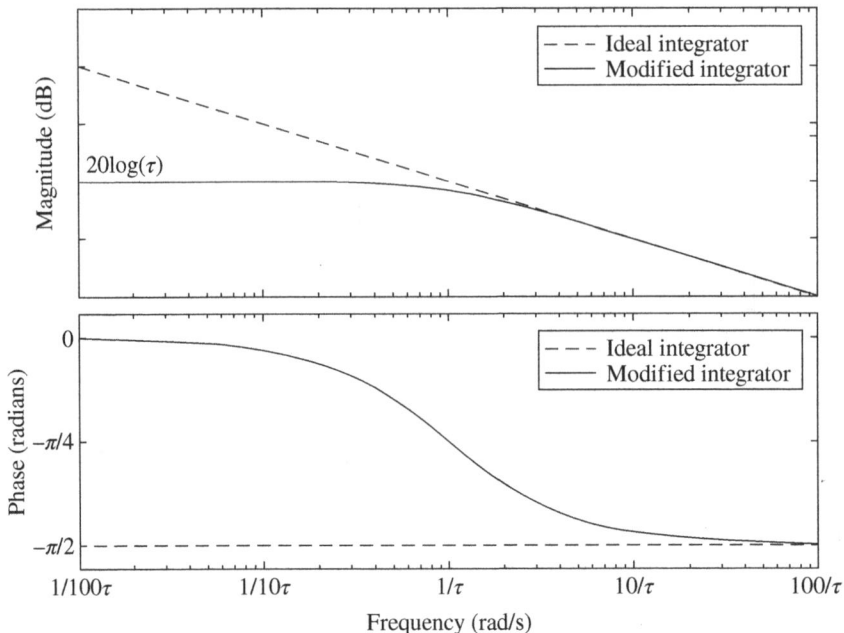

Fig. 14-11 Bode plot of an ideal and practical integrator.

The frequency response of an ideal integrator and the modified integrator is shown in Fig. 14-11. The choice of τ_i depends on what low-frequency signal needs to be suppressed from being integrated, and this would set the lower limit of ω_{mech} up to which sensorless vector control can be performed, in addition to the limit due to parameter variation.

The presence of noise in the measured voltage and current signals would cause the differentiators in the estimator model to generate randomly large signals, which could make the system unstable. Hence, the high-frequency noise signal is filtered using a low-pass filter before being fed to a differentiator. The combined transfer function is given in Eq. (14-22) and the frequency response is shown in Fig. 14-12.

$$D(s) = \frac{\dfrac{s}{\tau_d}}{s + \dfrac{1}{\tau_d}} \tag{14-22}$$

As seen in Fig. 14-11, at frequencies above $1/\tau_d$, the effect of the differentiator is canceled out by that of the low-pass filter. The choice of τ_d, depends on the lowest frequency component of the noise

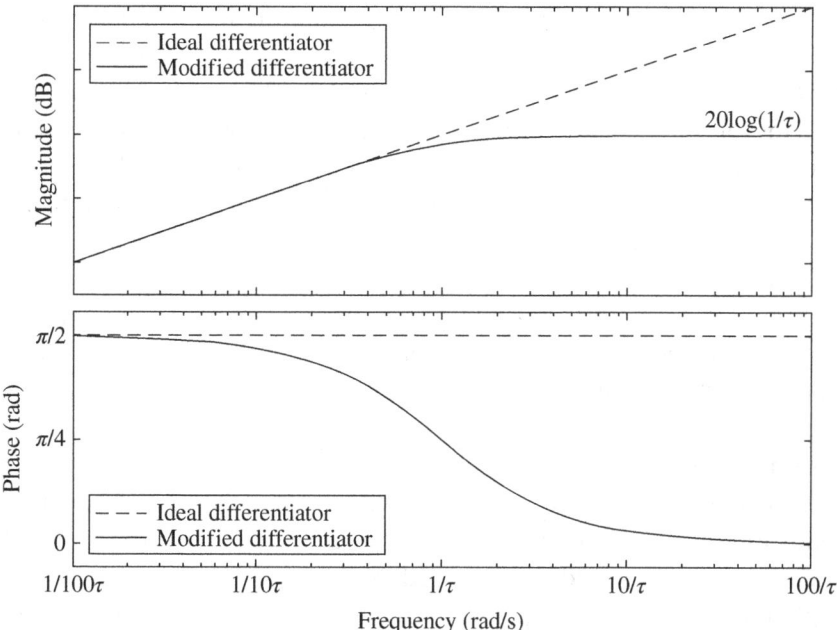

Fig. 14-12 Bode plot of an ideal and practical differentiator.

present in measured signal. This is usually the switching frequency or its second harmonics for electric drives that employ constant switching frequency-based modulation algorithms.

The vector control model developed in the previous chapter using Sciamble Workbench has been modified to include the MRAS estimator, with the changes discussed in this section to enable practical implementation of speed-sensorless vector control of induction motor. The procedure to test this system using actual hardware is available on the accompanying website.

14-6 SUMMARY

In this chapter, two different estimator models were presented to estimate the motor mechanical speed and the rotor flux-linkage space vector position. This information was utilized to perform vector control, developed in the previous chapter, without the use of a speed sensor based solely on applied terminal voltage and stator current measurements.

The operation of the open-loop estimator was verified using MATLAB in Example 14-1. This was followed by the design of an MRAS-based estimator, which utilizes the inherent redundancy in the motor model equations to form a closed-loop estimator. The functioning of this estimator was simulated using MATLAB in Example 14-2, and in the following section, the details of the controller design were discussed.

One of the major issues with both the estimator models discussed, is the deviation of estimated motor speed and the rotor flux-linkage space vector position from the actual values, in the presence of deviation in the measured motor parameter, compared to the actual motor. The effects of these parameter variations were observed in Example 14-4. The MRAS estimator, as seen, was insensitive to errors in the rotor time-constant, unlike the open-loop estimator. This leads to improved stability across a broader operating condition, but the error in estimated rotor speed of MRAS systems tends to be higher than an open-loop estimator, being especially more severe at lower speeds. It must be noted that both these methods would fail to estimate the motor speed when the applied stator voltage frequency tends to zero, and performance greatly deteriorates at lower speeds in the presence of parameter variations. Further analysis of these systems at low speeds, and a summary of alternative methods to overcome this limitation is presented in [2].

As the last step in this chapter, the modifications necessary to make this system practically implementable were discussed and the procedure to implement this hardware is made available on the accompanying website.

REFERENCES

1. Schauder, C. (1992). Adaptive speed identification for vector control of induction motors without rotational transducers. *IEEE Transaction on Industry Application* 28 (5): 1359–1394.
2. Holtz, J. (2002). Sensorless control of induction motor drives. *Proceedings of the IEEE* 90 (8): 1359–1394.

FURTHER READING

Mohan, N. (2001). *Electric Drives – An Integrative Approach*. MNPERE http://www.MNPERE.com.

Landau, Y.D. (1979). *Adaptive Control – The Model Reference Approach*. Marcel Dekker.

PROBLEMS

14-1 Modify Example 14-1 to start from the following steady-state operating condition: mechanical speed $\omega_{mech} = 185$ rad/s, load torque $T_{load} = 12$ Nm, and stator d-axis $I_{sd} = 3.1$ A.

14-2 Derive Eq. (14-5) from Eqs. (14-4a) and (14-4b).

14-3 Prove that Eqs. (14-16) and (14-17) are equivalent, i.e.

$$\hat{\lambda}^s_{r\_dq}\hat{\lambda}^r_{r\_dq}\sin\alpha = \lambda^r_{r\_d}\lambda^s_{r\_q} - \lambda^r_{r\_q}\lambda^s_{r\_d}$$

14-4 Redesign the PI controller in Example 14-2 for a bandwidth of 250 rad/s and a phase margin of $\pi/3$.

14-5 Compare the error in the estimated and actual of open-loop and MRAS estimators designed in Examples 14-1 and 14-2 for the following two cases where the estimated parameter is different from the actual motor parameters:

(a) $R_{s,est} = 1.8 \times R_s$
(b) $R_{r,est} = 1.8 \times R_r$

14-A APPENDIX

14-A-1 MRAS Linearized Error Function

From Eq. (14-11), for a small perturbation around the steady-state operating point,

$$\frac{d}{dt}\Delta\vec{\lambda}_{r\_dq} + \frac{1}{\tau_r}\Delta\vec{\lambda}_{r\_dq} = \frac{L_m}{\tau_r}\Delta\vec{i}_{s\_dq} - j\Delta\left(\omega_{dA}\vec{\lambda}_{r\_dq}\right)$$

$$\Rightarrow \left(s + \frac{1}{\tau_r} + j\omega_{dA,ss}\right)\Delta\vec{\lambda}_{r\_dq} = \left(\frac{L_m}{\tau_r}\Delta\vec{i}_{s\_dq} - j\Delta\omega_{dA}\vec{\lambda}_{r\_dq,ss}\right)$$

(14-23)

With a synchronous reference frame under steady-state $\omega_{dA} = \omega_{slip}$, which is the electrical rotor slip speed. From the above equation, the dynamic response of the stator model and rotor model of MRAS estimator is

$$\left(s + \frac{1}{\tau_r} + j\omega_{dA,ss}\right)\Delta\vec{\lambda}_{r\_dq}^s = \left(\frac{L_m}{\tau_r}\Delta\vec{i}_{s\_dq}\right)$$

$$\Rightarrow \Delta\vec{\lambda}_{r\_dq}^s = \frac{\left(\left(s + \frac{1}{\tau_r}\right) - j\omega_{dA,ss}\right)}{\left(\left(s + \frac{1}{\tau_r}\right)^2 + \omega_{dA,ss}^2\right)}\frac{L_m}{\tau_r}\Delta\vec{i}_{s\_dq}$$

(14-24)

$$\left(s + \frac{1}{\tau_r} + j\omega_{dA,ss}\right)\Delta\vec{\lambda}_{r\_dq}^r = \left(\frac{L_m}{\tau_r}\Delta\vec{i}_{s\_dq}\right)$$

$$\Rightarrow \Delta\vec{\lambda}_{r\_dq}^r = \frac{\left(\left(s + \frac{1}{\tau_r}\right) - j\omega_{dA,ss}\right)}{\left(\left(s + \frac{1}{\tau_r}\right)^2 + \omega_{dA,ss}^2\right)}\left(\frac{L_m}{\tau_r}\Delta\vec{i}_{s\_dq} - j\Delta\omega_{dA}\vec{\lambda}_{r\_dq,ss}^r\right)$$

(14-25)

Note that only the rotor model is affected by perturbations in the rotor speed. Thus $\Delta\omega_{dA} = 0$, in the stator model equation. Linearizing the error function in Eq. (14-12),

$$e = \Delta\lambda_{r\_d}^r\lambda_{r\_q,ss}^s - \Delta\lambda_{r\_q}^r\lambda_{r\_d,ss}^s + \lambda_{r\_d,ss}^r\Delta\lambda_{r\_q}^s - \lambda_{r\_q,ss}^r\Delta\lambda_{r\_d}^s \quad (14\text{-}26)$$

where $\lambda_{r\_d,ss}^s$ and $\lambda_{r\_q,ss}^s$ are the d-axis and q-axis rotor flux-linkages as estimated by the MRAS stator model and, $\lambda_{r\_d,ss}^r$ and $\lambda_{r\_q,ss}^r$ are the d-axis and q-axis rotor flux-linkages as estimated by the MRAS rotor model. At steady-state condition, the rotor flux-linkage estimator from the stator model and the rotor model are the same, leading to the simplifying assumption:

$$\lambda^s_{r\_d,ss} = \lambda^r_{r\_d,ss}$$
$$\lambda^s_{r\_q,ss} = \lambda^r_{r\_q,ss} \quad (14\text{-}27)$$

From the above assumption and substituting $\Delta\lambda^s_{r\_d}$ and $\Delta\lambda^s_{r\_q}$ from Eq. (14-26), and $\Delta\lambda^r_{r\_d}$ and $\Delta\lambda^r_{r\_q}$ from Eq. (14-25) into Eq. (14-26),

$$\Delta e = \hat{\lambda}^2_{r\_dq,ss} \frac{\left(s + \dfrac{1}{\tau_r}\right)}{\left(s + \dfrac{1}{\tau_r}\right)^2 + \omega^2_{slip,ss}} \Delta\omega_m \quad (14\text{-}28)$$

where $\hat{\lambda}_{r\_dq,ss} = \sqrt{\left(\lambda^r_{r\_d,ss}\right)^2 + \left(\lambda^r_{r\_q,ss}\right)^2}$ is the steady-state magnitude of $\vec{\lambda}_{r\_dq}$.

15 Analysis of Doubly Fed Generators (DFIGs) in Steady State and Their Vector Control

15-1 INTRODUCTION

Doubly fed generators (DFIGs) are used in harnessing wind energy. In utility-scale wind turbines, common configurations are to use squirrel-cage induction generators or PMAC generators, shown by a block diagram in Fig. 15-1a and in more detail in Fig. 15-1b for wind-turbine applications [1].

The advantage of this configuration, is that there is no need for mechanical contacts, i.e. slip-rings and brushes, which are needed for the configuration described in this section. Moreover, there is total flexibility of the speed of turbine rotation, which is decoupled by the power-electronics interface from the synchronous speed dictated by the grid frequency. In this arrangement, the power-electronics interface can also supply (or draw) reactive power to the grid for voltage stability. However, on the negative side, the entire power flows through the power-electronics interface, which is still expensive, but is declining in its relative cost.

(Adapted from chapter 11 of *Electric Machines and Drives: A First Course* ISBN: 978-1-118-07481-7 by Ned Mohan, January 2012 and adapted from chapter 7 of *Advanced Electric Drives: Analysis, Control, and Modeling Using MATLAB/Simulink* ISBN: 978-1-118-48548-4 by Ned Mohan, August 2014)

Analysis and Control of Electric Drives: Simulations and Laboratory Implementation, First Edition. Ned Mohan and Siddharth Raju.
© 2021 John Wiley & Sons, Inc. Published 2021 by John Wiley & Sons, Inc.
Companion website: www.wiley.com/go/Mohan/Vectorcontrolinelectricdrives

427

428 ANALYSIS OF DOUBLY FED GENERATORS (DFIGs)

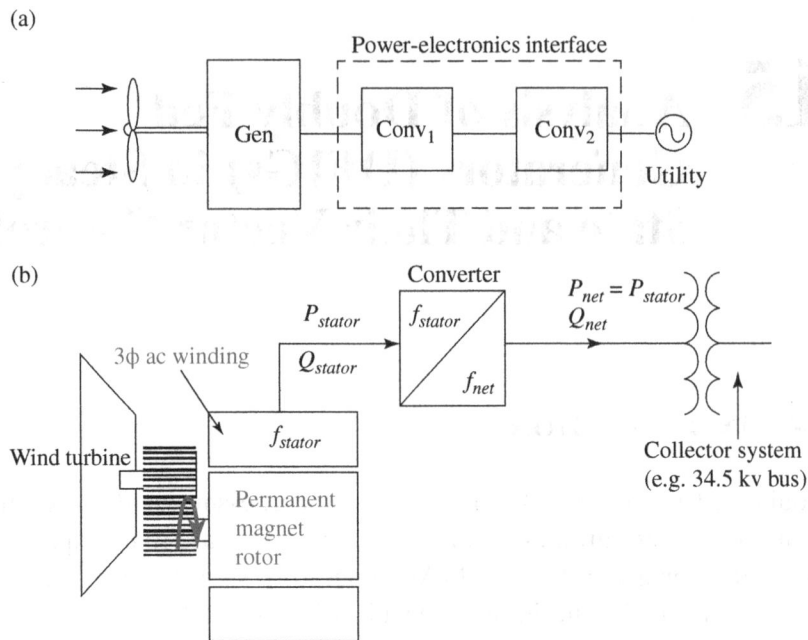

Fig. 15-1 Wind turbines with a complete power-electronics interface [1].

Another typical configuration uses generators with wound rotors, as shown in Fig. 15-2a in a block-diagram form in Fig. 15-2b in greater detail for wind-turbine applications [1]. The stator of such generators is directly connected to the three-phase grid voltages, but the three-phase windings on the rotor are fed appropriate currents through power electronics and the combination of slip-rings and brushes.

Since these generators are connected to the grid voltages on the stator-side, and are supplied by currents through a power-electronics interface on the rotor-side, they are called doubly fed generators and will be referred to as DFIGs from here on.

The benefits of using a DFIG in wind applications are as follows:

1. The speed can be controlled over a sufficiently wide range to make the turbine operate at its optimum coefficient of performance C_p.
2. The stator is directly connected to the grid. Only the rotor is supplied through power electronics that is approximately rated at 30% of the rated power of the wind turbine.

INTRODUCTION   429

3. The reactive power supplied to the rotor is controllable, and it is amplified on the stator-side.

A significant disadvantage of DFIGs is the periodic maintenance required of slip-rings and brushes.

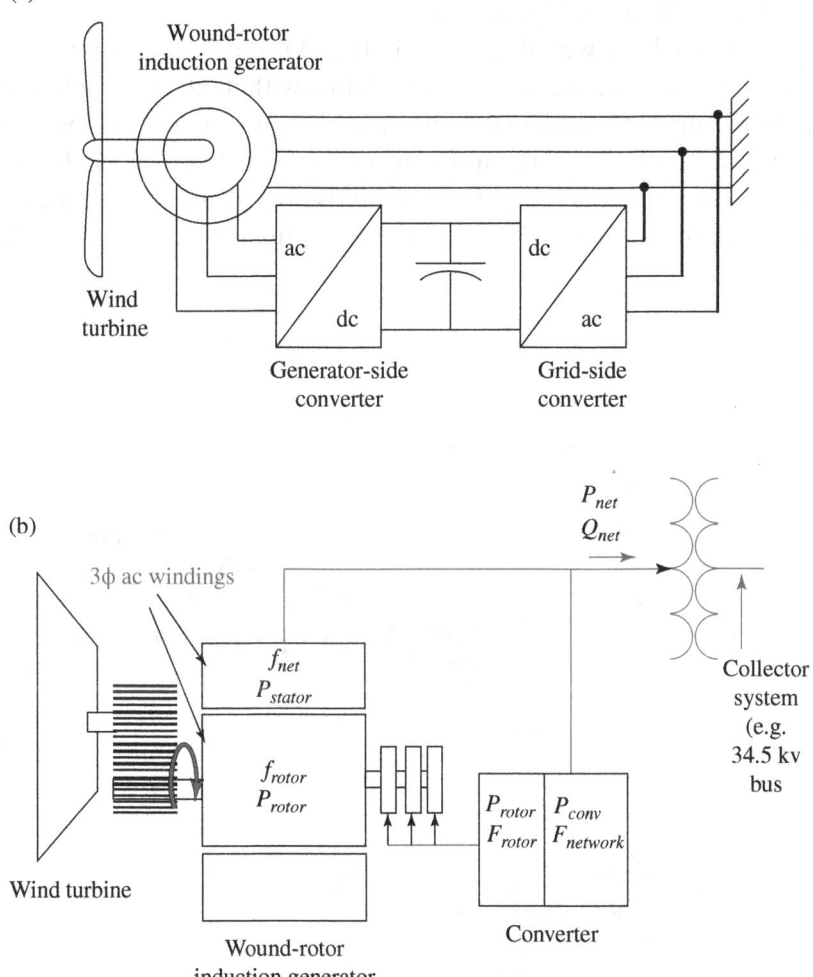

Fig. 15-2 Doubly fed generators (DFIGs) [1].

15-2 STEADY-STATE ANALYSIS

The cross-section of a DFIG is shown in Fig. 15-3. It consists of a stator, similar to the squirrel-cage induction machines, with a three-phase winding, each having N_s turns per-phase that are assumed to be distributed sinusoidally in space. The rotor consists of a wye-connected three-phase winding, each having N_r turns per-phase that are assumed to be distributed sinusoidally in space. Its terminals A, B, and C are supplied appropriate currents through slip-rings and brushes, as shown in Fig. 15-2b.

For our analysis, we will assume that this DFIG is operating under a balanced sinusoidal steady-state condition, with its stator supplied by the 60-Hz grid frequency voltages. In this simplified analysis, we will assume a 2-pole machine and neglect the stator resistance R_s and the leakage inductance $L_{\ell s}$. In this analysis, we will assume the motoring convention, where the currents are defined to be entering the

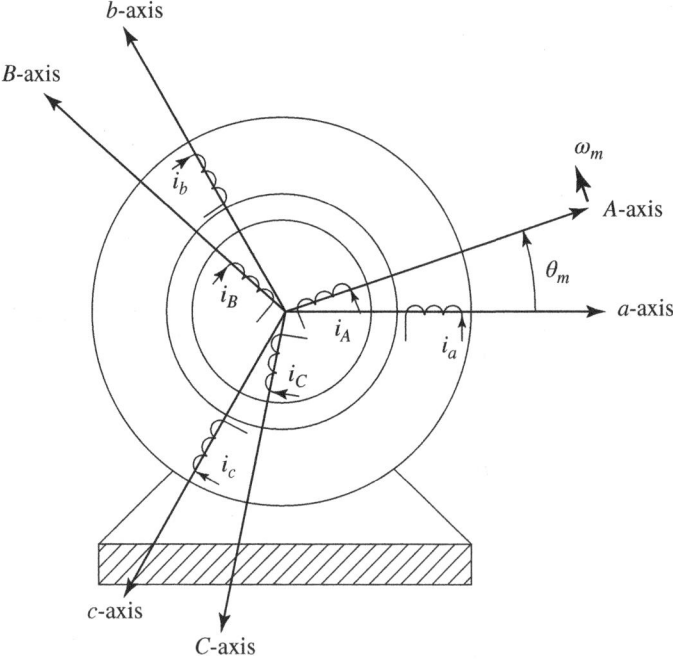

Fig. 15-3 Cross-section of a DFIG.

STEADY-STATE ANALYSIS 431

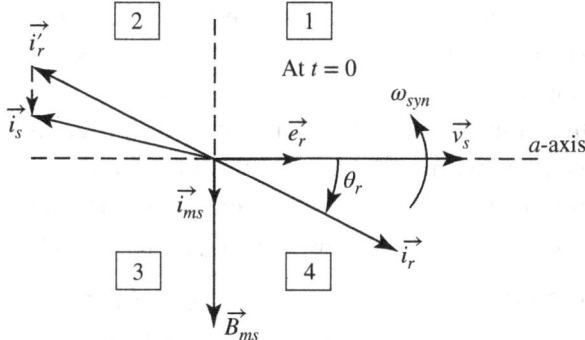

Fig. 15-4 DFIG space vectors at the time $t = 0$; drawn with $\omega_{slip} = +$.

stator- and the rotor-winding terminals, and the electromagnetic torque delivered to the shaft of the machine, is defined to be positive.

Let us also assume that the voltage in phase-a peaks at $\omega t = 0$. At this instant, as shown in Fig. 15-4, $\vec{v}_s$ is along the phase-a axis, and the resulting $\vec{i}_{ms}$ and $\vec{B}_{ms}$ space vectors are vertical.

All space vectors, with respect to the stationary stator windings, rotate at the synchronous speed ω_{syn} in the CCW direction. The rotor of the DFIG is turning at speed ω_m in the CCW direction, where the slip speed $\omega_{slip}(=\omega_{syn}-\omega_m)$ is positive in the sub-synchronous ($\omega_m < \omega_{syn}$) mode, and negative in the super-synchronous ($\omega_m > \omega_{syn}$) mode.

Based on Eq. (6-41), the induced back-emf in the stator windings can be represented by the following space vector (neglecting the stator winding resistances and the leakage inductances, the induced back-emfs are the same as the applied voltages) at the time $t = 0$ in Fig. 15-4:

$$\vec{e}_s = \vec{v}_s = \left(k_e \hat{B}_{ms} N_s \omega_{syn}\right) \angle 0° \quad (15\text{-}1)$$

The same flux-density distribution is cutting the rotor windings, however, at the slip speed. Therefore, the induced back-emfs in the rotor windings can be represented by the following space vector, where the subscript "r" designates the rotor:

$$\vec{e}_r = \left(k_e \hat{B}_{ms} N_r \omega_{slip}\right) \angle 0° \quad (15\text{-}2)$$

432 ANALYSIS OF DOUBLY FED GENERATORS (DFIGs)

Note that $\vec{e}_r$ is made up of the slip-frequency voltages $e_A(t)$, $e_B(t)$, and $e_C(t)$. At sub-synchronous speeds, when ω_{slip} is positive, it rotates at the slip-speed ω_{slip} relative to the rotor in the CCW direction, the same direction as the rotor is rotating in; otherwise, at super-synchronous speed with ω_{slip} negative, it rotates in the opposite direction. Since the rotor itself is rotating at ω_m, with respect to the stator, $\vec{e}_r$ rotates at $\omega_{syn}(=\omega_m + \omega_{slip})$, similar to $\vec{v}_s$. At $\omega t = 0$ in Fig. 15-4, $\vec{e}_r$ is also along the same axis as $\vec{e}_s \left(= \vec{v}_s \right)$ if ω_{slip} is positive (otherwise, opposite), regardless of where the rotor A-axis may be in Fig. 15-3 (why? see the Problems).

We should note that when $\vec{e}_r$ rotates in the CCW direction (at sub-synchronous speeds), the same as $\vec{v}_s$, and the phase sequence of the slip-frequency voltages induced in the rotor windings is A-B-C, same as the a-b-c sequence applied to the stator windings. However, at super-synchronous speeds, $\vec{e}_r$ rotates in the CW direction, opposite to $\vec{v}_s$, and the phase sequence of the slip-frequency voltages induced in the rotor windings is A-C-B, negative of the a-b-c sequence applied to the stator windings.

Appropriate slip-frequency voltages $\vec{v}_r$ are applied from the power-electronic converter, through slip-rings/brushes, as shown in the one-line diagram of Fig. 15-5, to control the current $\vec{i}_r$ to be as desired in Fig. 15-4.

Assuming the current direction to be going into the rotor windings as shown,

$$\vec{i}_r = \frac{\left(\vec{v}_r - \vec{e}_r\right)}{\left(R_r + j\omega_{slip}L_{\ell r}\right)} \text{ at slip} - \text{frequency} \quad (15\text{-}3)$$

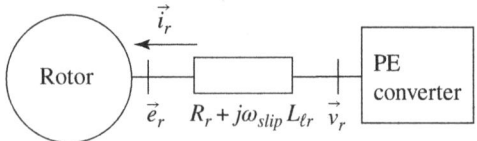

Fig. 15-5 Rotor circuit one-line diagram at slip-frequency.

STEADY-STATE ANALYSIS   433

In Fig. 15-4, to nullify the mmf produced by the rotor currents, additional currents drawn from the stator result in

$$\vec{i}_r' = -\left(\frac{N_r}{N_s}\right)\vec{i}_r \quad (15\text{-}4)$$

The current into the stator is

$$\vec{i}_s = \vec{i}_{ms} + \vec{i}_r' \quad (15\text{-}5)$$

Based on the space vectors shown in Fig. 15-4, the complex power $S_s(=P_s+jQ_s)$ *into* the stator is

$$S_s = P_s + jQ_s = \frac{2}{3}\vec{v}_s\vec{i}_s^* = \frac{2}{3}\vec{v}_s\left(\vec{i}_{ms}^* + \vec{i}_r'^*\right) \quad (15\text{-}6)$$

From Fig. 15-4,

$$\vec{i}_r = \hat{I}_r \angle -\theta_r \text{ (has a positive value as shown in Fig. 15–4)} \quad (15\text{-}7)$$

Using Eqs. (15-4) and (15-7) into Eq. (15-6), and recognizing that in Fig. 15-4 $\vec{v}_s = \hat{V}_s \angle 0°$, the real power into the stator is

$$P_s = \frac{2}{3}\hat{V}_s \operatorname{Re}\left[\vec{i}_r'^*\right] = -\frac{2}{3}\hat{V}_s\hat{I}_r'\frac{N_r}{N_s}\cos\theta_r \quad (15\text{-}8)$$

The reactive power into the stator is due to the magnetizing currents $\vec{i}_{ms}$ and $\vec{i}_r'$. Therefore, we can write the reactive power Q_s *into* the stator as

$$Q_s = Q_{mag} + Q_r' \quad (15\text{-}9)$$

where

$$Q_{mag} = \frac{2}{3}\hat{V}_s\hat{I}_{ms} \quad (15\text{-}10)$$

and

$$Q'_r = \frac{2}{3}\hat{V}_s \text{Im}\left[\vec{i}_r^{\prime*}\right] = -\frac{2}{3}\hat{V}_s\hat{I}'_r \frac{N_r}{N_s}\sin\theta_r \qquad (15\text{-}11)$$

Similarly, the complex power $S_r(=P_r+jQ_r)$ *into* the rotor back-emfs is

$$S_r = P_r + jQ_r = \frac{2}{3}\vec{e}_r \vec{i}_r^{\,*} \qquad (15\text{-}12)$$

where, recognizing that in Fig. 15-4 $\vec{e}_r = \hat{E}_r \angle 0°$,

$$P_r = \frac{2}{3}\hat{E}_r \text{Re}\left[\vec{i}_r^{\,*}\right] = \frac{2}{3}\hat{E}_r \hat{I}_r \cos\theta_r \qquad (15\text{-}13)$$

and

$$Q_r = \frac{2}{3}\hat{E}_r \text{Im}\left[\vec{i}_r^{\,*}\right] = \frac{2}{3}\hat{E}_r \hat{I}_r \sin\theta_r \qquad (15\text{-}14)$$

From Eqs. (15-8) and (15-13), and making use of Eqs. (15-1) and (15-2),

$$\frac{P_s}{P_r} = -\frac{\omega_{syn}}{\omega_{slip}} = -\frac{1}{s} \qquad (15\text{-}15)$$

The total real electrical power into this doubly fed machine, which gets converted into the output mechanical power to the shaft, is

$$P_{em} = P_s + P_r = P_s(1-s) \qquad (15\text{-}16)$$

Comparing the reactive powers,

$$\frac{Q'_r}{Q_r} = -\frac{\omega_{syn}}{\omega_{slip}} = -\frac{1}{s} \qquad (15\text{-}17)$$

which shows that the reactive power input of Q_r into the rotor back-emfs is amplified by a factor of $(1/s)$ at the stator in magnitude. Therefore, from Eq. (15-9),

$$Q_s = Q_{mag} + Q'_r = Q_{mag} - \frac{Q_r}{s} \qquad (15\text{-}18)$$

Figure 15-6 shows the flow of real and reactive powers, where the real and reactive power losses associated with the resistances and the leakage inductances in the stator and the rotor circuits are not included, and a motoring convention is used to define the flows. It should be noted that Q_r is unrelated to the reactive power associated with the grid-side converter shown in Fig. 15-6.

Table 15-1 shows various operating conditions in the sub-synchronous (sub-syn) and the super-synchronous (super-syn) modes.

In the analysis above, real and reactive powers associated with the stator and the rotor resistances and leakage inductances must be added for complete analysis (see Problems).

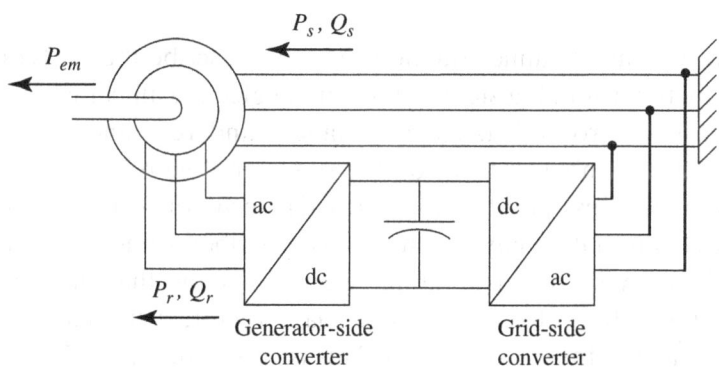

Fig. 15-6 Flows of real and reactive power in a DFIG using a motoring convention; the real and reactive power losses associated with the resistances and the leakage inductances in the stator and the rotor circuits are not included.

436 ANALYSIS OF DOUBLY FED GENERATORS (DFIGs)

TABLE 15-1 Various Operating Modes of DFIG

	ω_{slip}, s (speed)	$\vec{e}_r$ (at $t=0$)	$\vec{i}_r$ in quadrant	P_s (mode)	P_r	Q_r	Q'_r
1	+ (sub-syn)	+	4	$P_s = -$ (generating)	+	+	−
2	+ (sub-syn)	+	1	$P_s = -$ (generating)	+	−	+
3	+ (sub-syn)	+	3	$P_s = +$ (motoring)	−	+	−
4	+ (sub-syn)	+	2	$P_s = +$ (motoring)	−	−	+
5	− (super-syn)	−	4	$P_s = -$ (generating)	−	−	−
6	− (super-syn)	−	1	$P_s = -$ (generating)	−	+	+
7	− (super-syn)	−	3	$P_s = +$ (motoring)	+	−	−
8	− (super-syn)	−	2	$P_s = +$ (motoring)	+	+	+

Note that Row Number 1 Corresponds to the Space Vectors in Fig. 15-4.

15-3 UNDERSTANDING DFIG OPERATION IN dq AXIS

Before writing dynamic equations, we will describe the DFIG operation by first assuming steady state and neglecting all parasitics, such as stator and rotor leakage inductances and resistances. We will assume the number of turns on the stator and the rotor windings to be the same. This operation is described in terms of dq axis, compared to the steady-state analysis, which was described without the help of dq-axis analysis. In this analysis, we will assume that the d-axis is aligned with the rotor flux-linkage space vector, such that the rotor flux-linkage in the q-axis is zero. This is shown in Fig. 15-7.

It should be noted that having neglected the leakage fluxes, the flux-linkage in the rotor d-axis is the same as the stator flux in the d-axis (refer to Eqs. (12-19) to (12-22)). We will write all the necessary equations in steady state under the assumptions indicated earlier and

UNDERSTANDING DFIG OPERATION IN dq AXIS

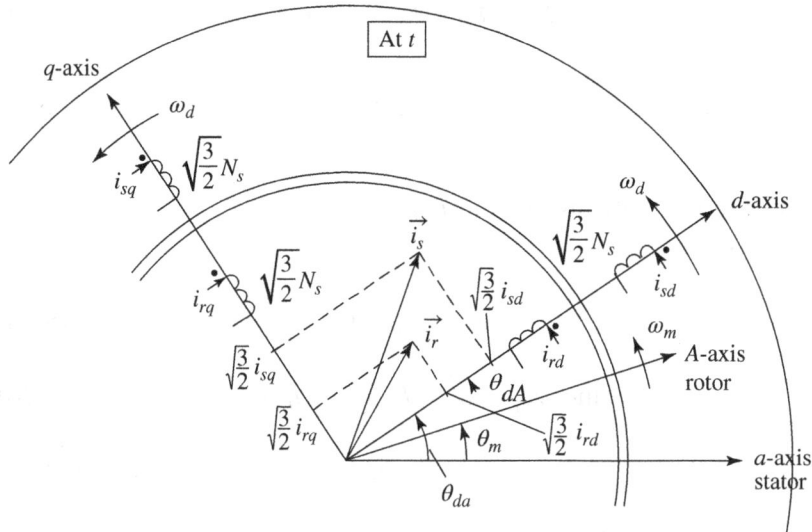

Fig. 15-7 d-axis aligned with the rotor flux; stator and rotor current vectors are shown for definition purposes only.

where P and Q inputs are defined into the stator and the rotor. Using the equations in Chapter 12, the following equations can be written.

15-3-1 Stator Voltages

$$v_{sd} = -\omega_d \lambda_{sq} = 0 \ (\text{since } \lambda_{sq} = 0) \qquad (15\text{-}19)$$

$$v_{sq} = \omega_d \lambda_{sd} \qquad (15\text{-}20)$$

$$v_{sq} = \omega_d \lambda_{sd} = \sqrt{\frac{2}{3}} \hat{V}_s \equiv \text{constant (since } v_{sd} = 0) \qquad (15\text{-}21)$$

where $\hat{V}_s$ is the amplitude of the stator voltage space vector.

15-3-2 Flux Linkages and Currents

d-axis:

$$\lambda_{rd} = \lambda_{sd} = \frac{v_{sq}}{\omega_d} \equiv \text{constant} \qquad (15\text{-}22)$$

$$\lambda_{sd} = \lambda_{rd} = L_m(i_{sd} + i_{rd}) \qquad (15\text{-}23)$$

$$i_{sd} = \frac{\lambda_{sd}}{L_m} - i_{rd} = i_{md} - i_{rd} \qquad (15\text{-}24)$$

where the magnetizing current $i_{md} = \frac{\lambda_{sd}}{L_m}$.
q-axis:

$$\lambda_{rq} = \lambda_{sq} = 0 \qquad (15\text{-}25)$$

Since $\lambda_{sq} = \lambda_{rq} = L_m(i_{sq} + i_{rq}) = 0 \qquad (15\text{-}26)$

$$i_{sq} = -i_{rq} \qquad (15\text{-}27)$$

$$\vec{i}_{s\_dq} = i_{sd} + ji_{sq} = (i_{md} - i_{rd}) - j\,i_{rq} \qquad (15\text{-}28)$$

15-3-3 Rotor Voltages

$$v_{rd} = -\omega_{dA}\lambda_{rq} = 0 \; (\text{since } \lambda_{rq} = 0) \qquad (15\text{-}29)$$

$$v_{rq} = \omega_{dA}\lambda_{rd} \qquad (15\text{-}30)$$

$$v_{rq} = s\omega_d v_{sq} \qquad (15\text{-}31)$$

where s is the slip.

15-3-4 Stator and Rotor Power Inputs

Stator:

$$P_s + jQ_s = \overline{V}_s \overline{I}_s^* = \Big(\underbrace{v_{sd}}_{(=0)} + jv_{sq}\Big)(i_{sd} - ji_{sq}) = v_{sq}i_{sq} + jv_{sq}i_{sd} \qquad (15\text{-}32)$$

$$P_s = v_{sq}i_{sq} = \omega_d \lambda_{sd} i_{sq} \qquad (15\text{-}33)$$

UNDERSTANDING DFIG OPERATION IN dq AXIS

$$Q_s = v_{sq}i_{sd} = \omega_d \lambda_{sd} i_{sd} \qquad (15\text{-}34)$$

Rotor:

$$P_r + jQ_r = \overline{V}_r \overline{I}_r^* = \left(\underbrace{v_{rd}}_{(=0)} + jv_{rq}\right)\left(i_{rd} - ji_{rq}\right) = v_{rq}i_{rq} + jv_{rq}i_{rd} \qquad (15\text{-}35)$$

$$P_r = v_{rq}i_{rq} = s\omega_d \lambda_{rd} i_{rq} \qquad (15\text{-}36)$$

$$Q_r = v_{rq}i_{rd} = s\omega_d \lambda_{rd} i_{rd} \qquad (15\text{-}37)$$

15-3-5 Electromagnetic Torque

$$T_{em} = -\frac{p}{2}\lambda_{rd}i_{rq} \; (\text{using Eq.(12-6) and } \lambda_{rq} = 0) \qquad (15\text{-}38)$$

15-3-6 Relationships of Stator and Rotor Real and Reactive Powers

$$\frac{P_s}{P_r} = -\frac{1}{s} \qquad (15\text{-}39)$$

$$Q_s = \left(\omega_d L_m i_{md}^2\right) - \frac{Q_r}{s} = Q_{mag} - \frac{Q_r}{s} \qquad (15\text{-}40)$$

EXAMPLE 15-1

A DFIG is operating in the motoring mode at a sub-synchronous speed at a lagging power factor (drawing Q_s from the grid). Calculate the signs of various quantities in this mode of operation (Fig. 15-8).

(Continued)

440 ANALYSIS OF DOUBLY FED GENERATORS (DFIGs)

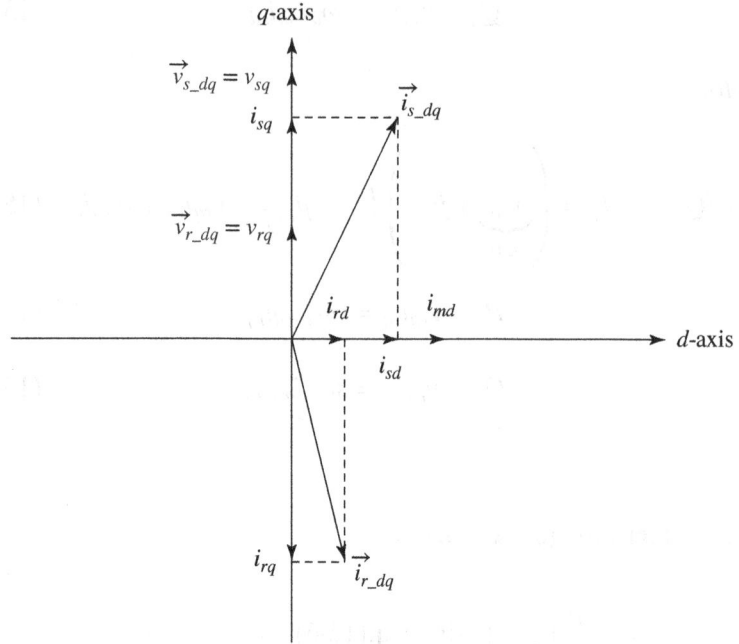

Fig. 15-8 Space vector diagram for Example 15-1.

Solution

$$\omega_{slip} = \omega_{dA} = +$$

$$s = \frac{\omega_{slip}}{\omega_{syn}} = +$$

$$T_{em} = +$$

$$P_s = v_{sq}i_{sq} = \omega_d \lambda_{sd} i_{sq} = +$$

$$\therefore i_{sq} = +$$

$$i_{rq} = -$$

$$Q_s = v_{sq}i_{sd} = \omega_d \lambda_{sd} i_{sd} = \underbrace{+}_{given}$$

$$\therefore i_{sd} = +$$

UNDERSTANDING DFIG OPERATION IN dq AXIS

$$P_r = v_{rq}i_{rq} = s\omega_d\lambda_{rd}i_{rq} = -$$

$$Q_r = v_{rq}i_{rd} = s\omega_d\lambda_{rd}i_{rd}$$

$$i_{rq} = -\text{ (since } i_{sq} = +\text{)}$$

$$i_{rd} = +\text{ taken as positive but } i_{rd} < i_{md}$$

$$v_{rd} = 0$$

$$v_{rq} = s\omega_d\lambda_{rd} = +$$

EXAMPLE 15-2

A DFIG is operating in the generator mode at a super-synchronous speed at a leading power factor (supplying Q_s to the grid). Calculate the signs of various quantities in this mode of operation (Fig. 15-9).

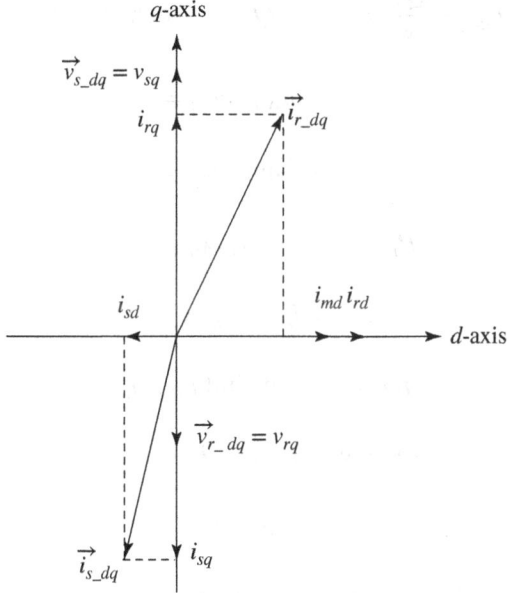

Fig. 15-9 Space vector diagram for Example 15-2.

(*Continued*)

Solution

$$\omega_{slip} = \omega_{dA} = -$$

$$s = \frac{\omega_{slip}}{\omega_{syn}} = -$$

$$T_{em} = -$$

$$P_s = -$$

$$P_s = v_{sq}i_{sq} = \omega_d \lambda_{sd} i_{sq}$$

$$i_{sq} = -$$

$$Q_s = v_{sq}i_{sd} = \omega_d \lambda_{sd} i_{sd} = \underbrace{-}_{given}$$

$$i_{sd} = -$$

$$Q_s = Q_{mag} - \frac{Q_r}{s} = - \quad \therefore Q_r = - \text{ such that } \frac{Q_r}{s} > Q_{mag}$$

$$\omega_{dA} = \omega_{syn} - \omega_m = -$$

$$i_{rq} = + \left(\text{since } i_{sq} = -\right)$$

$$P_r = v_{rq}i_{rq} = \omega_{dA} \lambda_{rd} i_{rq} = -$$

$$i_{sd} = i_{md} - i_{rd} = -$$

$$i_{rd} = + \text{ such that } i_{rd} > i_{md}$$

$$Q_r = v_{rq}i_{rd} = \omega_{dA} \lambda_{rd} i_{rd} = -$$

$$v_{rd} = 0$$

$$v_{rq} = \omega_{dA} \lambda_{rd} = -$$

15-4 DYNAMIC ANALYSIS OF DFIG

Equations for DFIG in terms of dq windings are the same as described in Chapter 12, where it is assumed that the rotor windings have the same number of turns as the stator windings, that is, $N_r = N_s$. However, for $n = N_r/N_s$, these equations can be rewritten, left as Problems.

15-5 VECTOR CONTROL OF DFIG

In Chapter 13, vector control was described by aligning the d-axis with the rotor flux. However, in controlling DFIG, it is common to align the d-axis with the stator flux-linkage vector since this leads to significant simplification in the controller design [2]. The overall block diagram of DFIG vector control is as shown in Fig. 15-10.

15-5-1 Rotor Current Controller

As discussed earlier, these are controlled by controlling the rotor currents i_{rd} and i_{rq}. In this section, the equations necessary for the rotor current controller design is analyzed. For this analysis, the d-axis is aligned with the stator flux-linkage space vector

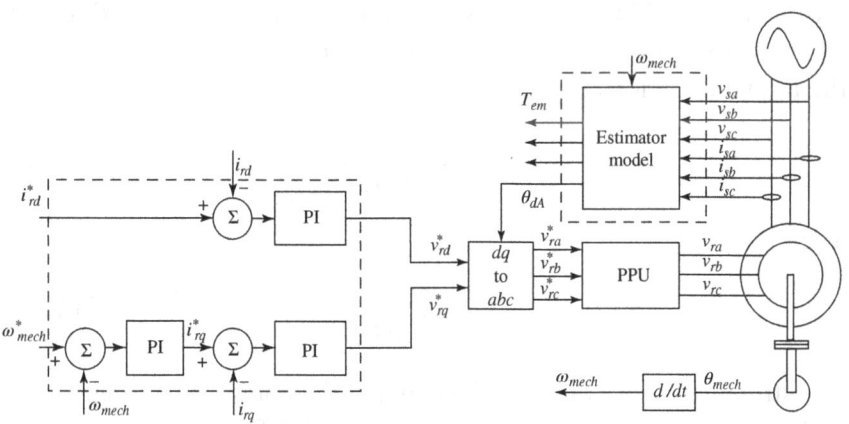

Fig. 15-10 Vector control DFIG drive.

444 ANALYSIS OF DOUBLY FED GENERATORS (DFIGs)

$$\lambda_{sq} = 0 \qquad (15\text{-}41)$$

Substituting the flux-linkage Eqs. (12-21) and (12-22) in the rotor voltage Eqs. (12-31) and (12-32),

$$v_{rd} = R_r i_{rd} + \sigma L_r \frac{di_{rd}}{dt} - \omega_{dA} \sigma L_r i_{rq} + \frac{L_m}{L_s} \frac{d\lambda_{sd}}{dt} \qquad (15\text{-}42)$$

and

$$v_{rq} = R_r i_{rq} + \sigma L_r \frac{di_{rq}}{dt} - \omega_{dA} \left(\sigma L_r i_{rd} + \frac{L_m}{L_s} \lambda_{sd} \right) \qquad (15\text{-}43)$$

Under the typical application of DFIG, i.e. MW level wind-turbine generator, the voltage drop due to the stator resistance is negligible compared to the applied stator terminal voltage. Hence, it is a reasonable assumption that the stator flux-linkage is mostly dictated by the stator voltage. Under normal operating conditions, the grid voltage magnitude is primarily constant, and hence the stator flux-linkage (and v_{sd} since $v_{sq} = 0$) is also a constant.

From the above reasoning, the last term in Eq. (15-42) can be considered to be zero, and the last term in Eq. (15-43) can be considered as a constant external offset, and hence both terms can be ignored for controller design. Similarly, the cross-coupling terms, $\omega_{dA}\sigma L_r i_{rq}$ in the v_{rd} equation and $\omega_{dA}\sigma L_r i_{rd}$ in the v_{rq} equation, can both be ignored as disturbances for the purposes of rotor current PI controller design, as they can be added finally as decompensation terms as done in Chapter 13. The inner current loop system transfer functions from the above equations are

$$i_{rd}(s) = \frac{1}{R_r + s\sigma L_r} v'_{rd}(s) \qquad (15\text{-}44)$$

and

$$i_{sq}(s) = \frac{1}{R_s + s\sigma L_s} v'_{sq}(s) \qquad (15\text{-}45)$$

The structure of the controller is similar to the one shown in Fig. 13-13, and the procedure to design the PI controller for the above system transfer function is the same as that presented in Chapter 13.

15-5-2 Rotor Speed Controller

In a typical wind-turbine generator application [3], the speed of the turbine is adjusted based on the wind speed to optimize the amount the energy collected from the wind. Thus, the current loop is extended by cascading a PI controller to control the rotor speed.

From rotor flux-linkage Eqs. (12-31) and (12-32),

$$i_{rd} = \frac{\lambda_{rd} - L_m i_{sd}}{L_r} \tag{15-46}$$

and

$$i_{rq} = \frac{\lambda_{rq} - L_m i_{sq}}{L_r} \tag{15-47}$$

Substituting the above two equations into the torque expressing in Eq. (12-47),

$$T_{em} = \frac{P}{2} \lambda_{sd} i_{sq} \tag{15-48}$$

Since the d-axis is aligned to the stator flux-linkage space vector, i.e. $\lambda_{sq} = 0$, stator q-axis flux-linkage expressing in Eq. (12-20) can be rewritten as

$$i_{rq} = -\frac{L_m}{L_s} i_{sq} \tag{15-49}$$

From the above two equations,

$$T_{em} = -\frac{P}{2}\frac{L_m}{L_s} \lambda_{sd} i_{rq} = k_t i_{rq} \tag{15-50}$$

Since the stator flux-linkage magnitude under normal operating conditions is constant, from the above expression, it is evident that the electromagnetic torque and, in turn, the rotor speed can be controlled through the rotor q-axis current.

The structure of the controller is the same as the one shown in Fig. 13-9, and the procedure to design the PI controller for the above system transfer function is the same as that presented in Chapter 13.

15-5-3 Stator Reactive Power Controller

The d-axis rotor current controller can be extended further to control the flow of reactive power. This allows not only for the DFIG drive to compensate for the reactive power drawn by its magnetizing circuit, but also to supply reactive power to the grid, which is crucial for voltage stability. The reactive power injected into the grid is directly proportional to the rotor current injected along the d-axis. This relationship is derived in this section.

The stator reactive power is given by

$$Q_s = \text{Im}\{V_s I_s^*\} = v_{sq} i_{sd} - v_{sd} i_{sq}$$

$$= v_{sq}\left(\frac{\lambda_{sd} - L_m i_{rd}}{L_s}\right) - v_{sd} i_{sq} \qquad (15\text{-}51)$$

$$= -\frac{L_m}{L_s} v_{sq} i_{rd} + \frac{\lambda_{sd}}{L_s} v_{sq} - v_{sd} i_{sq}$$

In the above expression, the last term can be ignored, as v_{sd} is negligible compared to v_{sq}. v_{sq} is closely proportional to the grid voltage, which has a constant magnitude and frequency under ideal operation. From the stator voltage Eq. (12-29), λ_{sd} is also a constant. Thus, $\frac{\lambda_{sd}}{L_s} v_{sq}$ can be ignored for the sake of controller design as a constant disturbance.

From the above simplification, the linearized system transfer function for the reactive power controller is given by

$$\Delta Q_s(s) = -\frac{L_m}{L_s} v_{sq} \Delta i_{rd}(s) = k_q \Delta i_{rd}(s) \qquad (15\text{-}52)$$

It must be noted that all these simplifications are valid only when the voltage drop across the stator resistance is negligible. The PI controller design for this is the same as that of the speed loop PI controller.

15-5-4 Rotor Position Estimator

From Eqs. (12-28) and (12-29),

$$\lambda_{sd} = \int (v_{sd} - R_s i_{sd}) \cdot dt \qquad (15\text{-}53)$$

$$\omega_{da} = \frac{v_{sq} - R_s i_{sq}}{\lambda_{sd}} \quad (15\text{-}54)$$

It must be noted that Eq. (15-53) involves open integration, which is not preferred in practical implementations, for reasons highlighted at the end of Chapter 14. Alternative methods to obtain the rotor position are available in various literatures. Once the stator flux-linkage speed is known, θ_{dA} is computed by

$$\theta_{dA} = \int (\omega_{da} - \omega_{mech}) \cdot dt \quad (15\text{-}55)$$

This completes all the information necessary to perform DFIG vector control, as illustrated in Fig. 15-10.

EXAMPLE 15-3

Consider a DFIG with the following parameters:

Voltage	460 V (L-L, rms)
Frequency	60 Hz
Phases	3
Number of poles	6
Full-load slip	1%
Moment of inertia	70 kg·m^2

Per-Phase Circuit Parameters:

$$R_s = 2\,m\Omega$$

$$R_r = 1.5\,m\Omega$$

$$L_m = 2.3\,mH$$

$$L_{ls} = 0.132\,mH$$

$$L_{lr} = 0.124\,mH$$

Design the controller and show the output results.

(*Continued*)

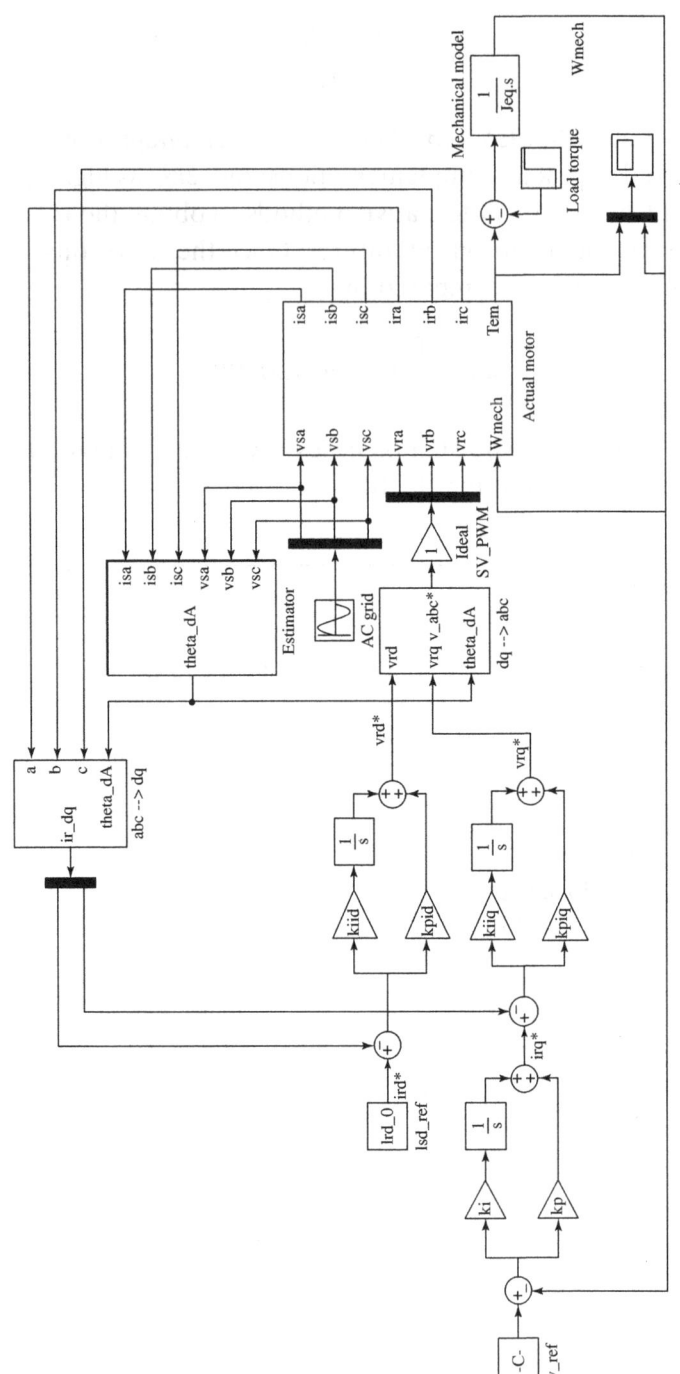

Fig. 15-11 Simulation of Example 15-3.

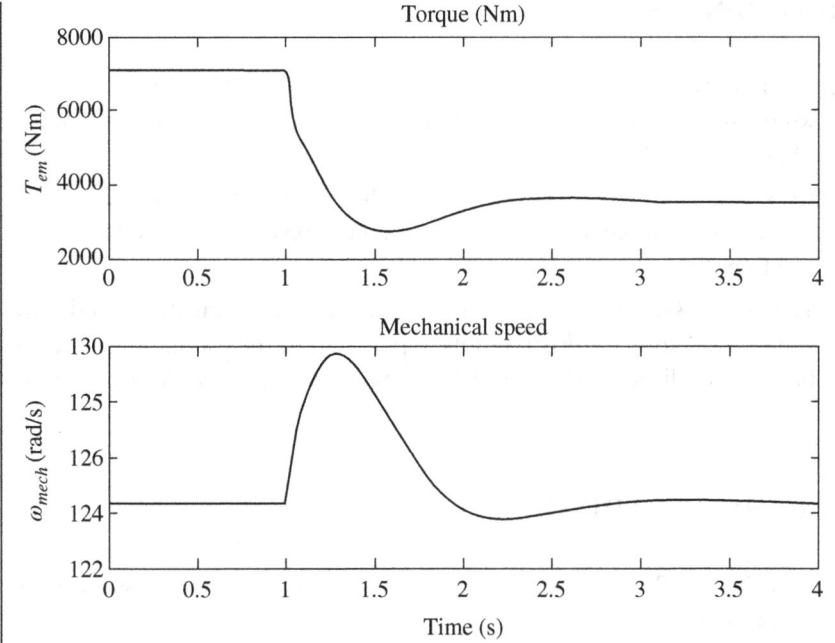

Fig. 15-12 Simulation results of Example 15-3.

Solution

Calculations for the initial conditions are repeated in the MATLAB file EX15_3calc.m, which is included on the accompanying website. It also shows the procedure for calculating the gain constants of the PI controller of the current loop. The simulation diagram of the SIMULINK file EX15_3.mdl (included on the accompanying website) is shown in Fig. 15-11 and the simulation results are plotted in Fig. 15-12.

15-6 SUMMARY

DFIGs are used in harnessing wind energy. In this chapter, the principle of operation of doubly fed machines is described mathematically in order to apply vector control.

REFERENCES

1. Clark, K., Miller, N.W., and Sanchez-Gasca, J.J. (2009). Modeling of GE wind turbine-generators for grid studies. GE energy report, version 4.4 (9 September 2009).
2. Brekken, T. (2005). A novel control scheme for a doubly-fed wind generator under unbalanced grid voltage conditions. PhD thesis. University of Minnesota.
3. Tewari, S., Geyer, C.J., and Mohan, N. (2011). A statistical model for wind power forecast error and its application to the estimation of penalties in liberalized markets. *IEEE Transactions on Power Systems* 26 (4).

FURTHER READING

Mohan, N. (2012). *Electric Machines and Drives*. Wiley. http://www.wiley.com/college/mohan.

PROBLEMS

15-1 In Fig. 15-4, explain why $\vec{e}_r$ is in phase with (or 180 opposite to) $\vec{v}_s$.

15.2 Draw the appropriate space vectors and the phasors corresponding to the generator-mode of operation in a sub-synchronous mode to confirm the entries in Table 15-1.

15-3 Draw the appropriate space vectors and the phasors corresponding to the generator-mode of operation in a super-synchronous mode to confirm the entries in Table 15-1.

15-4 A 6-pole, three-phase DFIG is rated at $V_{LL}(rms) = 480$ V at 60 Hz. In the per-phase equivalent circuit of Fig. 9-21, its parameters are as follows: $R_s = 0.008\,\Omega$, $X_{\ell s} = 0.1\,\Omega$, $X_m = 2.3\,\Omega$, $R'_r = 0.125\,\Omega$, and $X'_{\ell r} = 0.15\,\Omega$. Assume the friction, windage, and iron losses to be negligible. The rotor to the stator winding ratio is 2.5 to 1.0. This generator is supplying real power of

100 kW and the reactive power of 50 kVAR to the grid at a rotational speed of 1320 rpm. Calculate the voltages that should be applied to the rotor circuit from the power-electronics converter. Calculate the power and the reactive power losses within the machine.

15-5 A DFIG is operating in the motoring mode at a sub-synchronous speed at a leading power factor (supplying Q_s to the grid). Calculate the signs of various quantities in this mode of operation.

15-6 A DFIG is operating in the generator mode at a sub-synchronous speed at a leading power factor (supplying Q_s to the grid). Calculate the signs of various quantities in this mode of operation.

15-7 Equations for DFIG in terms of dq windings are the same as described in Chapter 12 where it is assumed that the rotor windings have the same number of turns as the stator windings, that is $N_r = N_s$. However, write these equations for $n = \frac{N_r}{N_s}$ and draw dq-winding equivalent circuits

1. "seen" from the stator-side.
2. "seen" from the rotor-side.

16 Direct Torque Control (DTC) and Encoder-Less Operation of Induction Motor Drives

16-1 INTRODUCTION

Unlike vector-control techniques described in previous chapters, in the direct torque control (DTC) scheme, no dq-axis transformation is needed, and the electromagnetic torque and the stator flux are estimated and directly controlled by applying the appropriate stator voltage vector. It is possible to estimate the rotor speed, thus eliminating the need for a rotor speed encoder.

16-2 SYSTEM OVERVIEW

Figure 16-1 shows the block diagram of the overall system, which includes the speed and the torque feedback loops, without a speed encoder. The estimated speed $\omega_{mech,est}$ is subtracted from the reference (desired) speed ω^*_{mech}, and the error between the two acts on a PI-controller to generate the torque reference signal T^*_{em}. The estimated speed generates the reference signal for the stator flux linkage

---

(Adapted from chapter 9 of *Advanced Electric Drives: Analysis, Control, and Modeling Using MATLAB/Simulink* ISBN: 978-1-118-48548-4 by Ned Mohan, August 2014)

Analysis and Control of Electric Drives: Simulations and Laboratory Implementation, First Edition. Ned Mohan and Siddharth Raju.
© 2021 John Wiley & Sons, Inc. Published 2021 by John Wiley & Sons, Inc.
Companion website: www.wiley.com/go/Mohan/Vectorcontrolinelectricdrives

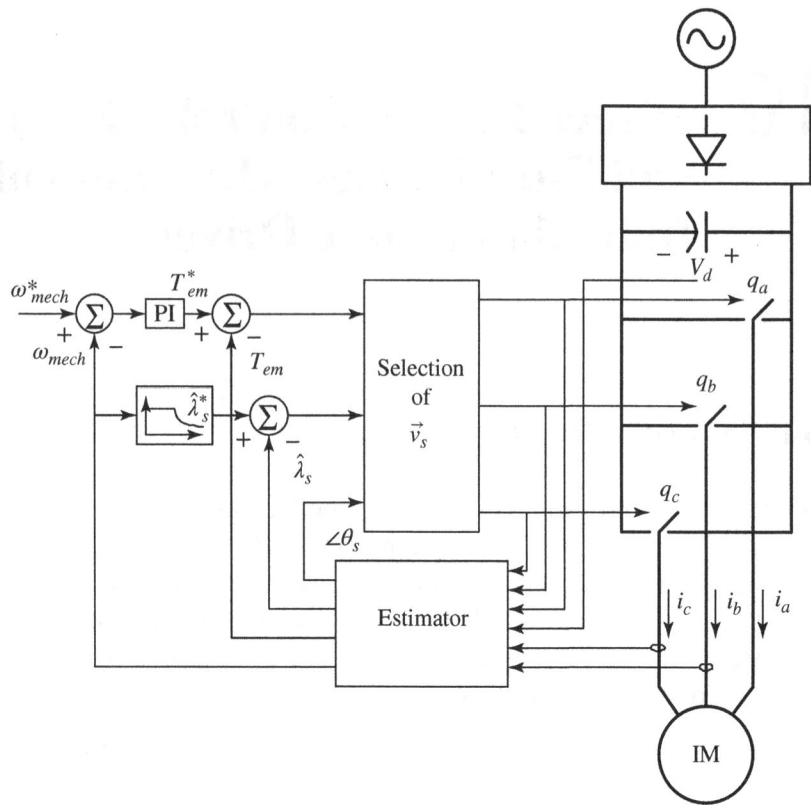

Fig. 16-1 Block diagram of DTC.

$\hat{\lambda}_s^*$ (thus allowing flux weakening for extended range of speed operation), which is compared with the estimated stator flux linkage $\hat{\lambda}_{s,est}$. The errors in the electromagnetic torque and the stator flux, combined with the angular position $\angle\theta_s$ of the stator flux-linkage space vector, determine the stator voltage space vector $\vec{v}_s$ that is applied to the motor during each sampling interval ΔT, for example, equal to 25 µs.

Estimating the electromagnetic torque, and the stator flux-linkage vector, requires measuring the stator currents and the stator phase voltages – the latter, as shown in Fig. 16-1, are indirectly calculated by measuring the dc-bus voltage and knowing within the digital controller, the status of the inverter switches.

16-3 PRINCIPLE OF ENCODER-LESS DTC OPERATION

Prior to detailed derivations, we can enumerate the various steps in the estimator block of Fig. 16-1 as follows, where all space vectors are implicitly expressed in electrical radians with respect to the stator a-axis as the reference axis (unless explicitly mentioned otherwise):

1. From the measured stator voltages and currents, calculate the stator flux-linkage space vector $\vec{\lambda}_s$:

$$\vec{\lambda}_s(t) = \vec{\lambda}_s(t - \Delta T) + \int_{t-\Delta T}^{t} \left(\vec{v}_s - R_s \vec{i}_s \right) \cdot d\tau = \hat{\lambda}_s e^{j\theta_s}$$

2. From $\vec{\lambda}_s$ and $\vec{i}_s$, calculate the rotor flux space vector $\vec{\lambda}_r$ and hence the speed of the rotor flux-linkage vector, where ΔT_ω is a sampling time for speed calculation:

$$\vec{\lambda}_r = \frac{L_r}{L_m} \left(\vec{\lambda}_s - \sigma L_s \vec{i}_s \right) = \hat{\lambda}_r e^{j\theta_r} \text{ and}$$

$$\omega_r = \frac{d}{dt} \theta_r = \frac{\theta_r(t) - \theta_r(t - \Delta T_\omega)}{\Delta T_\omega}$$

3. From $\vec{\lambda}_s$ and $\vec{i}_s$, calculate the estimated electromagnetic torque T_{em}:

$$T_{em} = \frac{2}{3}\frac{p}{2} \text{Im} \left(\vec{\lambda}_s^{conj} \vec{i}_s \right)$$

4. From $\vec{\lambda}_r$ and $T_{em,est}$, estimate the slip speed ω_{slip} and the rotor speed ω_m:

$$\omega_{slip} = \frac{2}{p} \left(\frac{3}{2} R_r \frac{T_{em}}{\hat{\lambda}_r^2} \right) \text{ and } \omega_m = \omega_r - \omega_{slip}$$

In the stator voltage selection block of Fig. 16-1, an appropriate stator voltage vector is calculated to be applied for the next sampling interval ΔT, based on the errors in the torque and the stator flux, in order to keep them within a hysteretic band.

16-4 CALCULATION OF $\vec{\lambda}_s$, $\vec{\lambda}_r$, T_{em}, AND ω_m

16-4-1 Calculation of the Stator Flux $\vec{\lambda}_s$

The stator voltage equation with the stator a-axis as the reference is

$$\vec{v}_s = R_s \vec{i}_s + \frac{d}{dt}\vec{\lambda}_s \qquad (16\text{-}1)$$

From Eq. (16-1), the stator flux-linkage space vector at a time t can be calculated in terms of the flux linkage at the previous sampling time as

$$\vec{\lambda}_s(t) = \vec{\lambda}_s(t - \Delta T) + \int_{t-\Delta T}^{t} \left(\vec{v}_s - R_s \vec{i}_s\right) d\tau = \hat{\lambda}_s e^{j\theta_s} \qquad (16\text{-}2)$$

where τ is the variable of integration, the applied stator voltage remains constant during the sampling interval ΔT, and the stator current value is that measured at the previous time step.

16-4-2 Calculation of the Rotor Flux $\vec{\lambda}_r$

From Chapter 3,

$$\vec{\lambda}_s = L_s \vec{i}_s + L_m \vec{i}_r \qquad (16\text{-}3)$$

and

$$\vec{\lambda}_r = L_r \vec{i}_r + L_m \vec{i}_s \qquad (16\text{-}4)$$

Calculating $\vec{i}_r$ from Eq. (16-3),

$$\vec{i}_r = \frac{\vec{\lambda}_s}{L_m} - \frac{L_s}{L_m}\vec{i}_s \qquad (16\text{-}5)$$

and substituting it into Eq. (16-4),

$$\begin{aligned}\vec{\lambda}_r &= \frac{L_r}{L_m}\vec{\lambda}_s - \frac{L_s L_r}{L_m}\vec{i}_s + L_m \vec{i}_s \\ &= \frac{L_r}{L_m}\left[\vec{\lambda}_s - L_s \vec{i}_s \underbrace{\left(1 - \frac{L_m^2}{L_s L_r}\right)}_{(=\sigma)}\right]\end{aligned} \qquad (16\text{-}6)$$

where the leakage factor σ is defined as (similar to Eq. (5-15))

$$\sigma = 1 - \frac{L_m^2}{L_s L_r} \qquad (16\text{-}7)$$

Therefore, the rotor flux-linkage space vector in Eq. (16-6) can be written as

$$\vec{\lambda}_r = \frac{L_r}{L_m}\left(\vec{\lambda}_s - \sigma L_s \vec{i}_s\right) = \hat{\lambda}_r e^{j\theta_r} \qquad (16\text{-}8)$$

We should note that similar to Eq. (16-2), for the stator flux-linkage vector, the rotor flux-linkage space vector can be expressed as follows, recognizing that the rotor voltage in a squirrel-cage rotor is zero:

$$\vec{\lambda}_r^A(t) = \vec{\lambda}_r^A(t - \Delta T) + \int_{t-\Delta T}^{t}\left(-R_r \vec{i}_r^A\right)\cdot d\tau = \hat{\lambda}_r e^{j\theta_r^A} \qquad (16\text{-}9)$$

where the space vectors and angles (in electrical radians) are expressed with respect to the rotor A-axis shown in Fig. 16-2. The above equation shows that the rotor flux changes very slowly with

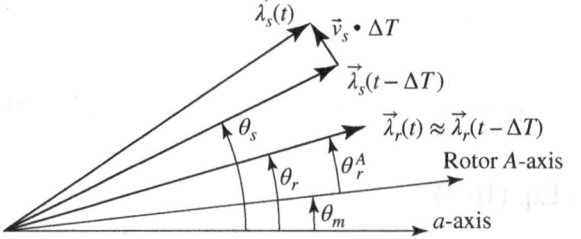

Fig. 16-2 Changing the position of stator flux-linkage vector.

time (in amplitude and in phase angle θ_r^A with respect to the rotor A-axis) only due to a small voltage drop across the rotor resistance.

16-4-3 Calculation of the Electromagnetic Torque T_{em}

The electromagnetic torque developed by the motor can be estimated in terms of the stator flux and the stator current, or in terms of the stator flux and the rotor flux. We will derive both expressions in Appendix 16.A; the final expressions that we need are given below.

Torque depends on the magnitude of the stator and the rotor fluxes and the angle between the two space vectors. As derived in Appendix 16.A, in terms of the machine leakage inductance L_σ (also defined in Appendix 16.A),

$$T_{em} = \frac{2p}{3}\frac{L_m}{2 L_\sigma^2}\hat{\lambda}_s \hat{\lambda}_r \sin\theta_{sr} \qquad (16\text{-}10a)$$

where

$$\theta_{sr} = \theta_s - \theta_r \qquad (16\text{-}10b)$$

The angles in Eq. (16-10) are expressed in electrical radians with respect to the stator a-axis, as shown in Fig. 16-2.

Equation (16-2) and Fig. 16-2, show that the torque can be controlled quickly by rapidly changing the position of the stator flux-linkage space vector (that is θ_s, hence θ_{sr}), by applying an appropriate voltage space vector during the sampling interval ΔT, while the rotor flux space vector position, $\theta_r\left(=\theta_r^A+\theta_m\right)$ changes relatively slowly.

CALCULATION OF $\vec{\lambda}_s$, $\vec{\lambda}_r$, T_{em}, AND ω_m

Thus, in accordance with Eq. (16-10a), a change in θ_{sr} results in the desired change in torque.

For torque estimation, it is better to use the expression below (derived in Appendix 16.A) in terms of the estimated stator flux linkage and the measured stator currents:

$$T_{em} = \frac{2}{3}\frac{p}{2}\operatorname{Im}\left(\vec{\lambda}_s^{conj}\vec{i}_s\right) \quad (16\text{-}11)$$

which, unlike the expression in Eq. (16-10a), does not depend on the rotor flux linkage (note that the rotor flux linkage in Eq. (16-8) depends on correct estimates of L_s, L_r, and L_m).

16-4-4 Calculation of the Rotor Speed ω_m

A much slower sampling rate with a sampling interval ΔT_ω, for example, equal to 1 ms, may be used for estimating the rotor speed. The speed of the rotor flux in electrical rad/s is calculated from the phase angle of the rotor flux space vector in Eq. (16-8) as follows:

$$\omega_r = \frac{d}{dt}\theta_r = \frac{\theta_r(t) - \theta_r(t - \Delta T_\omega)}{\Delta T_\omega} \quad (16\text{-}12)$$

The slip speed is calculated as follows: in Chapter 5, the torque and the speed expressions are given by Eqs. (5-7) and (5-5), where in the motor model, the d-axis is aligned with the rotor flux-linkage space vector. These equations are repeated below:

$$T_{em} = \frac{p}{2}\lambda_{rd}\left(\frac{L_m}{L_r}i_{sq}\right) \quad (16\text{-}13)$$

and

$$\omega_{slip} = R_r\frac{1}{\lambda_{rd}}\left(\frac{L_m}{L_r}i_{sq}\right) \quad (16\text{-}14)$$

where ω_{slip} is the slip speed, the same as ω_{dA} in Eq. (5-5). Calculating i_{sq} from Eq. (16-13) and substituting it into Eq. (16-14) (and recognizing that, in the model with the d-axis aligned with the rotor flux linkage, $\lambda_{rd} = \sqrt{2/3}\hat{\lambda}_r$), the slip speed in electrical radians per second is

$$\omega_{slip} = \frac{2}{p}\left(\frac{3}{2}R_r \frac{T_{em}}{\hat{\lambda}_r^2}\right) \quad (16\text{-}15)$$

Therefore, the rotor speed can be estimated from Eqs. (16-12) and (16-15) as

$$\omega_m = \omega_r - \omega_{slip} \quad (16\text{-}16)$$

where all speeds are in electrical radians per second. In a multipole machine with $p \geq 2$,

$$\omega_{mech} = (2/p)\,\omega_m \quad (16\text{-}17)$$

16-5 CALCULATION OF THE STATOR VOLTAGE SPACE VECTOR

A common technique in DTC is to control the torque and the stator flux amplitude with a hysteretic band around their desired values. Therefore, at a sampling time (with a sampling interval of ΔT), the decision to change the voltage space vector is implemented only if the torque and/or the stator flux amplitude are outside their range. The selection of the new voltage vector depends on the signs of the torque and the flux errors and the sector in which the stator flux-linkage vector lies, as explained below.

The plane of the stator voltage space vector is divided into six sectors, as shown in Fig. 16-3. We should note that these sectors are different from those defined for the stator voltage space vector-PWM in Chapter 7. The central vectors for each sector, which lie in the middle of a sector, are the basic inverter vectors as shown in Fig. 16-3.

The choice of the voltage space vector for sector 1 is explained below with the help of Fig. 16-4 and Eq. (16-10). Assuming that

CALCULATION OF THE STATOR VOLTAGE SPACE VECTOR

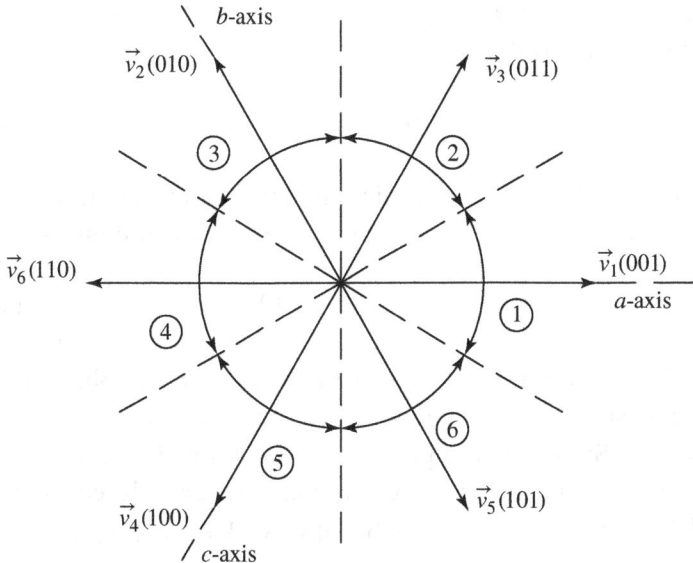

Fig. 16-3 Inverter basic vectors and sectors.

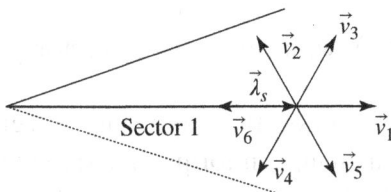

Fig. 16-4 Stator voltage vector selection in sector 1.

the stator flux-linkage space vector is along the central vector, the roles of various voltage vectors can be tabulated as follows:

There are some additional observations: The voltage vectors would have the same effects as tabulated above, provided the stator flux-linkage space vector is anywhere in sector 1. The use of voltage vectors $\vec{v}_1$ and $\vec{v}_6$ is avoided because their effect depends on where the stator flux-linkage vector is in sector 1. A similar table can be generated for all other sectors.

Use of zero vectors $\vec{v}_0(000)$ and $\vec{v}_7(111)$, results in the stator flux-linkage vector essentially unchanged in amplitude and in the angular position θ_s. In the torque expression of Eq. (16-10b), for small values of θ_{sr} in electrical radians,

$$\sin\theta_{sr} \approx (\theta_s - \theta_r) \qquad (16\text{-}18)$$

With a zero voltage vector applied, assuming that the amplitudes of the stator and the rotor flux-linkage vectors remain constant,

$$T_{em} \simeq k(\theta_s - \theta_r) \qquad (16\text{-}19)$$

where k is a constant. With the zero voltage vector applied, the position of the stator flux-linkage vector remains essentially constant, thus $\Delta\theta_s \simeq 0$. Similarly, the position of the rotor flux-linkage vector, with respect to the rotor A-axis, remains essentially constant, that is, $\Delta\theta_r^A \simeq 0$. However, as can be observed from Fig. 16-2, $\Delta\theta_r = \Delta\theta_m + \Delta\theta_r^A$. Therefore, the position of the rotor flux-linkage vector changes, albeit slowly, and the change in torque in Eq. (16-19) can be expressed as

$$\Delta T_{em} \simeq -k(\Delta\theta_m) \text{ (with zero voltage vector applied)} \qquad (16\text{-}20)$$

Equation (16-20) shows that applying a zero stator voltage space vector causes a change in torque in a direction opposite to that of ω_m. Therefore, with the rotor rating in a positive (counterclockwise) direction, for example, it may be preferable to apply a zero voltage vector to decrease torque in order to keep it with a hysteretic band.

In literature, there is no uniformity on the logic of space vector selection to keep the stator flux amplitude and the electromagnetic torque within their respective hysteretic bands. One choice of space vectors is illustrated by means of Example 16-1 below.

EXAMPLE 16-1

Implement simulation of encoder-less DTC for the "test" induction motor described after the-end-of-chapter.

CALCULATION OF THE STATOR VOLTAGE SPACE VECTOR

It is initially operating in steady state. At time $t = 0.4$ s, a load-torque disturbance causes it to reduce to one-half of its initial value. The objective is to keep the load speed constant at its initial value. The crossover-frequency for the speed controller is 25 rad/s and the phase margin is 60°.

Simulate the above system using Simulink, assuming the following additional parameters:

Maximum simulation time $T_{max} = 1.0$ s.
Sampling time in the stator voltage vector selection portion $\Delta T = 25$ ms.
Sampling time in the speed estimation portion $\Delta T_w = 1$ ms.
Hysteresis band for torque $\Delta T_{em} = 0.05 \times T_{em}(0)$.
Hysteresis band for flux $\Delta \hat{\lambda}_s = 0.005 \times \hat{\lambda}_s(0)$.
dc-bus voltage of the inverter $V_d = 1000$ V.
Switching frequency $f_s = 10$ kHz.
Triangular waveform peak $V_{tri} = 5$ V.

Note that tables in Simulink have zero-based indices. However, the tables in MATLAB have unity-based indices.

Selection of Stator Voltage Vector: To increase torque in the torque controller, a nonzero voltage vector is applied whose selection is based on the output of the flux controller (see Table 9-1). However, a zero voltage vector (v_0 or v_7) is applied to decrease torque, unless the estimated torque exceeds the reference torque by $2^* \Delta T_{em}$, in which case, a nonzero voltage vector is applied to decrease torque, taking the output of the flux controller into account (see Table 16-1). If the torque controller determines to apply a zero voltage vector, this is done regardless of the output of the flux controller.

Plot various results such as:

1. torque (reference and estimated)
2. speed (reference, actual, and estimated)
3. stator flux (reference and estimated)

(Continued)

TABLE 16-1 Effect of Voltage Vector on the Stator Flux-Linkage Vector in Sector 1

$\vec{v}_s$	T_{em}	$\hat{\lambda}_s$
$\vec{v}_3$	Increase	Increase
$\vec{v}_2$	Increase	Decrease
$\vec{v}_4$	Decrease	Decrease
$\vec{v}_5$	Decrease	Increase

Solution

Figure 16-5 shows the top-level diagram of the system simulation. Torque waveforms are shown in Fig. 16-6. Figure 16-7 shows the speed waveforms. Figure 16-8 shows the stator flux.

16-6 DIRECT TORQUE CONTROL USING dq-AXES

It is possible to perform the same type of control by aligning the d-axis with the stator flux-linkage vector. The amplitude of the stator flux-linkage vector is controlled by applying v_{sd} along the d-axis, and the torque is controlled by applying v_{sq} along the q-axis. The advantage of this type is control over the hysteretic control described earlier is that it results in a constant switching frequency. This is described by Example 16-2 on the accompanying website.

The procedure to implement the same in a hardware prototype is given in [1].

16-7 SUMMARY

This chapter discusses the DTC scheme where, unlike the vector control, no dq-axis transformation is needed, and the electromagnetic torque and the stator flux are estimated and directly controlled by applying the appropriate stator voltage vector. It is possible to

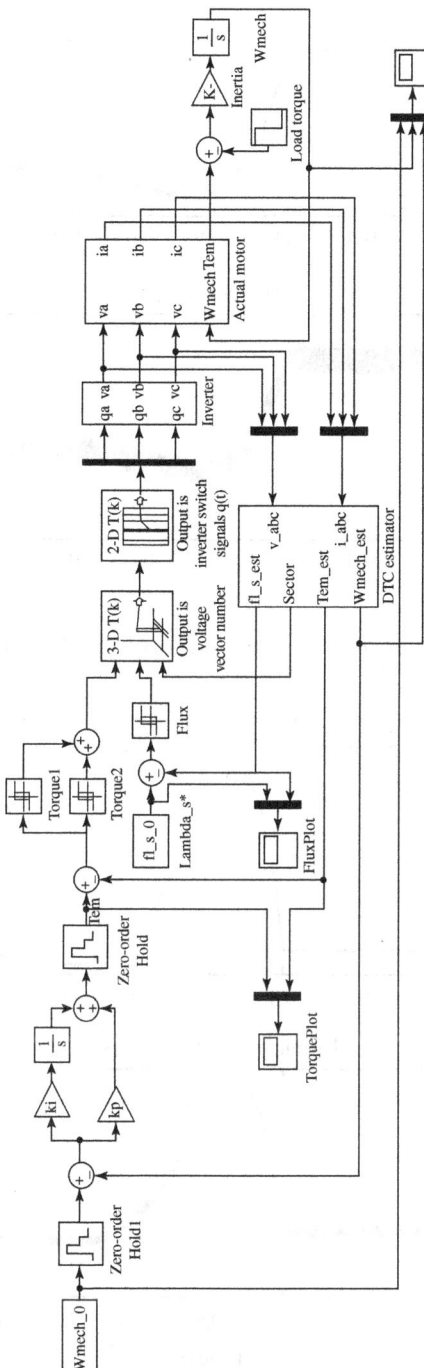

Fig. 16-5 Simulation of Example 16-1.

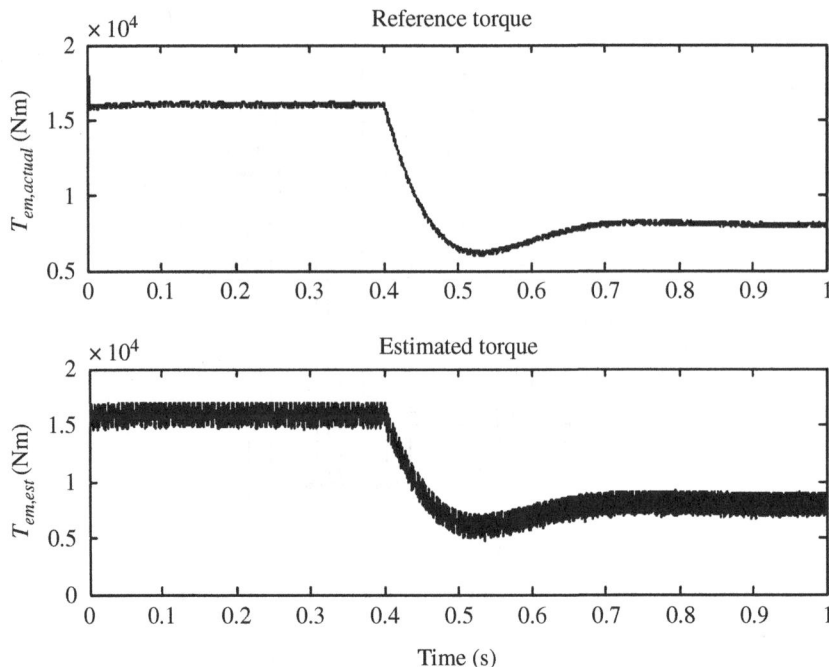

Fig. 16-6 Reference and estimated torque of Example 16-1.

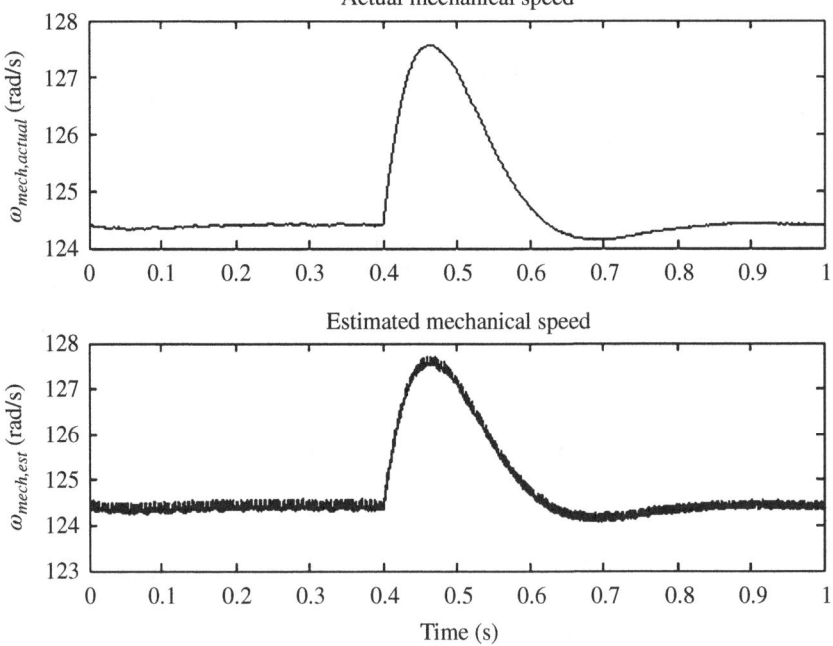

Fig. 16-7 Actual and estimated motor speed of Example 16-1.

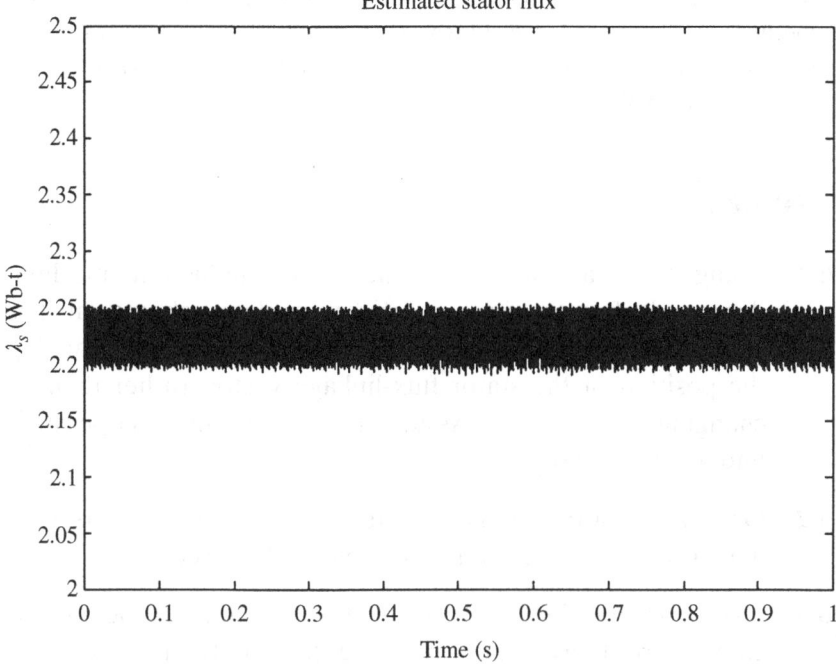

Fig. 16-8 Estimated stator flux of Example 16-1.

estimate the rotor speed, thus eliminating the need for rotor speed encoder.

REFERENCE

1. https://sciamble.com/Resources/pe-drives-lab/advanced-drives/im-dtc.

FURTHER READING

Depenbrock, M. (1988). Direct self control (DSC) of inverter-fed induction machines. *IEEE Transactions on Power Electronics* 13 (3): 420–429.

Takahashi, I. and Noguchi, T. (1986). A new quick response and high efficiency strategy of an induction motor. *IEEE Transactions on Industry Applications* 22 (7): 820–827.

Tiitinen, P. and Surendra, M. (1996). The next generation motor control method, DTC direct torque control. *Proceedings of International Conference on Power Electronics, Drives and Energy Systems, PEDES'96 New Delhi (India)*, vol. 1, 37–43.

PROBLEMS

16-1 Using the parameters of the "test" induction machine described after the-end-of-chapter problems, show that it is much faster to change electromagnetic torque by changing the position of the stator flux-linkage vector, rather than by changing its amplitude. Assume that the machine is operating under rated conditions.

16-2 Obtain the stator voltage vectors needed in other sectors, similar to what has been done in Table 16-1 for sector 1.

16-3 Assuming that the "test" machine described after the-end-of-chapter problems is operating under the rated conditions, compute the effect of applying a zero voltage space vector on the flux-linkage space vectors and on the electromagnetic torque produced.

16-4 In the system of Example 16-1, how can the modeling be simplified if the speed is never required to reverse.

16-5 Experiment with other schemes for selecting voltage space vector and compare results with that in Example 16-1.

16-6 In Example 16-1, include the field-weakening mode of operation.

TEST MACHINE

The parameters for a 1.5 MW wind-turbine induction machine are as follows:

Power	1.5 MW
Voltage	575 V (L-L, rms)

Frequency	60 Hz
Phases	3
Number of poles	6
Full-load slip	1%
Moment of inertia	75 kg·m²

Per-Phase Circuit Parameters:

$$R_s = 0.0014 \, \Omega$$

$$R_r = 0.000991 \, \Omega$$

$$X_{\ell s} = 0.034 \, \Omega$$

$$X_{\ell r} = 0.031 \, \Omega$$

$$X_m = 0.0576 \, \Omega$$

16-A APPENDIX

16-A-1 Derivation of Torque Expressions

The electromagnetic torque in terms of the stator flux and the stator current can be expressed as follows:

$$T_{em} = \frac{2}{3}\frac{p}{2} \operatorname{Im}\left(\vec{\lambda}_s^{conj} \vec{i}_s \right) \qquad (16\text{-}21)$$

To derive the above expression, it is easiest to assume it be correct and substitute the components to prove it. Taking the complex conjugate on both sides of Eq. (16-3),

$$\vec{\lambda}_s^{conj} = L_s \vec{i}_s^{conj} + L_m \vec{i}_r^{conj} \qquad (16\text{-}22)$$

Substituting in Eq. (16-21),

$$T_{em} = \frac{2p}{32}\left\{ \underbrace{L_s \text{Im}\left(\vec{i}_s^{conj}\vec{i}_s\right)}_{(=0)} + L_m \text{Im}\left(\vec{i}_r^{conj}\vec{i}_s\right)\right\} = \frac{p}{2}L_m \text{Im}\left(\vec{i}_r^{conj}\vec{i}_s\right)$$

(16-23)

Even although the *dq* transformation is not used in DTC, we can make use of *dq* transformations to prove our expressions. Therefore, in terms of an arbitrary *dq* reference set and the corresponding components substituted in Eq. (16-23),

$$T_{em} = \frac{2p}{32}L_m \text{Im}\left\{\left(i_{rd} - ji_{rq}\right)\left(i_{sd} + ji_{sq}\right)\right\} = \frac{p}{2}L_m\left(i_{sq}i_{rd} - i_{sd}i_{rq}\right)$$

(16-24)

which is identical to Eq. (12-47), thus proving the torque expression of Eq. (16-21) to be correct.

Another torque expression, which we will not use directly but which is the basis on which the selection of the stator voltage vector is made, is as follows:

$$T_{em} = \frac{2p}{32}\frac{L_m}{L_\sigma^2}\text{Im}\left(\vec{\lambda}_s \vec{\lambda}_r^{conj}\right)$$

(16-25)

where the machine leakage inductance is defined as

$$L_\sigma = \sqrt{L_s L_r - L_m^2}$$

(16-26)

Again, assuming the above expression in Eq. (16-25) to be correct, and substituting the expressions for the fluxes from Eqs. (16-3) and (16-4),

APPENDIX 471

$$T_{em} = \frac{2}{3}\frac{p}{2}\frac{L_m}{L_\sigma^2}\text{Im}\left\{\left(L_s\vec{i}_s + L_m\vec{i}_r\right)\left(L_r\vec{i}_r^{conj} + L_m\vec{i}_s^{conj}\right)\right\}$$

$$= \frac{2}{3}\frac{p}{2}\frac{L_m}{L_\sigma^2}L_sL_r\text{Im}\left(\vec{i}_s\vec{i}_r^{conj}\right) + \frac{2}{3}\frac{p}{2}\frac{L_m}{L_\sigma^2}L_m^2\text{Im}\left(\vec{i}_r\vec{i}_s^{conj}\right) \quad (16\text{-}27)$$

Note that $\text{Im}\left(\vec{i}_r\vec{i}_s^{conj}\right) = -\text{Im}\left(\vec{i}_s\vec{i}_r^{conj}\right)$. Therefore, in Eq. (16-26),

$$T_{em} = \frac{2}{3}\frac{p}{2}\frac{L_m}{L_\sigma^2}\underbrace{(L_sL_r - L_m^2)}_{(=L_\sigma^2)}\text{Im}\left(\vec{i}_s\vec{i}_r^{conj}\right) = \frac{2}{3}\frac{p}{2}L_m\text{Im}\left(\vec{i}_s\vec{i}_r^{conj}\right)$$

$$= \frac{p}{2}L_m\text{Im}\left\{\left(i_{sd} + ji_{sq}\right)\left(i_{rd} - ji_{rq}\right)\right\}$$

$$= \frac{p}{2}L_m\left(i_{sq}i_{rd} - i_{sd}i_{rq}\right)$$

(16-28)

which is identical to Eq. (12-47), thus proving the torque expression of Eq. (16-25) to be correct.

In Eq. (16-25), expressing flux linkages in their polar form:

$$T_{em} = \frac{2}{3}\frac{p}{2}\frac{L_m}{L_\sigma^2}\text{Im}\left(\hat{\lambda}_s e^{j\theta_s} \cdot \hat{\lambda}_r e^{-j\theta_r}\right) = \frac{2}{3}\frac{p}{2}\frac{L_m}{L_\sigma^2}\hat{\lambda}_s\hat{\lambda}_r\text{Im}\left(e^{j\theta_{sr}}\right)$$

$$= \frac{2}{3}\frac{p}{2}\frac{L_m}{L_\sigma^2}\hat{\lambda}_s\hat{\lambda}_r \sin\theta_{sr}$$

(16-29)

where

$$\theta_{sr} = \theta_s - \theta_r \quad (16\text{-}30)$$

is the angle between the two flux-linkage space vectors.

17 Vector Control of Permanent-Magnet Synchronous Motor Drives

17-1 INTRODUCTION

In Chapter 8, we looked at permanent-magnet synchronous-motor drives, also known as "brushless-dc motor" drives in steady state; without the help of dq-analysis, it was not possible to discuss dynamic control of such drives. In this chapter, we will make use of the dq-analysis of induction machines, which is easily extended to analyze and control synchronous machines.

17-2 dq-ANALYSIS OF PERMANENT-MAGNET SYNCHRONOUS MACHINES

Permanent-magnet synchronous motors are commonly classified under two categories, namely, surface permanent magnet (SPM) and interior permanent magnet (IPM) synchronous motor. In SPM, the magnets are attached to the surface of the rotor, as shown in Fig. 17-1a, while they are buried inside the rotor in IPM, as shown in Fig. 17-1b.

(Adapted from chapter 10 of *Advanced Electric Drives: Analysis, Control, and Modeling Using MATLAB/Simulink* ISBN: 978-1-118-48548-4 by Ned Mohan, August 2014)

Analysis and Control of Electric Drives: Simulations and Laboratory Implementation, First Edition. Ned Mohan and Siddharth Raju.
© 2021 John Wiley & Sons, Inc. Published 2021 by John Wiley & Sons, Inc.
Companion website: www.wiley.com/go/Mohan/Vectorcontrolinelectricdrives

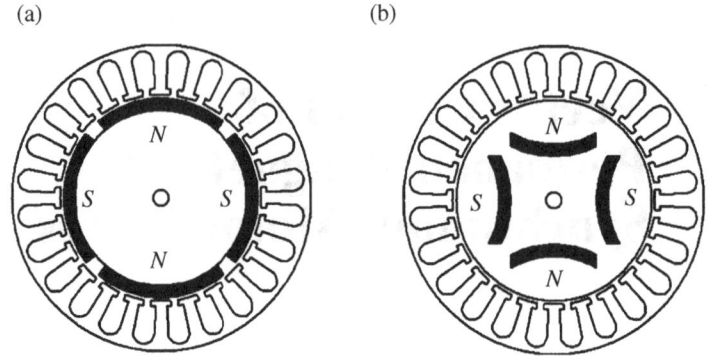

Fig. 17-1 Permanent-magnet synchronous motor rotor structure.

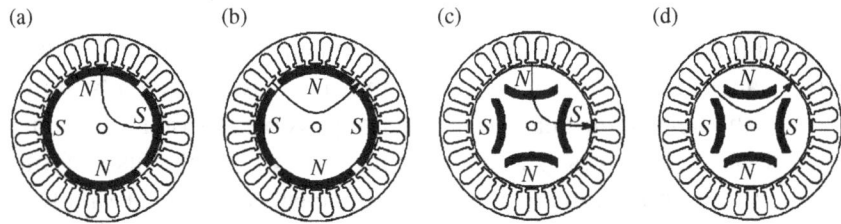

Fig. 17-2 Reluctance variation of permanent-magnet synchronous machine.

Since the magnetic permeability of permanent magnet material is close to that of air, the rotor can be considered magnetically round (non-salient) in synchronous motors with surface-mounted permanent magnets. As shown in Fig. 17-2a and b, the reluctance along any axis through the center of the machine is the same. In synchronous motors with buried permanent magnets, the reluctance along the magnetic axis, as shown in Fig. 17-2c, is higher than the reluctance along the electrically perpendicular axis, as shown in Fig. 17-2d.

A simplified representation of the rotor magnets is shown in Fig. 17-3a. The three-phase stator windings are sinusoidally distributed in space, like in an induction machine, with N_s number of turns per phase. In Fig. 17-3b, d-axis is always aligned with the rotor magnetic axis, with the q-axis 90° ahead in the direction of rotation, assumed to be counterclockwise. The stator three-phase windings are represented by equivalent d- and q-axis windings; each winding has $\sqrt{3/2}N_s$ turns that are sinusoidally distributed.

dq-ANALYSIS OF PERMANENT-MAGNET   475

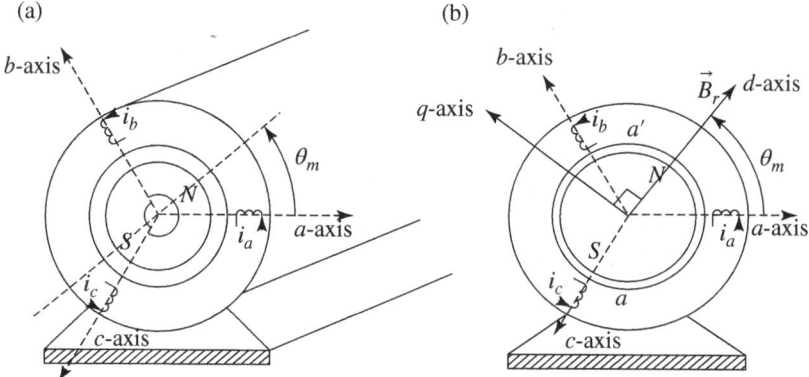

Fig. 17-3 Permanent-magnet synchronous machine (shown with $p = 2$).

17-2-1   Flux Linkages

The stator d- and q-winding flux linkages can be expressed as follows:

$$\lambda_{sd} = L_{sd}i_{sd} + \lambda_{fd} \qquad (17\text{-}1)$$

and

$$\lambda_{sq} = L_{sq}i_{sq} \qquad (17\text{-}2)$$

where in Eqs. (17-1) and (17-2), L_{sd} and L_{sq} are the stator inductance along the d-axis and q-axis, respectively, and λ_{fd} is the flux linkage of the stator d-winding due to flux produced by the rotor magnets (recognizing that the d-axis is always aligned with the rotor magnetic axis).

17-2-2   Stator dq-Winding Voltages

Using Eqs. (12-28) and (12-29) developed for induction machines, in dq windings,

$$v_{sd} = R_s i_{sd} + \frac{d}{dt}\lambda_{sd} - \omega_d \lambda_{sq} \qquad (17\text{-}3)$$

and

$$v_{sq} = R_s i_{sq} + \frac{d}{dt}\lambda_{sq} + \omega_d \lambda_{sd} \qquad (17\text{-}4)$$

where the speed of the equivalent dq windings is $\omega_d = \omega_m$ (in electrical rad/s) in order to keep the d-axis always aligned with the rotor magnetic axis [1]. The speed ω_m is related to the actual rotor speed ω_{mech} as

$$\omega_m = \frac{p}{2}\omega_{mech} \qquad (17\text{-}5)$$

17-2-3 Electromagnetic Torque

Using the analysis for induction machines in Chapter 12, and Eqs. (12-46) and (12-47), we can derive the following equation (see Problem 12-9a), which is also valid for synchronous machines:

$$T_{em} = \frac{p}{2}\left(\lambda_{sd} i_{sq} - \lambda_{sq} i_{sd}\right) \qquad (17\text{-}6)$$

Substituting for flux linkages in the above equation,

$$T_{em} = \frac{p}{2}\left(\lambda_{fd} - \left(L_{sq} - L_{sd}\right)i_{sd}\right)i_{sq} \qquad (17\text{-}7)$$

17-2-4 Electrodynamics

The acceleration is determined by the difference of the electromagnetic torque and the load torque (including friction torque) acting on J_{eq}, the combined inertia of the load and the motor,

$$\frac{d}{dt}\omega_{mech} = \frac{T_{em} - T_L}{J_{eq}} \qquad (17\text{-}8)$$

where ω_{mech} is in radiance per second and is related to ω_m as shown in Eq. (17-5).

17-3 NON-SALIENT POLE SYNCHRONOUS MACHINES

For a non-salient pole machine such as a SPM synchronous motor where $L_{sd} = L_{sq} = L_s$, the electromagnetic torque in the above equation can be rewritten as

$$T_{em} = \frac{p}{2}\lambda_{fd}i_{sq} \tag{17-9}$$

17-3-1 Relationship Between the *dq* Circuits and the Per-Phase Phasor-Domain Equivalent Circuit in Balanced Sinusoidal Steady State

In this section, we will see that under a balanced sinusoidal steady-state condition, the two *dq*-winding equivalent circuits combine to result in the per-phase equivalent circuit of a synchronous machine, that we have derived in the previous course. Note that in a synchronous motor used in a "brush-less dc" drive, the synchronous speed equals the rotor speed on an instantaneous basis. Therefore, our choice of $\omega_d = \omega_m$ also results in $\omega_d = \omega_m = \omega_{syn}$. Under a balanced sinusoidal steady-state condition, *dq*-winding quantities are dc, and their time derivatives are zero. In Eqs. (17-3) and (17-4) for stator voltages, substituting flux linkages from Eqs. (17-1) and (17-2) results in

$$v_{sd} = R_s i_{sd} - \omega_m L_s i_{sq} \tag{17-10}$$

and

$$v_{sq} = R_s i_{sq} + \omega_m L_s i_{sd} + \omega_m \lambda_{fd} \tag{17-11}$$

Multiplying both sides of Eq. (17-11) by (j) and adding to Eq. (17-10) (and multiplying both sides of the resulting equation by $\sqrt{3/2}$) leads to the following space vector equation, with the *d*-axis as the reference axis:

$$\vec{v}_s = R_s \vec{i}_s + j\omega_m L_s \vec{i}_s + \underbrace{j\sqrt{3/2}\omega_m \lambda_{fd}}_{\vec{e}_{fs}} \tag{17-12}$$

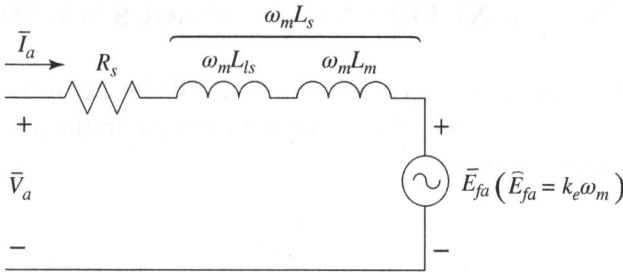

Fig. 17-4 Per-phase equivalent circuit in steady state (ω_m in elec. rad/s).

noting that $\vec{v}_s = \sqrt{3/2}\left(v_{sd} + jv_{sq}\right)$ and so on. Dividing both sides of the above space vector equation by 3/2, we obtain the following phasor equation for phase a in a balanced sinusoidal steady state:

$$\overline{V}_a = R_s \overline{I}_a + j\omega_m L_s \overline{I}_a + \underbrace{j\omega_m \sqrt{\frac{2}{3}}\lambda_{fd}}_{\overline{E}_{fa}} \qquad (17\text{-}13)$$

The above equation corresponds to the per-phase equivalent circuit of Fig. 17-4, that was derived in the previous course under a balanced sinusoidal steady-state condition.

Relationship Between k_E and λ_{fd} From Eq. (17-13),

$$\hat{E}_{fa} = \underbrace{\sqrt{\frac{2}{3}}\lambda_{fd}}_{k_E} \omega_m = k_E \omega_m \qquad (17\text{-}14)$$

Therefore,

$$k_E = \sqrt{\frac{2}{3}}\lambda_{fd} \qquad (17\text{-}15)$$

17-3-2 *dq*-Based Dynamic Controller for "Brush-less dc" Drives

In the previous section, in the absence of the *dq*-analysis, a hysteretic converter was used, where the switching frequency did not remain

NON-SALIENT POLE SYNCHRONOUS MACHINES   479

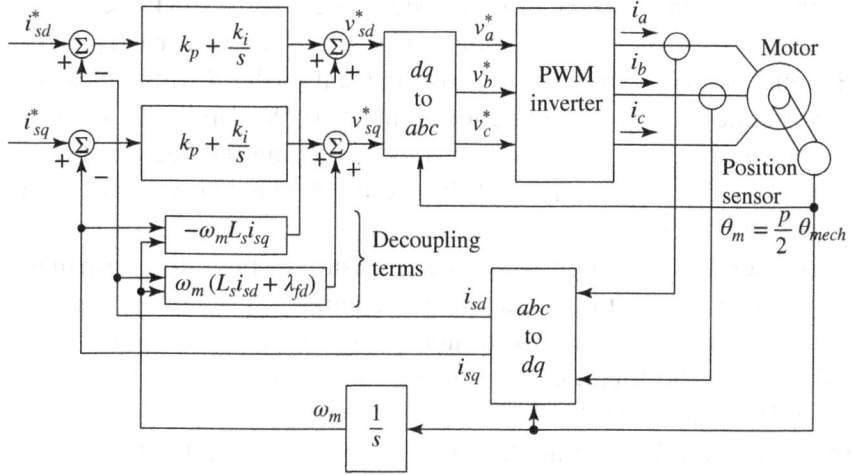

Fig. 17-5   Controller in the dq reference frame.

constant. In this section, we will see that it is possible to use a converter with a constant switching frequency. The block diagram of such a control system is shown in Fig. 17-5. In Eqs. (17-3) and (17-4), using the flux linkages of Eqs. (17-1) and (17-2), the voltage can be expressed as follows, recognizing that the time-derivative of the rotor-produced flux λ_{fd} is zero:

$$v_{sd} = R_s i_{sd} + L_s \frac{d}{dt} i_{sd} + \underbrace{\left(-\omega_m L_s i_{sq}\right)}_{comp_d} \quad (17\text{-}16)$$

and

$$v_{sq} = R_s i_{sq} + L_s \frac{d}{dt} i_{sq} + \underbrace{\omega_m \left(L_s i_{sd} + \lambda_{fd}\right)}_{comp_q} \quad (17\text{-}17)$$

In Fig. 17-5, the PI controllers in both channels are designed assuming that the inverter is ideal and the compensation (decoupling) terms in Eqs. (17-16) and (17-17) are utilized, to result in the desired phase margin at the chosen open-loop crossover frequency.

480 VECTOR CONTROL OF PERMANENT-MAGNET

Flux Weakening In the normal speed range below the rated speed, the reference for the d-winding current is kept at zero ($i_{ds} = 0$). Beyond the rated speed, a negative current in the d-winding causes flux weakening (a phenomenon similar to that in brush-type dc machines and induction machines), thus keeping the back-emf from exceeding the rated voltage of the motor. A negative value of i_{sd} in Eq. (17-11) causes v_{sq} to decrease.

To operate synchronous machines with surface-mounted permanent magnets at above the rated speed, requires a substantial negative d-winding current to keep the terminal voltage from exceeding its rated value. Since the total current into the stator cannot exceed its rated value in steady state, the higher the magnitude of the d-winding current, the lower the magnitude of the q-winding current has to be, since

$$\sqrt{|i_{sd}|^2 + |i_{sq}|^2} \leq \hat{I}_{dq,rated} \left(= \sqrt{\frac{3}{2}} \hat{I}_{a,rated} \right) \quad (17\text{-}18)$$

Hence in Eq. (17-7), a lower limit on i_{sq} in the flux-weakening mode results in a reduced torque capability.

EXAMPLE 17-1

For analyzing the performance of the dynamic control procedure, a motor from a commercial vendor catalog [2] is selected, whose specifications are as follows:
Nameplate Data:

Continuous stall torque	3.2 Nm
Continuous current	8.74 A
Peak torque	12.8 Nm
Peak current	31.5 A
Rated voltage	200 V
Rated speed	6000 RPM
Phases	3
Number of poles	4

Per-Phase Motor Circuit Parameters:
$R_s = 0.416\,\Omega$
$L_s = 1.365\,\text{mH}$

Voltage Constant k_E (as in Eq. (17-14) and Fig. 17-2): 0.0957 V/(elect. rad/s).
The total equivalent inertia of the system (motor-load combination) is

$$J_{eq} = 3.4 \times 10^{-4}\,\text{kg}\cdot\text{m}^2$$

Initially, the drive is operating in steady-state at its rated speed, supplying its rated torque of 3.2 Nm to the mechanical load connected to its shaft.

At time $t = 0.1$ s, a load-torque disturbance occurs, which causes it to suddenly decrease by 50% (there is no change in load inertia). The feedback control objective is to keep the shaft speed at its initial steady-state value after the load-torque disturbance. Design the speed feedback controller with the open-loop crossover frequency of 2500 rad/s and a phase margin of 60°. The open-loop crossover frequency of the internal current feedback loop is ten times higher than that of the speed loop, and the phase margin is 60°.

Solution

The simulation block diagram is shown in Fig. 17-6, and the Simulink file EX17_1.mdl is included on the accompanying website to this textbook. The simulation results are shown in Fig. 17-7.

17-4 SALIENT-POLE SYNCHRONOUS MACHINES

Since the magnets of an IPM motor are buried, they tend to be more rugged, especially at high speeds, than SPM motors which have the magnets attached to the rotor surface. In addition, synchronous machines with IPMs result in unequal reluctance along the d- and the q-axis. This saliency, due to difference in reluctance, enables

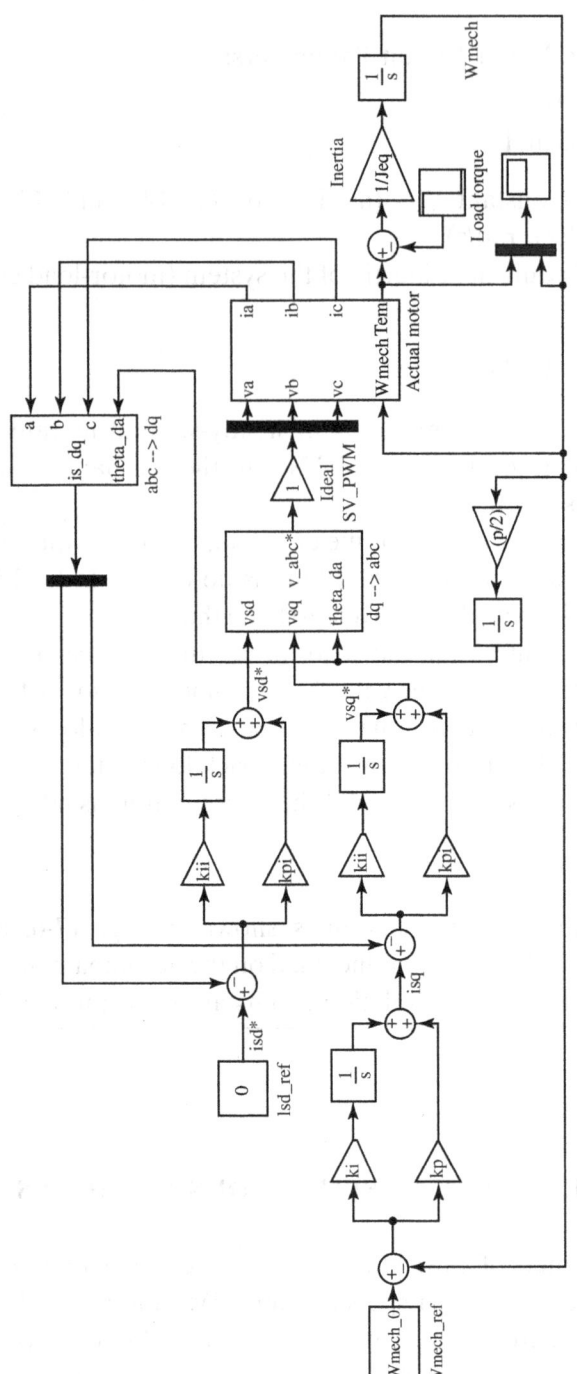

Fig. 17-6 Simulation of Example 17-1.

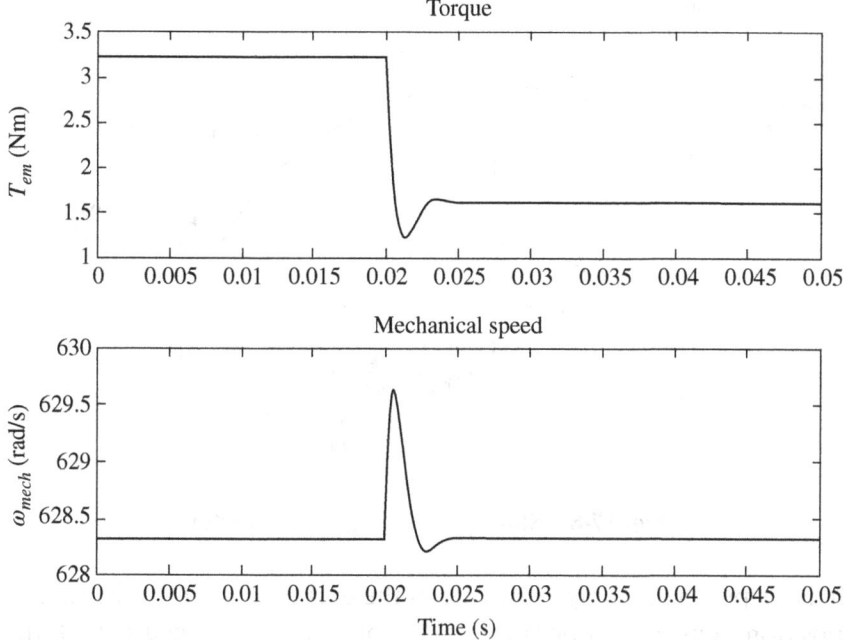

Fig. 17-7 Simulation results of Example 17-1.

IPM synchronous machines to generate reluctance torque in addition to magnet torque as given by Eq. (17-7).

In Chapter 14, speed-sensorless vector control of induction motor was presented based on an open-loop and MRAS speed estimator. Those methods can be easily extended to do speed-sensorless vector control for a permanent-magnet synchronous motor. Both these methods are highly sensitive to parameter variations, especially at low speeds, and are incapable of estimating rotor speed as the frequency of the applied voltage tends towards zero.

In this section, we shall explore an alternative method of speed estimation proposed in [3] that leverages the saliency of IPM motors. This method is insensitive to motor parameter variation and can estimate speed over a wide range, including at zero speed.

17-4-1 Rotor Position Estimation Using High-Frequency Injection

Consider at time t, the rotor is at an angle of θ_m from the stator a-axis, as shown in Fig. 17-8. The d-axis is aligned with the rotor magnetic

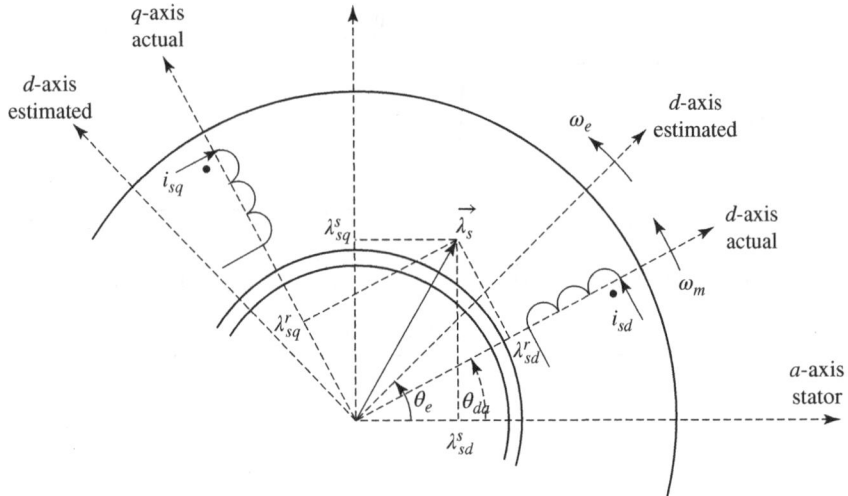

Fig. 17-8 Stator flux-linkage space vector.

axis, as per the assumption made in the earlier section. In the case of position sensorless operation, the objective is to ensure that the estimated d-axis position θ_e converges towards the actual rotor position θ_{da}.

The stator d- and q-winding flux linkage given by Eqs. (17-1) and (17-2) can be expressed as

$$\begin{bmatrix} \lambda_{sd}^r \\ \lambda_{sq}^r \end{bmatrix} = \begin{bmatrix} L_d & 0 \\ 0 & L_q \end{bmatrix} \begin{bmatrix} i_{sd}^r \\ i_{sq}^r \end{bmatrix} + \begin{bmatrix} \lambda_{fd} \\ 0 \end{bmatrix} \qquad (17\text{-}19)$$

In this section, a superscript s refers to stationary reference frame along stator a-axis, superscript r refers to the synchronously rotating reference frame aligned with the rotor magnetic axis, and superscript e refers to components along the estimated λ d- and q-axis.

Transforming the stator flux-linkage components, in the previous equation, along the stator a-axis,

$$\begin{bmatrix} \lambda_{sd}^s \\ \lambda_{sq}^s \end{bmatrix} = \begin{bmatrix} L - \Delta L \cos(2\theta_{da}) & -\Delta L \sin(2\theta_{da}) \\ -\Delta L \sin(2\theta_{da}) & L + \Delta L \cos(2\theta_{da}) \end{bmatrix} \begin{bmatrix} i_{sd}^s \\ i_{sq}^s \end{bmatrix} + \begin{bmatrix} \cos(\theta_{da}) \\ \sin(\theta_{da}) \end{bmatrix} \lambda_{fd}$$

$$(17\text{-}20)$$

where $L = \frac{L_{sd} + L_{sq}}{2}$ and $\Delta L = \frac{L_{sd} - L_{sq}}{2}$. The proof of this is given in Appendix 17.A.

Consider a flux-linkage vector of frequency ω_i and magnitude V_{si}/ω_i injected along the estimated q-axis,

$$\begin{bmatrix} \lambda_{sdi}^e \\ \lambda_{sqi}^e \end{bmatrix} = \frac{V_{si}}{\omega_i} \sin(\omega_i t) \begin{bmatrix} 0 \\ 1 \end{bmatrix} \qquad (17\text{-}21)$$

If the estimated d-axis was aligned with the actual d-axis, the injected flux-linkage vector would cause currents to flow in the q-axis alone. The presence of error in the estimated rotor angle θ_e, would lead to currents at the injected frequency to flow in the estimated d-axis as well. The estimated position can be made to follow the actual rotor position by designing a controller that drives the current at the injected frequency along the d-axis to zero, as detailed below. The frequency of the injected signal is chosen, such that it is sufficiently higher than the current controller bandwidth, so that the current controller does not actively correct the injected current.

Transforming the injected flux-linkage components in Eq. (17-21) to the stationary reference frame along stator a-axis,

$$\begin{bmatrix} \lambda_{sdi}^s \\ \lambda_{sqi}^s \end{bmatrix} = \frac{V_{si}}{\omega_i} \sin(\omega_i t) \begin{bmatrix} -\sin(\theta_e) \\ \cos(\theta_e) \end{bmatrix} \qquad (17\text{-}22)$$

Substituting Eq. (17-22) in Eq. (17-20) and considering only current components in the injected frequency,

$$\begin{bmatrix} i_{sdi}^s \\ i_{sqi}^s \end{bmatrix} = \frac{V_{si}}{\omega_i} \frac{1}{L^2 - \Delta L^2} \sin(\omega_i t) \begin{bmatrix} -L\sin(\theta_e) & \Delta L \sin(2\theta_{da} - \theta_e) \\ L\cos(\theta_e) & -\Delta L \cos(2\theta_{da} - \theta_e) \end{bmatrix} \qquad (17\text{-}23)$$

The stator current along the estimated d-axis is

$$i_{sdi}^e = i_{sdi}^s \cos(\theta_e) + i_{sqi}^s \sin(\theta_e) \qquad (17\text{-}24)$$

Substituting Eq. (17-23) in Eq. (17-24),

$$i_{sdi}^e = \frac{V_{si}}{\omega_i} \frac{\Delta L}{L^2 - \Delta L^2} \sin(\omega_i t) \sin(2(\theta_{da} - \theta_e)) = I_{si} \sin(\omega_i t) \sin(2(\theta_{da} - \theta_e))$$
(17-25)

From the above equation, a controller that minimizes i_{sdi}^e would cause θ_e to converge to θ_{da}. Multiplying the above equation by $\sin(\omega_i t)$ yields Eq. (17-26), which contains a dc component independent of the frequency of the injected signal and an ac component, which varies at twice the frequency of the injected signal.

$$i_{sdi}^e \sin(\omega_i t) = \frac{I_{si}}{2} \sin(2(\theta_{da} - \theta_e)) - \frac{I_{si}}{2} \cos(2\omega_i t) \sin(2(\theta_{da} - \theta_e))$$
(17-26)

The dc component can be obtained using a low-pass filter, which yields the error signal

$$e = \frac{I_{si}}{2} \sin(2(\theta_{da} - \theta_e))$$
(17-27)

The stator voltage that needs to be applied to generate the high-frequency flux linkage is given in Eq. (17-27), and is obtained by substituting Eq. (17-22) into Eqs. (17-3) and (17-4). The drop due to R_s can be ignored, as it is relatively small compared to the impedance due to stator inductances at high injected frequency.

$$\begin{bmatrix} V_{sdi}^e \\ V_{sqi}^e \end{bmatrix} = V_{si} \begin{bmatrix} \sin(\omega_i t) \\ -\frac{\omega_e}{\omega_i} \cos(\omega_i t) \end{bmatrix}$$
(17-28)

where $\omega_e = d\theta_e/dt$ is the speed of the estimated d-axis.

17-4-2 Speed-Sensorless Dynamic Controller for IPM Motor

The controller block diagram in Fig. 17-6 is modified, as shown in Fig. 17-9, to use the high-frequency injection-based estimator to

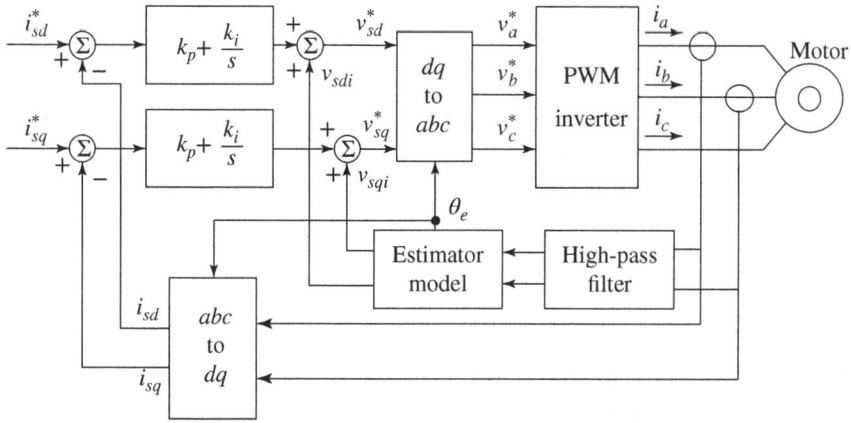

Fig. 17-9 IPM motor position estimator and controller.

estimate the rotor position instead of relying on a position sensor. For the sake of simplicity, the decompensation terms have been ignored but can be added to the output of the current controllers if desired.

The injection frequency current component in the measured current, $\tilde{i}^s_{sdi}$ and $\tilde{i}^s_{sdi}$, is obtained using a high-pass filter with a crossover frequency lower than the injection frequency. The injection frequency current components are transformed into the error signal using the method highlighted in Eqs. (17-24) through (17-27). The error signal is minimized using a PID controller, the output of which is fed to the mechanical model of the motor to obtain the rotor speed, as shown in Fig. 17-10.

It must be noted that even if the estimated motor mechanical parameters, rotor inertia J_{eq} and coefficient of friction B, in Fig. 17-10

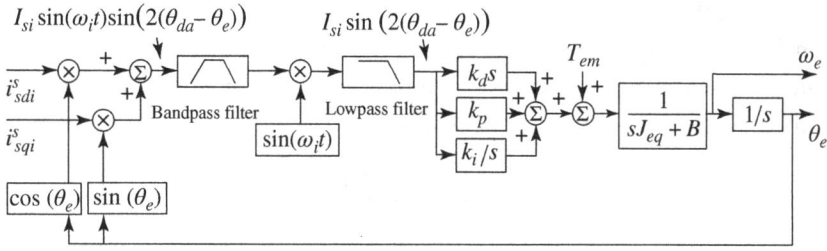

Fig. 17-10 Rotor position and speed estimation.

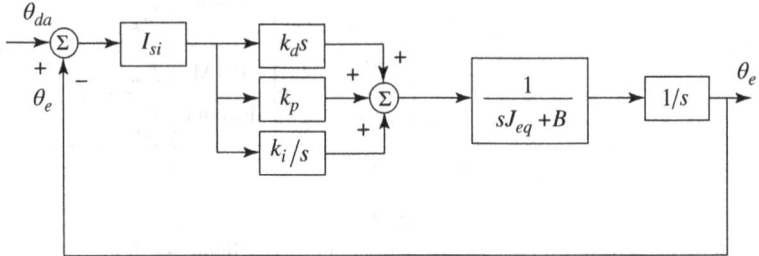

Fig. 17-11 Rotor position estimator linearized system transfer function.

are significantly different from the actual motor parameters, the estimated rotor position converges toward the actual rotor position. Only the transient response tracking is affected by any parameter mismatch.

17-4-3 Designing PID Controller

For a small perturbation in the rotor position $\Delta\theta_m$, the controller error from Eq. (17-27),

$$\Delta e = I_{si}(\Delta\theta_m - \Delta\theta_e) \qquad (17\text{-}29)$$

The system transfer function of the rotor position estimator under this condition is shown in Fig. 17-11. The method to design the controller is presented through an example.

EXAMPLE 17-2

Design a PID controller with a crossover frequency of $\omega_c = 100$ rad/s and phase margin $\phi_m = \pi/3$ for a motor similar to the one in Example 17-1, with the only difference being an IPM instead of an SPM motor with the following inductance:

$$L_d = 1.365 \text{ mH}$$
$$L_q = 2.72 \text{ mH}$$

Coefficient of viscous friction $B = 2 \times 10^{-4}$ N·m·s.
The magnitude of the injected voltage for position sensing is 20 V at a frequency of 1000 Hz.

Solution

From Eq. (17-25), the magnitude of the injected current is

$$I_{si} = \frac{V_{si}}{\omega_i} \frac{\Delta L}{L^2 - \Delta L^2} = \frac{20}{2\pi \times 1000} \frac{\left(\frac{2.72 - 1.365}{2}\right)}{\left(\frac{2.72 + 1.365}{2}\right)^2 - \left(\frac{2.72 - 1.365}{2}\right)^2} = 0.583 \text{ A}$$

From Fig. 17-11, the controller open-loop transfer function is

$$G(s) = I_{si} \frac{k_d s^2 + k_p s + k_i}{s} \frac{1}{Js + B} \frac{1}{s}$$

$$= 0.583 \times 10^4 \frac{k_d s^2 + k_p s + k_i}{s^2} \frac{1}{3.4s + 2}$$

$$\Rightarrow G(j\omega) = -0.583 \times 10^4 \frac{(k_p \omega)j + (k_i - k_d \omega^2)}{\omega^2} \frac{1}{(3.4\omega)j + 2}$$

$$\Rightarrow G(j\omega_c) = \left(-6.47 \times 10^{-9} + j6.91 \times 10^{-6}\right)$$
$$\times \left(628.3 k_p j + \left(k_i - 3.95 \times 10^5 k_d\right)\right)$$

At cross-over frequency

$$|G(j\omega_c)| = 1 \Rightarrow (628.3 k_p)^2 + \left(k_i - 3.95 \times 10^5 k_d\right)^2 = 1.45 \times 10^5$$

and

$$\angle G(j\omega_c) = -\pi + \phi_c \Rightarrow \frac{628.3 k_p}{k_i - 3.95 \times 10^5 k_d}$$

$$= \tan\left(-\pi + \frac{\pi}{3} + \tan^{-1}\left(\frac{-1.31 \times 10^{-4}}{4.65 \times 10^{-6}}\right)\right) = 0.5786$$

Solving the above two equations yields $k_p = 115.3$ and $k_i - 3.95 \times 10^5 k_d = -1.25 \times 10^5$.

(Continued)

From the above solution, k_i and k_d can assume a wide range of value. The only constraint is that the phase angle of the open-loop transfer function should at no point exceed $-\pi$.

$$\angle G(j\omega) = -\pi + \tan^{-1}\left(\frac{k_p\omega}{k_i - k_d\omega^2}\right) - \tan^{-1}\left(\frac{J\omega}{B}\right) > -\pi$$

$$\Rightarrow k_i - k_d\omega^2 < \frac{k_p B}{J}$$

For the above condition to hold true for all values of $\omega < \omega_c$,

$$k_i < k_p B / J$$

It would be preferable to maximize k_i, to bring the steady-state error to zero as quickly as possible, while also not exceeding the above constraints accounting for the possibility that measured motor parameters might not be accurate or changed over time. Thus, k_i is chosen to be half of the maximum possible value.

Solving the above equation yields $k_p = 115.3$, $k_i = 33.9$, and $k_d = 0.317$.

The PID controller in Fig. 17-11 consists of a differentiator which is not preferred in practical implementations in which measured signals tend to be noisy. This measured noise gets amplified by the differentiator. To avoid this, the controller in Fig. 17-11 is restructured, as shown in Fig. 17-12.

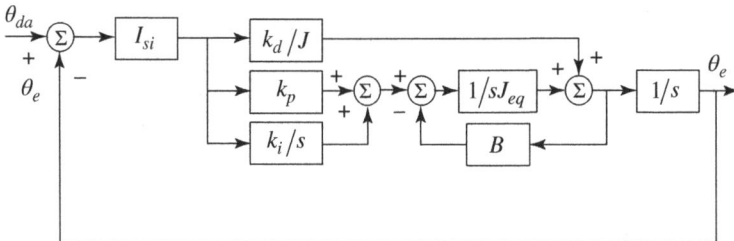

Fig. 17-12 Restructured rotor position estimator transfer function.

SALIENT-POLE SYNCHRONOUS MACHINES 491

EXAMPLE 17-3

Simulate the controller designed in Example 17-2 under the following condition. The motor starts from zero steady-state condition. The reference speed is stepped from 0 to 100 rad/s at time $t = 0.1$ s. The load torque is stepped from 0 to half the rated torque at time $t = 0.5$ s.

Solution

The simulation block diagram is shown in Fig. 17-13, and the Simulink file EX17_3.mdl is included on the accompanying website to this textbook. The simulation results are shown in Fig. 17-14.

17-4-4 Electromagnetic Torque

One of the advantages of IPM synchronous motors is the production of reactance torque in addition to the magnet torque as given by Eq. (17-7). Noting that $\vec{i}_s = \sqrt{3/2}\left(i_{sd} + ji_{sq}\right)$,

$$i_{sd} = \sqrt{\frac{2}{3}} I_s \cos(\delta) \qquad (17\text{-}30a)$$

and

$$i_{sq} = \sqrt{\frac{2}{3}} I_s \sin(\delta) \qquad (17\text{-}30b)$$

where $\vec{i}_s = I_s \angle \delta$.
Substituting Eq. (17-30) in Eq. (17-7),

$$\begin{aligned}T_{em} &= \frac{P}{2}\frac{2}{3}\left(\lambda_f I_s \sin(\delta) - (L_q - L_d)I_s^2 \sin(\delta)\cos(\delta)\right) \\ &= \frac{P}{2}\frac{2}{3} I_s\left(\lambda_f \sin(\delta) - (L_q - L_d)I_s \sin(2\delta)\right)\end{aligned} \qquad (17\text{-}31)$$

where $\lambda_f = \sqrt{\frac{3}{2}}\lambda_{fd}$.

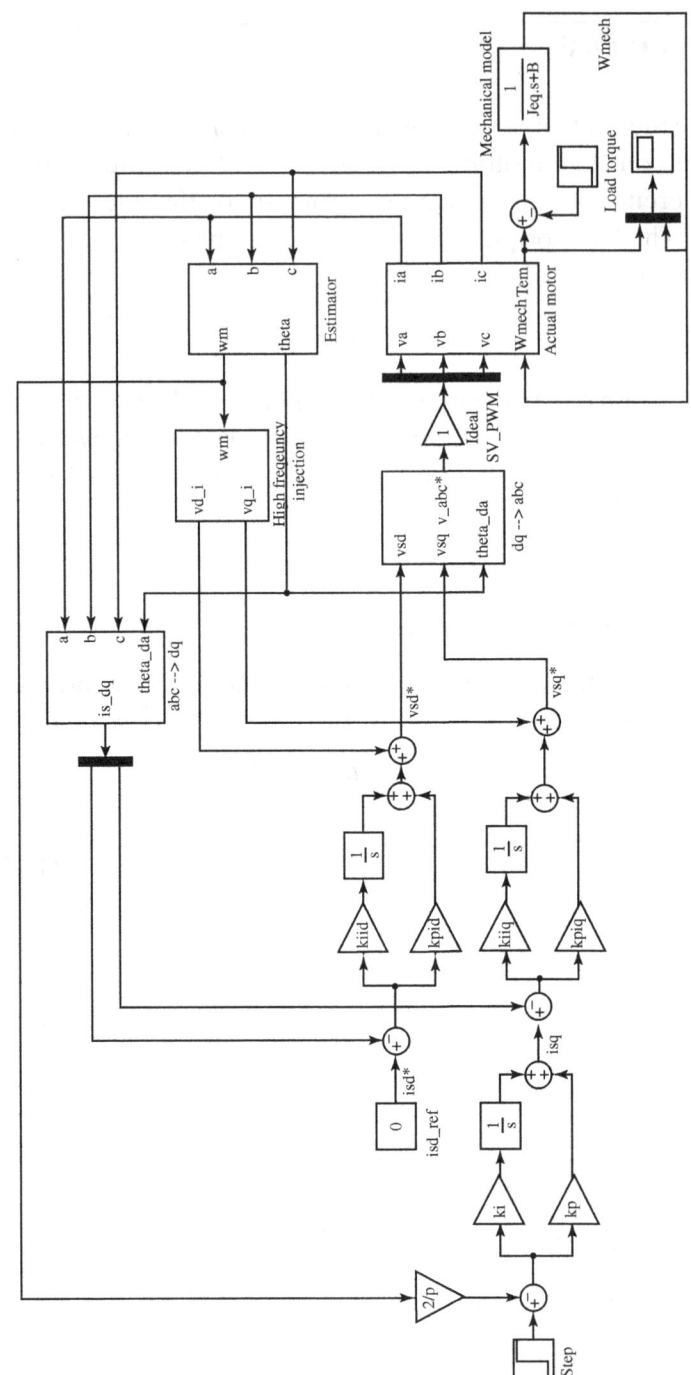

Fig. 17-13 Simulation of Example 17-3.

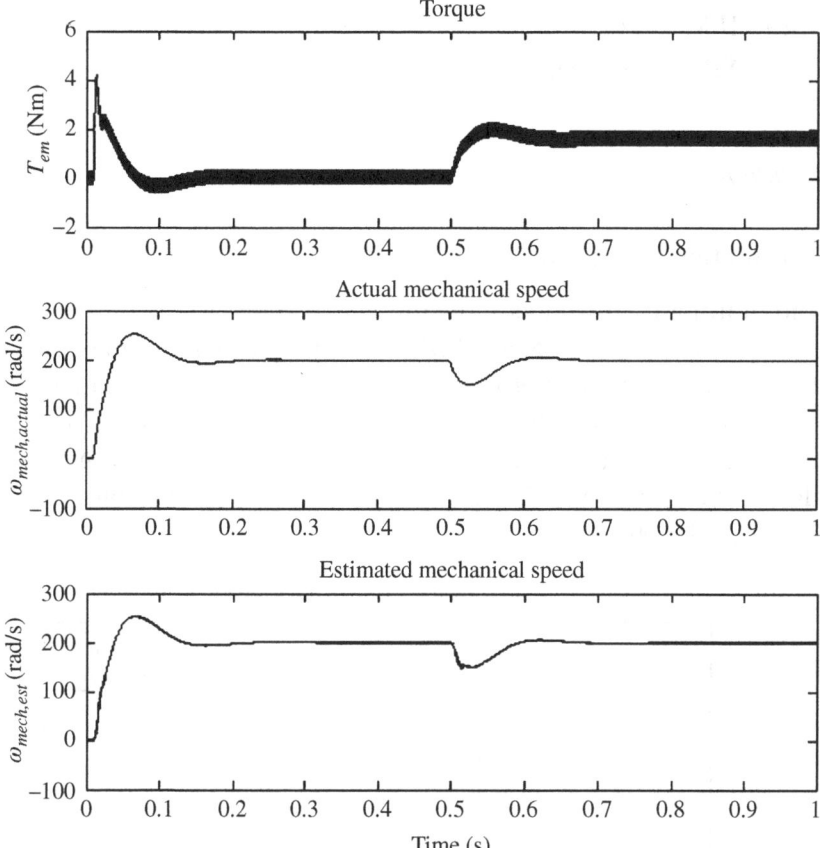

Fig. 17-14 Simulation result of Example 17-3.

EXAMPLE 17-4

Consider a synchronous motor with the following operating steady-state condition:

Stator current space vector magnitude $I_s = 15$ A.
Flux linkage due to rotor magnet $\lambda_{fd} = 0.0956$ Wb.
Plot the electromagnetic torque as a function of stator current space vector angle for the following two cases:

(*Continued*)

(a) IPM motor with $L_d = 1.4\,\text{mH}$ and $L_q = 2.7\,\text{mH}$.
(b) SPM motor with $L_d = 1.4\,\text{mH}$ and $L_q = 1.4\,\text{mH}$.

Solution

Substituting $\lambda_f = \sqrt{\dfrac{3}{2}}\lambda_{fd} = 0.1171\,\text{Wb}$ and $I_s = 15\,\text{A}$ into Eq. (17-31) and plotting the result as shown in Fig. 17-15.

As seen in Fig. 17-15, an IPM motor generates more maximum torque than an SPM motor of similar design and steady-state condition. This is due to additional reluctance torque due to saliency. It must be noted that unlike SPM, the maximum torque for an IPM does not occur when the current space vector leads the rotor magnetic axis by 90°.

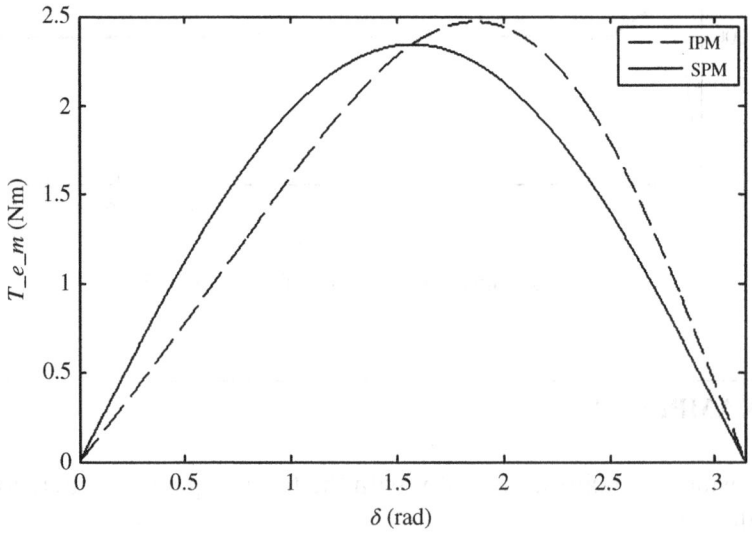

Fig. 17-15 Electromagnetic torque as a function of current space vector position.

17-5 HARDWARE PROTOTYPING OF VECTOR CONTROL OF SPM SYNCHRONOUS MOTOR

In Section 17.3, vector control of SPM synchronous motor drives was explained in detail, along with accompanying simulation results. In this section, hardware results of vector control of SPM synchronous motor are analyzed.

For the purposes of this comparison, a commercially available tabletop 36 V (L-L, rms) 3ϕ SPM synchronous motor and its vector control is modeled in Workbench. Although the exact motor parameters of the motor are not necessary, this is available on the accompanying website. The Workbench file of the modeled system is shown in Fig. 17-16. The speed controller has a step reference input from 0 to 100 rad/s at time $t = 1$ s. A step-load torque of 0.04 Nm is applied at time $t = 3$ s by means of a coupled dc generator.

The speed result from running this model in hardware is shown in Fig. 17-17. The hardware results closely match that of the simulation, and the speed settles down rapidly to the desired speed even after a load torque is applied. The d-axis current remains 0. Since no field weakening is performed, the q-axis increases proportional to the load torque (including the applied load torque as well as torque due to friction).

The step-by-step procedure for recreating the above experiment is presented in [4].

17-6 SUMMARY

In this chapter, we have extended the dq-analysis of induction machines to analyze and control the SPM and the IPM synchronous machines. A speed-sensorless rotor position and speed estimation methodology using high-frequency voltage injection were analyzed in detail. As the last step in this chapter, we implemented a real-time hardware vector control of a laboratory SPM motor – dc generator coupled system.

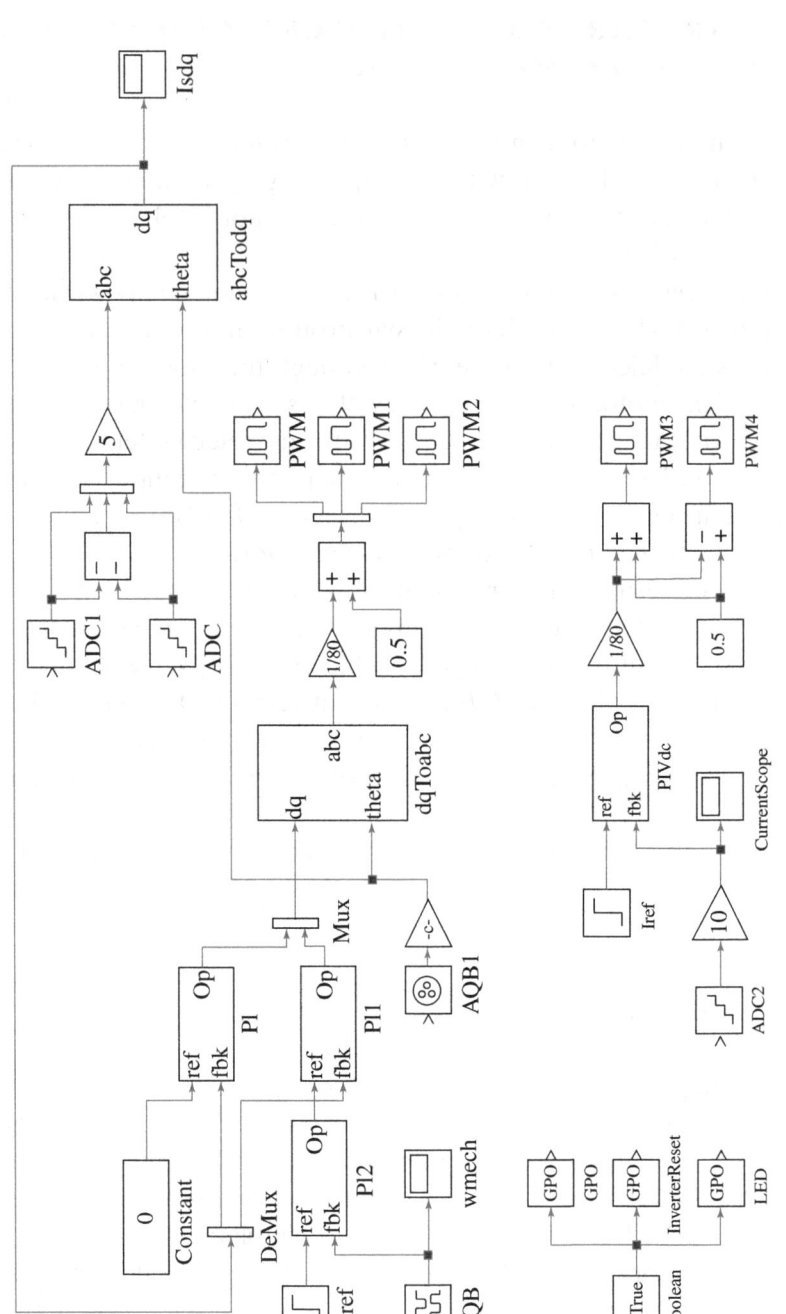

Fig. 17-16 Real-time SPM synchronous motor vector control.

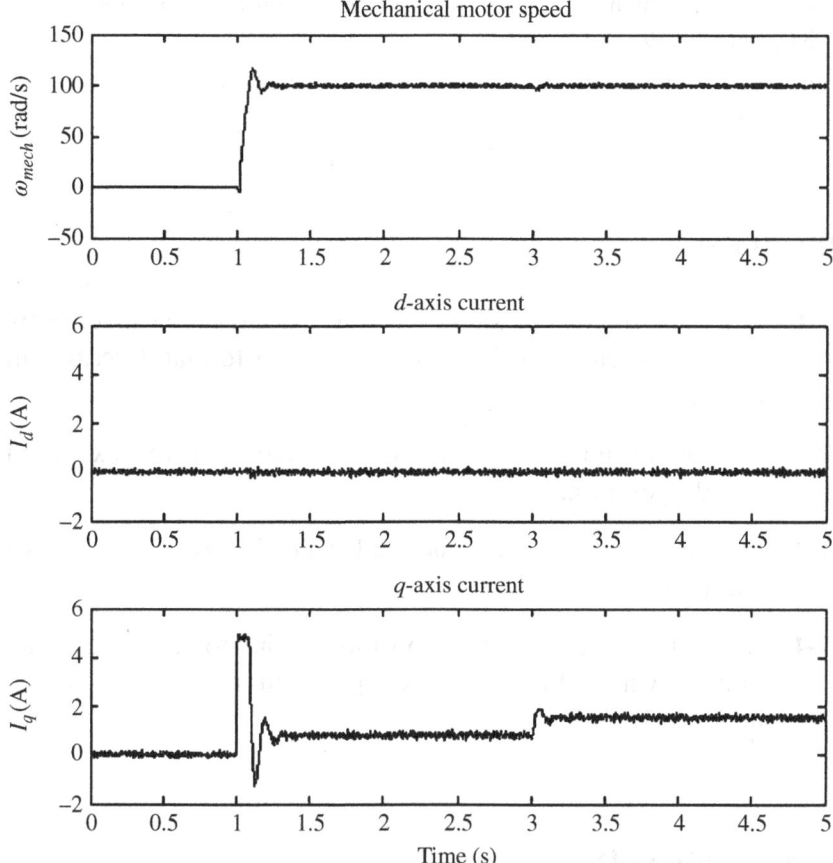

Fig. 17-17 Hardware speed and dq-current result.

REFERENCES

1. Kimbark, E.W. (1995). *Power System Stability: Synchronous Machines*. IEEE Press. ISBN: 0-7803-1135-3.
2. http://www.baldor.com.
3. Corley, M.J. and Lorenz, R.D. (1998). Rotor position and velocity estimation for a salient-pole permanent magnet synchronous machine at

standstill and high speeds. *IEEE Transactions on Industry Applications* 34 (4): 784–789.

4. https://sciamble.com/Resources/pe-drives-lab/advanced-drives/pm-vector-control.

PROBLEMS

17-1 In the simulation of Example 17-1, replace the ideal inverter by an appropriate SV-PWM inverter, similar to that described in Chapter 7.

17-2 Implement flux-weakening in Example 17-1 for extended speed operation.

17-3 Derive the current expression in Eq. (17-23) from Eqs. (17-20) and (17-22).

17-4 Prove that the linearized position estimator transfer function shown in Fig. 17-11 is equivalent to the one shown in Fig. 17-12.

17-A APPENDIX

17-A-1 Transformation of Stator Flux-Linkage From Rotating dq Frame to Stationary Frame

From Fig. 17-8, the projection of stator flux-linkage in the stationary frame to the synchronously rotating dq frame, is achieved by multiplying the flux linkage and currents in Eq. (17-19) by the rotation matrix R:

$$R = \begin{bmatrix} \cos(\theta_{da}) & \sin(\theta_{da}) \\ -\sin(\theta_{da}) & \cos(\theta_{da}) \end{bmatrix} \qquad (17\text{-}32)$$

i.e.,

$$\begin{bmatrix} \cos(\theta_{da}) & \sin(\theta_{da}) \\ -\sin(\theta_{da}) & \cos(\theta_{da}) \end{bmatrix} \begin{bmatrix} \lambda_{sd}^s \\ \lambda_{sq}^s \end{bmatrix} = \begin{bmatrix} L_d & 0 \\ 0 & L_q \end{bmatrix} \begin{bmatrix} \cos(\theta_{da}) & \sin(\theta_{da}) \\ -\sin(\theta_{da}) & \cos(\theta_{da}) \end{bmatrix}$$
$$\begin{bmatrix} i_{sd}^s \\ i_{sq}^s \end{bmatrix} + \begin{bmatrix} \lambda_{fd} \\ 0 \end{bmatrix}$$

$$\Rightarrow \begin{bmatrix} \lambda_{sd}^s \\ \lambda_{sq}^s \end{bmatrix} = \begin{bmatrix} \cos(\theta_{da}) & -\sin(\theta_{da}) \\ \sin(\theta_{da}) & \cos(\theta_{da}) \end{bmatrix} \begin{bmatrix} L_d & 0 \\ 0 & L_q \end{bmatrix}$$
$$\begin{bmatrix} \cos(\theta_{da}) & \sin(\theta_{da}) \\ -\sin(\theta_{da}) & \cos(\theta_{da}) \end{bmatrix} \begin{bmatrix} i_{sd}^s \\ i_{sq}^s \end{bmatrix}$$
$$+ \begin{bmatrix} \cos(\theta_{da}) & -\sin(\theta_{da}) \\ \sin(\theta_{da}) & \cos(\theta_{da}) \end{bmatrix} \begin{bmatrix} \lambda_{fd} \\ 0 \end{bmatrix}$$

$$= \begin{bmatrix} L_d \cos^2(\theta_{da}) + L_q \sin^2(\theta_{da}) & -(L_q - L_d)\cos(\theta_{da})\sin(\theta_{da}) \\ -(L_q - L_d)\cos(\theta_{da})\sin(\theta_{da}) & L_d \sin^2(\theta_{da}) + L_q \cos^2(\theta_{da}) \end{bmatrix}$$
$$\begin{bmatrix} i_{sd}^s \\ i_{sq}^s \end{bmatrix} + \begin{bmatrix} \cos(\theta_{da}) \\ \sin(\theta_{da}) \end{bmatrix} \lambda_{fd}$$

$$= \begin{bmatrix} \dfrac{L_q + L_d}{2} - \dfrac{L_q - L_d}{2}\cos(2\theta_{da}) & -\dfrac{L_q - L_d}{2}\sin(2\theta_{da}) \\ -\dfrac{L_q - L_d}{2}\sin(2\theta_{da}) & \dfrac{L_q + L_d}{2} + \dfrac{L_q - L_d}{2}\cos(2\theta_{da}) \end{bmatrix}$$
$$\begin{bmatrix} i_{sd}^s \\ i_{sq}^s \end{bmatrix} + \begin{bmatrix} \cos(\theta_{da}) \\ \sin(\theta_{da}) \end{bmatrix} \lambda_{fd}$$

$$= \begin{bmatrix} L - \Delta L \cos(2\theta_{da}) & -\Delta L \sin(2\theta_{da}) \\ -\Delta L \sin(2\theta_{da}) & L + \Delta L \cos(2\theta_{da}) \end{bmatrix} \begin{bmatrix} i_{sd}^s \\ i_{sq}^s \end{bmatrix} + \begin{bmatrix} \cos(\theta_{da}) \\ \sin(\theta_{da}) \end{bmatrix} \lambda_{fd}$$
(17-33)

18 Reluctance Drives: Stepper-Motors and Switched-Reluctance Drives

18-1 INTRODUCTION

Reluctance machines operate on principles that are different from those associated with all of the machines discussed so far. Reluctance drives can be broadly classified into three categories: stepper-motor drives, switched-reluctance drives, and synchronous-reluctance-motor drives. Only the stepper-motor and the switched-reluctance-motor (SRM) drives are discussed in this chapter.

Stepper-motor drives are widely used for position control in many applications, for example, computer peripherals, textile mills, integrated-circuit fabrication processes, and robotics. A stepper-motor drive can be considered as a digital electromechanical device, where each electrical pulse input results in a movement of the rotor by a discrete angle called the step-angle of the motor, as shown in Fig. 18-1. Therefore, for the desired change in position, the corresponding number of electrical pulses is applied to the motor, without the need for any position feedback.

(Adapted from chapter 13 of *Advanced Electric Drives: Analysis, Control, and Modeling Using MATLAB/Simulink* ISBN: 978-1-118-48548-4 by Ned Mohan, August 2014)

Analysis and Control of Electric Drives: Simulations and Laboratory Implementation, First Edition. Ned Mohan and Siddharth Raju.
© 2021 John Wiley & Sons, Inc. Published 2021 by John Wiley & Sons, Inc.
Companion website: www.wiley.com/go/Mohan/Vectorcontrolinelectricdrives

502 RELUCTANCE DRIVES: STEPPER-MOTORS

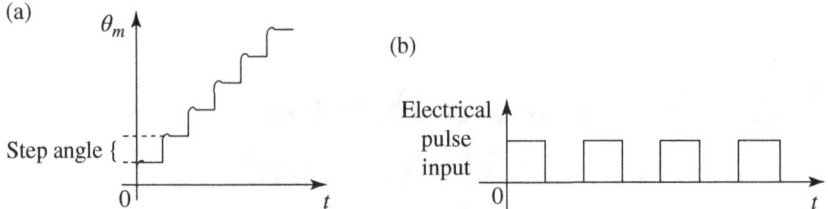

Fig. 18-1 Position change in a stepper-motor.

SRM drives are operated with controlled currents, using feedback. They are being considered for a large number of applications discussed later in this chapter.

18-2 THE OPERATING PRINCIPLE OF RELUCTANCE MOTORS

Reluctance motors operate by generating reluctance torque. This requires that the reluctance in the magnetic-flux path be different along the various axes. Consider the cross-section of a primitive machine shown in Fig. 18-2a in which the rotor has no electrical

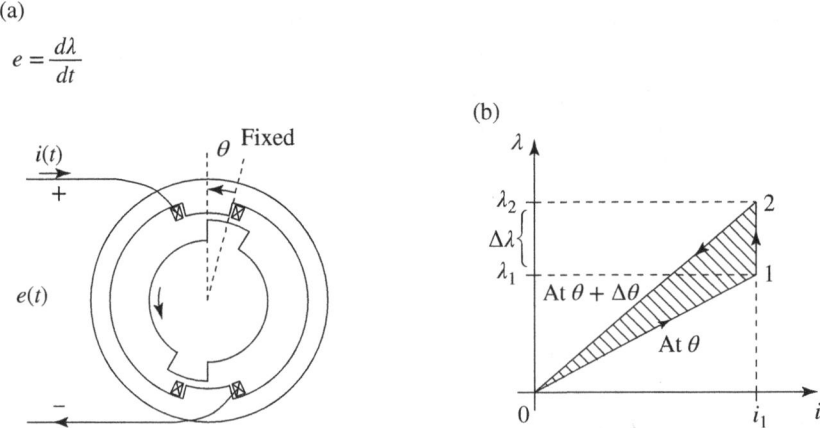

Fig. 18-2 (a) Cross-section of a primitive machine and (b) $\lambda - i$ trajectory during motion.

THE OPERATING PRINCIPLE OF RELUCTANCE MOTORS   503

excitation, and the stator has a coil excited by a current $i(t)$. In the following analysis, we will neglect the losses in the electrical and the mechanical systems, but these losses can also be accounted for. In the machine of Fig. 18-2a, the stator current would produce a torque on the rotor in the counterclockwise direction, due to fringing fluxes, in order to align the rotor with the stator pole. This torque can be estimated by the principle of energy conservation; this principle states that

Electrical energy input = increase in stored energy + mechanical output

(18-1)

Assuming that magnetic saturation is avoided, the stator coil has an inductance $L(\theta)$, which depends on the rotor position θ. Thus, the flux-linkage λ of the coil can be expressed as

$$\lambda = L(\theta)\, i \qquad (18\text{-}2)$$

The flux-linkage λ depends on the coil inductance, as well as the coil current. At any time, the voltage e across the stator coil, from Faraday's Law, is

$$e = \frac{d\lambda}{dt} \qquad (18\text{-}3)$$

The polarity of the induced voltage is indicated in Fig. 18-2a. Based on Eqs. (18-2) and (18-3), the voltage in the coil may be induced due to the time-rate of change of the current and/or the coil inductance. Using Eq. (18-3), the energy supplied by the electrical source from a time t_1 (with a flux linkage of λ_1) to time t_2 (with a flux linkage of λ_2) is

$$W_{el} = \int_{t_1}^{t_2} e \cdot i \cdot dt = \int_{t_1}^{t_2} \frac{d\lambda}{dt} \cdot i \cdot dt = \int_{\lambda_1}^{\lambda_2} i \cdot d\lambda \qquad (18\text{-}4)$$

In order to calculate the torque developed by this motor, we will consider the counterclockwise movement of the rotor in

Fig. 18-2a by a differential angle $d\theta$ in the following steps shown in Fig. 18-2b:

- Keeping θ constant, the current is increased from zero to a value i_1. The current follows the trajectory from 0 to 1 in the $\lambda - i$ plane in Fig. 18-2b. Using Eq. (18-4), we find that the energy supplied by the electrical source is obtained by integrating with respect to λ in Fig. 18-2b; thus, the energy supplied equals $Area\,(0-1-\lambda_1)$:

$$W_{el}(0 \to 1) = Area\,(0-1-\lambda_1) = \frac{1}{2}\lambda_1 i_1 \qquad (18\text{-}5)$$

This is the energy that gets stored in the magnetic field of the coil since there is no mechanical output.

- Keeping the current constant at i_1, the rotor angle is allowed to be increased by a differential angle, from θ to $(\theta + \Delta\theta)$, in the counterclockwise direction. This follows the trajectory from 1 to 2 in the $\lambda - i$ plane of Fig. 18-2b. The change in flux linkage of the coil is due to the increased inductance. From Eq. (18-2),

$$\Delta\lambda = i_1 \Delta L \qquad (18\text{-}6)$$

Using Eq. (18-4) and integrating with respect to λ, we find that the energy supplied by the electrical source during this transition in Fig. 18-2b is

$$W_{el}(1 \to 2) = Area\,(\lambda_1 - 1 - 2 - \lambda_2) = i_1(\lambda_2 - \lambda_1) \qquad (18\text{-}7)$$

- Keeping the rotor angle constant at $(\theta + \Delta\theta)$, the current is decreased from i_1 to zero. This follows the trajectory from 2 to 0 in the $\lambda - i$ plane in Fig. 18-2b. Using Eq. (18-4), we see that the energy is now supplied to the electrical source. Therefore, in Fig. 18-2b,

$$W_{el}(2 \to 0) = -Area\,(2-0-\lambda_2) = -\frac{1}{2}\lambda_2 i_1 \qquad (18\text{-}8)$$

During these three transitions, the coil current started with a zero value and ended at zero. Therefore, the increase in the energy storage term in Eq. (18-1) is zero. The net energy supplied by the electric source is

$$W_{el,net} = Area(0-1-\lambda_1) + Area(\lambda_1-1-2-\lambda_2) - Area(2-0-\lambda_2)$$
$$= Area(0-1-2) \qquad (18\text{-}9)$$

$Area(0-1-2)$ is shown hatched in Fig. 18-2b. This triangle has a base of $\Delta\lambda$ and a height of i_1. Thus, we can find its area:

$$W_{el,net} = Area(0-1-2) = \frac{1}{2}i_1(\Delta\lambda) \qquad (18\text{-}10)$$

Using Eqs. (18-6) and (18-10),

$$W_{el,net} = \frac{1}{2}i_1(\Delta\lambda) = \frac{1}{2}i_1(i_1 \cdot \Delta L) = \frac{1}{2}i_1^2 \Delta L \qquad (18\text{-}11)$$

Since there is no change in the energy stored, the electrical energy has been converted into mechanical work by the rotor, which is rotated by a differential angle $\Delta\theta$ due to the developed torque T_{em}. Therefore,

$$T_{em}\Delta\theta = \frac{1}{2}i_1^2 \Delta L \text{ or } T_{em} = \frac{1}{2}i_1^2 \frac{\Delta L}{\Delta\theta} \qquad (18\text{-}12)$$

Assuming a differential angle,

$$T_{em} = \frac{1}{2}i_1^2 \frac{dL}{d\theta} \qquad (18\text{-}13)$$

This shows that the electromagnetic torque in such a reluctance motor depends on the current squared. Therefore, the counterclockwise torque in the structure of Fig. 18-2a is independent of the direction of the current. This torque, called the reluctance torque, forms the basis of operation for stepper-motors and SRMs.

18-3 STEPPER-MOTOR DRIVES

Stepper-motors come in a large variety of constructions, with three basic categories: variable-reluctance motors, permanent-magnet motors, and hybrid motors. Each of these is briefly discussed.

18-3-1 Variable-Reluctance Stepper-Motors

Variable-reluctance stepper-motors have double saliency; that is, both the stator and the rotor have different magnetic reluctances along various radial axes. The stator and the rotor also have a different number of poles. An example is shown in Fig. 18-3, in which the stator has six poles, and the rotor has four poles. Each phase winding in this three-phase machine is placed on the two diametrically opposite poles.

Exciting phase-a with a current i_a results in a torque that acts in a direction to minimize the magnetic reluctance to the flux produced by i_a. With no load connected to the rotor, this torque will cause the rotor to align at $\theta = 0°$, as shown in Fig. 18-3a. This is the no-load equilibrium position. If the mechanical load causes a small deviation in θ, the motor will develop an opposing torque in accordance with Eq. (18-13).

To turn the rotor in a clockwise direction, i_a is reduced to zero and phase-b is excited by i_b, resulting in the no-load equilibrium position shown in Fig. 18-3b. The point z on the rotor moves by the step-angle of the motor. The next two transitions with i_c and back to i_a are shown in Fig. 18-3c and d. Following the movement of point z, we see that the rotor has moved by one rotor-pole-pitch for three changes in excitation ($i_a \rightarrow i_b$, $i_b \rightarrow i_c$, and $i_c \rightarrow i_a$). The rotor-pole-pitch equals ($360°/N_r$), where N_r equals the number of rotor poles. Therefore, in a q-phase motor, the step-angle of rotation for each change in excitation will be

$$\text{Step-angle} = \frac{360°}{qN_r} \qquad (18\text{-}14)$$

In the motor of Fig. 18-3 with $N_r = 4$ and $q = 3$, the step-angle equals 30°. The direction of rotation can be made counterclockwise by excitation in the sequence a-c-b-a.

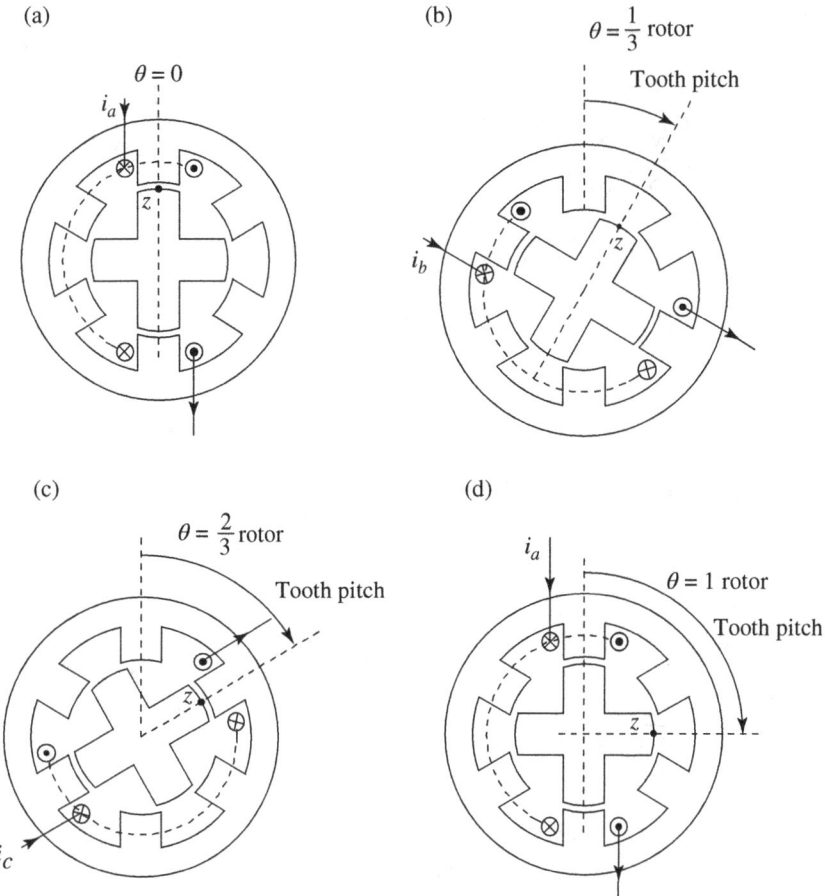

Fig. 18-3 Variable-reluctance motor; excitation sequence a-b-c-a. (a) Phase-a excited, (b) phase-b excited, (c) phase-c excited, and (d) phase-a excited.

18-3-2 Permanent-Magnet Stepper-Motors

In permanent-magnet stepper-motors, permanent magnets are placed on the rotor, as in the example shown in Fig. 18-4. The stator has two phase windings. Each winding is placed on four poles, the same as the number of poles on the rotor. Each phase winding produces the same number of poles as the rotor. The phase currents

508 RELUCTANCE DRIVES: STEPPER-MOTORS

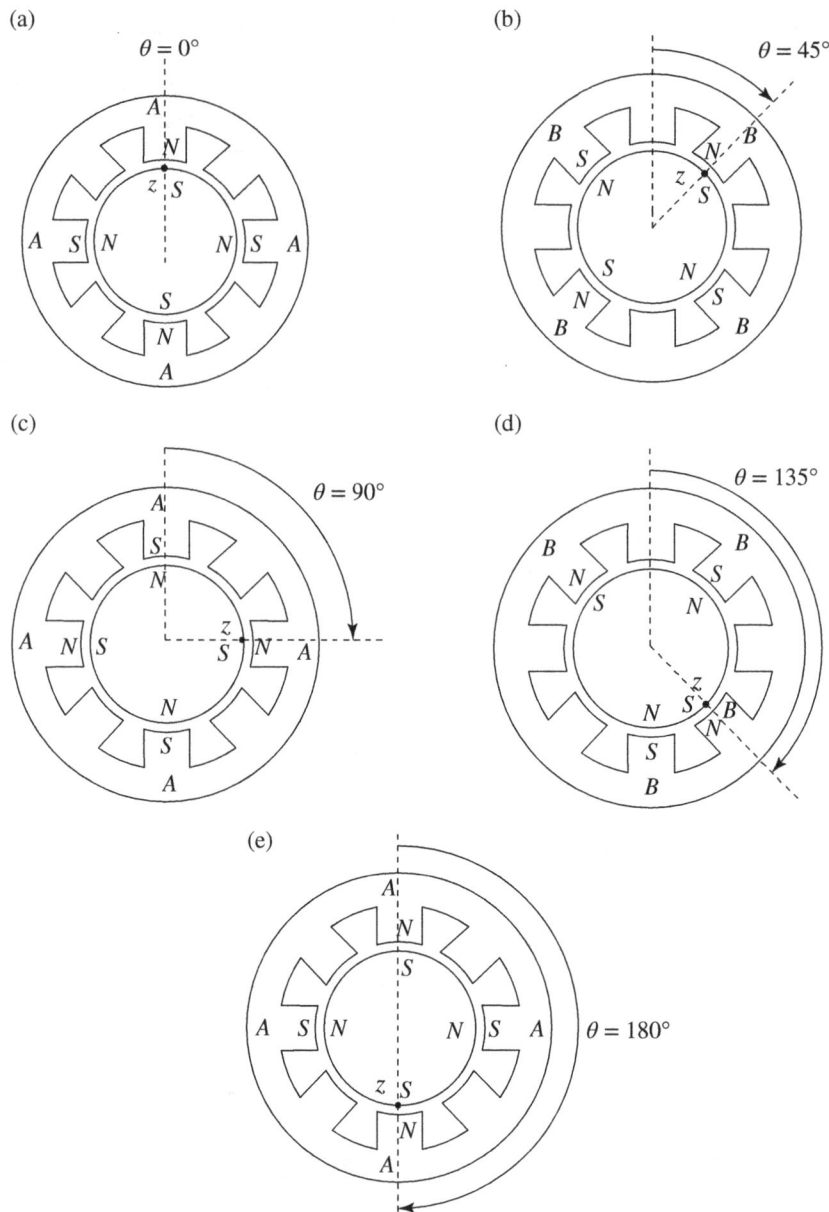

Fig. 18-4 Two-phase permanent-magnet stepper-motor; excitation sequence $i_{a+}, i_{b+}, i_{a-}, i_{b-}, i_{a+}$. (a) i_{a+}, (b) i_{b+}, (c) i_{a-}, (d) i_{b-}, (e) i_{a+}.

are controlled to be positive or negative. With a positive current i_a^+, the resulting stator poles and the no-load equilibrium position of the rotor are as shown in Fig. 18-4a. Reducing the current in phase-a to zero, a positive current i_b^+ in phase-b results in a clockwise rotation (following the point z on the rotor), shown in Fig. 18-4b. To rotate further, the current in phase-b is reduced to zero, and a negative current i_a^- causes the rotor to be in the position shown in Fig. 18-4c. Figure 18-4 illustrates that an excitation sequence $(i_a^+ \rightarrow i_b^+, i_b^+ \rightarrow i_a^-, i_a^- \rightarrow i_b^-, i_b^- \rightarrow i_a^+)$ produces a clockwise rotation. Each change in excitation causes rotation by one-half of the rotor-pole-pitch, which yields a step-angle of 45° in this example.

18-3-3 Hybrid Stepper-Motors

Hybrid stepper-motors utilize the principles of both the variable-reluctance and the permanent-magnet stepper-motors. An axial cross-section is shown in Fig. 18-5. The rotor consists of permanent magnets with a north and a south pole at the two opposite ends. In addition, each side of the rotor is fitted with an end cap with N_r teeth; N_r is equal to 10 in this figure. The flux produced by the permanent

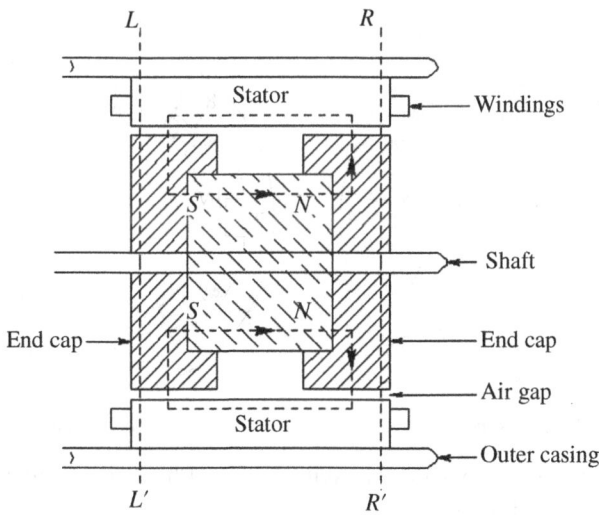

Fig. 18-5   Axial view of a hybrid stepper-motor.

magnets is shown in Fig. 18-5. All of the end-cap teeth on the left act like south poles, while all of the end-cap teeth on the right act like north poles.

The left and the right cross-sections, perpendicular to the shaft, along $L - L'$ and $R - R'$, are shown in Fig. 18-6. The two rotor end caps are intentionally displaced, with respect to each other, by one-half of the rotor-tooth-pitch. The stator in this figure consists of 8 poles in which the slots run parallel to the shaft axis.

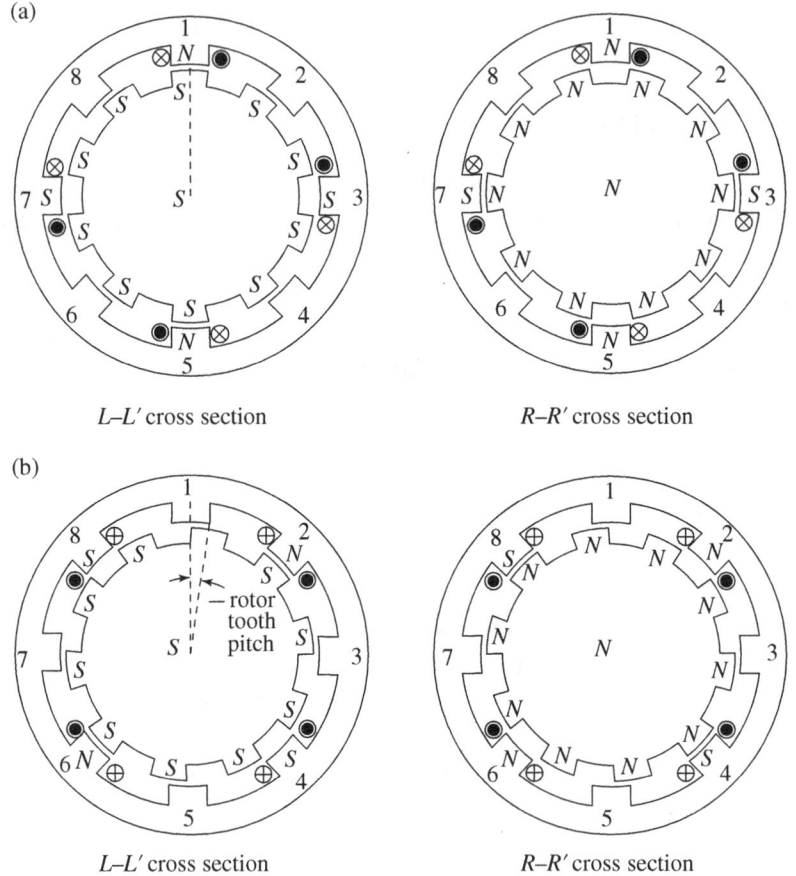

Fig. 18-6 Hybrid stepper-motor excitation. (a) Phase-a excited with i_{a+} and (b) phase-b excited with i_{b+}.

The stator consists of two phases; each phase winding is placed on 4 alternate poles, as shown in Fig. 18-6. Excitation of phase-a by a positive current i_a^+ results in north and south poles, as shown in both cross-sections in Fig. 18-6a. In the no-load equilibrium position, shown in Fig. 18-6a, on both sides, the opposite stator and rotor poles align while the similar poles are as far apart as possible. For clockwise rotation, the current in phase-a is brought to zero and phase-b is excited by a positive current i_b^+, as shown in Fig. 18-6b. Again, on both sides, the opposite stator and rotor poles align while the similar poles are as far apart as possible. This change of excitation $(i_a^+ \to i_b^+)$ results in clockwise rotation by one-fourth of the rotor-tooth-pitch. Therefore, in a two-phase motor,

$$\text{Step-angle} = \frac{360°/N_r}{4} \qquad (18\text{-}15)$$

which, in this example with $N_r = 10$, equals 9°.

18-3-4 Equivalent-Circuit Representation of a Stepper-Motor

Similar to other machines discussed previously, stepper-motors can be represented by an equivalent circuit on a per-phase basis. Such an equivalent circuit for phase-a is shown in Fig. 18-7 and consists of a back-emf, a winding resistance R_s, and a winding inductance L_s.

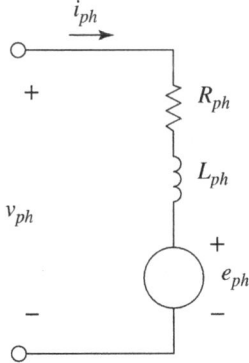

Fig. 18-7 Per-phase equivalent circuit of a stepper-motor.

The magnitude of the induced emf depends on the speed of rotation, and the polarity of the induced emf is such that it absorbs power in the motoring mode.

18-3-5 Half-Stepping and Micro-Stepping

It is possible to get smaller angular movement for each transition in the stator currents. For example, consider the variable-reluctance motor, for which the no-load equilibrium positions with i_a and i_b, shown in Fig. 18-3a and b, respectively. Exciting phases a and b simultaneously cause the rotor to be in the position shown in Fig. 18-8, which is one-half of a step-angle away from the position in Fig. 18-3a with i_a. Therefore, if "half-stepping" in the clockwise direction is required in the motor of Fig. 18-3, the excitation sequence will be as follows:

$$i_a \rightarrow (i_a,i_b) \rightarrow i_b \rightarrow (i_b,i_c) \rightarrow i_c \rightarrow (i_c,i_a) \rightarrow i_a \qquad (18\text{-}16)$$

By precisely controlling the phase currents, it is possible to achieve micro-step angles. For example, there are hybrid stepper-motors in which a step-angle can be divided into 125 micro-steps. This results in 25 000 micro-steps/revolution in a two-phase hybrid motor with a step-angle of 1.8°.

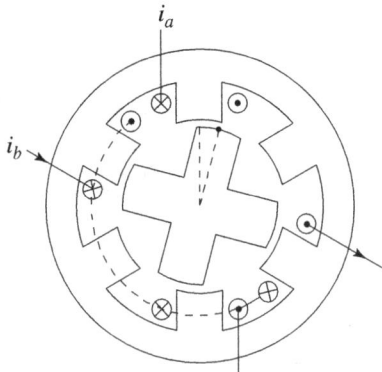

Fig. 18-8 Half-excitation by exciting two phases.

18-3-6 Power Electronic Converters for Stepper-Motors

In variable-reluctance drives, the phase currents need not reverse direction. A unidirectional-current converter for such motors is shown in Fig. 18-9a. Turning both switches on simultaneously causes the phase current to build up quickly, as shown in Fig. 18-9b. Once the current builds up to the desired level, it is maintained at that level by pulse-width-modulating one of the switches (for example T_1) while keeping the other switch on. By turning both switches off, the current is forced to flow into the dc-side source through the two diodes, thus decaying quickly.

Bidirectional currents are needed in permanent-magnet and hybrid stepper-motors. Supplying these currents requires a converter such as that shown in Fig. 18-10. This converter is very similar to those used in dc-motor drives discussed in Chapter 7.

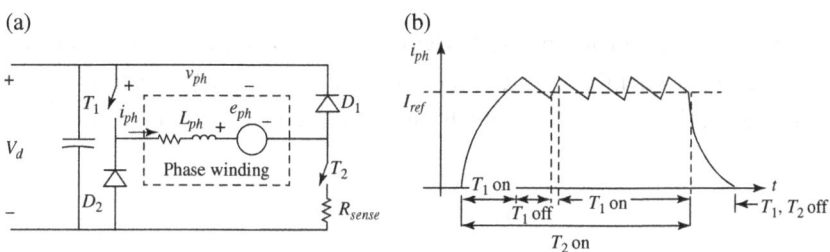

Fig. 18-9 Unipolar voltage drive for variable-reluctance motor: (a) circuit and (b) current waveform.

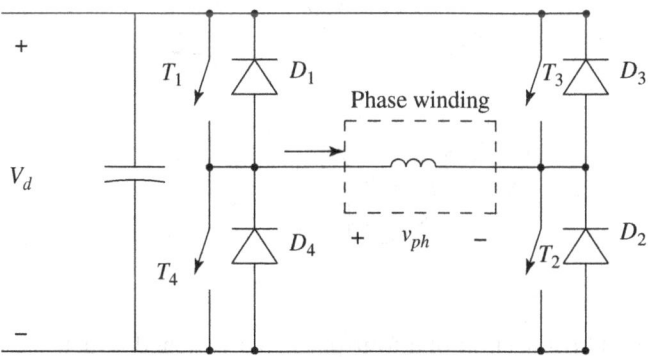

Fig. 18-10 Bipolar voltage drive.

18-4 SRM DRIVES

As discussed above, reluctance stepper-motors are generally used for position control in an open-loop manner, where by counting the number of electrical pulses supplied and knowing the step-angle of the motor, it is possible to rotate the shaft by the desired angle without any feedback. In contrast, SRM drives, also doubly salient in construction, are intended to provide continuous rotation, and compete with induction motor and PMAC motor drives in certain applications such as washing machines, automobiles, earthmoving equipment, and others. In this section, we will briefly look at the basic principles of SRM operation and how it is possible to control them in an encoder-less manner.

18-4-1 Switched-Reluctance Motor

Cross-section of a four-phase SRM is shown in Fig. 18-11, which looks identical to a variable-reluctance stepper-motor. It has a four-phase winding on the stator. In order to achieve a continuous rotation, each

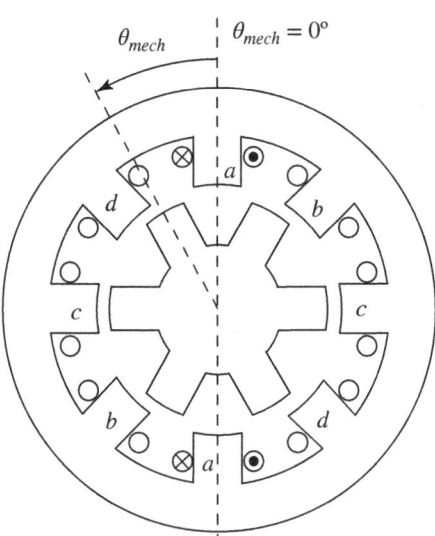

Fig. 18-11 Cross-section of a four-phase 8/6 switched reluctance machine.

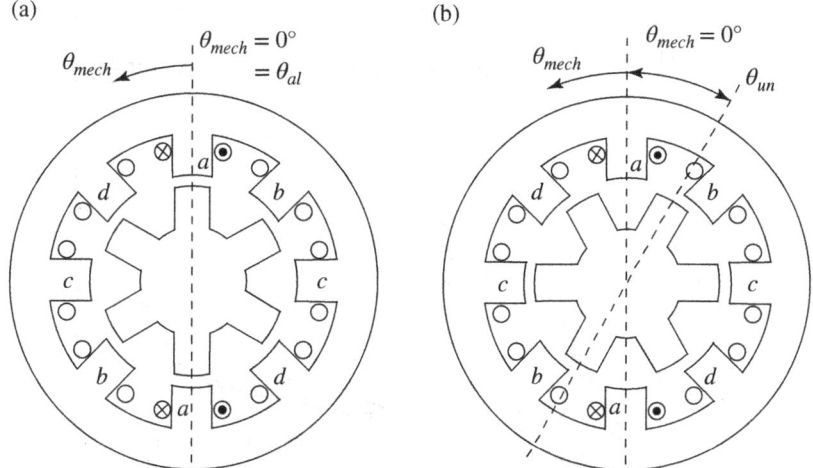

Fig. 18-12 (a) Aligned position for phase-a and (b) unaligned position for phase-a.

phase winding is excited by an appropriate current, at an appropriate rotor angle, as well as de-excited at a proper angle.

An SRM must be designed to operate with magnetic saturation, and the reason to do so will be discussed later on in this chapter. Figure 18-12 shows the aligned and the unaligned rotor positions for phase-a. For phase-a, the flux-linkage λ_a as a function of phase current i_a is plotted in Fig. 18-13, for various values of the rotor position. In the unaligned position, where the rotor pole is midway between two stator poles (see Fig. 18-12b, where θ_{mech} equals θ_{un}), the flux path includes a large air gap; thus, the reluctance is high. Low flux density keeps the magnetic structure in its linear region and the phase inductance has a small value. As the rotor moves toward the aligned position of Fig. 18-12a (where θ_{mech} equals zero), the characteristics become progressively more saturated at higher current values.

18-4-2 Electromagnetic Torque T_{em}

With the current built up to a value I_1 as shown in Fig. 18-14, holding the rotor at a position θ_1 between the unaligned and the aligned positions, the instantaneous electromagnetic torque can be calculated

516 RELUCTANCE DRIVES: STEPPER-MOTORS

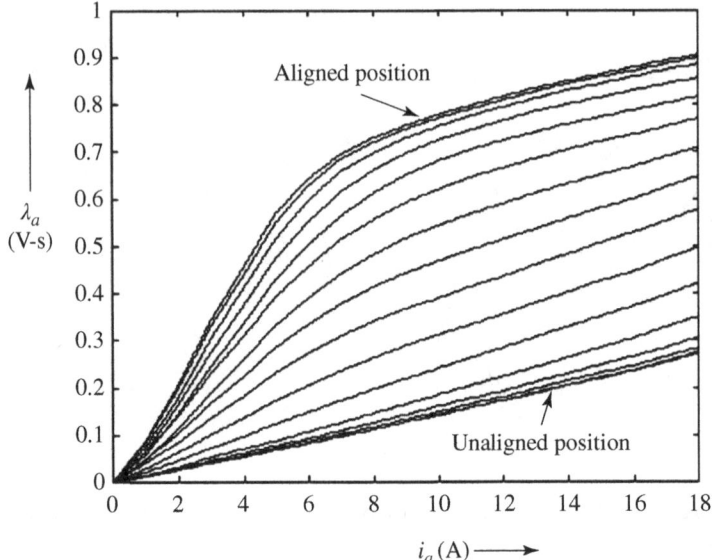

Fig. 18-13 Typical flux-linkage characteristics of an SRM.

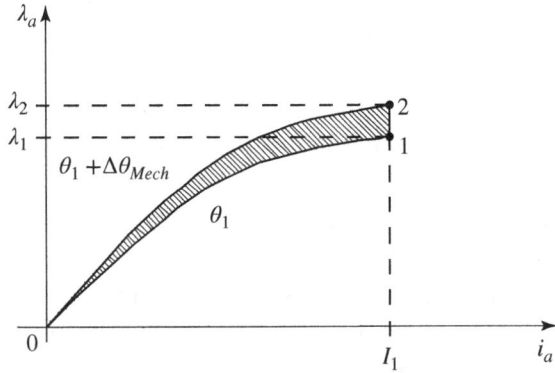

Fig. 18-14 Calculation of torque.

as follows: allowing the rotor to move incrementally under the influence of the electromagnetic torque from position θ_1 to $\theta_1 + \Delta\theta_{mech}$, keeping the current constant at I_1, the incremental mechanical work done is

$$\Delta W_{mech} = T_{em}\Delta\theta_{mech} \qquad (18\text{-}17)$$

The increment of energy supplied by the electrical source is

$$\Delta W_{elec} = area(1 - \lambda_1 - \lambda_2 - 2 - 1) \tag{18-18}$$

and the incremental increase in energy storage associated with the phase-*a* winding is

$$\Delta W_{storage} = area(0 - 2 - \lambda_2 - 0) - area(0 - 1 - \lambda_1 - 0) \tag{18-19}$$

The mechanical work performed is the difference of the energy supplied by the electrical source minus the increase in energy storage:

$$\Delta W_{mech} = \Delta W_{elec} - \Delta W_{storage} \tag{18-20}$$

Therefore, in Eq. (18-20),

$$\begin{aligned}
T_{em}\Delta\theta &= area(1 - \lambda_1 - \lambda_2 - 2 - 1) - \{area(0 - 2 - \lambda_2 - 0) - area(0 - 1 - \lambda_1 - 0)\} \\
&= \underbrace{\{area(1 - \lambda_1 - \lambda_2 - 2 - 1) + area(0 - 1 - \lambda_1 - 0)\}}_{area(0-1-2-\lambda_2-0)} - area(0 - 2 - \lambda_2 - 0) \\
&= area(0 - 1 - 2 - 0)
\end{aligned} \tag{18-21}$$

which is shown shaded in Fig. 18-14. Therefore,

$$T_{em} = \frac{area(0 - 1 - 2 - 0)}{\Delta\theta_{mech}} \tag{18-22}$$

which is in the direction to increase this area. The area between the $\lambda - i$ characteristic and the horizontal current axis is usually defined as the co-energy W'. Therefore, the area in Eq. (18-22), shown shaded in Fig. 18-14, represents an increase in co-energy. Thus, on a differential basis, we can express the instantaneous electromagnetic torque developed by this motor as the partial derivative of co-energy, with respect to the rotor angle, keeping the current constant:

$$T_{em} = \left.\frac{\partial W'}{\partial \theta_{mech}}\right|_{i_a = \text{constant}} \tag{18-23}$$

18-4-3 Induced Back-EMF e_a

With phase-a excited by i_a, the movement of the rotor results in a back-emf e_a and the voltage across the phase-a terminals includes the voltage drop across the resistance of the phase winding

$$v_a = Ri_a + e_a \qquad (18\text{-}24)$$

and

$$e_a = \frac{d}{dt}\lambda_a(i_a, \theta_{mech}) \qquad (18\text{-}25)$$

where the phase-winding flux linkage is a function of the phase current and the rotor position, as shown in Fig. 18-3. In terms of partial derivatives, we can rewrite the back-emf in Eq. (18-25) as

$$e_a = \left.\frac{\partial \lambda_a}{\partial i_a}\right|_{\theta_{mech}} \frac{d}{dt}i_a + \left.\frac{\partial \lambda_a}{\partial \theta_{mech}}\right|_{i_a} \frac{d}{dt}\theta_{mech} \qquad (18\text{-}26)$$

where it is important to recognize that a partial derivative with respect to one variable is obtained by keeping the other variable constant.

In Fig. 18-3, the movement of the rotor by an angle $\Delta\theta_{mech}$, keeping the current constant results in a back-emf, which from Eq. (18-26) can be written as

$$e_a = \left.\frac{\partial \lambda_a}{\partial \theta_{mech}}\right|_{i_a} \underbrace{\frac{d}{dt}\theta_{mech}}_{\omega_{mech}} = \left.\frac{\partial \lambda_a}{\partial \theta_{mech}}\right|_{i_a} \omega_{mech} \quad \left(\frac{d}{dt}i_a = 0\right) \qquad (18\text{-}27)$$

where ω_{mech} is the instantaneous rotor speed. However, the instantaneous power $(e_a i_a)$ is not equal to the instantaneous mechanical output, due to the change in stored energy in the phase winding.

18-5 INSTANTANEOUS WAVEFORMS

For a clear understanding, we will initially assume an idealized condition where it is possible to supply the phase winding with a current i_a

that has a rectangular waveform as a function of θ_{mech}, as shown in Fig. 18-15. The current is assumed to be built up instantaneously (this will require infinite voltage) at the unaligned position θ_{un} and instantaneously goes to zero at the aligned position θ_{al}. The corresponding waveforms for the electromagnetic torque $T_{em,a}$ and the induced back-emf e_a are also plotted by means of Eqs. (18-23) and (18-27), respectively, with the current held constant.

The objectives in selecting these two rotor positions θ_{un} and θ_{al} for current flow are twofold: (1) to maximize the average torque per ampere, and (2) to build up the current to its desired level while the back-emf is small. We can appreciate that with the current flow prior to the unaligned position, and after the aligned position, the instantaneous torque would be negative, which would be counter

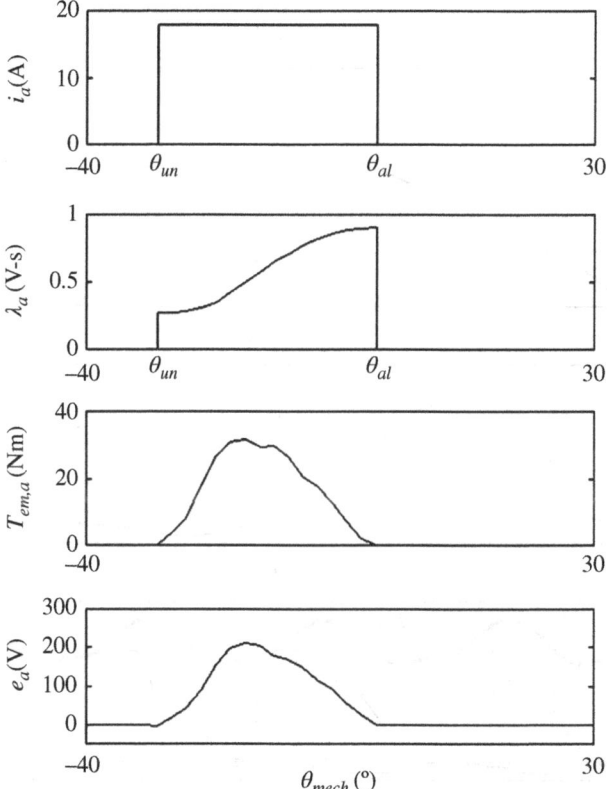

Fig. 18-15 Performance assuming idealized current waveform.

520 RELUCTANCE DRIVES: STEPPER-MOTORS

to our objective of maximizing the average torque per ampere. At the unaligned position, the winding inductance is the lowest and it is easier to build up current in that position, compared to other rotor positions.

To achieve instantaneous buildup and decay of phase current assumed in the plots of Fig. 18-15, would require that an infinite phase voltage (positive and negative) is available. In reality, with a finite voltage available from the power-processing unit to the motor, the phase-current waveform for a four-phase motor may look as shown in Fig. 18-16, with the corresponding flux linkage and torque

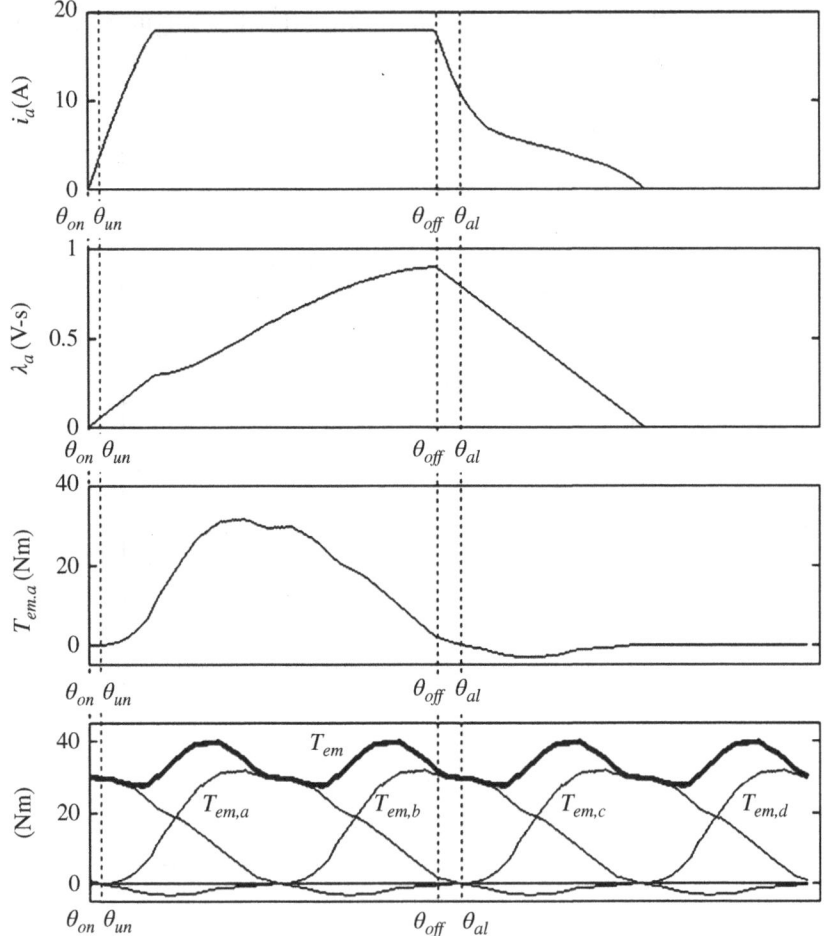

Fig. 18-16 Performance with a power-processing unit.

waveforms. The phase-current buildup is started at an angle θ_{on} (prior to the unaligned position), and the current decay is started at an angle θ_{off} (prior to the aligned position). In order to reduce torque ripple, there is generally overlapping of phases where during a short duration, two of the phases are simultaneously excited. The bottom part in Fig. 18-16, shows the resultant electromagnetic torque T_{em} by summing the torque developed by each of the four phases.

18-6 ROLE OF MAGNETIC SATURATION [1]

Magnetic saturation plays an important role in SRM drives. During each excitation cycle, a large ratio of the energy supplied to a phase winding should be converted into mechanical work, rather than returned to the electrical source at the end of the cycle. We will call this an energy conversion factor. In Fig. 18-17, assuming magnetic saturation and a finite voltage is available from the power-processing unit, this factor is as follows:

$$\text{Energy conversion factor} = \frac{W_{em}}{W_{em} + W_f} \qquad (18\text{-}28)$$

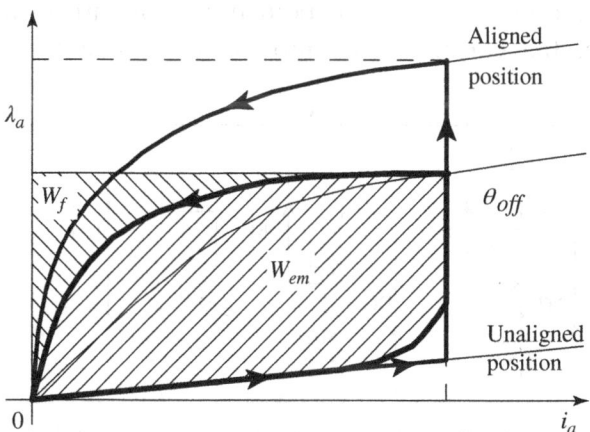

Fig. 18-17 Flux-linkage trajectory during motoring.

522 RELUCTANCE DRIVES: STEPPER-MOTORS

As can be seen from Fig. 18-17, this factor would be clearly higher in the idealized case where an instantaneous buildup and decay of phase current is assumed. However, without saturation, this energy conversion factor is limited to a value of nearly 50%. Magnetic saturation also keeps the rating of the power processor unit from becoming unacceptable.

It should be noted that the energy conversion factor is *not* the same as energy efficiency of the motor, although there is a correlation – a lower energy conversion factor means that a larger fraction of energy sloshes back-and-forth between the power-processing unit and the machine, resulting in power losses in the form of heat and thus in a lower energy efficiency.

18-7 POWER ELECTRONIC CONVERTERS FOR SRM DRIVES

A large number of topologies for SRM converters have been proposed in the literature. Figure 18-18 shows a topology that is most versatile. For current buildup, both transistors are turned on simultaneously. (This also shows the robustness of the SRM drive power-processing unit where turning on both transistors simultaneously is normal, which can be catastrophic in other drives.) To maintain the current within an hysteretic band around the reference value, either one of the transistors is turned off, thus making the current freewheel through the opposite diode, or both transistors are turned off, in which case, the current flows into the dc bus and decreases in

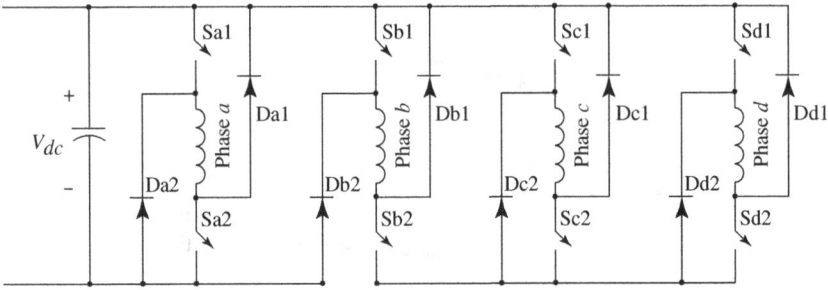

Fig. 18-18 Power converter for a four-phase switched reluctance drive.

CONTROL IN MOTORING MODE

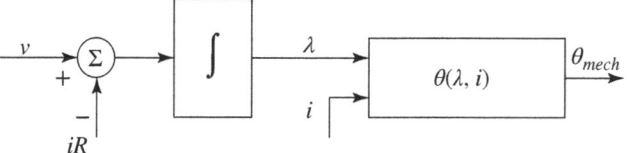

Fig. 18-19 Estimation of rotor position.

magnitude. The latter condition is also used to quickly de-energize a phase winding.

18-8 DETERMINING THE ROTOR POSITION FOR ENCODER-LESS OPERATION

It is necessary to determine the rotor position so that the current buildup and decay can be started at rotor positions θ_{on} and θ_{off}, respectively, for each phase. There are various methods proposed in the literature to determine the rotor position. One of the easiest methods to explain is shown by means of Figs. 18-13 and 18-19. The motor can be characterized to achieve the family of curves shown in Fig. 18-13, where the flux linkage of a phase is plotted as a function of the phase current for various values of the rotor position. From this information, knowing the flux linkage and the current allows the determination of the rotor position. In Fig. 18-9, the flux linkage of an excited phase is computed by integrating the difference of the applied phase voltage and the voltage drop across the winding resistance (see Eq. (18-24)). The combination of the measured phase current and the estimated flux linkage then determines the rotor position, using the information of Fig. 18-13.

18-9 CONTROL IN MOTORING MODE

A simple block diagram for speed control is shown in Fig. 18-20, where the actual rotor position is either sensed or estimated using the method described in the previous section, or some other technique. The speed error between the reference speed and the actual speed is amplified by means of a proportional-integral (PI) controller

524 RELUCTANCE DRIVES: STEPPER-MOTORS

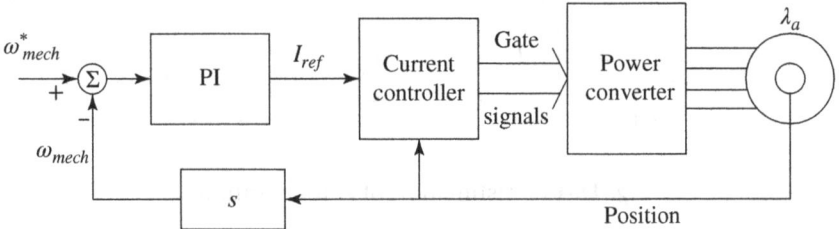

Fig. 18-20 Control block diagram for motoring.

to generate a current reference. The rotor angle determines which phases are to be excited, and their current is controlled to equal the reference current as much as possible, in view of the limited dc-bus voltage of the power-processing unit shown in Fig. 18-18.

18-10 SUMMARY/REVIEW QUESTIONS

This chapter discusses stepper-motors that are often used in an open-loop manner and SRM drives, doubly salient in construction, which are intended to provide continuous rotation, and compete with induction motors and PMAC motor drives.

1. What are the three broad categories of reluctance drives?

2. How is the principle on which reluctance drives operate differently than that seen earlier with other drives?

3. Write down the reluctance torque expression. What does the direction of torque depend on?

4. Describe the operating principle of a variable-reluctance stepper-motor.

5. Describe the operating principle of a permanent-magnet stepper-motor.

6. Describe the operating principle of a hybrid stepper-motor.

7. What is the equivalent-circuit representation of a stepper-motor?

8. How are half-stepping and micro-stepping achieved in stepper-motors?
9. What is the nature of power-processing units in stepper-motor drives?
10. Describe the operating principles of switched-reluctance drives.
11. What are the application areas of switched-reluctance drives?

REFERENCE

1. Miller, T.J.E. (ed.) (2001). *Electronic Control of Switched Reluctance Machines*. Newnes. ISBN: 0-7506-50737.

FURTHER READING

Kenjo, T. (1985). *Stepping Motors, and Their Microprocessor Control*. Oxford: Oxford Science Publications.

Acarnley, P.P. (1984). *Stepping Motors: A Guide to Modern Theory and Practice*. IEE Control Engineering Series 19, Revised 2nd Ed.

Slemon, G.R. (1997). Electrical machines for drives. In: *Power Electronics and Variable Frequency Drives* (ed. B.K. Bose). IEEE Press.

PROBLEMS

18-1 Determine the phase excitation sequence, and draw the rotor positions, in a variable-reluctance step drive for counterclockwise rotation.

18-2 Repeat Problem 18-1 for a permanent-magnet stepper-motor drive.

18-3 Repeat Problem 18-1 for a hybrid stepper-motor drive.

526 RELUCTANCE DRIVES: STEPPER-MOTORS

18-4 Describe the half-step operation in a permanent-magnet stepper-motor drive.

18-5 Describe the half-step operation in a hybrid stepper-motor drive.

18-6 Show that without magnetic saturation, the energy conversion factor in Fig. 18-7 would be limited to 50%.

18-7 What would the plot of the phase inductance be as a function of the rotor angle (between the unaligned and the aligned rotor positions) for various values of the phase current.

18-8 Although only the motoring mode is discussed in this chapter, SRM drives (like all other drives) can also be operated in a generator mode. Explain how this mode of operation is possible in SRM drives.

INDEX

Ampere's Law, 52
Analysis of doubly-fed generators (DFIG) in steady-state and their vector control, 427
   dynamic analysis of DFIG, 443
   rotor current controller, 443
   rotor position estimator, 446
   rotor speed controller, 445
   stator reactive power controller, 446
   steady-state analysis, 430
   understanding DFIG operation in dq-axis, 436
   vector control of DFIG, 443
Application of the basic principles, 80
Applied voltage amplitudes to keep $\hat{B}_{ms} = \hat{B}_{ms,rated}$, 291

Balanced sinusoidal steady-state excitation (Rotor Open-Circuited), 186
Basic structure of AC machines, 71
Bi-directional switching power-pole, 99
Blocked-rotor test to estimate R'_r and the leakage inductances, 272

Calculating the stator voltages to be applied, 388
Calculation of $\vec{\lambda}_s$, $\vec{\lambda}_r$, T_{em}, and ω_m, 456
Calculation of ω_{dA}, 379
Calculation of initial conditions, 364
Calculation of the electromagnetic torque T_{em}, 380, 458
Calculation of the reference values $i_a^*(t)$, $i_b^*(t)$, and $i_c^*(t)$ of the stator currents, 225
Calculation of the rotor flux $\vec{\lambda}_r$, 456
Calculation of the rotor speed ω_m, 459
Calculation of the stator flux $\vec{\lambda}_s$, 456
Calculation of the stator voltage space vector, 460
Capability to operate below and above the rated speed, 298
Choice of the dq-winding speed ω_d, 355
Climate crisis, 4
Computer simulation, 363
Computer simulation of SV-PWM inverter, 208

Analysis and Control of Electric Drives: Simulations and Laboratory Implementation, First Edition. Ned Mohan and Siddharth Raju.
© 2021 John Wiley & Sons, Inc. Published 2021 by John Wiley & Sons, Inc.
Companion website: www.wiley.com/go/Mohan/Vectorcontrolinelectricdrives

Conditions for efficient speed control over a wide range, 286
Control in electric drives
  average representation of the power-processing unit (PPU), 140
  cascade control structure, 139
  controller design, 143
  controller design example, 145
  control objectives, 134
  DC motors, 130
  designing feedback controllers for motor drives, 134
  hardware prototyping of DC motor speed control, 156
  modeling of the DC machine and the mechanical load, 141
  PI controllers, 143
    anti-windup (non-windup) integration, 155
    feed-forward, 154
    limits, 154
    position control loop, 151
    speed loop, 148
    torque (current) control loop, 145
  requirements imposed by DC machines on the PPU, 134
  steps in designing the feedback controller, 139
  system representation for small-signal analysis, 140
Control in motoring mode, 523
Controllable switches, 120
Conversion between linear and rotary motion, 39
Converters for DC motor drives $(-V_d < \bar{v}_o < V_d)$, 104
Coupling mechanisms, 39

d-and q-axis equivalent circuits, 360
d-axis rotor flux linkage dynamics, 380
DC-resistance test to estimate R_s, 270
Derivation of torque expressions, 469
Designing PID controller, 488
Designing the PI controller, 391, 414
Determining the rotor position for encoder-less operation, 523
Device ratings, 119
Direct torque control (DTC) and encoder-less operation of induction motor drives, 453
  calculation of $\vec{\lambda}_s$, $\vec{\lambda}_r$, T_{em}, and ω_m, 456
  calculation of the electromagnetic torque T_{em}, 458
  calculation of the rotor flux $\vec{\lambda}_r$, 456
  calculation of the rotor speed ω_m, 459
  calculation of the stator flux $\vec{\lambda}_s$, 456
  calculation of the stator voltage space vector, 460
  derivation of torque expressions, 469
Direct torque control using dq-axes, 464
dq analysis of permanent magnet synchronous machines, 473
dq-based dynamic controller for "brush-less DC" drives, 478

INDEX

dq-winding representation, 341
dq-winding voltage equations, 351
Dynamic analysis of DFIG, 443
Dynamic analysis of induction machines in terms of dq-windings, 341
  calculation of initial conditions, 364
  choice of the dq-winding speed ω_d, 355
  computer simulation, 363
  d-and q-axis equivalent circuits, 360
  dq-winding representation, 341
  dq-winding voltage equations, 351
  electrodynamics, 360
  electromagnetic torque, 357
  flux linkages of dq windings in terms of their currents, 350
  mathematical relationships of the dq windings (at an arbitrary speed ω_d), 347
  mutual inductance between dq windings on the stator and the rotor, 346
  net electromagnetic torque T_{em} on the rotor, 359
  obtaining fluxes and currents with voltages as inputs, 355
  phasor analysis, 365
  relating dq winding variables to phase winding variables, 349
  relationship between the dq windings and the per-phase phasor-domain equivalent circuit in balanced sinusoidal steady state, 361
  rotor dq-windings (along the same dq-axes as in the stator), 345
  stator dq winding representation, 342
  torque on the rotor d-axis winding, 357
  torque on the rotor q-axis winding, 358

Electrical analogy of mechanical systems, 36
Electrically open-circuited rotor, 243
Electric transportation, 10
Electrodynamics, 360, 476
Electromagnetic force, 76
Electromagnetic torque T_{em}, 256, 357, 476, 491, 515
Emf induction, 76
Encoder-less vector control, 401, 486
Energy conversion, 81
Energy efficiency, 83
Energy saving opportunities, 4–5, 7, 9
Equivalent-circuit representation of a stepper-motor, 511
Equivalent windings in a squirrel-cage rotor, 324
Estimation of ω_{dA} and θ_{da}, 411

Faraday's Law, 65
Ferromagnetic materials, 54
Flux φ, 56
Flux density B, 54
Flux linkage, 57, 330, 475
Flux linkages of dq windings in terms of their currents, 350
Four-quadrant operation, 44
Friction, 33

Gear ratio, 42
Gears, 41
Generator (regenerative braking) mode of operation, 260

530 INDEX

Generator-mode of operation of PMAC drives, 233
Half-stepping and micro-stepping, 512
Hardware Prototyping
  closed-loop speed control of induction motor, 309
  induction motor parameter estimation, 277
  PMAC motor hysteresis current control, 235
  pulse width modulation, 122
  space vector pulse width modulation, 213
  vector control of induction motor, 393
  vector control of SPM synchronous motor, 495
Harmonics in the PPU output voltages, 305
High-frequency injection, 483
Hybrid stepper-motors, 509

Induced back-emf e_a, 518
Induced emf, 76, 78
Induced emf in the stator windings due to rotating $\overrightarrow{B_r}(t)$, 228
Induced emf in the stator windings due to rotating $\overrightarrow{i_s}(t)$: armature reaction, 229
Induced emfs in the stator windings during balanced sinusoidal steady state, 228
Induced voltages in stator windings, 193
Inductances, 61
Induction-generator drives, 301
Induction machine equations in phase quantities: assisted by space vectors, 317
  equivalent windings in a squirrel-cage rotor, 324
  flux linkages, 330
  making a case for a dq-winding analysis, 334
  mutual inductances between the stator and the rotor phase windings, 326
  per-phase magnetizing-inductance L_m, 323
  relationship between phasors and space vectors in sinusoidal steady state, 329
  review of space vectors, 327
  rotor flux linkage (stator open-circuited), 331
  rotor-winding inductances (stator open-circuited), 325
  sinusoidally-distributed stator windings, 318
  stator and rotor flux linkages (simultaneous stator and rotor currents), 332
  stator and rotor voltage equations in terms of space vectors, 333
  stator flux linkage (rotor open-circuited), 330
  stator-inductance L_s, 324
  stator inductances (rotor open-circuited), 320
  stator mutual-inductance L_{mutual}, 322
  stator single-phase magnetizing inductance $L_{m,1-phase}$, 320

INDEX   **531**

three-phase, sinusoidally distributed stator windings, 319
Induction motor characteristics at rated voltages in magnitude and frequency, 272
Induction motors in sinusoidal steady state, 241
blocked-rotor test to estimate R'_r and the leakage inductances, 272
DC-resistance test to estimate R_s, 270
electrically open-circuited rotor, 243
electromagnetic torque, 256
generator (regenerative braking) mode of operation, 260
hardware prototyping of induction motor parameter estimation, 277
including the rotor leakage inductance, 261
including the stator winding resistance R_s and leakage inductance L_{ls}, 269
induction motor characteristics at rated voltages in magnitude and frequency, 272
induction motors of NEMA design A, B, C, and D, 275
line start, 277
the no-load test to estimate L_m, 271
per-phase steady-state equivalent circuit (including rotor leakage), 265
principles of induction motor operation, 242
reversing the direction of rotation, 261
short-circuited rotor, 245
slip frequency, f_{slip}, in the rotor circuit, 254
structure of three-phase, squirrel-cage induction motors, 241
tests to obtain the parameters of the per-phase equivalent circuit, 270
transformer analogy, 245
Induction motors of NEMA design A, B, C, and D, 275
Initial startup, 388
Instantaneous waveforms, 518
Insulated-gate bipolar transistors (IGBTs), 121

Leakage and magnetizing inductances, 68
Limit on the amplitude $\hat{V}_s$ of the stator voltage space vector $\vec{v}^a_s$, 211
Linear motion, 23
Line start, 277
Load types, 43–44

Magnetic circuit concepts, 52
Magnetic energy storage in inductors, 63
Magnetic field produced by current-carrying conductors, 52
Magnetics, 51
  Ampere's Law, 52
  application of the basic principles, 80
  basic structure of AC machines, 71
  electromagnetic force, 76

Magnetics (*cont'd*)
  emf induction, 76
  energy conversion, 81
  energy efficiency, 83
  Faraday's Law, 65
  ferromagnetic materials, 54
  flux φ, 56
  flux density B, 54
  flux linkage, 57
  induced emf, 76, 78
  inductances, 61
  leakage and magnetizing inductances, 68
  magnetic circuit concepts, 52
  magnetic energy storage in inductors, 63
  magnetic field produced by current-carrying conductors, 52
  magnetic shielding of conductors in slots, 79
  magnetic structures with air gaps, 58
  mutual inductances, 71
  power losses, 83
  production of magnetic field, 73
  regenerative braking, 82
  relating $e(t)$, $\varphi(t)$, and $i(t)$, 67
  torque production, 71, 76
  voltage induction, 71, 76
Magnetic shielding of conductors in slots, 79
Magnetic structures with air gaps, 58
Making a case for a dq-winding analysis, 334
Mathematical description of vector control in induction machines, 377
  calculating the stator voltages to be applied, 388
  calculation of ω_{dA}, 379
  calculation of T_{em}, 380
  d-axis rotor flux linkage dynamics, 380
  designing the PI controllers, 391
  hardware prototyping of vector control of induction motor, 393
  initial startup, 388
  motor model, 381
  motor model with the d-axis aligned along the rotor flux linkage $\vec{\lambda}_r$-axis, 378
  speed and position control loops, 385
  vector control, 384
Mathematical relationships of the dq windings (at an arbitrary speed ω_d), 347
Mechanical system of PMAC drives, 224
Mechanical system requirements, 21
  conversion between linear and rotary motion, 39
  coupling mechanisms, 39
  electrical analogy of mechanical systems, 36
  four-quadrant operation, 44
  friction, 33
  gear ratio, 42
  gears, 41
  linear motion, 23
  loads types
    centrifugal, 43
    constant power, 43
    constant torque, 43
    squared power, 43
  rotating systems, 25
  torsional resonances, 35

INDEX   533

Metal-oxide-semiconductor field-effect transistors (MOSFETs), 120
Modeling the PPU-supplied induction motors in steady state, 308
Model-reference adaptive system (MRAS) estimator, 404
Motivation, 3
  climate crisis, 4
  electric transportation, 10
  energy saving opportunities
    commercial sectors, 8
    end-use of electricity, 6
    generation of electricity, 5
    process industry, 7
    residential sector, 8
    wind energy, 6
  multi-disciplinary nature of drive systems
    control theory, 13
    interaction of drives with the utility grid, 14
    mechanical system modeling, 14
    power electronics, 12
    real-time control of DSP, 14
    sensors, 14
    theory of electric machines, 12
  precise speed and torque control applications
    drones, 10
    process industry, 10
    robotics, 10
  range of electric drives, 11
  structure of the textbook
    dynamic analysis and control, 16
    fundamental concepts, 16
    stead-state analysis, 16

Motor model with the d-axis aligned along the rotor flux linkage $\vec{\lambda}_r$-axis, 378
MRAS linearized error function, 423
Multi-disciplinary nature of drive systems, 12–14
Mutual inductance between dq windings on the stator and the rotor, 346
Mutual inductances, 71
Mutual inductances between the stator and the rotor phase windings, 326

The no-load test to estimate L_m, 271
Non-salient pole synchronous machines, 477

Obtaining fluxes and currents with voltages as inputs, 355
Open loop speed estimator, 402
The operating principle of reluctance motors, 502

Parameter sensitivity of open-loop estimator and MRAS estimator, 416
Permanent-magnet stepper-motors, 507
Per-phase magnetizing-inductance L_m, 323
Per-phase steady-state equivalent circuit (including rotor leakage), 265
Phase components of space vectors $\vec{i_s}(t)$ and $\vec{v_s}(t)$, 184
Physical interpretation of the stator current space vector $\vec{i_s}(t)$, 181

Power diodes, 119
Power electronic converters
  Bi-directional switching
    power-pole, 99
  controllable switches, 120
    insulated-gate bipolar
      transistors (IGBTs), 121
    MOSFETs, 120
    "smart power" modules
      including gate drivers and
      wide bandgap devices, 121
  converters for DC motor drives
    ($-V_d < \bar{v}_o < V_d$), 104
  device ratings, 119
  hardware prototyping of pulse
    width modulation, 122
  overview, 95
  power diodes, 119
  power semiconductor devices, 118
  pulse-width-modulation (PWM)
    of the bi-directional switching
      power-pole, 101
    of the switching power-pole
      (constant f_s), 98
  switching power-pole as the
    building block, 97
  switching waveforms
    converter for DC motor
      drives, 108
    three-phase inverter with
      sine-PWM, 117
  switch-mode conversion, 97
  switch-mode power electronic
    converters, 95
  synthesis of low-frequency
    AC, 112
  three-phase inverters, 113
Power electronic converters for
  SRM drives, 522

Power electronic converters for
  stepper-motors, 513
Power semiconductor devices, 118
Practical implementation, 417
Precise speed and torque control
  applications, 10–11
Principle of encoder-less DTC
  operation, 455
Principle of operation, 219
Principles of induction motor
  operation, 242
Production of magnetic field, 73
Pulse-width-modulation (PWM)
  of the bi-directional switching
    power-pole, 101
  power-processing unit, 305
  of the switching power-pole
    (constant f_s), 98

Rated power capability above the
  rated speed by
  flux-weakening, 300
Rated torque capability below
  the rated speed
  (with $\hat{B}_{ms,rated}$), 299
Reduction of $\hat{B}_{ms}$ at light loads, 308
Regenerative braking, 82
Relating $e(t)$, $\varphi(t)$, and $i(t)$, 67
Relating dq winding variables to
  phase winding variables, 349
Relationship between phasors and
  space vectors in sinusoidal
  steady state, 329
Relationship between space vectors
  and phasors in balanced
  three-phase sinusoidal steady
  state ($\vec{v_s}|_{t=0} \Leftrightarrow \bar{V}_a$ and
  $\vec{i_{ms}}|_{t=0} \Leftrightarrow \bar{I}_{ma}$), 191

INDEX 535

Relationship between the dq circuits and the per-phase phasor-domain equivalent circuit in balanced sinusoidal steady state, 477
Relationship between the dq windings and the per-phase phasor-domain equivalent circuit in balanced sinusoidal steady state, 361
Reluctance drives: stepper-motors and switched-reluctance drives, 501
  control in motoring mode, 523
  determining the rotor position for encoder-less operation, 523
  electromagnetic torque T_{em}, 515
  equivalent-circuit representation of a stepper-motor, 511
  half-stepping and micro-stepping, 512
  hybrid stepper-motors, 509
  induced back-emf e_a, 518
  instantaneous waveforms, 518
  the operating principle of reluctance motors, 502
  permanent-magnet stepper-motors, 507
  power electronic converters for SRM drives, 522
  power electronic converters for stepper-motors, 513
  role of magnetic saturation, 521
  stepper-motor drives, 506
  switched-reluctance motor (SRM) drives, 514
  variable-reluctance stepper motors, 506
Reversing the direction of rotation, 261
Review of space vectors, 327
Role of magnetic saturation, 521
Rotating stator MMF space vector, 187
  in multi-pole machines, 189
Rotating systems, 25
Rotor current controller, 443
Rotor dq-windings (along the same dq-axes as in the stator), 345
Rotor flux linkage (stator open-circuited), 331
Rotor position estimation using high-frequency injection, 483
Rotor position estimator, 446
Rotor-produced flux density distribution, 219
Rotor speed controller, 445
Rotor speed estimation, 407
Rotor-winding inductances (stator open-circuited), 325

Salient-pole synchronous machines, 481
Sensorless vector control, 401, 486
Short-circuited rotor, 245
Sinusoidally-distributed stator windings, 166, 318
Sinusoidal permanent magnet AC (PMAC) drives in steady state, 217
  calculation of the reference values $i_a^*(t)$, $i_b^*(t)$, and $i_c^*(t)$ and of the stator currents, 225
  the controller and the power-processing unit (PPU), 233
  generator mode, 224
  generator-mode of operation of PMAC drives, 233

Sinusoidal permanent magnet AC (PMAC) drives in steady state (*cont'd*)
  hardware prototyping of PMAC motor hysteresis current control, 235
  induced emf in the stator windings due to rotating $\vec{B_r}(t)$, 228
  induced emf in the stator windings due to rotating $\vec{i_s}(t)$: armature reaction, 229
  induced emfs in the stator windings during balanced sinusoidal steady state, 228
  mechanical system of PMAC drives, 224
  per-phase equivalent circuit, 231
  principle of operation, 219
  rotor-produced flux density distribution, 219
  structure of permanent-magnet AC (PMAC) machines, 219
  superposition of the induced emfs in the stator windings, 230
  torque production, 220
Slip frequency, f_{slip} in the rotor circuit, 254
Smart power modules including gate drivers and wide bandgap devices, 121
Space vector pulse-width-modulated (SV-PWM) inverters, 203
  computer simulation of SV-PWM inverter, 208
  hardware prototyping of space vector pulse width modulation, 213
  limit on the amplitude $\hat{V}_s$ of the stator voltage space vector $\vec{v}_s^a$, 211
  synthesis of stator voltage space vector $\vec{v}_s^a$, 203
Space-vector representation of combined terminal currents and voltages, 180
Space vectors to analyze AC machines
  balanced sinusoidal steady-state excitation (rotor open-circuited), 186
  induced voltages in stator windings, 193
  phase components of space vectors and $\vec{i_s}(t)$ and $\vec{v_s}(t)$, 184
  physical interpretation of the stator current space vector $\vec{i_s}(t)$, 181
  relationship between space vectors and phasors in balanced three-phase sinusoidal steady state $(\vec{v_s}|_{t=0} \Leftrightarrow \overline{V}_a$ and $\vec{i_{ms}}|_{t=0} \Leftrightarrow \overline{I}_{ma})$, 191
  rotating stator MMF space vector, 187
  rotating stator MMF space vector in multi-pole machines, 189
  sinusoidally-distributed stator windings, 166
  space-vector representation of combined terminal currents and voltages, 180
  space vectors to represent sinusoidal field distributions in the air gap, 175

three-phase, sinusoidally-distributed stator windings, 173
Space vectors to represent sinusoidal field distributions in the air gap, 175
Speed and position control loops, 385
Speed control of induction-motor drives, 285, 302
  applied voltage amplitudes to keep $\hat{B}_{ms} = \hat{B}_{ms,rated}$, 291
  capability to operate below and above the rated speed, 298
  conditions for efficient speed control over a wide range, 286
  hardware prototyping of closed-loop speed control of induction motor, 309
  harmonics in the PPU output voltages, 305
  induction-generator drives, 301
  modeling the PPU-supplied induction motors in steady state, 308
  pulse-width-modulated power-processing unit, 305
  rated power capability above the rated speed by flux-weakening, 300
  rated torque capability below the rated speed (with $\hat{B}_{ms,rated}$), 299
  reduction of $\hat{B}_{ms}$ at light loads, 308
  speed control of induction-motor drives, 302
  starting considerations in drives, 296

Speed-sensorless dynamic controller for IPM motor, 486
Speed-sensorless vector control of induction motor, 421, 483
  designing the PI controller, 414
  estimation of ω_{dA} and θ_{da}, 411
  introduction, 401
  model-reference adaptive system (MRAS) estimator, 404
  MRAS linearized error function, 423
  open loop speed estimator, 402
  parameter sensitivity of open-loop estimator and MRAS estimator, 416
  practical implementation, 417
  rotor speed estimation, 407
  stator d-and q-axis current reference, 410
Starting considerations in drives, 296
Stator and rotor flux linkages (simultaneous stator and rotor currents), 332
Stator and rotor voltage equations in terms of space vectors, 333
Stator d-and q-axis current reference, 410
Stator dq winding representation, 342
Stator flux linkage (rotor open-circuited), 330
Stator-inductance L_s, 324
Stator inductances (rotor open-circuited), 320
Stator mutual-inductance L_{mutual}, 322
Stator reactive power controller, 446
Stator single-phase magnetizing inductance $L_{m,1-phase}$, 320

Steady-state analysis, 430
Stepper-motor drives, 506
Structure of permanent-magnet AC (PMAC) machines, 219
Structure of the textbook, 16–17
Structure of three-phase, squirrel-cage induction motors, 241
Superposition of the induced emfs in the stator windings, 230
Switched-reluctance motor (SRM) drives, 514
Switching power-pole as the building block, 97
Switching waveforms in a converter for DC motor drives, 108
Switching waveforms in a three-phase inverter with sine-PWM, 117
Switch-mode conversion, 97
Switch-mode power electronic converters, 95
Synthesis of low-frequency AC, 112
Synthesis of stator voltage space vector $\vec{v}_s^{\,a}$, 203
System overview, 453

Test machine, 468
Tests to obtain the parameters of the per-phase equivalent circuit, 270
Three-phase inverters, 113
Three-phase, sinusoidally distributed stator windings, 319

Torsional resonances, 35
Transformer analogy, 245

Understanding DFIG operation in dq axis, 436

Variable-reluctance stepper motors, 506
Vector control of DFIG, 443
Vector control of permanent magnet synchronous motor drives, 473
 designing PID controller, 488
 dq analysis of permanent magnet synchronous machines, 473
 dq-based dynamic controller for "brush-less dc" drives, 478
 electrodynamics, 476
 electromagnetic torque, 476, 491
 flux linkages, 475
 hardware prototyping of vector control of SPM synchronous motor, 495
 non-salient pole synchronous machines, 477
 relationship between the dq circuits and the per-phase phasor-domain equivalent circuit in balanced sinusoidal steady state, 477
 rotor position estimation using high-frequency injection, 483
 salient-pole synchronous machines, 481
 speed-sensorless dynamic controller for IPM motor, 486

Printed in the USA/Agawam, MA
May 31, 2024

867067.007